SYSTEM DYNAMICS

SYSTEM DYNAMICS

Modeling and Simulation of Mechatronic Systems

Fourth Edition

DEAN C. KARNOPP
Department of Mechanical and Aeronautical Engineering
University of California
Davis, CA

DONALD L. MARGOLIS
Department of Mechanical and Aeronautical Engineering
University of California
Davis, CA

RONALD C. ROSENBERG
Department of Mechanical Engineering
Michigan State University
East Lansing, MI

WILEY

JOHN WILEY & SONS, INC.

Library of Congress Cataloging-in-Publication Data:

Karnopp, Dean.
    System dynamics : modeling and simulation of mechatronic systems /
Dean Karnopp, Donald L. Margolis, Ronald C. Rosenberg. — 4th ed.
        p.   cm.
    Includes bibliographical references and index.
    ISBN-13: 978-0-471-70965-7 (cloth : alk. paper)
    ISBN-10: 0-471-70965-4 (cloth : alk. paper)
        1. System engineering.   2. System analysis.   3. Bond graphs.   4. Mechatronics
    I. Margolis, Donald L.   II. Rosenberg, Ronald C.   III. Title.
    TA168.K362   2005
    620′.001′1—DC22   2005013732

Printed in the United States of America

10 9 8 7 6 5 4 3

CONTENTS

PREFACE

We are once again very pleased to have been asked to produce yet another edition of our text on the modeling and simulation of dynamic physical systems. We continue to believe that an understanding of physical systems is imperative for the design of modern engineering systems, and that the ability to represent mathematical models of systems in a uniform way that leads directly to analytical or computer response predictions is essential. In a real engineering environment, the engineer must work and interact in many different disciplines. Virtually all complex systems are subject to thermal problems, structural problems, vibration and noise problems, and control and stability issues, as well as other design problems that do not fit into a single discipline. An understanding of the interaction of these different disciplines is a valuable asset for any engineer. We continue to believe that bond graphs provide the best way to study and solve these multi-energy domain system problems.

Our book is about modeling and simulation of all types of engineering systems. We think of our book as one that truly deals with "mechatronic" issues in a generalized sense. Since the inception of the first edition, the word *mechatronics* has come into use in the world. The word is often interpreted to mean the field in which electronic controls are applied to mechanical systems. In this book, we deal with a wide variety of physical systems as well as purely mechanical systems. Our book shows how to model and simulate very complex systems in such a way as to expedite the design of electronic controls for complex engineering systems.

Since the introduction of the third edition, there has been a proliferation worldwide of courses, and even departments, dealing with mechatronics. An Internet search of "Bond Graph Courses" produces hundreds of courses worldwide that now use bond graphs to teach system modeling. Several of these courses are in diverse areas such as biology, economics, and physics, and as well as in traditional engineering. There was a time when we, the authors, knew practically every bond grapher in the world. That time has long past, as more and more disciplines recognize the value of a unified approach to modeling all types of physical systems, producing both linear and nonlinear mathematical models.

The earlier editions of the book have been divided into an undergraduate text in the first six chapters and a graduate text with advanced material in the succeeding chapters. In the third edition we concentrated primarily on improving the advanced material that covers multidimensional nonlinear mechanics, distributed systems typically represented with partial differential equations, magnetic circuits with application to electromagnetic actuators, and thermofluid systems in Chapters 7 through 12. We also added Chapter 13, devoted to simulation of complex nonlinear systems using commercially available software.

In this fourth edition we have concentrated on improving the basic material in the first six chapters. We have a combined experience of many decades of teaching this material to undergraduates and have rewritten the first six chapters to reflect how we best get this material into the heads of beginners. The topics are basically unchanged, but the presentation is all new. New problems have been added to each chapter and, as before, a solution manual is available. These chapters are suitable for use as an undergraduate text for a quarter or semester course.

Bond graphs are equally useful for representing linear and nonlinear systems. The introductory chapters emphasize linear systems, although reminders and examples are spread throughout the presentation to illustrate that real systems are actually nonlinear and that nonlinearities must ultimately be considered. It is made very clear how linear systems yield a standard starting point for analysis and simulation regardless of the energy domains involved. A section has been added that discusses automated simulation in which a bond graph can be drawn on a computer screen and software then derives all equations and sets them up for simulation using any of several available packages. The later chapters have been used for graduate-level courses and as a reference for advanced bond graph modeling techniques.

We have been rewarded many times over by the acceptance and growth of the material of our book. It has proved particularly satisfying to see that bond graph methods cannot only be used effectively to introduce undergraduate students to the mathematical modeling of dynamic physical systems, but can be extended to treat advanced topics in graduate-level courses and to deal with realistic problems in an industrial context. It is our opinion that bond graphs have an advantage in that although they can be presented to beginners in system modeling as easily as competing methods, they continue to retain their usefulness for more challenging system problems where less organized alternative methods are often of little help.

The topic of physical system modeling continues to be very dear to all of us. We hope that students will enjoy this book and find it most useful for their engineering careers and that professional engineers will find it to be of continuing benefit.

1

INTRODUCTION

This book is concerned with the development of an understanding of the dynamic physical systems that engineers are called upon to design. The type of systems to be studied can be described by the term *mechatronic*, which implies that while the elements of the system are mechanical in a general sense, electronic control will also be involved. For the design of a computer-controlled system, it is crucial that the dynamics of systems that exchange power and energy in various forms be thoroughly understood. Methods for modeling real systems will be presented, ways of analyzing systems in order to shed light on system behavior will be shown, and techniques for using computers to simulate the dynamic response of systems to external stimuli will be developed. Before beginning the study of physical systems, it is worthwhile to reflect a moment on the nature of the discipline that is usually called *system dynamics* in engineering.

The word *system* is used so often and so loosely to describe a variety of concepts that it is hard to give a meaningful definition of the word or even to see the basic concept that unites its diverse meanings. When the word *system* is used in this book, two basic assumptions are being made:

1. A system is assumed to be an entity separable from the rest of the universe (the environment of the system) by means of a physical or conceptual boundary. An animal, for example, can be thought of as a system that reacts to its environment (the temperature of the air, for example) and that interchanges energy and information with its environment. In this case the boundary is physical or spatial. An air traffic control system, however, is a complex, man-made system, the environment of which is not only the physical surroundings but also the fluctuating demands for air traffic, which ultimately come from human decisions about travel and the shipping of goods. The unifying element in these

1

two disparate systems is the ability to decide what belongs in the system and what represents an external disturbance or command originating from outside the system.

2. A system is composed of interacting parts. In an animal we recognize organs with specific functions, nerves that transmit information, and so on. The air traffic control system is composed of people and machines with communication links between them. Clearly, the *reticulation* of a system into its component parts is something that requires skill and art, since most systems could be broken up into so many parts that any analysis would be swamped with largely irrelevant detail.

These two aspects of systems can be recognized in everyday situations as well as in the more specific and technical applications that form the subject matter of most of this book. For example, when one hears a complaint that the transportation system in this country does not work well, one may see that there is some logic in using the word *system*. First of all, the transportation system is roughly identifiable as an entity. It consists of air, land, and sea vehicles and the human beings, machines, and decision rules by which they are operated. In addition, many parts of the system can be identified—cars, planes, ships, baggage handling equipment, computers, and the like. Each part of the transportation system could be further reticulated into parts (i.e., each component part is itself a system), but for obvious reasons we must exercise restraint in this division process.

The essence of what may be called the "systems viewpoint" is to concern oneself with the operation of a complete system rather than with just the operation of the component parts. Complaints about the transportation system are often real "system" complaints. It is possible to start a trip in a private car that functions just as its designers had hoped it would, transfer to an airplane that can fly at its design speed with no failures, and end in a taxi that does what a taxi is supposed to do, and yet have a terrible trip because of traffic jams, air traffic delays, and the like. Perfectly good components can be assembled into an unsatisfactory system.

In engineering, as indeed in virtually all other types of human endeavor, tasks associated with the design or operation of a system are broken up into parts that can be worked on in isolation to some extent. In a power plant, for example, the generator, turbine, boiler, and feed water pumps typically will be designed by separate groups. Furthermore, heat transfer, stress analysis, fluid dynamics, and electrical studies will be undertaken by subsets of these groups. In the same way, the bureaucracy of the federal government represents a splitting up of the various functions of government. All the separate groups working on an overall task must interact in some manner to make sure that not only will the parts of the system work, but also the system as a whole will perform its intended function. Many times, however, oversimplified assumptions about how a system will operate are made by those working on a small part of the system. When this happens, the results can be disappointing. The power plant may undergo damage during a full load rejection, or the economy of a country may collapse because of the unfavorable interaction of segments of government, each of which assiduously pursues seemingly reasonable policies.

In this book, the main emphasis will be on studying system aspects of behavior as distinct from component aspects. This requires a knowledge of the component parts of the systems of interest and hence some knowledge in certain areas of engineering that are taught and sometimes even practiced in splendid isolation from other areas. In the engineering systems of primary interest in this book, topics from vibrations, strength of materials, dynamics, fluid mechanics, thermodynamics, automatic control, and electrical circuits will be used. It is possible, and perhaps even common, for an engineer to spend a major part of his or her professional career in just one of these disciplines, despite the fact that few significant engineering projects concern a single discipline. Systems engineers, however, must have a reasonable command of several of the engineering sciences as well as knowledge pertinent to the study of systems per se.

Although many systems may be successfully designed by careful attention to static or steady-state operation in which the system variables are assumed to remain constant in time, in this book the main concern will be with *dynamic* systems, that is, those systems whose behavior as a function of time is important. For a transport aircraft that will spend most of its flight time at a nearly steady speed, the fuel economy at constant speed is important. For the same plane, the stress in the wing spars during steady flight is probably less important than the time-varying stress during flight through turbulent air, during emergency maneuvers, or during hard landings. In studying the fuel economy of the aircraft, a static system analysis might suffice. For stress prediction, a dynamic system analysis would be required.

Generally, of course, no system can operate in a truly static or steady state, and both slow evolutionary changes in the system and shorter time transient effects associated, for example, with startup and shutdown are important. In this book, despite the importance of steady-state analysis in design studies, the emphasis will be on dynamic systems. Dynamic system analysis is more complex than static analysis but is extremely important, since decisions based on static analyses can be misleading. Systems may never actually achieve a possible steady state due to external disturbances or instabilities that appear when the system dynamics are taken into account. Moreover, systems of all kinds can exhibit counterintuitive behavior when considered statically. A change in a system or a control policy may appear beneficial in the short run from static considerations but may have long-run repercussions opposite to the initial effect. The history of social systems abounds with sometimes tragic examples, and there is hope that dynamic system analysis can help avoid some of the errors in "static thinking" [1]. Even in engineering with rather simple systems, one must have some understanding of the dynamic response of a system before one can reasonably study the system on a static basis.

A simple example of a counterintuitive system in engineering is the case of a hydraulic power generating plant. In order to reduce power, wicket gates just before the turbine are moved toward the closed position. Temporarily, however, the power actually increases as the inertia of the water in the penstock forces the flow through the gates to remain almost constant, resulting in a higher velocity of flow through the smaller gate area. Ultimately, the water in the penstock slows down and power is reduced. Without an understanding of the dynamics of this system, one would be led to open the gates to *reduce* power. If this were done, the immediate result would be a gratifying decrease of power followed by a surprising and inevitable increase. Clearly, a good understanding of dynamic response is crucial to the design of a controller for mechatronic systems.

1.1   MODELS OF SYSTEMS

A central idea involved in the study of the dynamics of real systems is the idea of a *model* of a system. Models of systems are simplified, abstracted constructs used to predict their behavior. Scaled physical models are well known in engineering. In this category fall the wind tunnel models of aircraft, ship hull models used in towing tanks, structural models used in civil engineering, plastic models of metal parts used in photoelastic stress analysis, and the "breadboard" models used in the design of electric circuits.

The characteristic feature of these models is that some, but not all, of the features of the real system are reflected in the model. In a wind tunnel aircraft model, for example, no attempt is made to reproduce the color or interior seating arrangement of the real aircraft. Aeronautical engineers assume that some aspects of a real craft are unimportant in determining the aerodynamic forces on it, and thus the model contains only those aspects of the real system that are supposed to be important to the characteristics under study.

In this book, another type of model, often called a *mathematical model*, is considered. Although this type of model may seem much more abstract than the physical model, there are strong similarities between physical and mathematical models. The mathematical model also is used to predict only certain aspects of the system response to inputs. For example, a mathematical model might be used to predict how a proposed aircraft would respond to pilot input command signals during test maneuvers. But such a model would not have the capability of predicting every aspect of the real aircraft response. The model might not contain any information on changes in aerodynamic heating during maneuvers or about high-frequency vibrations of the aircraft structure, for example.

Because a model must be a simplification of reality, there is a great deal of art in the construction of models. An unduly complex and detailed model may contain parameters virtually impossible to estimate, may be practically impossible to analyze, and may cloud important results in a welter of irrelevant detail if it can be analyzed. An oversimplified model will not be capable of exhibiting important effects. It is important, then, to realize that *no system can be modeled exactly* and that any competent system designer needs to have a procedure for constructing a variety of system models of varying complexity so as to find the simplest model capable of answering the questions about the system under study.

The rest of this book deals with models of systems and with the procedures for constructing models and for extracting system characteristics from models. The models will be mathematical models in the usual meaning of the term even though they may be represented by stylized graphs and computer printouts rather than the more conventional sets of differential equations.

System models will be constructed using a uniform notation for all types of physical systems. It is a remarkable fact that models based on apparently diverse branches of engineering science can all be expressed using the notation of *bond graphs* based on energy and information flow. This allows one to study the *structure* of the system model. The nature of the parts of the model and the manner in which the parts interact can be made evident in a graphical format. In this way, analogies between various types of systems are made evident, and experience in one field can be extended to other fields.

Using the language of bond graphs, one may construct models of electrical, magnetic, mechanical, hydraulic, pneumatic, thermal, and other systems using only a rather small set of ideal elements. Standard techniques allow the models to be translated into differential equations or computer simulation schemes. Historically, diagrams for representing dynamic system models developed separately for each type of system. For example, parts *a*, *b*, and *c* of Figure 1.1 each represent a diagram of a typical model. Note that in each case the elements in the diagram seem to have evolved from sketches of devices, but in fact a photograph of the real system would not resemble the diagram at all. Figure 1.1*a* might well represent the dynamics of the heave motion of an automobile, but the masses, springs, and dampers of the model are not directly related to the parts of an automobile visible in a photograph. Similarly, symbols for resistors and inductors in diagrams such as Figure 1.1*b* may not correspond to separate physical elements called resistors and chokes but instead may correspond to the resistance and inductance effects present in a single physical device. Thus, even semipictorial diagrams are often a good deal more abstract than they might at first appear.

When mixed systems such as that shown in Figure 1.1*d* are to be studied, the conventional means of displaying the system model are less well developed. Indeed, few such diagrams are very explicit about just what effects are to be included in the model. The basic structure of the model may not be evident from the diagram. A bond graph is more abstract than the type of diagrams shown in Figure 1.1, but it is explicit and has the great advantage that all the models shown in Figure 1.1 would be represented using exactly the same set of symbols. For mixed systems such as that shown in Figure 1.1*d*, a universal language such as bond graphs provide is required in order to display the essential structure of the system model.

1.2   SYSTEMS, SUBSYSTEMS, AND COMPONENTS

In order to model a system, it is usually necessary first to break it up into smaller parts that can be modeled and perhaps studied experimentally and then to assemble the system model from the parts. Often, the breaking up of the system is conveniently accomplished in several stages. In this book major parts of a system will be called *subsystems* and primitive parts of subsystems will be called *components*. Of course, the hierarchy of components, subsystems, and systems can never be absolute, since

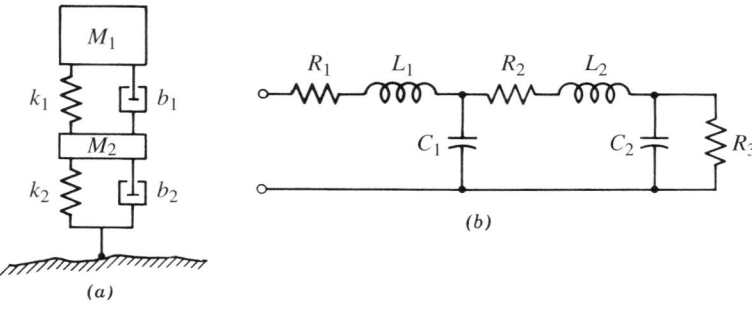

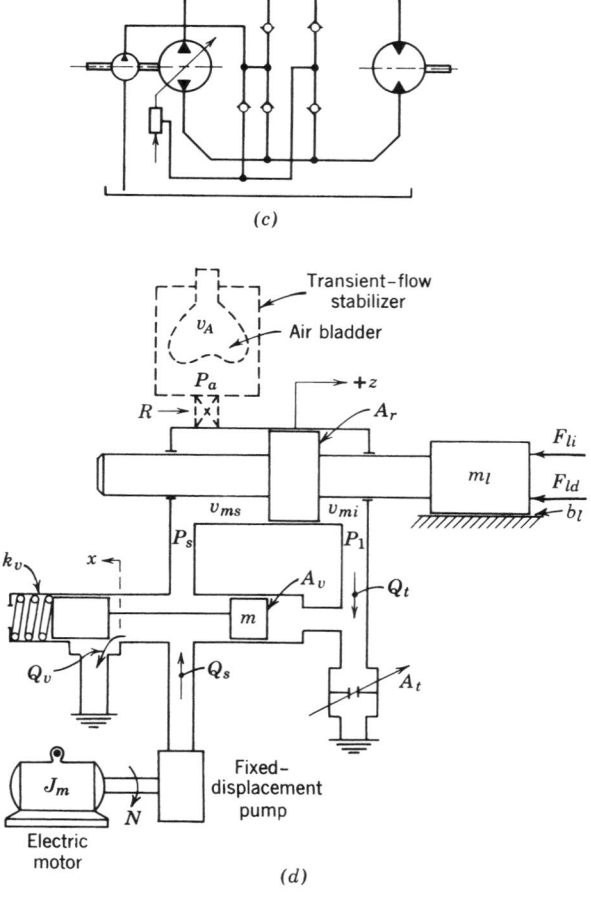

FIGURE 1.1. (*a*) Typical schematic diagram; (*b*) typical electric circuit diagram; (*c*) typical hydraulic diagram; (*d*) schematic diagram of system containing mechanical, electrical, and hydraulic components.

even the most primitive part of a system could be modeled in such detail that it would be a complex subsystem. Yet in many engineering applications, the subsystem and component categories are fairly obvious.

Basically, a subsystem is a part of a system that will be modeled as a system itself; that is, the subsystem will be broken into interacting component parts. A component, however, is modeled as a unit and is not thought of as composed of simpler parts. One needs to know how the component interacts with other components and one must have a characterization of the component, but otherwise a component is treated as a "black box" without any need to know what causes it to act as it does.

To illustrate these ideas, consider the vibration test system shown in Figure 1.2. The system is intended to subject a test structure to a vibration environment specified by a signal generator. For example, if the signal generator delivers a random-noise signal, it may be desired that the acceleration of the shaker table be a faithful reproduction of the electrical noise signal waveform. In a system that is assembled from physically separate pieces, it is natural to consider the parts that are assembled by connecting wires and hydraulic lines or by mechanical fasteners as subsystems. Certainly, the electronic boxes labeled signal generator, controller, and electrical amplifier are subsystems, as are the electrohydraulic valve, the hydraulic shaker, and the test structure. It may be possible to treat some of these subsystems as components if their interactions with the rest of the system can be specified without knowledge of the internal construction of the subsystem. The electrical amplifier is obviously composed of many components, such as resistors, capacitors, transistors, and the like, but if the amplifier is sized correctly so that it is not overloaded, then it may be possible to treat the amplifier as a component specified by the manufacturer's input–output data. Other subsystems may require a subsystem analysis in order to achieve a dynamic description suitable for the overall system study.

Consider, for example, the electrohydraulic valve. A typical servo valve is shown in Figure 1.3. Clearly, the valve is composed of a variety of electrical, mechanical, and hydraulic parts that work together to produce the dynamic response of the valve. For this subsystem the components might be the torque motor, the hydraulic

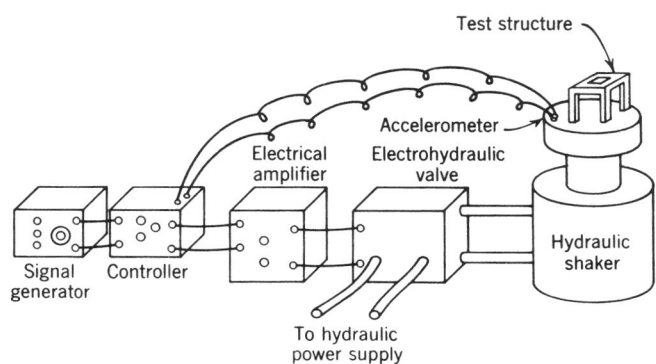

FIGURE 1.2. Vibration test system.

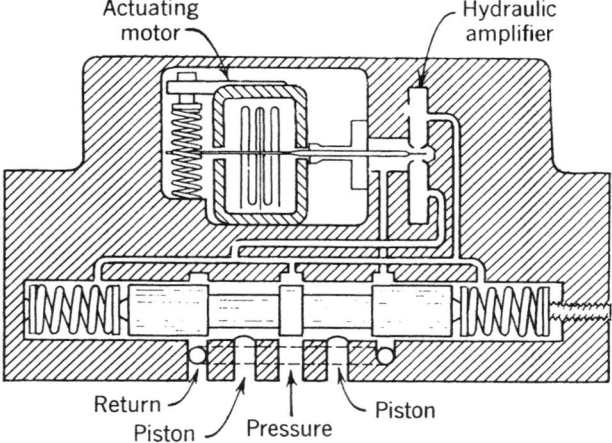

FIGURE 1.3. Electrohydraulic valve.

amplifier, mechanical springs, hydraulic passages, and the spool valve. A subsystem dynamic analysis can reveal weaknesses in the subsystem design that may necessitate the substitution of another subsystem or a reconfiguration of the overall system. Yet such an analysis may indicate that, from the point of view of the overall system, the subsystem may be adequately characterized as a simple component. A skilled and experienced system designer often makes a judgment on the appropriate level of detail for the modeling of a subsystem on an intuitive basis. A major purpose of the methods presented in this book is to show how system models can be assembled conveniently from component models. It is then possible to experiment with subsystem models of varying degrees of sophistication in order to verify or disprove initial modeling decisions.

1.3 STATE-DETERMINED SYSTEMS

The goal of this book is to describe means for setting up mathematical models for systems. The type of model that will be found is often described as a "state-determined system." In mathematical notation, such a system model is often described by a set of ordinary differential equations in terms of so-called *state variables* and a set of algebraic equations that relate other system variables of interest to the state variables. In succeeding chapters an orderly procedure, beginning with physical effects to be modeled and ending with state equations, will be demonstrated. Even though some techniques of analysis and computer simulation do not require that the state equations be written, from a mathematical point of view all the system models are state-determined systems.

The future of all the variables associated with a state-determined system can be predicted if (1) the state variables are known at some initial time and (2) the future time history of the input quantities from the environment is known.

Such models, which are virtually the only ones used in engineering, have some built-in philosophical implications. For example, events in the future do not affect the present state of the system. This implication is correlated with the assumption that time runs only in one direction—from past to future. That models should have these properties probably seems plausible, if not obvious, yet it is remarkably difficult to conceive of a demonstration that real systems always have these properties.

Clearly, past history can have an effect on a system; yet the influence of the past is exhibited in a special way in state-determined systems. All the past history of a state-determined system is summed up in the present values of its state variables. This means that many past histories could have resulted in the same present value of state variables and hence the same future behavior of the system. It also means that if one can condition the system to bring the state variables to some particular values, then the future system response is determined by the future inputs and nothing is important about the past except that the state variables were brought to those values.

Scientific experiments are run as if the systems to be studied were state determined. The system is always started from controlled conditions that are expressed in terms of carefully monitored variables. If the experiment is repeatable, then the assumption is that the state variables are properly initialized by the operations used to set up the experiment. If the experiment is not repeatable, then the assumption is that some important influence has not been controlled. This influence can be either a state variable that was not monitored and initialized properly or an unrecognized input quantity through which the environment influences the system.

State-determined system models have proved useful over centuries of scientific and technical work. For the usual macroscopic systems encountered in engineering, state-determined system models are nearly universal, and there is continuing interest in developing such models for social and economic systems. This book can be regarded as a textbook devoted to the establishment and study of state-determined system models using well-defined physical systems of interest to engineers as examples.

1.4 USES OF DYNAMIC MODELS

In Figure 1.4 a general dynamic system model is shown schematically. The system, S, is characterized by a set of state variables, indicated by X, that are influenced by a set of input variables, U, that represent the action of the system's environment on the system. The set of output variables, Y, are observable aspects of the system's

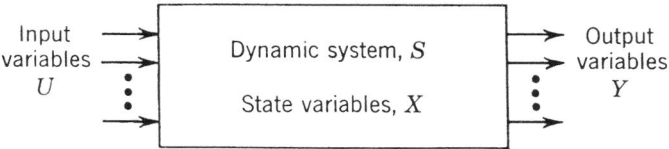

FIGURE 1.4. General dynamic system model.

response or back effects from the system onto the environment. This type of dynamic system model may be used in three quite distinct ways:

1. *Analysis.* Given U for the future, X at the present, and the model S, predict the future of Y. Assuming that the system model is an accurate representation of the real system, analysis techniques allow one to predict system behavior.
2. *Identification.* Given time histories of U and Y, usually by experimentation on real systems, find a model S and state variables X which are consistent with U and Y. This is the essence of scientific experimentation. Clearly, a "good" model is one that is consistent with a great variety of sets U and Y.
3. *Synthesis.* Given U and some desired Y, find S such that U acting on S will produce Y. Most of engineering deals with synthesis, but only in limited contexts are there direct synthesis methods. Often we must be content to accomplish synthesis of systems via a trial-and-error process of repetitive analysis of a series of candidate systems. In this regard, dynamic models pay a vital role, since progress would be slow indeed if one had to construct each candidate system "in the metal" in order to discover its properties.

In this book we will concentrate on setting up system models and predicting the behavior of the systems using analytical or computational techniques. Thus, we will concentrate on analysis, but it is important to remember that the techniques are useful for identification problems and that the major challenge to a systems engineer is to synthesize desirable systems. It may not be too much to say that analysis, except in the service of synthesis, is a rather sterile pursuit for an engineer.

1.5   LINEAR AND NONLINEAR SYSTEMS

An overall system model, consisting of subsystems and their components, requires modeling decisions as to what dynamic effects must be included in order to use the model for its intended purpose. The result of these modeling decisions is typically a system schematic that indicates the important dynamic effects. Figures 1.1 and 1.2 are examples of system schematics where modeling decisions have been made. In Figure 1.1d, the important dynamic effects at the component level are indicated by labeling inertial, compliance, and resistance effects. In Figure 1.2, modeling decisions are indicated at the subsystem level, while the detail modeling of each subsystem remains to be done. A very important aspect of the modeling process is whether components of subsystems behave linearly or nonlinearly. As we progress through the chapters, it will become very clear as to what is meant by linear or nonlinear behavior. For now it is simply stated that linear systems are represented by sets of linear, first-order differential equations, and nonlinear systems, while still state determined, are represented by sets of nonlinear, first-order differential equations.

If it is justified to assume that an overall system can be represented as linear, then there exist an abundance of analytical tools for obtaining exact analytical solutions to the linear equations and for extracting incredibly detailed information about the response of the system. Some of the analytical information that is covered in later chapters includes eigenvalues, transfer functions, and frequency response. If the systems have large numbers of state variables, then pencil and paper analysis may not be possible, and one must resort to computation to obtain the linear properties of the system.

If a single component in a system model is represented as a nonlinear element, then the system is nonlinear, and linear analysis tools will not work. Sliding friction is an example of a nonlinear component. There will not exist analytical eigenvalues, transfer functions, or frequency response. In order to extract information about the response of nonlinear systems, one must resort to time step simulation. Fortunately, there is an abundance of commercial programs to simulate nonlinear systems.

The fact is that there are virtually no physical systems that are linear. However, in order to introduce the concepts of constructing overall system models of interacting electrical, mechanical, hydraulic, and thermal components, it is easier to start with linear systems and then extend the procedures to nonlinear systems. In the following five chapters, the emphasis is on linear system models, but whenever possible, the reader is reminded that real physical systems are nonlinear and simulation tools must be used to obtain system responses.

1.6 AUTOMATED SIMULATION

Mathematical models of dynamic physical systems have been made ever since the invention of differential equations. But until the development of powerful computers, there were severe limitations on the analysis of these models. Practically speaking, the dynamic behavior could generally be predicted only for low-order linear models that often were not very accurate representatives of real systems.

There is a lot to be said for the study of low-order linear models in order to gain an appreciation of system dynamics, and the first six chapters of this book deal primarily with just such system models. However, computer simulation can now be used to gain experience with system dynamics even when the system models become large and when they contain nonlinear elements. Chapter 13 discusses some of the issues that arise when dealing with complex but realistic models.

The next Chapters, 2, 3, and 4, present techniques for representing elements of mechanical, electrical, and fluid systems (and combination systems) in the abstract form of bond graphs instead of the schematic diagrams usually used to show vibratory systems, electric circuits, or hydraulic systems. For some, this may seem to be an unnecessary step away from physical reality, but it has useful consequences.

First of all, a bond graph is a precise way to represent a mathematical model. Often schematic diagrams are not entirely clear about whether certain effects are to be included or neglected in the model. Second, for many systems involving two or more forms of energy, such as mechanical, electrical and hydraulic, there are no standard schematic diagrams that clearly indicate assumptions made in the modeling process. Finally, it is much easier to communicate a bond graph model unambiguously to a computer than a schematic diagram.

The bond graph uses only a few standard symbols, whereas typical schematic diagrams for the same system model drawn by different people are almost never identical. Just as computers more easily read bar codes than handwriting, they more easily interpret bond graphs than schematic diagrams.

Computer programs have been developed that recognize bond graphs and can process them in the same manner that a human would in order to extract differential equations for analysis or simulation. In the process, useful facts about the mathematics of the model are discovered even before any numerical parameters or laws have been supplied. Furthermore, when the parameters of the elements and the forcing functions for the system have been specified, the programs can then simulate the response of the system. In this process, only a minimum of human intervention is required.

Although it is important for a system engineer to understand the entire process of modeling and simulation, the use of bond graphs and bond graph simulation programs allows a beginner to start developing the skills associated with computer simulation even before all the bond graph modeling techniques have been learned. This has proved to be very effective in teaching. From the first day, the student can see that, given a bond graph model and a bond graph simulation program, it is possible to see how the model reacts to various input forcing functions and to variations in system parameters. Simple design studies on dynamic systems can be assigned without waiting until the student has learned to make models, derive equations, and use an equation solver. This provides motivation to learn about bond graph dynamic system modeling and numerical simulation techniques.

The fact that the simulation programs are effective for nonlinear as well as linear models, and for large models as well as small ones, may not be fully appreciated by a beginning student. However, in the course of time the significance of this fact should become apparent.

References [2–4] give the names of some of the more well known commercial bond graph processors, some of which are used in conjunction with simulation programs to solve the differential equations. A web search will reveal that there are a number of other bond graph processor programs.

Another category of program is based on a stored library of predetermined bond graph submodels, but in use, replaces bond graph submodels with icons. See Reference [5], for example. Such programs are useful for studying large engineering models but they are less useful for learning about bond graph modeling.

REFERENCES

[1] J. W. Forrester, *Urban Dynamics*, Cambridge, MA: MIT Press, 1969.

[2] CAMP-G, Cadsim Engineering, P.O. Box 4083, Davis, CA 95616.

[3] 20-SIM, Controllab Products B.V., Drienerlolaan 5, EL-CE, 7255 NB Enschede, Netherlands.

[4] SYMBOLS 2000, Hightech Consultants, STEP IIT Ktragpur—721302, WB, India.

[5] AMESim, IMAGINE SA, 5 rue Brison, F 42300 ROANNE, France.

PROBLEMS

1-1 Suppose you were a heating engineer and you wished to consider a house as a dynamic system. Without a heater, the average temperature in the house would clearly vary over a 24-h period. What might you consider for inputs, outputs, and state variables for a simple dynamic model? How would you expand your model so that it would predict temperatures in several rooms of the house? How does the installation of a thermostatically controlled heater change your model?

1-2 For a particular car operated on a level road at steady conditions there is a relation between throttle position and speed. Sketch the general shape you would expect for this curve. If recordings were made of instantaneous speed and throttle position while the car was driven normally for several miles on ordinary roads, do you think that the instantaneous values of speed and throttle position would fall on the steady-state curve? What inputs, outputs, and state variable might prove useful in trying to find a dynamic model useful in predicting dynamic speed variations?

1-3 A car is driven over a curb twice—once very slowly and once quite rapidly. What would you need to know about the car in the second case that you did not need to know in the first case if you were required to find the tire force that resulted from going over the curb?

1-4 In the steady state a good weather vane points into the wind, but when the wind shifts, the vane cannot always be trusted to be pointing into the wind. Identify inputs, outputs, and the parameters of the weather vane system that affect its response to the wind. Sketch your idea of how the position of the vane would change in time if the wind suddenly shifted $10°$.

1-5 The height of water in a reservoir fluctuates over time. If you had to construct a dynamic system model to help water resource planners predict variations in the height, what input quantities would you consider? How many state variables do you think you would need for your model?

1-6 A mass, M, and spring, k, are at rest in a gravity field, about to be struck by a mass, m, falling from a height, h. The mass, m, sticks to M, and the two move downward. The variable, x, keeps track of the displacement after impact. Sketch the general motion of x for some period of time after impact. How many equations do you think are needed to describe this system mathematically? If the system ever came to rest, what would be the deflection of the spring?

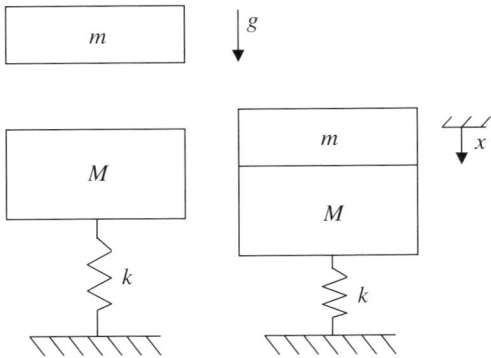

1-7   The system is an electric circuit consisting of an input voltage, $e(t)$, and a capacitor, resistor, and inductor, C, R, L. As will be seen in later chapters, if a voltage is applied to a capacitor, current flows easily at first and then slows as the capacitor becomes charged. Inductors behave just the opposite, in that they reluctantly pass current when a voltage is first applied, and then the current passes easily as time passes. If the input voltage is suddenly raised from zero to some constant value, sketch the current in the capacitor, i_C, and the inductor, i_L, as a function of time. What is the steady-state current in the capacitor and inductor?

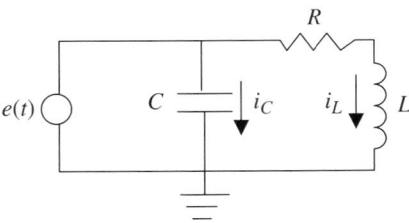

1-8   A hydraulic system consists of a supply pressure, P_s, and a long fluid-filled tube. Branching from the tube is an accumulator, C_a, consisting of a compliant gas separated from the fluid by a diaphragm. The long tube has inertia and resistance, I_f and R_f. It may be hard to believe at this point, but this hydraulic

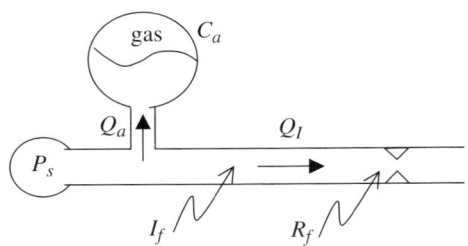

system exhibits identical behavior to that of the electric circuit of Problem 1-7. Armed with this information, sketch the volume flow rate of the fluid into the accumulator, Q_a, and in the tube, Q_1, as a function of time. Also construct a word bond graph of this hydraulic system.

1-9 Shown here is the hydraulic system in Problem 1-8 connected to a hydraulic cylinder of piston area A_p. The piston is connected to a mass, m, attached to the ground through the spring and damper, k and b. This is a "system" consisting of interacting hydraulic and mechanical components. How many variables do you think are needed to fully describe the motion of the system? Sketch how you think the volume flow rate in the accumulator and in the tube will respond to a sudden elevation of the supply pressure. Sketch the motion, x, of the mass. Construct a word bond graph for this system.

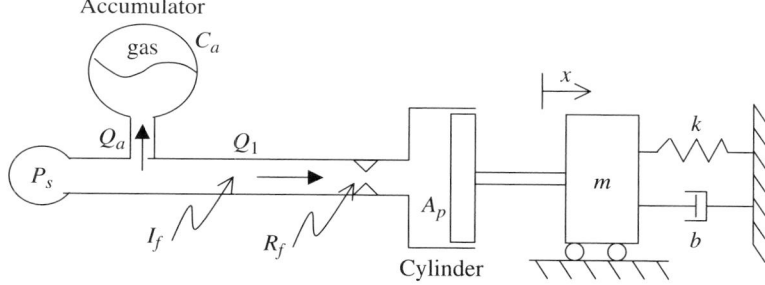

2

MULTIPORT SYSTEMS
AND BOND GRAPHS

In this chapter the first steps are taken toward the development of system-modeling techniques for engineering systems involving power interactions. First, major subsystems are identified, and the means by which the subsystems are interconnected are studied. The fact that interacting physical systems must transmit power then is used to unify the description of interconnected subsystems. A uniform classification of the variables associated with power and energy is established, and bond graphs showing the interconnection of subsystems are introduced. Finally, the notions of inputs, outputs, and pure signal flows are discussed.

2.1 ENGINEERING MULTIPORTS

In Figure 2.1 a representative collection of subsystems or components of engineering systems is shown. Although the subsystems sketched are quite elementary, they will serve to introduce the concept of an engineering multiport. "Engineering" is used to imply that the devices are used to build up systems such as automobiles, television sets, machine tools, or electric power plants that are designed to accomplish some specific objectives. "Multiport" refers to a point of view taken in the description of the subsystems.

Inspection of Figure 2.1 reveals that a number of variables have been labeled on the subsystems. These variables are torques, angular speeds, forces, velocities, voltages, currents, pressures, and volume flow rates. The variables occur in pairs associated with points at which the subsystems could be connected with other subsystems to form a system. It would be possible, for example, to couple the shaft of the electric motor (a) to one end of the drive shaft (c) and the hydraulic motor shaft (b) to the other end of the drive shaft. After the coupling, the motor torque and speed would be

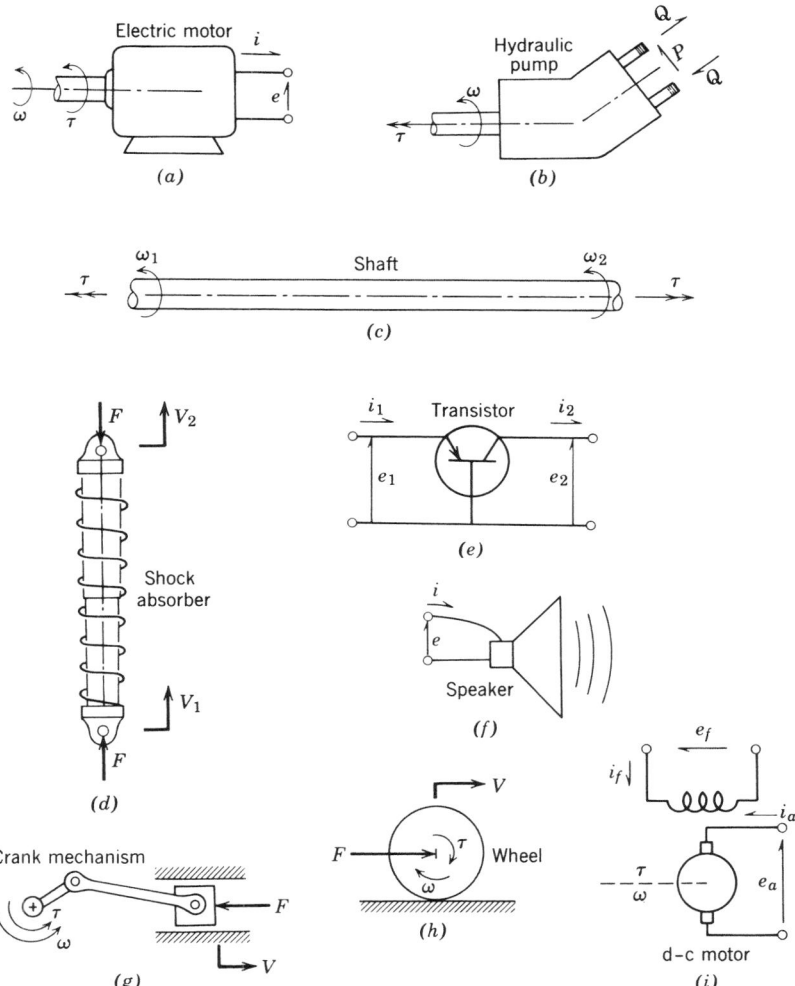

FIGURE 2.1. A collection of engineering multiports. (a) Electric motor: torque τ, angular speed ω, voltage e, current i; (b) hydraulic pump: torque τ, angular speed ω, pressure P, volume flow rate Q; (c) drive shaft: torque τ, angular speeds ω_1 and ω_2; (d) spring shock absorber unit: force F, velocities V_1 and V_2; (e) transistor: voltages e_1 and e_2, currents i_1 and i_2; (f) loudspeaker: voltage e, current i; (g) crank and slider mechanism: torque τ, angular speed ω, force F, velocity V; (h) wheel: force F, velocity V, torque τ, angular speed ω; (i) separately excited direct current (d-c) motor: torque τ, angular speed ω, voltages e_a and e_f, currents i_a and i_f.

identical to the torque and speed of one end of the drive shaft. Similarly, the torque and speed of the other end of the drive shaft would be identical to the torque and speed of the hydraulic pump. (If the drive shaft were not rigid, then the two angular speeds, ω_1 and ω_2, on the drive shaft ends would not necessarily be equal at all times.) Similarly, one could connect the two terminals of the transistor (e) associated with e_2 and i_2 to the terminals of the loudspeaker (f). After the connection, the voltage and current associated with one terminal pair of the transistor would be identical to the voltage and current associated with the loudspeaker terminals. Generally, when two subsystems or components are joined together physically, two complementary variables are simultaneously constrained to be equal for the two subsystems.

Places at which subsystems can be interconnected are places at which power can flow between the subsystems. Such places are called *ports*, and physical subsystems with one or more ports are called *multiports*. A system with a single port is called a *1-port*, a system with two ports is called a *2-port*, and so on. The multiports in Figures 2.1a–h are shown as 2-ports. Figure 2.1f is a 1-port as long as it is considered only as an electrical element and not as an element coupling electrical and acoustic subsystems. Figure 2.1i is shown as a 3-port.

The variables listed for the multiports in Figure 2.1 and the variables that are forced to be identical when two multiports are connected are called *power variables*, because the product of the two variables considered as functions of time is the instantaneous power flowing between the two multiports. For example, if the electrical motor (Figure 2.1a) were coupled to the hydraulic pump (Figure 2.1b), the power flowing from the motor to the pump would be given by the product of the angular speed and the torque. Since power could flow in either direction, a sign convention for the power variables will be established. Similarly, power can be expressed as the product of a force and a velocity for a multiport in which mechanical translation is involved, as the product of voltage and current for an electrical port, and as the product of pressure and volume flow rate for a port at which hydraulic power is interchanged.

Since power interactions are always present when two multiports are connected, it is useful to classify the various power variables in a universal scheme and to describe all types of multiports in a common language. In this book all power variables are called either *effort* or *flow*. Table 2.1 shows effort and flow variables for several types of power interchange.

As Table 2.1 indicates, in general discussions the symbols $e(t)$ and $f(t)$ are used to denote effort and flow quantities as functions of time. For specific applications, more traditional notation suggestive of the physical variable involved may be used.

TABLE 2.1. Some Effort and Flow Quantities

Domain	Effort, $e(t)$	Flow, $f(t)$
Mechanical translation	Force component, $F(t)$	Velocity component, $V(t)$
Mechanical rotation	Torque component, $\tau(t)$	Angular velocity component, $\omega(t)$
Hydraulic	Pressure, $P(t)$	Volume flow rate, $Q(t)$
Electric	Voltage, $e(t)$	Current, $i(t)$

A curse of system analysis that becomes evident as soon as problems involving several energy domains are studied is that it is hard to establish notation that does not conflict with conventional usage. In Table 2.1, for example, a force is an effort quantity, $e(t)$, even though the common use of the letter F to stand for a force might be confused with the $f(t)$, which stands for a flow quantity. These notational difficulties are bothersome but not fundamental, and cannot be avoided except by using entirely new notation. For example, the letter Q has been used for charge in electric circuits, volume flow rate in hydraulics, and heat in thermodynamics. In this book, both the generalized notation e and f and the physical notation in Figure 2.1 and Table 2.1 are used. The context in which the symbols are used will resolve any possible ambiguities in meaning. The power, $P(t)$, flowing into or out of a port can be expressed as the product of an effort and a flow variable, and thus in general notation is given by the following expression:

$$P(t) = e(t)f(t). \tag{2.1}$$

In a dynamic system the effort and the flow variables, and hence the power, fluctuate in time. Two other types of variables turn out to be important in describing dynamic systems. These variables, sometimes called *energy variables* for reasons that will become clearer later, are called the *momentum* $p(t)$ and the *displacement* $q(t)$ in generalized notation.

The momentum is defined as the time integral of an effort. That is,

$$p(t) \equiv \int^t e(t)\,dt = p_0 + \int_{t_0}^t e(t)\,dt, \tag{2.2}$$

in which either the indefinite time integral can be used or one may define p_0 to be the initial momentum at time t_0 and use the definite integral from t_0 to t. In the same way, a displacement variable is the time integral of a flow variable:

$$q(t) \equiv \int^t f(t)\,dt = q_0 + \int_{t_0}^t f(t)\,dt. \tag{2.3}$$

Again, the second integral expression in Eq. (2.3) indicates that at time t_0 the displacement is q_0.

Other ways of writing the definitions in Eqs. (2.2) and (2.3) follow by considering the differential rather than the integral forms:

$$\frac{dp(t)}{dt} = e(t), \quad dp = e\,dt; \tag{2.2a}$$

$$\frac{dq(t)}{dt} = f(t), \quad dq = f\,dt. \tag{2.2b}$$

The energy, $E(t)$, which has passed into or out of a port is the time integral of the power, $P(t)$. Thus,

$$E(t) \equiv \int^t P(t)\,dt = \int^t e(t)f(t)\,dt. \tag{2.4}$$

The reason p and q are sometimes called energy variables is that in Eq. (2.4), one may be able to write $e\,dt$ as dp or $f\,dt$ as dq by using Eq. (2.2a) or (2.3a). Alternative expressions for $\mathbf{E}$ then follow:

$$\mathbf{E}(t) = \int^t e(t)dq(t) = \int^t f(t)\,dp(t). \tag{2.5}$$

In the next chapter, cases will be encountered in which an effort is a function of a displacement or a flow is a function of a momentum. Then the energy can be expressed not only as a function of time but also as a function of one of the energy variables; thus,

$$\mathbf{E}(q) = \int^q e(q)\,dq \tag{2.5a}$$

or

$$\mathbf{E}(p) = \int^p f(p)\,dp. \tag{2.5b}$$

This provides the motivation for calling p and q energy variables in distinction to the power variables e and f.

In Figure 2.2 a mnemonic device, fancifully called the "tetrahedron of state," is shown. The four variable types, e, f, p, and q, are associated with the four vertices of a tetrahedron. Along two of the edges of the tetrahedron are indicated the relationships between e and p and between f and q. In Chapter 3 the same figure will be augmented to display the variables related by certain basic multiport elements.

It is an interesting fact that the only types of variables that will be needed to model physical systems are represented by the power and energy variables, e, f, p, and q. In order to make this statement more plausible, let us study the variables in several energy domains in more detail.

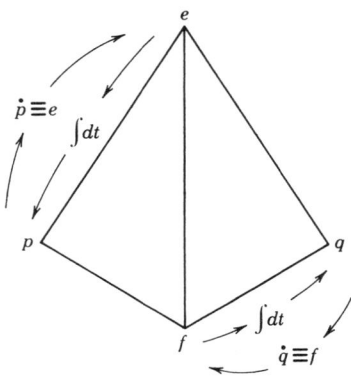

FIGURE 2.2. The tetrahedron of state.

TABLE 2.2. Power and Energy Variables for Mechanical Translational Systems

Generalized Variables	Mechanical Translation	SI Units
Effort, e	Force, F	Newtons (N)
Flow, f	Velocity, V	Meters per second (m/s)
Momentum, p	Momentum, P	N-s
Displacement, q	Displacement, X	m
Power, P	$F(t)V(t)$	N-m/s
Energy, E	$\displaystyle\int^{x} F\, dx, \int^{p} V\, dP$	N-m

Table 2.2 presents power and energy variables for mechanical translational ports. Since the power variables force and velocity are considered primitive, the units of the remaining variables follow from a choice in units for the power variables. Units are the shoals on which many a system analysis has foundered, and it is worthwhile to consider the great advantage that the metric or International System of Units (SI) has for system dynamic studies as compared with the many other unit systems that have been used in the past. In the SI system, power is always measured in newton-meters per second (N-m/s) or the equivalent watts (W), no matter what type of physical system is being studied. Similarly, energy will always be measured in newton-meters or the equivalent joules (J) for an electrical, mechanical, hydraulic, or any other type of physical system. Thus, if the e, f, p, and q variables are given SI units, no bothersome unit conversions will be necessary to properly account for power and energy interactions.

Anyone who has attempted to describe a complex system using traditional units such as pounds, slugs, feet, volts, pounds per square inch, gallons per hour, and the like will appreciate how difficult it is to ensure that the proper unit conversions have been incorporated. In fact, most computer programs for processing bond graphs into the equivalent differential equations for subsequent analysis and simulation are incapable of incorporating conversion factors and thus essentially assume that the SI system will be used. In this text, we will simply assume that the SI system will be used. After an analysis or simulation has been completed, it is a simple matter to convert some results to a traditional unit system. The power demand of an electric car in kilowatts (kW), for example, could readily be converted to horsepower if this would be better understood by consumers, but we believe that it is a mistake to create a mathematical model using units that require conversion factors internally.

Table 2.3 gives power and energy variables for ports involving mechanical rotation. The shafts of motors, pumps, gears, and many other useful devices represent such ports.

The entries in Table 2.4 for hydraulic power again are related to the variables used in solid mechanics, but some unusual quantities are defined. The momentum quantity is defined according to Eq. (2.2) as the integral of the effort, or in this case, the pressure. Not only is the pressure momentum a quantity not often encountered

TABLE 2.3. Power and Energy Variables for Mechanical Rotational Ports

Generalized Variables	Mechanical Rotation	SI Units
Effort, e	Torque, τ	Newton-meters (N-m)
Flow, f	Angular velocity, ω	Radians per second (rad/s)[a]
Momentum, p	Angular momentum, p_τ	N-m-s
Displacement, q	Angle, θ	rad[a]
Power, P	$\tau(t)\omega(t)$	N-m/s
Energy, E	$\displaystyle\int^{\theta} \tau\,d\theta,\ \int^{p_\tau} \omega\,dp_\tau$	N-m

[a]Radians and other angular measures are dimensionless, but there are scale factors between, say, radians, revolutions, and degrees which can cause errors not discoverable by dimensional analysis. The formulas used in this book all are based on the radian as the unit of angular measure.

TABLE 2.4. Power and Energy Variables for Hydraulic Ports

Generalized Variables	Hydraulic Variables	SI Units
Effort, e	Pressure, P	Newtons per square meter (N/m^2)[a]
Flow, f	Volume flow rate, Q	Cubic meters per second (m^3/s)
Momentum, p	Pressure momentum, p_p	$N\text{-}s/m^2$
Displacement, q	Volume, V	m^3
Power, P	$P(t)Q(t)$	N-m/s
Energy, E	$\displaystyle\int^{v} P\,dV,\ \int^{p_p} Q\,dp_p$	N-m

[a]In subsequent tables, when pressure is involved, the units will be given as N/m^2 rather than the equivalent pascals (Pa) for clarity.

in conventional fluid mechanics, but it is also a quantity without an obvious symbol. The symbol p_p is meant to indicate a momentum quantity that is the integral of $P(t)$, just as in Table 2.3 p_τ was a momentum quantity defined as the time integral or $\tau(t)$. Fortunately, the lack of a commonly accepted symbol for certain variables is not a serious handicap. When some facility in system modeling has been developed, the generalized variables, e, f, p, and q, can be used for variables in all the energy domains, if desired.

Finally, Table 2.5 gives power and energy variables for electrical ports. The only new quantity that needs to be defined is the unit of electrical charge, the coulomb. It is common to use *volts* and *amperes* for the units of voltage and current rather than their equivalents in terms of coulombs and SI units. Most of the variables in Table 2.5 should be familiar, with the possible exception of the momentum or flux linkage variable λ. The usefulness of this variable will become evident when inductors are studied in Chapter 3.

TABLE 2.5. Power and Energy Variables for Electrical Ports

Generalized Variable	Electrical Variable	Units
Effort, e	Voltage, e	Volt (V) = newton-meter per coulomb (N-m/C)
Flow, f	Current, i	Ampere (A) = coulomb per second (C/s)
Momentum, p	Flux linkage variable, λ	V-s
Displacement, q	Charge, q	C = A-s
Power, P	$e(t)i(t)$	V-A = W = N-m/s
Energy, E	$\int^{q} e\, dq, \int^{\lambda} i\, d\lambda$	J = V-A-s = W-s = N-m

The tables of variables presented give at least preliminary evidence that the variables associated with a variety of physical systems can be fitted into the scheme of Figure 2.2. The usefulness of this viewpoint will become increasingly evident as systems are modeled in detail. In the next sections the ways in which subsystem interconnections can be indicated graphically using the e, f, p, q classification will be shown.

2.2   PORTS, BONDS, AND POWER

The devices sketched in Figure 2.1 can all be treated as multiport elements with ports that can be connected to other multiports to form systems. Further, when two multiports are connected, power can flow through the connected ports and the power can be expressed as the product of an effort and a flow quantity, as given in Tables 2.2–2.5. We now develop a universal way to represent multiports and systems of interconnected multiports based on the variable classifications in the tables.

Consider the separately excited d-c motor shown in Figure 2.3. Physically, such motors have three obvious ports. The two electrical ports are represented by armature and field terminal pairs, and the shaft is a rotary mechanical port as sketched in Figure 2.3a. Figure 2.3b is a conventional schematic diagram in which the mechanical shaft is represented by a dashed line, the field coils are represented by a symbol similar to the circuit symbol for an inductance, and the armature is represented by a highly schematic sketch of a commutator and brushes. Note that the schematic diagram does not indicate what the detailed internal model of this subsystem or component will be. To write down equations describing the motor, an analyst must decide how detailed a model is necessary.

Figure 2.3c represents a further step in simplifying the representation of this engineering multiport. The name *d-c motor* is used to stand for the device, and the ports are simply indicated by single lines emanating from the word representing the device. As a convenience, the effort and flow variables are written next to the lines representing the ports. Whenever the port lines are either horizontal or vertical, it is useful to use the following convention:

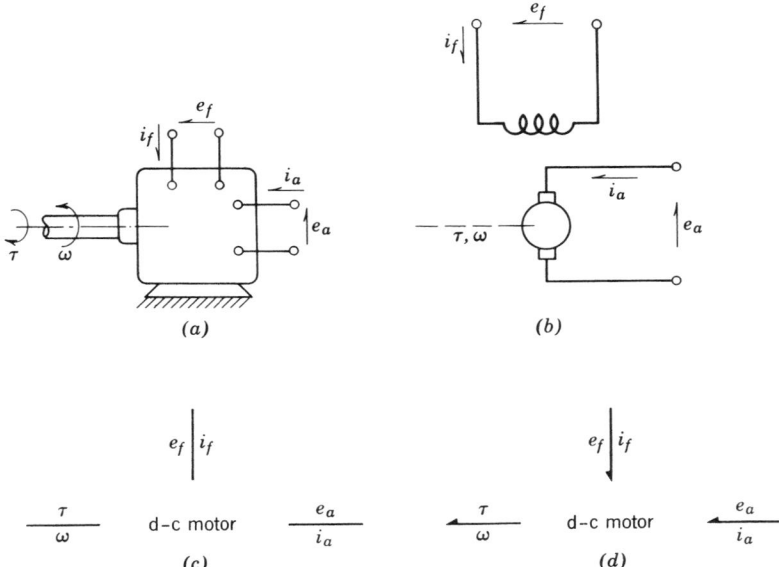

FIGURE 2.3. Separately excited d-c motor. (*a*) Sketch of motor; (*b*) conventional schematic diagram; (*c*) multiport representation; (*d*) multiport representation with sign convention for power.

- Efforts are placed either *above* or to the *left* of the port lines.

- Flows are placed either *below* or to the *right* of port lines.

- When diagonal lines are used, some judgment is required for placement of the effort and flow variables.

Note that Figures 2.3*a*, *b*, and *c* all contain the same information, namely, that the d-c motor is a 3-port with power variables τ, ω, e_f, i_f, e_a, and i_a. In Figure 2.3*d*, a sign convention has been added: The *half arrow* on a port line indicates the direction of power flow at any instant of time when the effort and flow variables both happen to be positive.

For example, if ω is positive in the direction shown in Figure 2.3*a* and if τ is interpreted to be the torque on the motor shaft resulting from a connection to some other multiport and is positive in the direction shown in Figure 2.3*a*, then when τ and ω are both positive (or for that matter, both negative), the product $\tau\omega$ is positive and represents power flowing *from* the motor to some other multiport coupled to the motor shaft. Thus, the half arrow in Figure 2.3*d* points *away* from the d-c motor. Similarly, when e_f, i_f, e_a, and i_a are positive, power flows to the motor from whatever other multiports are connected to the field and armature terminals. Hence, the half arrows associated with the field and armature ports point *toward* the motor.

Anytime one desires to be specific about the characteristics of a multiport—for instance, in equation form or in the form of tabulated data—then a sign convention is necessary. The establishment of sign conventions is fairly straightforward for electric circuits or for the circuit-like parts of representations of multiports such as those of Figures 2.3a and b. Anyone who has struggled with the definition of forces and moments on interconnected rigid bodies using "free-body diagrams," however, knows that the establishment of sign conventions in mechanical systems is not trivial. The problem is that the action and reaction forces show up as oppositely directed in most representations. Thus, in Figure 2.3a, one must decide whether τ represents the torque *on* the motor shaft or *from* the motor onto some other multiport. On diagrams such as Figure 2.3b, the mechanical signs are often not indicated at all, and it is up to the analyst to insert plus or minus signs in the equations without much help from the schematic diagram of the system.

When two multiports are coupled together so that the effort and flow variables become identical, the two multiports are said to have a common *bond*, in analogy to the bonds between component parts of molecules. Figure 2.4 shows part of a system consisting of three multiports bonded together. The motor and pump have a common angular speed, ω, and torque at the coupling, τ. The battery and the motor have a common voltage and current defined at the terminals at which the battery leads connect to the motor armature. To represent this type of subsystem interconnection in the manner of Figure 2.3c or d is very straightforward; the joined ports are represented by a single line or bond between the multiports. This has been done in Figure 2.5. The line between the pump and motor in Figure 2.5 implies that a port of the motor and a port of the pump have been connected, and hence a single torque and a single angular

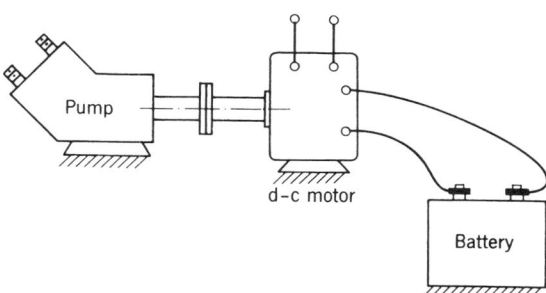

FIGURE 2.4. Partially assembled system.

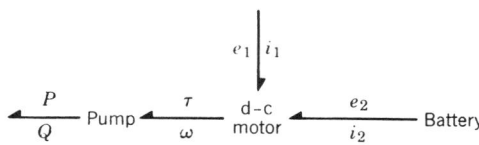

FIGURE 2.5. Word bond graph for system of Figure 2.4.

speed pertain to both the pump and the motor. The half arrow on the bond means that the torque and the speed are defined in such a way that when their product, $\tau\omega$, is positive, power is flowing from the motor to the pump. Thus, lines associated with isolated multiports indicate ports or potential bonds. For interconnected multiports, a line represents the conjunction of two ports, that is, a bond.

2.3 BOND GRAPHS

The mechanism for studying dynamic systems to be used subsequently in this book is the *bond graph*. A bond graph simply consists of subsystems linked together by lines representing power bonds, as in Figure 2.5. When major subsystems are represented by words, as in Figure 2.5, then the graph is called a *word bond graph*. Such a bond graph establishes multiport subsystems, the way in which the subsystems are bonded together, effort and flow variables at the ports of the subsystems, and sign conventions for power interchanges.

Since the word bond graph serves to make some initial decisions about the representation of dynamic systems, it is worthwhile to consider some example systems even before the details of dynamic systems have been presented. In Figure 2.6 part of a positioning system for a radar antenna is shown. The word bond graph indicates the major subsystems to be considered, and the bonds with the effort and flow variables indicated introduce some variables that will be useful in characterizing the subsystems at a later stage in the analysis. You should be able to associate all the efforts and flows on the bond graph with physical quantities associated with the physical system being modeled. Try it.

In Figure 2.7 another example system is shown. Again, it is instructive to try to understand the effort and flow quantities associated with the bonds in the word bond graph. For example, what are the three efforts and three flows associated with the 3-port—Diff—? Can you see that—Wheel—is a 2-port that relates a torque and an angular speed to a force and a velocity? Do not be surprised if the construction of a word bond graph for a dynamic system seems less than obvious at this stage. As you progress in the details of system modeling, it will become easier to recognize ports and bonds and, hence, multiport subsystems.

In Figure 2.7 you will notice that the influences of throttle position, clutch linkage position, and gear selector position are indicated using a bond with a full arrowhead. This notation, which is discussed in more detail in the next section, indicates that an influence on the system from its environment occurs at essentially zero power flow. In the present example, the driver of the car is part of the environment of the car and can control the car using the accelerator pedal, clutch pedal, and gear shift using low power, as compared with the power present in the drive train. A bond with a full arrow is an *active bond*, and it indicates a signal flow at very low power. In the present case, we assume that the controls of the car can be moved by the driver at will, and our dynamic model need not concern itself with the forces required to move the controls. A word bond graph is useful for sorting true power interactions from the one-way influences of active bonds.

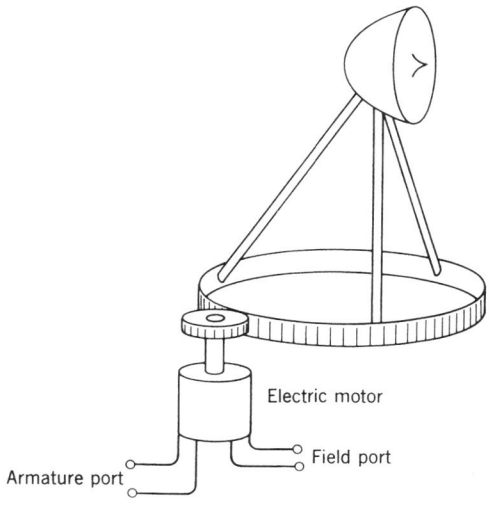

Field
supply

v_2 | i_2

Armature $\dfrac{v_1}{i_1}$ — Motor $\dfrac{\tau_3}{\omega_3}$ — Shaft $\dfrac{\tau_4}{\omega_4}$ — Gears $\dfrac{\tau_5}{\omega_5}$ — Pedestal
supply   load

τ = torque
ω = angular velocity
v = voltage
i = current

FIGURE 2.6. Schematic diagram and word bond graph for radar antenna pedestal drive system.

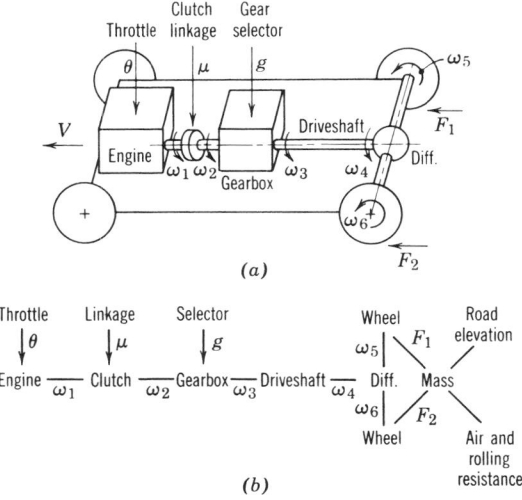

FIGURE 2.7. Automotive drive train example. (*a*) Schematic diagram; (*b*) word bond graph.

Bond graphs will subsequently be used to model subsystems in detail internally. For this purpose, a set of basic multiport elements denoted not by words, but by letters and numbers, will be developed in the next chapter. Ultimately, detailed bond graphs must be substituted for the multiports designated by words in a word bond graph. From a sufficiently detailed bond graph, state equations may be derived using standard techniques or computer simulations of the system can be made. Several computer programs will accept a wide variety of bond graphs directly and produce either state equations for subsequent analysis or system response predictions. In addition, some types of analyses can be performed on a bond graph without either writing the state equations or using a computer.

2.4 INPUTS, OUTPUTS, AND SIGNALS

Multiport subsystem characteristics typically are determined by a combination of experimental and theoretical methods. It might be fairly easy to compute the moment of inertia of a rotor, for example, merely by knowing the density of the material of which the rotor was made and having a drawing of the part, but to predict the port characteristics of a fan in great detail by theoretical means would be much more difficult than by measuring the characteristics. In performing experiments on a subsystem, the notions of *input* and *output* or, equivalently, *excitation* and *response*, arise. The same concepts will carry over when "mathematical" models of subsystems are assembled into a system model.

In performing experiments on a multiport, one must make a decision about what is to be done at the ports. At each port, both an effort and a flow variable exist, and one can control either one but not both of these variables simultaneously. As an example, consider the problem of determining the steady-state characteristics of a d-c motor such as the one shown in Figures 2.3–2.5.

Figure 2.8*a* shows a sketch of equipment that could be used in experimenting on the motor. The dynamometer is supposed to be capable of setting the speed of the motor regardless of the torque delivered by the motor. This speed, ω, is then an *input variable* to the motor. The torque being delivered by the motor is then measured by means of a torque gage. The torque is thus an output variable of the motor. Note that, in general, it is not possible to adjust the dynamometer for both torque and speed. The nature of the experiment is to discover what the motor torque is at a given speed.

Similarly, if voltages are supplied to the two electrical ports, that is, if voltages are input variables, then the motor responds with measurable currents that are output variables of the motor. Figure 2.8*b* is an attempt to use lines and arrows to show which quantities are inputs to the motor and which are outputs. Figure 2.8*b* is a simple example of a *block diagram*, in which lines with arrows indicate the direction of flow of *signals*. For multiports each port or bond has both an effort and a flow, and when these two types of variables are represented as paired signals, it is possible for only one of these signals to be an input and the other to be an output.

To know which of the effort and flow signals at a port is the input of the multiport, only one piece of information must be supplied to Figures 2.3*c*, 2.3*d*, or 2.5. This is

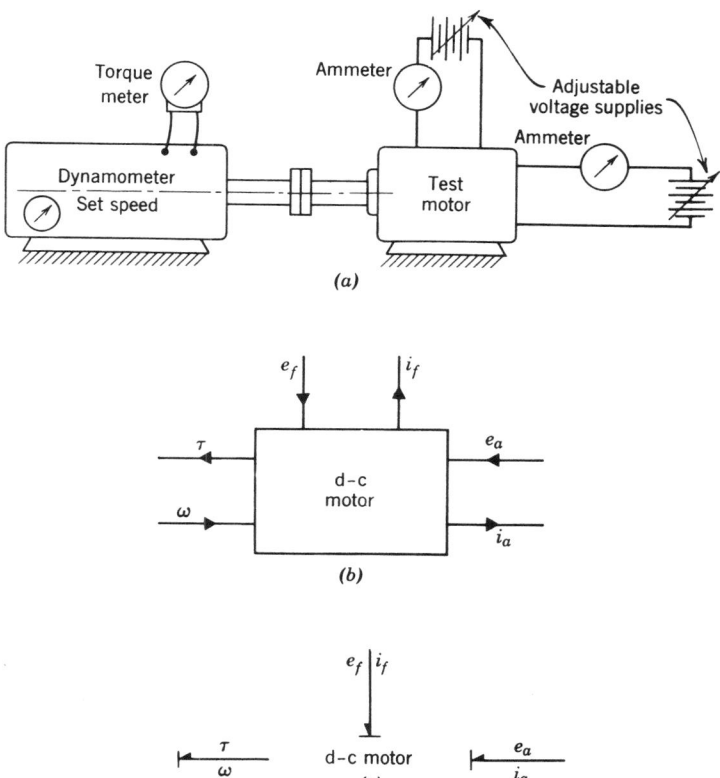

FIGURE 2.8. Experimental testing of a d-c motor. (*a*) Sketch of the test apparatus; (*b*) block diagram showing input and output signal flow; (*c*) causal strokes added to multiport representation.

because if either the effort or flow variable is an input, the other is an output. In bond graphs the way in which inputs and outputs are specified is by means of the *causal stroke*. The causal stroke is a short, perpendicular line made at one end of a bond or port line. It indicates the direction in which the effort signal is directed. (By implication, the end of a bond that does not have a causal stroke is the end toward which the flow signal arrow points.) In Figure 2.8*c* causal strokes have been added to the multiport representation of Figure 2.3*d*. By comparing Figures 2.8*a–c*, all of which contain the same information regarding input and output variables, the meaning of causal strokes may be appreciated. It is summarized in Figure 2.9, in which both bond graphs and block diagrams are shown. Note that the half-arrow sign convention for power flow and the causal stroke are completely independent. Thus, using A and B to stand for subsystems as in Figure 2.9, all the following combinations of sign convention and causal strokes are possible: $A \vdash\!\!\rightarrow B$, $A \vdash\!\!\leftarrow B$, $A \rightarrow\!\!\dashv B$, $A \leftarrow\!\!\dashv B$. The study of input–output *causality*, which is a uniquely useful feature of bond graphs, will be dealt with at length in succeeding chapters.

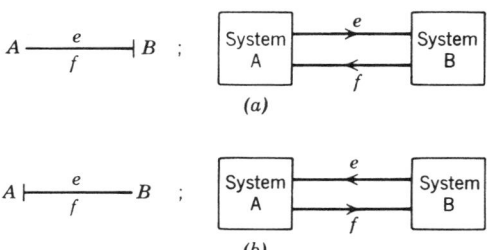

FIGURE 2.9. The meaning of causal strokes. (a) Effort is output of A, input to B; flow is output of B, input to A; (b) effort is output of B, input to A; flow is output of A, input to B.

Finally, we come to the question of pure signal flow, or the transfer of information with negligible power flow, which we already encountered in the example of Figure 2.7. Multiports in principle all transmit finite power when interconnected. This is correlated with the fact that both an effort and a flow variable exist when multiports are coupled. Thus, systems are interconnected by the matching of a *pair* of signals representing the power variables.

In many cases, however, systems are so designed that only one of the power variables is important, that is, so that a single signal is transmitted between two subsystems. For example, an electronic amplifier may be designed so that the voltage from a circuit influences the amplifier, but the current drawn by the amplifier has virtually no effect on the circuit. Essentially, the amplifier reacts to a voltage but extracts negligible power in doing so, as compared with the rest of the power levels in the circuit. No information can really be transmitted at zero power, but, practically speaking, information can be transmitted at power levels that are negligible, as compared with other system power levels. Every instrument is designed to extract information about some system variable without seriously disturbing the system to which the instrument is attached. An ideal ammeter indicates current but introduces no voltage drop, an ideal voltmeter reads a voltage while passing no current, an ideal pressure gage reads pressure with no flow, an ideal tachometer reads angular speed with no added torque, and the like. When an instrument reads an effort or flow variable, but with negligible power, there is a signal connection between subsystems without the back effect associated with power interaction.

The block diagrams of control engineering or the signal flow graphs that were developed first for electrical systems ideally show signal coupling. As Figure 2.8b shows, when multiports are considered, power interactions require a pair of bilaterally oriented signals. The bond graph, in which each bond implies the existence of both an effort and a flow signal, is a more efficient way of describing multiports than are block diagrams or signal flow graphs. Yet when the system is dominated by signal interactions due to the presence of instruments, isolating amplifiers, and the like, then either an effort or a flow signal may be suppressed at many interconnection points. In such a case, a bond degenerates to a single signal and may be shown as an *active bond*. The notation for an active bond is identical to that for a signal in a block diagram; for example, $A \xrightarrow{e} B$ indicates that the effort, e, is determined by subsystem A and is an input to subsystem B. Normally, this situation would be indicated by $A \xrightarrow[f]{e} B$, in which the flow, f, is determined by B and is an input to A. When e is shown as a signal (by means of the full arrow on the bond) or, in other words, an activated bond, the implication is that the flow, f, has a negligible effect on A.

When automatic control systems are added to physical systems, the control systems usually receive signals by means of nearly ideal instruments and affect the systems through nearly ideal amplifiers. The use of active bonds for such cases simplifies the analysis of the systems. Notice that in using bond graphs, one always assumes that multiports are coupled with both forward and backward effects unless a specific modeling decision has been made that a back effect is negligible.

In the following chapter we begin the detailed modeling of subsystems by considering a basic idealized set of multiports that can be assembled to model the pertinent physical effects in a subsystem. At this detailed level, physical parameters must be estimated and the rules of causality among ideal multiports must be discovered and obeyed in assembling the subsystem model from elemental multiports. As this process goes on, the notation and concepts briefly introduced in this chapter will become more familiar and useful.

PROBLEMS

2-1 Construct four tetrahedra of state similar to that shown in Figure 2.2 for the following four physical domains: mechanical translation, mechanical rotation, hydraulic systems, electrical systems. Replace e, f, p, and q with their physical counterpart variables, and list the dimensions of each variable.

2-2 For each multiport in Figure 2.1 construct a word bond graph similar to that shown in Figure 2.3. Construct several systems by bonding several multiports together.

2-3 Suppose a pump was tested by running it at various speeds and measuring the volume flow rate and torque for various pressures at the pump outlet. Draw a schematic diagram, block diagram, and bond graph for the pump test analogous to those shown in Figure 2.8 for an electric motor test.

2-4 If the system of Figure 2.5 had the causality

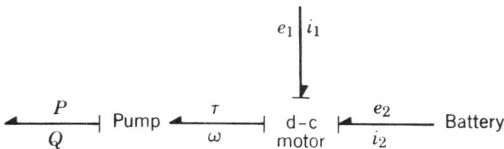

show how the signals flow by using a block diagram of the type used in Figure 2.8b for each multiport. See also Figure 2.9.

2-5 Apply causal strokes in an arbitrary manner to each bond in the bond graph of Figure 2.6. Construct an equivalent block diagram for this system using one block for each multiport as in Figure 2.8b. Indicate the signal flow directions that correspond to your causal marks as in Figure 2.9.

2-6 Repeat Problem 2-5 for the system of Figure 2.7. (Note that active bonds act just like one-way signal flows in a block diagram, but that normal bonds each result in two signal flows for e and f.)

2-7 Consider the system of Problem 2-4. Identify the system input variables that come from the environment of the system, given the causal stroke pattern indicated. What variables are indicated as system outputs (and hence inputs to the environment)?

2-8 How long would a 100-W light bulb have to burn to use up the same energy that would be required to raise a 10-kg mass 30 m up in the earth's gravity field?

2-9 Represent an electric drill as a multiport. Consider the switch position influence as occurring on an active bond. Apply causal strokes to your bond graph, assuming that the drill is plugged into a 100-V outlet and that the torque is determined by the material being drilled. Show a block diagram for the drill corresponding to your choice of causality at the ports.

2-10 If a positive-displacement hydraulic pump is 100% efficient (so that the mechanical power is always instantaneously equal to the hydraulic power) and if a torque of 5 N-m produces a pressure of 7.0 MPa, what is the relationship between volume flow and angular speed? (7.0 MPa $= 7.0 \times 10^6$ N/m^2.)

2-11 The slider-crank mechanism is the fundamental kinematic device in virtually all internal combustion engines. This device relates the rotational motion of the crankshaft to the reciprocating motion of the piston. In its most idealized representation, the slider-crank is massless, frictionless, and constructed from rigid components. Under these assumptions, the device is power conserving, in that $\tau\omega = Fv$, where τ is the torque on the crankshaft, F is the force on the end of the connecting rod, ω is the angular velocity of the crank, and v is the velocity of the rod end. If we can derive how are related, then we automatically know how F and τ are related.

As a word bond graph, the slider-crank will be represented as indicated in the figure. We will soon learn that this device is a modulated transformer.

Derive the relationship between v and ω. Here is some help.

$$x = R\cos\theta + l\cos\alpha$$

$$l\sin\alpha = R\sin\theta$$

Solve the second equation for $\sin\alpha$ and then use $\cos\alpha = \sqrt{1 - \sin^2\alpha}$. Substitute into the first equation. Then differentiate the result to relate $\dot{x} = -v$ to $\dot{\theta} = \omega$. If you complete these steps, you will have derived $v = m(\theta)\omega$,

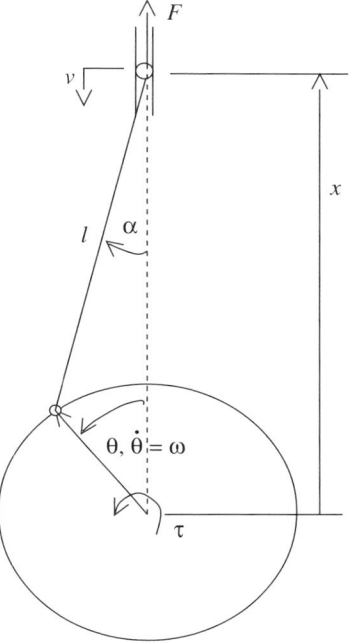

(*a*) Slider-crank device

$$\frac{\tau}{\omega} \quad \text{Slider-crank device} \quad \frac{F}{v}$$

(*b*) Word bond graph

where $m(\theta)$ is a function of the crank angle. Since this device is power conserving, we immediately know that $\tau = m(\theta)F$.

Now try to derive the relationship between τ and F by using force and moment equilibrium conditions. You will find this far more difficult than deriving the velocity-angular velocity relationship.

2-12 The hydraulic system from Problem 1-9 has the word bond graph shown in the following figure. Causality has been assigned to the bonds. Identify the inputs and outputs for each element.

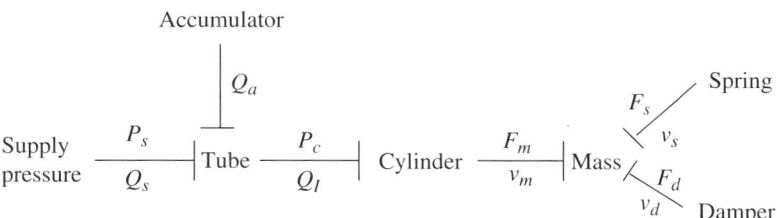

3

BASIC COMPONENT MODELS

In Chapter 2 real devices were considered as subsystems from the point of view of power exchanges and external port variables. In this chapter a basic set of multiports is defined that can be used to model subsystems in detail. These multiports function as components of subsystem and system models and are, in many cases, idealized mathematical versions of real components such as resistors, capacitors, masses, springs, pipes, and so on. In other cases, however, the basic multiports are used to model *physical effects* in a device and cannot be put into a one-to-one correspondence with physical components of the device. For example, one might create a model of an electrical or fluid transmission line using a finite collection of resistance, capacitance, and inertia elements, even though in the real device the effects being modeled are distributed along the transmission line and not concentrated into lumps as in the model.

Using bond graphs and the classification of power and energy variables presented in the previous chapter, it turns out that only a few basic types of multiport elements are required in order to represent models in a variety of energy domains. The bond graph notation often allows one to visualize aspects of the system more easily than would be possible with just the state equations or with some other graphical notation designed for a single energy domain or for signal flow rather than power flow. The search for a bond graph model of a complex system frequently increases one's physical understanding of the system.

3.1 BASIC 1-PORT ELEMENTS

A 1-port element is addressed through a single power port, and at the port a single pair of effort and flow variables exists. Generally, a 1-port can be a very complex subsystem. An ordinary electrical wall outlet can represent the port of a 1-port in a system analysis. The port actually connects to a vast network of power generation and distribution equipment, yet from the point of view of a system model, a relatively simple characterization of what is behind the wall outlet as a 1-port may suffice.

Here we deal with the most primitive 1-ports. We consider, in order, elements that dissipate power, store energy, and supply power.

The *1-port resistor* is an element in which the effort and flow variables at the single port are related by a static function. Figure 3.1 shows the bond graph symbol for the resistor, a typical graph of the constitutive relation between e and f, and sketches of resistors in several energy domains. The electrical resistor is an R-element because it is characterized by the linear volt–current constitutive relationship,

$$e = Ri.$$

Since e is an effort variable and i is a flow variable, this constitutive relationship exactly duplicates our definition of a linear 1-port resistor. The mechanical dashpot is a 1-port resistor for the same reason as in the electrical resistor. The ideal dashpot is characterized by the linear force–velocity relationship

$$F = bV,$$

where b is the dashpot constant. Since F is an effort and is V a flow, the constitutive relationship fits exactly with our definition of a 1-port resistor. The hydraulic device is also a 1-port resistor because it is characterized by a pressure–volume flow rate relationship, although in most cases the relationship is not linear. This is dealt with in great detail in Chapter 12; however, here we just specify that for turbulent flow

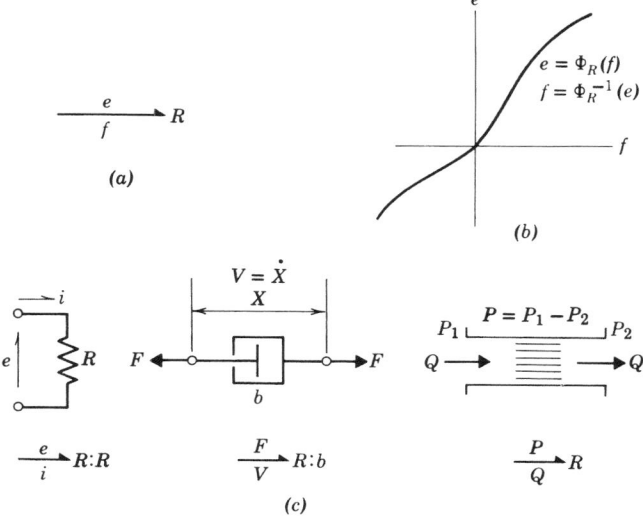

FIGURE 3.1. The 1-port resistor. (*a*) Bond graph symbol; (*b*) defining relation; (*c*) representations in several physical domains.

through a restriction, the pressure–flow relationship is

$$P_1 - P_2 = \frac{1}{2A^2}\rho Q|Q|$$

or

$$Q = A\sqrt{\frac{2}{\rho}|P_1 - P_2|}\,\text{sgn}(P_1 - P_2),$$

where A is the flow area. These relationships are nonlinear effort–flow relationships and again fit our definition of a one-port resistor, given in generalized form at the top of Table 3.1 and in particular for hydraulic systems listed lower in the table.

Usually, resistors dissipate energy. This must be true for simple electrical resistors, mechanical dampers or dashpots, porous plugs in fluid lines, and other analogous passive elements. Noting from Figure 3.1a that power flows *into* the port when the product of e and f is positive according to the sign convention shown, we may deduce that power is always dissipated if the defining constitutive relation between e and f lies in the first and third quadrants of the e–f plane as shown in Figure 3.1b, for then the product ef is positive.

When the relation between e and f for a 1-port resistor plots as a curved line as in Figure 3.1b, then the resistor is a *nonlinear element*. If the relation is a straight line, then it is a *linear* resistor. In the special case of a linear element, a coefficient, the *resistance*, or its inverse, the *conductance*, may be defined. When a resistive element is assumed to be linear, it is conventional to indicate this on the bond graph by appending a colon (:) next to the $-R$ and noting the physical symbol for the resistance parameter. This is done in Figure 3.1 for the electrical resistance and the mechanical dashpot. For the hydraulic resistor, no parameter is indicated since this is a nonlinear element and no resistance parameter can be identified.

TABLE 3.1. The 1-Port Resistor, $\dfrac{e}{f}\,R$

	General Relation	Linear Relation	SI Units for Linear Resistance Parameter
Generalized variables	$e = \Phi_R(f)$ $f = \Phi_R^{-1}(e)$	$e = Rf$ $f = Ge = e/R$	$R = e/f$
Mechanical translation	$F = \Phi_R(V)$ $V = \Phi_R^{-1}(F)$	$F = bV$	$b = \text{N} - \text{s/m}$
Mechanical rotation	$\tau = \Phi_R(\omega)$ $\omega = \Phi_R^{-1}(\tau)$	$\tau = c\omega$	$c = \text{N} - \text{m} - \text{s}$
Hydraulic systems	$P = \Phi_R(Q)$ $Q = \Phi_R^{-1}(P)$	$P = RQ$	$R = \text{N} - \text{s/m}^5$
Electrical systems	$e = \Phi_R(i)$ $i = \Phi_R^{-1}(e)$	$e = Ri$ $i = Ge$	$R = \text{V/A} = \Omega$ (ohm)

The resistance relationships are summarized in the first lines of Table 3.1. Note that for power-dissipating resistors, with the sign convention shown in Figure 3.1 and Table 3.1, the resistance and conductance parameters, R and G, respectively, are positive.

For simplicity, we establish the following arbitrary but useful rule: For passive resistors, establish the power sign convention by means of a half arrow pointing *toward the resistor*. Then linear resistance parameters will be positive, and nonlinear relations will fall in the first and third quadrants of the $e-f$ plane.

Since linear models are of great usefulness in certain fields (vibrations and electric circuits, for example), the linear versions of resistance relations in various energy domains are shown in Table 3.1 with the same notation employed in Chapter 2. The units of the linear resistance parameter are simply the units of effort divided by the units of flow. The units displayed in Table 3.1 are worth studying, since many of them may not be familiar. The only resistance unit dignified with its own name is the electrical ohm.

Next consider a 1-port device in which a static constitutive relation exists between an effort and a displacement. Such a device stores and gives up energy without loss. In bond graph terminology, an element that relates e to q is called a *1-port capacitor* or *Compliance*. In physical terms, a capacitor is an idealization of such devices as springs, torsion bars, electrical capacitors, gravity tanks, and accumulators. The bond graph symbol, the defining constitutive relation, and some physical examples are shown in Figure 3.2.

As with the 1-port resistor, there are idealized linear compliance elements as well as nonlinear ones. In Figure 3.2b, a general nonlinear constitutive e, q relationship is shown. If the element can be assumed linear, then the e, q curve will be a straight line and a compliance parameter can be defined such that $e = q/C$. Note that it is customary to define the linear compliance relationship using the inverse of the slope of the e, q curve. The reason for this will become clear when more physical elements are presented in the next chapter. For the linear case, it is customary to indicate the compliance parameter on the bond graph as shown in Figure 3.2.

The electrical capacitor, of capacitance C farads, is a compliance element because its idealized behavior is

$$e = \frac{q}{C},$$

where $q = \int i\, dt$ is the charge on the capacitor. This fits perfectly with our definition of a linear 1-port capacitor. The spring of stiffness, k, is a 1-port capacitor because it is characterized by

$$F = kx,$$

where $x = \int V\, dt$ is the relative displacement across the spring. This definition fits the general definition of a 1-port capacitor and the specific definition of a linear 1-port C-element as indicated in Table 3.2. In this case, as indicated on the bond graph, the compliance parameter is $C = 1/k$. A water storage tank is discussed in the next chapter, and it is ideally a linear compliance element. The torsional spring of

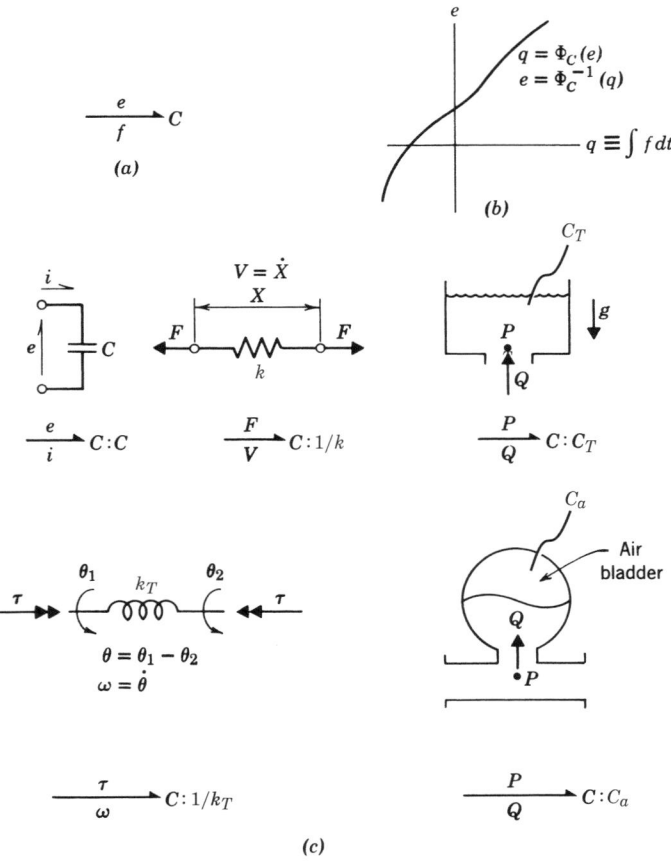

FIGURE 3.2. The 1-port capacitor. (*a*) Bond graph symbol; (*b*) defining relation; (*c*) representation in several physical domains.

TABLE 3.2. The 1-Port Capacitor, $\overset{e}{\underset{f=\dot{q}}{\longrightarrow}} C$

	General Relation	Linear Relation	SI Units for Linear Capacitance Parameter
Generalized	$q = \Phi_C(e)$	$q = Ce$	$C = q/e$
	$e = \Phi_C^{-1}(q)$	$e = q/C$	$1/C = e/q$
Mechanical	$X = \Phi_C(F)$	$X = CF$	$C = \text{m/N}$
translation	$F = \Phi_C^{-1}(X)$	$F = kX$	$k = \text{N/m}$
Mechanical	$\theta = \Phi_C(\tau)$	$\theta = C\tau$	$C = \text{rad/N} - \text{m}$
rotation	$\tau = \Phi_C^{-1}(\theta)$	$\tau = k\theta$	$k = \text{N} - \text{m/rad}$
Hydraulic	$V = \Phi_C(P)$	$V = CP$	$C = \text{m}^5/\text{N}$
systems	$P = \Phi_C^{-1}(V)$	$P = V/C$	
Electrical	$q = \Phi_C(e)$	$q = Ce$	$C = \text{A} - \text{s/V}$
systems	$e = \Phi_C^{-1}(q)$	$e = q/C$	$= \text{farad (F)}$

stiffness, k_τ, N-m/rad, is a linear 1-port compliance for the identical reason as in the linear spring. The compliance parameter for the torsional spring is $1/k_\tau$, as indicated on the bond graph of Figure 3.2. An air bladder is also discussed in the next chapter. There are circumstances in which it may be linear. However, in general, compressing air is a nonlinear process. If the process was isentropic, then the behavior of the air bladder could be characterized by

$$P = \frac{P_0 V_0^\gamma}{V^\gamma},$$

where $V = \int Q\,dt$, P_0, V_0 are initial pressure and volume in the bladder, and γ is the ratio of specific heats for air. This behavior is nonlinear, but still fits our general definition of a 1-port capacitor as indicated in Table 3.2. Thus, for the case here, the air bladder is a compliance element, but it is not a linear one, and no compliance parameter can be identified nor indicated on the bond graph.

Note that when a sign convention similar to that used for the resistor, namely, $\rightarrow C$, is used for the C-element, then ef represents power flowing *to* the capacitor and

$$\mathbf{E}(t) = \int_0^t e(t) f(t)\, dt + \mathbf{E}_0 \tag{3.1}$$

represents the energy stored in the capacitor at any time t. The energy stored initially at $t = 0$ (if any) is called $\mathbf{E}_0$.

Since from Eq. (2.2b) the displacement q is defined so that $f\,dt \equiv dq$, and the constitutive relation of a C-element implies that e is a function of q, $e = e(q)$, then Eq. (3.1) can be rewritten as

$$\mathbf{E}(q) = \int_{q_0}^q e(q)\, dq + \mathbf{E}_0, \tag{3.2}$$

where $\mathbf{E}_0$ is the energy stored when $q = q_0$. Usually, it is convenient to define the energy stored to be zero when the effort is zero. Then, if q_0 is that value of q at which $e = 0$, and $\mathbf{E}_0 = 0$, Eq. (3.2) may be written as

$$\mathbf{E}(q) = \int_{q_0}^q e(q)\, dq. \tag{3.2a}$$

The operation indicated in Eq. (3.2a) may be interpreted graphically as shown in Figure 3.3. As q varies, the area under the curve of e versus q varies, and this area is equal to $\mathbf{E}$. The *conservation of energy* for $-C$ is almost obvious. If q goes from q_0 to q as in Figure 3.3a, then energy is stored; if q then ever returns to q_0, the shaded area disappears and all the stored energy disappears. The power flow into the port, which resulted in the storage of energy, reverses and power flows out of the port. During the process, no energy is lost.

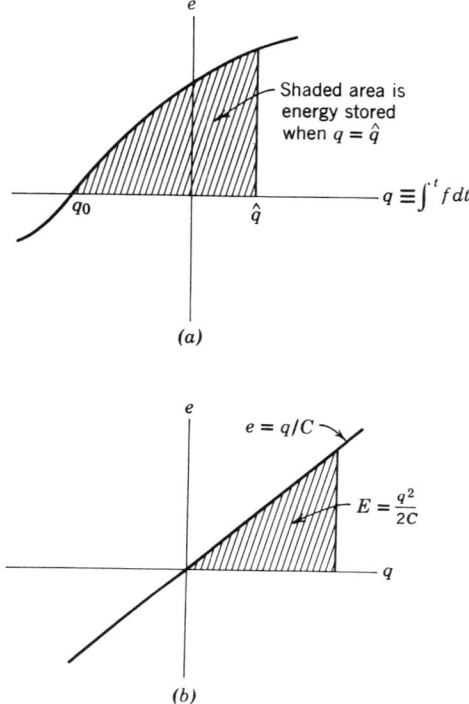

$$q \equiv \int^t f\, dt$$

(a)

(b)

FIGURE 3.3. Area interpretation of stored energy for 1-port capacitor. (*a*) Nonlinear case; (*b*) linear case.

Table 3.2 summarizes the relationships characterizing capacitors. The units for linear capacitance parameters are given, and again it may be noted that only the electrical unit is given a name, the farad. For linear mechanical systems, it is common to use the *spring constant*, k, rather than the *compliance*, $C \equiv 1/k$, which is analogous to the electrical capacitance, C, and the parameter C in generalized variables. In mixed electrical-mechanical systems, one must simply be careful to note whether a numerical parameter corresponds to C or the inverse of C in a bond graph. Once again, the reader is urged to study the units shown, since some units will probably be unfamiliar.

A second energy-storing 1-port arises if the momentum p is related by a static constitutive law to the flow f. Such an element is called an *inertia* in bond graph terminology. The bond graph symbol for an inertia, the constitutive relation, and several physical examples are shown in Figure 3.4. The inertia is used to model inductance effects in electrical systems, and mass or inertia effects in mechanical or fluid systems.

The 1-port inertia is characterized by an f, p relationship as indicated in Figure 3.4*b*. If the relationship is linear, then it will plot as a straight line and the constitutive relationship will have the form $f = p/I$, where I is the inertia parameter

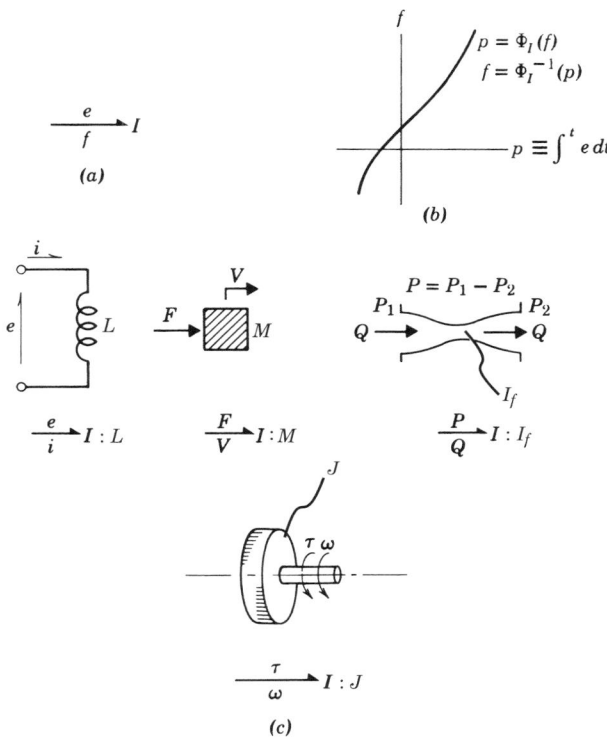

FIGURE 3.4. The 1-port inertia. (*a*) Bond graph symbol; (*b*) defining relation; (*c*) representation in several physical domains.

and $p = \int e \, dt$. Note that, as with the linear compliance element, it is customary to define the inertance parameter as the inverse of the slope of the linear relationship. Shown in Figure 3.4 is an electrical inductor with inductance L, a mass m, a section of fluid-filled pipe with fluid inertia, I_f, and a rotating disk with moment of inertia, J. These are all examples of linear 1-port inertia elements. The inductor is ideally represented by the constitutive relationship $i = \lambda/L$, where $\lambda = \int e \, dt$, the mass is represented by $V = p/m$, where $p = \int F \, dt$, and the rotating disk has the ideal behavior, $\omega = p_\tau/J$, where $p_\tau = \int \tau \, dt$. All these elements fit exactly our definition of a 1-port inertia as shown in Table 3.3. The fluid inertia is also a linear 1-port inertia, which is covered thoroughly in Chapter 4.

Using the sign convention $\longrightarrow I$, the power flowing into the inertia is given by the expression in Eq. (3.1). In the present case, Eq. (2.2a) allows us to write $e \, dt \equiv dp$, and if $f = f(p)$, then Eq. (3.1) can be written thus:

$$\mathbf{E}(p) = \int_{p_0}^{p} f(p) \, dp + \mathbf{E}_0. \tag{3.3}$$

TABLE 3.3. The 1-Port Inertia, $\overset{e=\dot{p}}{\underset{f}{\rightharpoonup}} I$

	General Relation	Linear Relation	SI Units for Linear Inertance Parameter
Generalized	$p = \Phi_I(f)$	$p = If$	$I = p/f$
variables	$f = \Phi_I^{-1}(p)$	$f = p/I$	$1/I = f/p$
Mechanical	$p = \Phi_I(V)$	$p = mV$	$m = $ N-s^2/m $= $ kg
translation	$V = \Phi_I^{-1}(p)$	$V = p/m$	
Mechanical	$p_\tau = \Phi_I(\omega)$	$p_\tau = J\omega$	$J = $ N-m-s$^2 = $ kg $-$ m^2
rotation	$\omega = \Phi_I^{-1}(p_\tau)$	$\omega = p_\tau/J$	
Hydraulic	$p_p = \Phi_I(Q)$	$p_p = IQ$	$I = $ N-s^2/m^5
systems	$Q = \Phi_I^{-1}(p_p)$	$Q = p_p/I$	
Electrical	$\lambda = \Phi_I(i)$	$\lambda = Li$	$L = $ V-s/A
systems	$i = \phi_I^{-1}(\lambda)$	$i = \lambda/L$	$= $ henrys (H)

If the energy is defined to vanish when f vanishes and if p_0 corresponds to that point in the plot f of p at which $f = 0$, then

$$\mathbf{E}(p) = \int_{p_0}^{p} f(p)\, dp. \tag{3.3a}$$

The similarities between Eqs. (3.2) and (3.3) should be noted. Often the energy associated with a capacitor is called *potential energy*, whereas the energy associated with an inertia is called *kinetic energy*. These names are applied primarily to mechanical systems. In electrical systems, the corresponding two forms of stored energy are sometimes called *electric* and *magnetic* energy.

As in the case of the capacitor, if the constitutive relation of the inertia is plotted, then there is an area interpretation of the stored energy. This interpretation is shown in Figure 3.5, and again, you should be able to demonstrate that any energy stored in an $—I$ can be recovered without loss.

Table 3.3 shows the constitutive relations for inertias and gives units for inertance parameters for the linear case. Since most engineering work is accomplished successfully using Newton's law rather than the postulates of relativity, the relation between velocity and momentum is linear and the mass or moment of inertia is the inertia parameter. Although it is common to think of mass as a ratio of force to acceleration a from the equation

$$F = ma, \qquad a \equiv \dot{V}, \tag{3.4}$$

the table gives the fundamental definition of a mass according to

$$p \equiv mV \tag{3.5}$$

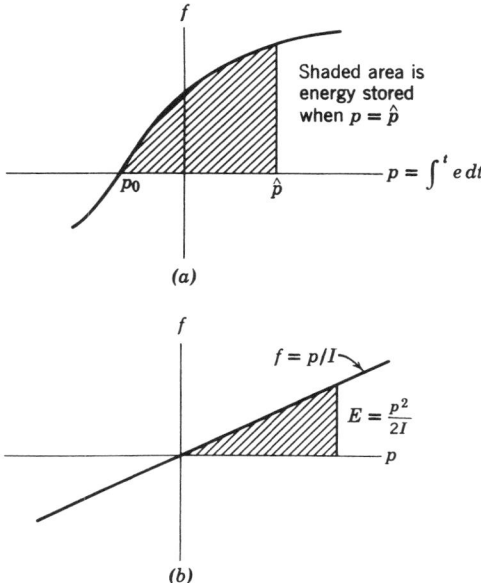

FIGURE 3.5. Area interpretation of stored energy for 1-port inertia. (*a*) Nonlinear case; (*b*) linear case.

with

$$\dot{p} \equiv F. \tag{3.6}$$

Clearly, when Eq. (3.5) is differentiated with respect to time, and Eq. (3.6) is used, then Eq. (3.4) can be derived. If, on the other hand, Eq. (3.5) is replaced with a nonlinear relation,

$$p = \Phi_I(V) = \frac{mV}{(1 - V^2/c^2)^{1/2}}, \tag{3.7}$$

where m is the rest mass and c is the velocity of light, then Eqs. (3.5) and (3.6) hold for the special theory of relativity. See Reference [1, p. 19], for example. Thus, there is some justification for the general constitutive relations given for mechanical systems, even though engineering is overwhelmingly concerned with the linear case. For electrical systems, however, the relation between the flux linkage variable (the time integral of the voltage) and the current in an inductor is nonlinear in typical cases. The use of the linear parameter L is then the result of a modeling decision. It is more satisfactory to generalize $\lambda = Li$ to $\lambda = \Phi_I(i)$ with $\dot{\lambda} = e$ than to try to generalize $e = L\,di/dt$ to the nonlinear case.

As an aid in remembering the three 1-port relationships, the tetrahedron of state introduced in Figure 2.2 may be used. See Figure 3.6. We now know something about five of the six edges of the tetrahedron. The sixth edge, which stretches between the

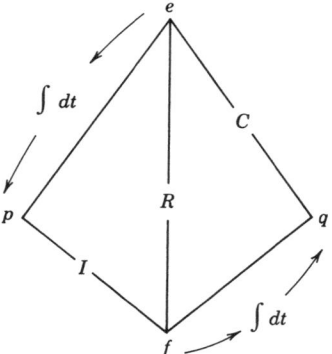

FIGURE 3.6. The three 1-ports placed on the tetrahedron of state according to the variables to which they relate.

vertices representing p and q, is hidden from view in Figure 3.6. This is just as well, since no element will relate p and q.*

Finally, two useful and rather simple 1-ports must be defined—the *effort source* and the *flow source*: the 1-port sources are idealized versions of voltage supplies, pressure sources, vibration shakers, constant-flow systems, and the like. In each case, an effort or flow is either maintained reasonably constant, independent of the power supplied or absorbed by the source, or constrained to be some particular function of time. As an example of a constant-effort source, consider the gravity force on a mass. Near the surface of the earth, this force is essentially independent of the velocity of the mass. As an example of a time-varying source, the electrical wall outlet will serve. The wall outlet enforces a sinusoidal voltage across the power cord wires of most small appliances. Over a reasonable range of currents, the voltage is independent of fluctuations in the current. Of course, the voltage is actually affected by large currents, and a fuse will blow to protect the circuits if very large currents build up, but this simply means that the real outlet is not modeled exactly by an ideal source of effort.

Table 3.4 presents bond graph symbols and the constitutive relations for sources. In this table, physical names are given for the respective energy domains. There are sources of velocity, S_V, sources of force, S_F, sources of pressure, S_P, and so forth. In general, it is best to use the generalized name for the effort or flow source, S_e or S_f, and denote next to the generalized name the specific energy domain being represented. This is done by example in most of the following chapters. It should be further noted that the symbols SE and SF for effort and flow source are often used in computer programs.

---

*One can, in fact, define an element corresponding to the hidden edge, the "memristor." While interesting and occasionally useful, memristors can be represented in terms of other elements to be introduced later, so the memristor will not be considered to be a basic element. See G. F. Oster and D. M. Auslander, "The Memristor: A New Bond Graph Element," *Trans. ASME, J. Dynamic Systems, Measurement, and Control*, **94**, Ser. G, no. 3 pp. 249–252 (Sept. 1972).

TABLE 3.4. The 1-Port Source Elements

	Bond Graph Symbol	Defining Relation
Generalized variables	$S_e \rightharpoonup$	$e(t)$ given, $f(t)$ arbitrary
	$S_f \rightharpoonup$	$f(t)$ given, $e(t)$ arbitrary
Mechanical translation	$S_F \rightharpoonup$	$F(t)$ given, $V(t)$ arbitrary
	$S_V \rightharpoonup$	$V(t)$ given, $F(t)$ arbitrary
Mechanical rotation	$S_\tau \rightharpoonup$	$\tau(t)$ given, $\omega(t)$ arbitrary
	$S_\omega \rightharpoonup$	$\omega(t)$ given, $\tau(t)$ arbitrary
Hydraulic systems	$S_P \rightharpoonup$	$P(t)$ given, $Q(t)$ arbitrary
	$S_Q \rightharpoonup$	$Q(t)$ given, $P(t)$ arbitrary
Electrical systems	$S_e \rightharpoonup$	$e(t)$ given, $i(t)$ arbitrary
	$S_i \rightharpoonup$	$i(t)$ given, $e(t)$ arbitrary

Typically, source elements are thought of as supplying power to a system. This accounts for the sign-convention half arrow shown, which implies that when $e(t)f(t)$ is positive, power flows from the source to whatever system is connected to the source. Since a source maintains one of the power variables constant or a specified function of time no matter how large the other variable may be, a source can supply an indefinitely large amount of power. This is, of course, not a realistic assumption, and real devices are not really sources even though they may be modeled approximately by sources. As an example, consider the problem of predicting the current flowing from a 12-V automotive battery into a variable resistor connected to the battery. Figure 3.7 shows a circuit diagram, a bond graph, and a plot of voltage versus current. This is a static system operating at points at which the source characteristic intersects the resistor characteristics. For small currents (or high values of the resistance R), the battery is almost a constant-voltage source. When the resistance is lowered toward zero, the predicted current approaches infinity. Actually, when the current gets large, the internal resistance in the battery reduces the voltage below the nominal 12 V. In fact, if the resistance approaches zero, as it will when a shorting bar is put across the battery terminals, the battery current will approach a finite value, labeled "short-circuit current" in Figure 3.7c. When more basic multiports have been defined, it will be possible to model the battery with an ideal source and a resistor in such a way that the actual characteristics in Figure 3.7 will be reproduced by the model. For now, we simply note that ideal sources are useful in modeling real devices but should not be expected to be realistic models in all power ranges unless supplemented by other multiports.

A universe made up only of 1-ports would be very simple, since bond graphs more complicated than that in Figure 3.7b would be impossible. This leads one to anticipate that 1-ports are not the whole story. A logical next step is to consider 2-ports.

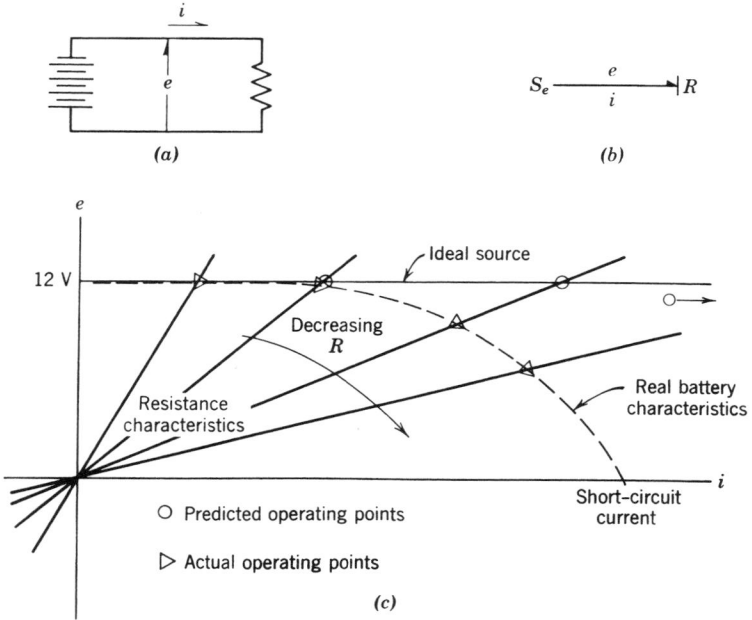

FIGURE 3.7. Study of a battery connected to a variable resistance. (*a*) Electric circuit diagram; (*b*) bond graph; (*c*) plot of source, real battery, and resistance characteristics.

3.2   BASIC 2-PORT ELEMENTS

One might expect that it would be necessary to define more basic types of 2-ports than 1-ports, but, in fact, only two basic types of 2-ports are required. There are, of course, an unlimited number of 2-port subsystems, but we need to discuss here only those that cannot be modeled using the basic 1-ports of the previous section and other elements to be defined later.

The 2-ports to be discussed here are ideal in the specific sense that *power is conserved*. If any 2-port, —*T P*—, has the sign convention

$$\stackrel{e_1}{\rightharpoonup} T P \stackrel{e_2}{\rightharpoonup},$$
$$f_1 f_2$$

then power conservation means that at every instant of time

$$e_1(t)f_1(t) = e_2(t)f_2(t). \tag{3.8}$$

The power sign convention implied in Eq. (3.8) and shown in the bond graph just above is a *through power* sign convention in the sense that power is thought of as flowing through the 2-port. Equation (3.8) states that whatever power is flowing into one side of the 2-port is simultaneously flowing out of the other side.

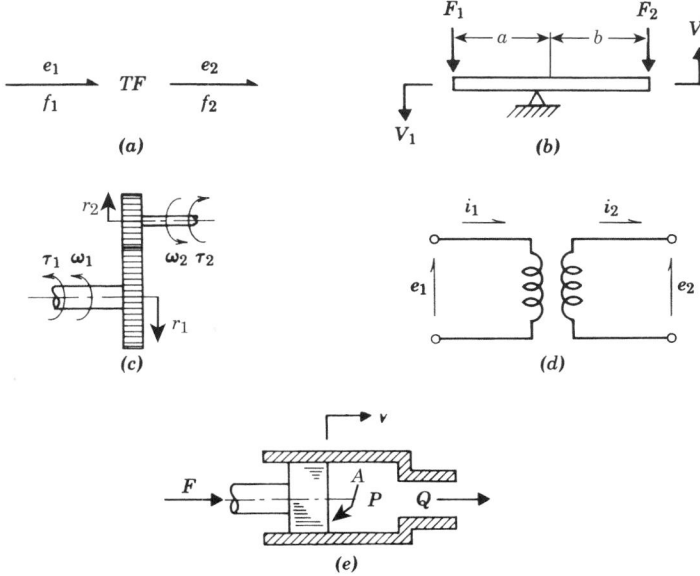

FIGURE 3.8. Transformers. (*a*) Bond graph; (*b*) ideal rigid lever; (*c*) gear pair; (*d*) electrical transformer; (*e*) hydraulic ram.

One way in which Eq. (3.8) can be satisfied is found in the 2-port known as a transformer and given in the bond graph symbol $\rightharpoonup TF \rightharpoonup$. The constitutive laws of the ideal 2-port transformer are

$$e_1 = me_2, \qquad mf_1 = f_2, \tag{3.9}$$

in which the parameter m is called the *transformer modulus** and the subscripts 1 and 2 correspond to the two ports, as shown in Figure 3.8*a*. Note that Eqs. (3.9) and (3.8) both imply the use of the through sign convention shown in the figure. Also shown in Figure 3.8 are a number of devices which in idealized form are modeled by transformers. In no case is the physical device exactly a transformer. For example, the lever in Figure 3.8*b* would only be a $\rightharpoonup TF \rightharpoonup$ if it were massless, rigid, and frictionless. Similar restrictions can be made on the validity of the transformer as a model for the other physical devices. Actual models of the devices can be made using the ideal transformer and other multiports to account for nonideal effects if these effects are important to the system under study.

The lever is an ideal transformer because kinematics dictates that $(b/a)V_1 = V_2$ and moment equilibrium requires $F_1 = (b/a)F_2$. This is exactly the definition for the ideal transformer from Eq. (3.9). If the velocity relationship had been derived first, then there is no need to derive the force relationship, as it comes for free due

---

*There is an ambiguity here since one could just as well write Eq. (3.9) as, $me_1 = e_2$, $f_1 = mf_2$, in which case the modulus would be defined as the inverse of the modulus in Eq. (3.9).

to the power conserving nature of the transformer. Similarly, the gear set is an ideal transformer because kinematics dictates that $(r_1/r_2)\omega_1 = \omega_2$ and moment equilibrium requires $\tau_1 = (r_1/r_2)\tau_2$. This also fits exactly the definition from Eq. (3.9). If the angular velocity relationship had been derived first, then there is no need to derive the torque relationship, as it comes for free due to the power conserving nature of the transformer. The electrical transformer from Figure 3.8d is a transformer in the bond graph sense because the voltage is stepped up or down as the current is stepped down or up according to the turns ratio of the windings of the transformer. In these first three cases, the transformer relates similar power variables on each side of the device. An extremely important use of the transformer is to cross from one energy domain to another. The hydraulic ram is a first example of this.

The hydraulic ram is shown in Figure 3.8e. Hydraulic power is transduced into mechanical power. The constitutive laws of the ideal version of this device are

$$F = AP, \qquad AV = Q, \tag{3.10}$$

in which the area of the piston, A, functions as the transformer modulus, m, as it appears in Eq. (3.9). The two equations of (3.10) can be derived separately from physical considerations, or if one is derived, the other follows because of power conservation. The fact that only a single modulus exists serves as a useful check on constitutive equations such as (3.10).

Another way in which the power balance of Eq. (3.8) may be satisfied is embodied in the *gyrator*, which is symbolized thus: $\rightharpoonup GY \rightharpoonup$. The constitutive laws of the gyrator are

$$e_1 = rf_2, \qquad rf_1 = e_2, \tag{3.11}$$

in which r is the *gyrator modulus* and the through sign convention of Figure 3.9a is implied. The modulus is called r because Eq. (3.11) reminds one of the 1-port linear

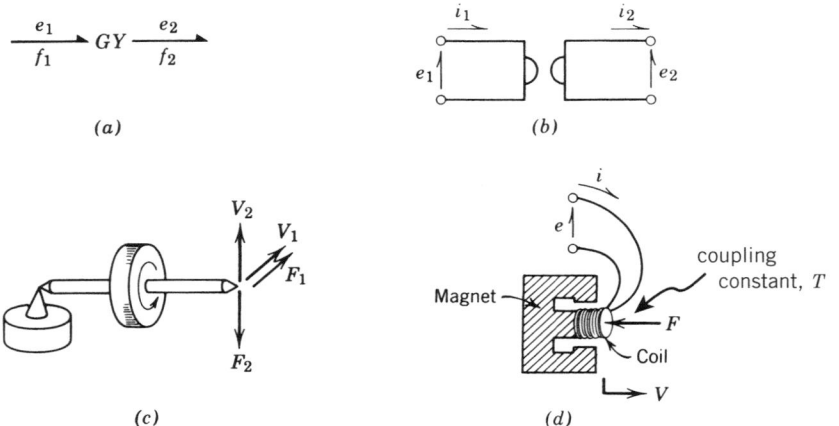

FIGURE 3.9. Gyrators. (a) Bond graph; (b) symbol for electrical gyrator; (c) mechanical gyrator; (d) voice coil transducer.

resistance law, as shown in Table 3.1. In Eq. (3.11), however, the effort and flow at two *different* ports are statically related. Thus, a 1-port resistor dissipates power, whereas the 2-port gyrator conserves power, as can be seen by multiplying the two equations in (3.11) together.

Figure 3.9 shows some physical devices that are at least approximately gyrators. The electric circuit symbol is used to represent gyrators in electric network diagrams. Electric gyrators can be made using the Hall effect, and the gyrator is needed to model effects at microwave frequencies even though in that case it cannot be identified as a separate physical device. Anyone who has played with a toy gyroscope has observed a gyrator. If the rotor of Figure 3.9c spins very rapidly, a gentle push in the direction of F_1 will yield a proportional velocity V_2. Similarly, a force F_2 will result in a velocity V_1. The counterintuitive behavior of the gyroscope is predicted by Eq. (3.11). For example, if the gravity force is in the direction of F_2, then the device precesses in a horizontal path. If a gyroscope rotates slowly, or if large disturbances are applied, the gyroscope must be modeled as a multi-dimensional rigid body. The bond graph, then, is much more complex than a simple $\rightarrow GY \rightarrow$, but it does contain gyrators. For a restricted range of spin speeds and forces, the gyroscope is approximately a gyrator and gives the gyrator its name.

Figure 3.9d shows a useful transducer, which is a gyrator if certain nonideal effects may be neglected. It is the *voice coil* used in electrodynamic loudspeakers, vibration shakers, seismic mass accelerometers, and many other devices. The voice coil and other similar electromechanical conversion devices are covered thoroughly in Chapter 4. Here we simply specify its behavior as an example of a gyrator.

The constitutive laws of the device are

$$e = TV, \qquad Ti = F, \tag{3.12}$$

in which T plays the role of r in Eq. (3.11) and is the coupling constant for the device. The units of T might be V/(ft/s) from one equation and lb/A from the other equation, and one might be tempted to measure the two versions of T that have such different units in two separate experiments. Actually, there is only a single T for a gyrator. The different units merely arise because power is measured in volt-amperes (watts) on one side of the transducer and in foot-pounds per second on the other. Thus, one may measure T using one of the equations (3.12), invoke power conservation using the conversion factor between W and ft-lb/s, and deduce the T in the other equation. The two values of T are numerically different *only* because of the choice of units. The recognition that the device is power conserving and representable as a gyrator helps one avoid the mistake of using numerical values of the two Ts in Eq. (3.12), which would allow the model to create power out of nothing. An important advantage of the SI system (International System of Units) is that the numerical values of T in the two parts of Eq. (3.12) are identical. We recommend that the SI system always be used in order to avoid any unit inconsistencies, especially when multiple energy domains are being modeled.

The gyrator always seems to be a more mysterious element than the transformer. Before the significance of the gyrator was recognized, it was common to make equivalent electrical network diagrams for electromechanical or electrohydraulic systems using only transformers. This is not possible, in general, but in many special cases one may apply the analogy between electrical and mechanical or hydraulic variables so that a gyrator is treated as a transformer. In bond graph terms, the voice coil is a transformer if, for example, we call current an effort and voltage a flow. This switching of the identification of effort–flow variables is entirely unnecessary if one only recognizes that gyrators are really necessary for devices such as those shown in Figures 3.9b and c, in any case.

In fact, a gyrator is a more fundamental element than a transformer. Two gyrators cascaded are equivalent to a transformer:

$$\frac{e_1}{f_1} GY_1 \frac{e_2}{f_2} GY_2 \frac{e_3}{f_3} = \frac{e_1}{f_1} TF_3 \frac{e_3}{f_3}$$

$$e_1 = r_1 f_2, \quad r_2 f_2 = e_3 \quad \rightarrow \quad e_1 = (r_1/r_2)e_3,$$

$$r_1 f_1 = e_2, \quad e_2 = r_2 f_3 \quad \rightarrow \quad (r_1/r_2)f_1 = f_3.$$

In contrast, cascaded transformers are equivalent only to another transformer:

$$\frac{e_1}{f_1} TF_1 \frac{e_2}{f_2} TF_2 \frac{e_3}{f_3} = \frac{e_1}{f_1} TF_3 \frac{e_3}{f_3}$$

$$e_1 = m_1 e_2, \quad e_2 = m_2 e_3 \quad \rightarrow \quad e_1 = m_1 m_2 e_3,$$

$$m_1 f_1 = f_2, \quad m_2 f_2 = f_3 \quad \rightarrow \quad m_1 m_2 f_1 = f_3.$$

Thus, one could, in principle, consider every transformer as a cascade combination of two gyrators and dispense with $\rightarrow TF \rightarrow$ as a basic 2-port. It is more convenient, however, to retain $\rightarrow TF \rightarrow$ as a basic bond graph element.

It is also important to realize that the gyrator essentially interchanges the roles of effort and flow. This may be seen by replacing r in Eq. (3.11) with unity. Then the effort at one port of the $\rightarrow GY \rightarrow$ is just the flow at the other, and vice versa. Thus, the combination $\rightarrow GY \rightarrow I$ is equivalent to $\rightarrow C$. To see this, recall that $\rightarrow I$ relates f to the integral of e, or p. After the gyrator is added and the roles of e and f are interchanged, the combination relates e and the integral of f or q at the external port. The element relating e to q is $\rightarrow C$. Similarly, $\rightarrow GY \rightarrow C$ is equivalent to $\rightarrow I$. Thus, one could in principle dispense with either $\rightarrow C$ or $\rightarrow I$ as a basic 1-port as long as $\rightarrow GY \rightarrow$ was available. Again, it is more convenient and natural to retain both $\rightarrow C$ and $\rightarrow I$ as basic 1-ports.

As an example of a deduction about a system based purely on its bond graph representation, consider the equivalence $I_1 \rightarrow GY \rightarrow I_2 = I_1 \rightarrow C$, in which $\rightarrow GY \rightarrow I_2$ has been replaced by $\rightarrow C$. An $\rightarrow I$ bonded to a $\rightarrow C$ is an oscillator. In physical terms, it could be a mass–spring or inductor–capacitor system. Thus, we see that $I_1 \rightarrow GY \rightarrow I_2$ (or, for that matter, $C_1 \rightarrow GY \rightarrow C_2$) will act just like an inertia–capacitor system.

Finally, there is a generalization of the transformers and gyrators discussed above, based on the curious fact that in both Eqs. (3.9) and (3.11) the power conservation between the two ports is maintained even when the moduli m and r are not constant. This gives rise to the *modulated transformer* and the *modulated gyrator*, denoted in bond graph symbolism by

$$\overset{\downarrow m}{\underset{f_1}{\xrightarrow{e_1}}} MTF \overset{e_2}{\underset{f_2}{\xrightarrow{\quad}}} \quad \text{and} \quad \overset{\downarrow r}{\underset{f_1}{\xrightarrow{e_1}}} MGY \overset{e_2}{\underset{f_2}{\xrightarrow{\quad}}} .$$

Note that m and r are shown as *signals* on an *activated bond*. This means that no power is associated with the changes in m and r, and $e_1 f_1$ is always exactly equal to $e_2 f_2$, as in the case of the constant modulus, $\rightarrow TF \rightarrow$ and $\rightarrow GY \rightarrow$.

Many physical devices may be modeled by the modulated 2-ports. For example, the electrical autotransformer contains a mechanical wiper which, when moved, alters the turns ratio between the primary and secondary coils and thus changes the transformer ratio. This alteration takes no power (if we can assume mechanical friction is negligible), and for any wiper position the device essentially conserves electrical power.

Both gyrators in Figures 3.9c and d have moduli that can be changed without changing the fact that power is conserved at the two ports. For the gyroscope, a motor can be made to change the spin speed of the rotor, which changes r. Similarly, for the voice coil, if an electromagnet is substituted for a permanent magnet, then the transduction coefficient T in Eqs. (3.12) may be varied. At every instant, the power is conserved at the two ports, but the characteristics of the device change.

In mechanics, the *MTF* is particularly important and may be used to represent geometric transformations or kinematic linkages. As a simple example, consider the rotating arm shown in Figure 3.10. The arm is in equilibrium under the action of the torque τ and force F, and it provides a relation between θ and y or $\dot{\theta} \equiv \omega$ and $\dot{y} \equiv V_y$. Writing the displacement relation first,

$$y = l \sin \theta, \tag{3.13}$$

we can differentiate this to yield a constitutive relation for velocities,

$$\dot{y} = (l \cos \theta) \dot{\theta},$$

or

$$V_y = (l \cos \theta) \omega. \tag{3.14}$$

The equilibrium relation between τ and F is

$$(l \cos \theta) F = \tau. \tag{3.15}$$

The occurrence of $l \cos \theta$ in both Eqs. (3.14) and (3.15) might appear to be coincidental until we remember that the device must conserve power. Equations (3.14) and (3.15) are embodied in the bond graph of Figure 3.10b. The *MTF* is called a *displacement*-modulated transformer because the modulus m is a function of the displacement variable, θ. Such transformers allow one to create a bond graph for the

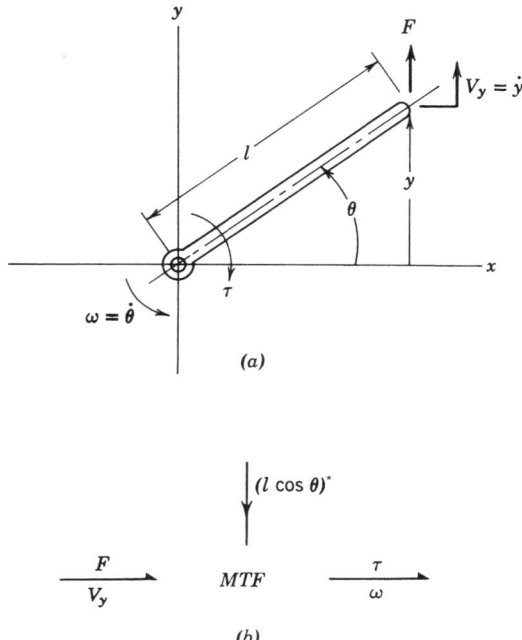

FIGURE 3.10. A displacement-modulated transformer. (*a*) Sketch of rigid, massless rotating arm; (*b*) bond graph.*

extremely complex dynamic systems associated with three-dimensional rigid-body motion.

Although some useful 2-port components have been defined, still only very simple bond graphs can be assembled from 2-ports and 1-ports. Only chains of 2-ports with 1-ports at the ends can be made. In order to create the complex models used in engineering, it turns out that 3-ports are required, but only two basic 3-ports are required to model a very rich variety of systems.

3.3   THE 3-PORT JUNCTION ELEMENTS

Imagine that you are surrounded with electrical components such as resistors, capacitors, inductors, motors, and so forth, mechanical components such as springs, shock absorbers, flywheels, and the like, and hydraulic components such as pipes, tees, accumulators, and so on. Further imagine that you are to connect these components any way you desire. You might imagine that there are hundreds of ways these diverse

---

*If one wishes to follow the convention of Figure 3.8 and Eq. (3.9), one might wish to indicate the modulus as the inverse of $l \cos \theta$. Equations (3.14) and (3.15) show unambiguously how the modulus $l \cos \theta$ is to be used.

components could be connected. In fact, there are only two ways that all components can be connected together, and this brings us to the 3-port junction elements that allow all energy domains to be assembled into overall system models.

We now introduce two 3-port components which, like the 2-ports of the previous section, are power conserving. These 3-ports are called *junctions*, since they serve to interconnect other multiports into subsystem or system models. These 3-ports represent one of the most fundamental ideas behind the bond graph formalism. The idea is to represent in multiport form the two types of connections which, in electrical terms, are called the *series* and *parallel* connections. As we shall see, such connections really occur in all types of systems, even though traditional treatments may not recognize the existence of the junctions as multiports.

First, consider the *flow junction*, *0-junction*, or *common effort junction*. The symbol for this junction is a zero with three bonds emanating from it (as will become evident, it is easy to extend the definition to a 4-, 5-, or more-port version of this 3-port):

$$\big|_{} \quad \big|2 \quad 2\big| $$
$$-0-, \qquad -0-, \qquad \rightarrow 0 \leftarrow.$$
$$\quad\;\; 1 \quad 3 \qquad 1 \quad 3$$

This element is ideal in that power is neither dissipated nor stored. Using the inward power sign convention shown in the last version of the junction, this implies

$$e_1 f_1 + e_2 f_2 + e_3 f_3 = 0. \tag{3.16}$$

The 0-junction is defined such that all efforts are the same, thus,

$$e_1(t) = e_2(t) = e_3(t). \tag{3.17}$$

Combining (3.16) and (3.17) yields

$$f_1(t) + f_2(t) + f_3(t) = 0. \tag{3.18}$$

In words, the efforts on all bonds of a 0-junction are always identical, and the algebraic sum of the flows always vanishes. In other words, if power is flowing in on two ports of the three, then it must be flowing out of the third port.

The use of the 0-junction is suggested by Figure 3.11a. The most obvious examples of 0-junctions are the electrical conductors connected as shown to provide three terminal pairs and the pipe tee junction, which is an idealized version of the hardware store variety. The mechanical example may seem obscure, and it is contrived. Mechanical 0-junctions are just as necessary as electrical or hydraulic ones, but they do not appear so readily in gadget form. The two carts in the mechanical example of Figure 3.11a are supposed to be rigid and massless. Note that $V_3 = -V_1 - V_2$, which conforms with Eq. (3.18). If F is the force across the gap, X_3, then F is the port effort for V_1, V_2, and V_3, in accordance with Eq. (3.17). Such a force would, in fact, exist if F were due to a massless spring connected between the two

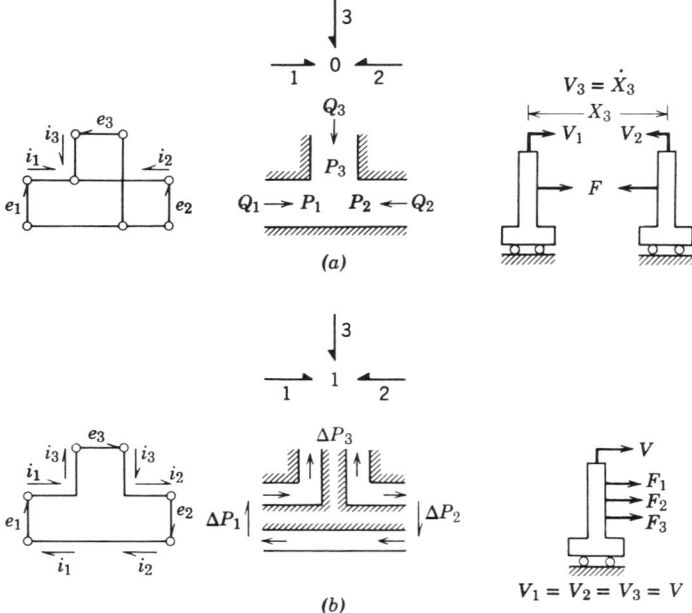

FIGURE 3.11. Basic 3-ports in various physical domains. (a) 0-junction; (b) 1-junction.

carts. With the spring connected, a bond graph of the system would be*

$$
\begin{array}{c}
C \\
F_3 \uparrow V_3 \\
\dfrac{F_1}{V_1}\ 0\ \dfrac{F_2}{V_2}
\end{array}
$$

Before considering more examples in which the 0-junction is used, consider the dual of the 0-junction, that is, a multiport in which the roles of effort and flow are interchanged. Such an element is an *effort junction*, a *1-junction*, or a *common flow junction*. The symbol for this multiport is a 1 with three bonds:

$$
-\!1\!-, \qquad -\!1\!\underset{1\quad 3}{\overset{|2}{-}}, \qquad \underset{f_1}{\overset{e_2\,|\,f_2}{\underset{}{e_1}}}\,1\,\underset{f_3}{\overset{e_3}{}}
$$

This element is again power conserving according to Eq. (3.16); however, this time the element is defined such that every bond has the identical flow, thus,

$$
f_1(t) = f_2(t) = f_3(t), \tag{3.19}
$$

---

*In order to point the half arrow toward the C, we have redefined V_3 to be $V_1 + V_2$.

which, when combined with the power conserving idealization, requires

$$e_1(t) + e_2(t) + e_3(t) = 0. \tag{3.20}$$

As with the 0-junction, the constitutive equations for the 1-junction combine to ensure power conservation in the form of Eq. (3.16).

The 1-junction has a single flow, and the sum of the effort variables on the bonds vanishes. Figure 3.11*b* shows some instances in which a 1-junction can be used to model physical situations. Both the electrical conductors and the hydraulic passages are arranged so that if 1-port components were attached to the ports, one could describe the resulting connection as a series connection. A single current or volume flow would circulate, and the voltages and pressures at the ports would sum algebraically to zero. In the mechanical example, the three forces are all associated with a common velocity, and the forces must sum to zero, since the cart is assumed to be massless.

An understanding of the meaning of the 0- and 1-junctions is important for anyone learning bond graph techniques, and it may be helpful to give some physical interpretation for these multiports in several physical domains.

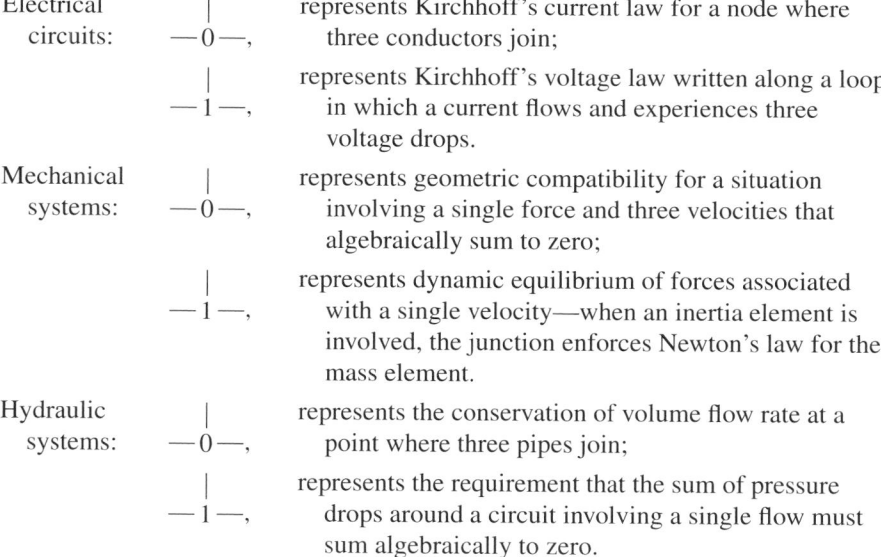

Electrical circuits:	$-0-,$	represents Kirchhoff's current law for a node where three conductors join;
	$-1-,$	represents Kirchhoff's voltage law written along a loop in which a current flows and experiences three voltage drops.
Mechanical systems:	$-0-,$	represents geometric compatibility for a situation involving a single force and three velocities that algebraically sum to zero;
	$-1-,$	represents dynamic equilibrium of forces associated with a single velocity—when an inertia element is involved, the junction enforces Newton's law for the mass element.
Hydraulic systems:	$-0-,$	represents the conservation of volume flow rate at a point where three pipes join;
	$-1-,$	represents the requirement that the sum of pressure drops around a circuit involving a single flow must sum algebraically to zero.

As might be expected, the existence of 0- and 1-junctions within complex systems is not always obvious, but in succeeding chapters formal techniques for modeling systems using these basic elements are presented.

To make clear the utility of the junctions, four elementary example systems are displayed in Figure 3.12. Note that only two bond graphs are involved. The series and parallel aspects of the junctions are more obvious in the electrical than in the mechanical cases. The reader should study these examples to understand how the

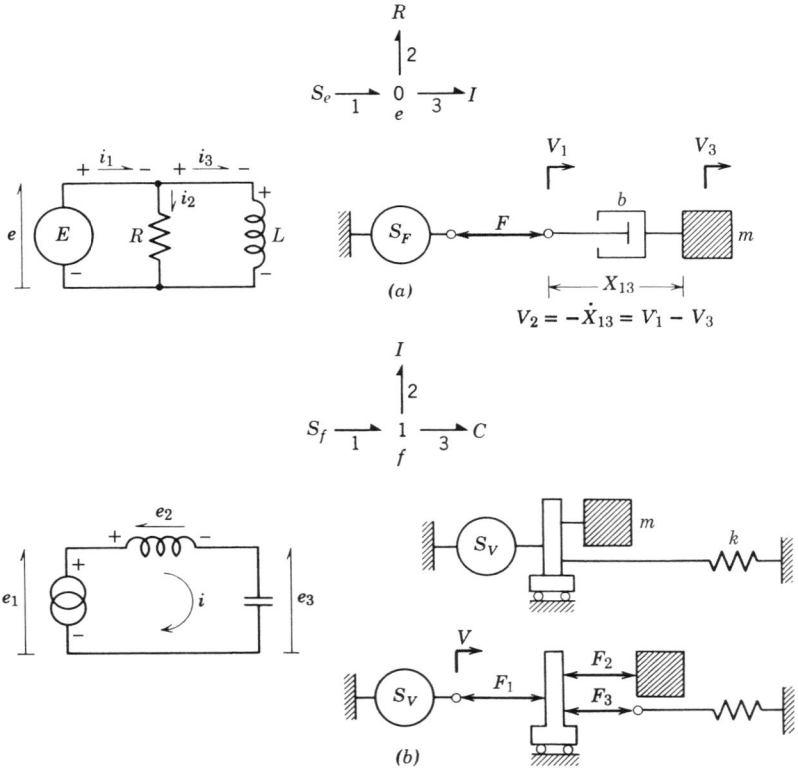

FIGURE 3.12. Example systems involving basic 3-ports. (*a*) Systems using 0-junctions; (*b*) systems using 1-junctions.

sign conventions are transferred from the physical sketches to the bond graph. Note that the 1-ports have the sign conventions as they were presented in the tables at the beginning of this chapter, but the junction signs are not all inward pointing. When the sign convention arrows are changed from the inward-pointing convention used to introduce the 3-ports, Eqs. (3.18) and (3.20) must be modified with a minus sign for each port with an outward-pointing sign. Equations (3.17) and (3.19) remain invariant, however, under changes in a sign convention. *A 0-junction has only a single effort and a 1-junction has only a single flow, independent of the sign convention.* As an example, consider

$$\begin{array}{c} e_2 \uparrow f_2 \\ \overset{e_1}{\longrightarrow} 1 \overset{e_3}{\longrightarrow} \\ f_1 \quad f_3 \end{array}$$

the equations of which are

$$f_1 = f_2 = f_3, \qquad e_1 - e_2 - e_3 = 0. \tag{3.21}$$

The reader should verify that the systems and bond graphs are consistent by writing equations such as (3.21).

The slight generalization from 3-port junctions to 4- or n-port junctions is worth emphasizing. In bond graph symbolism, two similar 3-ports may be combined into a 4-port thus:

$$-0-0-=-0-$$

$$-1-1-=-1-.$$

An n-port 0- or 1-junction has a common effort or flow on all bonds, and the algebraic sum of the complementary power variables on the bonds vanishes. Occasionally, 2-port junctions arise, and, in some cases, these are precisely equivalent to a single bond. The following bond graph identities are always valid:

$$\rightarrow 0 \rightarrow \; = \; \rightarrow, \qquad \rightarrow 1 \rightarrow \; = \; \rightarrow.$$

However, with some sign patterns, the 2-port 0- and 1-junctions serve to reverse the sign definition of an effort or flow. For example,

$$\frac{e_1}{f_1} 0 \frac{e_2}{f_2} \quad \text{implies} \quad e_1 = e_2, \, f_1 = -f_2,$$

and

$$\frac{e_1}{f_1} 1 \frac{e_2}{f_2} \quad \text{implies } f_1 = f_2, e_1 = -e_2.$$

Such 2-ports are sometimes necessary when two multiports are to be joined by a bond, but the two multiports have been defined with signs that are not compatible with a single bond. In connecting a spring, $-C$, to a mass, $-I$, one could define a common velocity but use a 2-port 1-junction to express the fact that the spring force is the negative of the force on the mass. The resulting bond graph would then be $C \leftarrow 1 \rightarrow I$, in which the passive 1-ports have the convenient inward sign convention.

The constitutive relations for 0- and 1-junctions are summarized in Table 3.5.

TABLE 3.5. Summary of Basic 3-Ports

Flow junction, or 0-junction	$\frac{e_1}{f_1} \; 0 \; \frac{e_3}{f_3}$ over $e_2 \mid f_2$	$e_1 = e_2 = e_3,$ $f_1 + f_2 + f_3 = 0$
Effort junction, or 1-junction	$\frac{e_1}{f_1} \; 1 \; \frac{e_3}{f_3}$ over $e_2 \mid f_2$	$f_1 = f_2 = f_3,$ $e_1 + e_2 + e_3 = 0$

3.4   CAUSALITY CONSIDERATIONS FOR THE BASIC MULTIPORTS

The concept of causality was discussed in general terms in Section 2.4, and we may now consider some more specific uses of the idea with respect to the basic multiports. Some of the causal properties developed here will be applied in later chapters. For now, we simply note that some of the basic multiports are heavily constrained with respect to possible causalities, some are relatively indifferent to causality, and some exhibit their constitutive laws in quite different forms for different causalities.

3.4.1   Causality for Basic 1-Ports

The effort and flow sources are the most easily discussed from a causal point of view, since, by definition, a source impresses either an effort or flow time history upon whatever system is connected to it. Thus, if we use the symbols S_e— and S_f— for the abstract effort and flow sources, the only permissible causalities for these elements are

$$S_e \dashv \quad \text{and} \quad S_f \vdash.$$

in which the causal stroke indicates the direction that the effort signal is oriented. The causal forms for effort and flow sources are summarized in the first two rows of Table 3.6.

In contrast to the sources, the 1-port resistor is normally indifferent to the causality imposed upon it. The two possibilities may be represented in equation form as follows:

$$e = \Phi_R(f), \qquad f = \Phi_R^{-1}(e),$$

TABLE 3.6. Causal Forms for Basic 1-Ports

Element	Acausal Form	Causal Form	Causal Relation
Effort source	$S_e \rightharpoonup$	$S_e \dashv$	$e(t) = E(t)$
Flow source	$S_f \rightharpoonup$	$S_f \vdash$	$f(t) = F(t)$
Resistor	$R \leftharpoonup$	$R \dashv$	$e = \Phi_R(f)$
		$R \vdash$	$f = \Phi_R^{-1}(e)$
Capacitor	$C \leftharpoonup$	$C \dashv$	$e = \Phi_C^{-1}\left(\int^t f\,dt\right)$
		$C \vdash$	$f = \dfrac{d}{dt}\Phi_C(e)$
Inertia	$I \leftharpoonup$	$I \vdash$	$f = \Phi_I^{-1}\left(\int^t e\,dt\right)$
		$I \dashv$	$e = \dfrac{d}{dt}\Phi_I(f)$

where we use the convention that the variable on the left of the equality sign represents the output of the resistor (the dependent variable), and that appearing in the function of the right side is the input (independent) variable for the element. This convention is used commonly, but not universally, in writing equations and corresponds to the notation used in computer programming.

The correspondences between the causally interpreted equations and the causal strokes on the bond of the R— element are shown in the third row of Table 3.6. As long as both the functions Φ_R and Φ_R^{-1} exist and are known, there is no reason for preferring one causality over the other. It is possible, however, that the static relation between e and f shown in Figure 3.1 is multiple valued in one direction or the other; that is, either Φ_R or Φ_R^{-1} might be multiple valued. In such a case, the single-valued causality would be clearly preferable. In the linear case, with a finite slope of the $e-f$ characteristic, the 1-port resistor is indifferent to the causality imposed upon it, although the resistance law would be written in two forms:

$$e = Rf \quad \text{or} \quad f = (1/R)e.$$

The constitutive laws of the C— and I— elements are expressed as static relations between e and $q = \int^t f \, dt$ and f and $p = \int^t e \, dt$, respectively. In expressing causal relations between es and fs, we will find that the choice of causality has an important effect. Taking the capacitor, we may rewrite the relations from Table 3.2 as follows:

$$e = \Phi_C^{-1} \left(\int^t f \, dt \right), \qquad f = \frac{d}{dt} \Phi_C(e), \tag{3.22}$$

in which causality is implied by the form of the equation. Note that when f is the input to the C—, e is given by a static function of the time integral of f, but when e is the input, f is the time derivative of a static function of e. The correspondences between these causal equations and the causal stroke notation for the capacitance are shown in the fourth row of Table 3.6. The implications of the two types of causality, which are called *integral causality* and *derivative causality*, respectively, will be discussed in some detail in later chapters.

Since inertia is the dual* of the capacitor, similar effects occur with the two choices of causality. Rewriting the inertia element relations from Table 3.3, we have

$$f = \Phi_I^{-1} \left(\int^t e \, dt \right), \qquad e = \frac{d}{dt} \Phi_I(f). \tag{3.23}$$

In this case, integral causality exists when e is the input to the inertia, and derivative causality exists when f is the input. These observations are summarized in the fifth row of Table 3.6. Equations (3.22) and (3.23) are written in a form suitable for nonlinear C— and I— elements, but the distinction between integral and derivative causality remains for the special case of linear elements.

---

*Dual elements have identical constitutive laws except that the roles of effort and flow are interchanged.

3.4.2   Causality for Basic 2-Ports and 3-Ports

Proceeding now to the basic 2-ports, one might think initially that there would be a total of four possibilities for the assignment of causality of a transformer, namely, any combination of the two possible causalities for each of the two ports. However, there are only two possible causality assignments, as the defining relations (3.9) and (3.11) show. As soon as one of the e's or f's has been assigned as an input to the $-TF-$, the other e or f is constrained to be an output by Eq. (3.9). Thus, in fact, the only two possible choices for causality for the transformer are $\vdash TF\vdash$ and $\dashv TF\dashv$. The possible causalities are tabulated in the first row of Table 3.7. In Table 3.7, a simplified naming of the efforts and flows has been achieved by simply numbering the bonds. This technique will be explored in more detail in subsequent chapters. Again, causal equation equivalents to the causal stroke notation are given for all elements in Table 3.7.

For the gyrator, Eqs. (3.11) show that as soon as the causality for one bond has been determined, the causality for the other is also. Thus, the only permissible causal choices for the $-GY-$ are $\dashv GY\vdash$ and $\vdash GY\dashv$. The choices for the causality for the gyrator are summarized in the second row of Table 3.7.

The causal properties of 3-port 0- and 1-junctions are somewhat similar to those of the basic 2-ports. Although each bond of the 3-ports, considered alone, could have either of the two possible causalities assigned, not all combinations of bond causalities are permitted by the constitutive relations of the element. For example, the constitutive relations for the 0-junction given in Table 3.5 indicate that all efforts on all the bonds are equal and the flows must sum to zero. Thus, if on any bond the effort is an input to a 0-junction, then all other efforts are determined, and on all other

TABLE 3.7.   Causal Forms for Basic 2-Ports and 3-Ports

Element	Acausal Graph	Causal Graph	Causal Relations
Transformer	$\overset{1}{-}TF\overset{2}{-}$	$\overset{1}{\vdash}TF\overset{2}{\vdash}$	$e_1 = me_2$ $f_2 = mf_1$
		$\overset{1}{-\!\!\!\rightarrow}TF\overset{2}{-\!\!\!\rightarrow}$	$f_1 = f_2/m$ $e_2 = e_1/m$
Gyrator	$\overset{1}{-}GY\overset{2}{-}$	$\overset{1}{\vdash}GY\overset{2}{-\!\!\!\rightarrow}$	$e_1 = rf_2$ $e_2 = rf_1$
		$\overset{1}{-\!\!\!\rightarrow}GY\overset{2}{\vdash}$	$f_1 = e_2/r$ $f_2 = e_1/r$
0-Junction	$\overset{1}{-}0\overset{2}{-}$ $3\uparrow$	$\overset{1}{-\!\!\!\rightarrow}0\overset{2}{-\!\!\!\vdash}$ $3\downarrow$	$e_2 = e_1$ $e_3 = e_1$ $f_1 = -(f_2 + f_3)$
1-Junction	$\overset{1}{-}1\overset{2}{-}$ $3\uparrow$	$\overset{1}{\vdash}1\overset{2}{\vdash}$ $3\uparrow$	$f_2 = f_1$ $f_3 = f_1$ $e_1 = -(e_2 + e_3)$

bonds they must be outputs of the 0-junction. Conversely, if all the flows on all bonds except one are inputs to the 0-junction, the flow on the remaining bond is determined and must be an output of the junction. A typical permissible causality for a 0-junction is shown in the third row of Table 3.7. Here the causal stroke on the end of bond 1 nearest the 0 indicates that e_1 is an input to the junction and that all other bonds must have causal strokes at the end away from the 0. To interpret the diagram another way, the flows on bonds 2 and 3 are inputs to the 0-junction. These considerations are also expressed by the causal equations shown in Table 3.7. For a 3-port 0-junction, then, there are only three different permissible causalities in which each of the three bonds in succession plays the role assigned to bond 1 in the example shown in the table. For an n-port 0-junction this description of the constraints on causality is still valid, and there are exactly n different permissible causal assignments.

For a 1-junction the same considerations apply as for a 0-junction except that the roles of the efforts and flows are interchanged. Table 3.5 indicates that flows on all the bonds are equal and the efforts sum to zero. Thus, if the flow on any single bond is an input to the 1-junction, the flows on all other bonds are determined and must be considered outputs of the junction. Yet when the efforts on all bonds except one are inputs to the 1-junction, the effort on the remaining bond is determined and must be an output of the junction. A typical permissible causality is shown in the fourth row of Table 3.7. In this example, bond 1 plays the special role of determining the common flow at the junction, and the remaining bonds supply effort inputs that suffice to determine the effort on bond 1. Clearly, there are three permissible causalities for a 3-port 1-junction, and there are n permissible different causal assignments for an n-port 1-junction.

Although the causal considerations have been stated for all the basic multiports defined so far (summarized in Tables 3.6 and 3.7), it can hardly be clear as to what all the implications of causality are. The study of causality is very important, and bond graphs are uniquely suited to this study. However, only when some real system models have been assembled is it clear why causal information is so important. In the next chapter, system models are built up using the basic multiports just discussed. Using the rules of causality, it is then possible to predict features of these systems even before the exact characterization of the multiports has been decided. For instance, it will be possible to predict the order of the system model before any equations are written and before a firm decision has been made about whether the model should be linear or nonlinear. In addition, causal considerations will prove invaluable in writing state equations or setting up computational block diagrams.

3.5   CAUSALITY AND BLOCK DIAGRAMS

Block diagrams indicate input and output quantities for each block and thus are inherently causal. When causal strokes are added to a bond graph, one may represent the information by a block diagram. For example, the block-diagram versions of the causal forms for the R, C, and I 1-ports shown in Table 3.6 are given in Figure 3.13. Similarly, block diagrams for 2-ports and 3-ports corresponding to entries

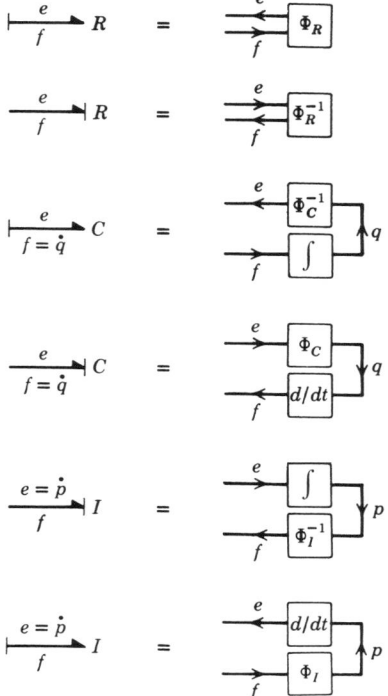

FIGURE 3.13. Block diagrams for 1-ports.

in Tables 3.7 are shown in Figures 3.14 and 3.15. It should be possible to correlate the signal flow paths in the block diagrams with the equations in the tables and with the bond graph representation. Note that when one rigorously maintains the spatial arrangements with efforts above and to the left of bonds and flows below and to the right, the block diagrams have fixed patterns.

It may also be seen that block diagrams are more complex graphically than bond graphs because a single bond implies two signal flows on a block diagram. Initially, block diagrams are easier to understand than bond graphs because they contain re-dundant information. For systems with some complexity, however, block diagrams rapidly become so complex that the conciseness of bond graphs is an advantage. For example, Figure 3.16 shows a block diagram equivalent to a bond graph model of the automotive drive train system of Figure 2.7. Note that the sign-convention half arrows have yet to be put on the bond graph, and the corresponding + and − signs do not appear in the block diagram near the circles representing signal summation.

After procedures for constructing bond graph models and adding causal strokes to them have been discussed in the following chapters, one option is to construct a block diagram from the bond graph.

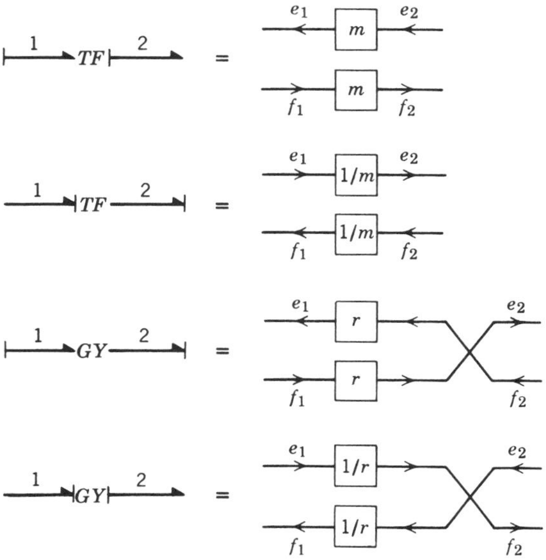

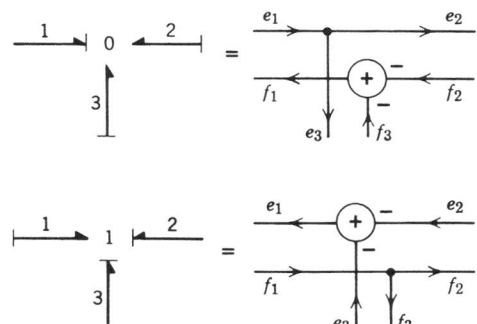

FIGURE 3.14. Block diagrams for 2-ports.

FIGURE 3.15. Block diagrams for 3-ports.

3.6   PSEUDO–BOND GRAPHS AND THERMAL SYSTEMS

In these introductory chapters, the number of physical domains to be discussed is purposely limited. Later, when more sophisticated modeling concepts have already been discussed, the range of physical systems is broadened considerably. Because thermal systems are so important, we briefly introduce some bond graph representations of thermal elements. Traditionally, thermal systems have been presented as analogous to electric circuits, usually with temperature analogous to voltage and heat flow analogous to current. With this analogy there are then thermal resistors,

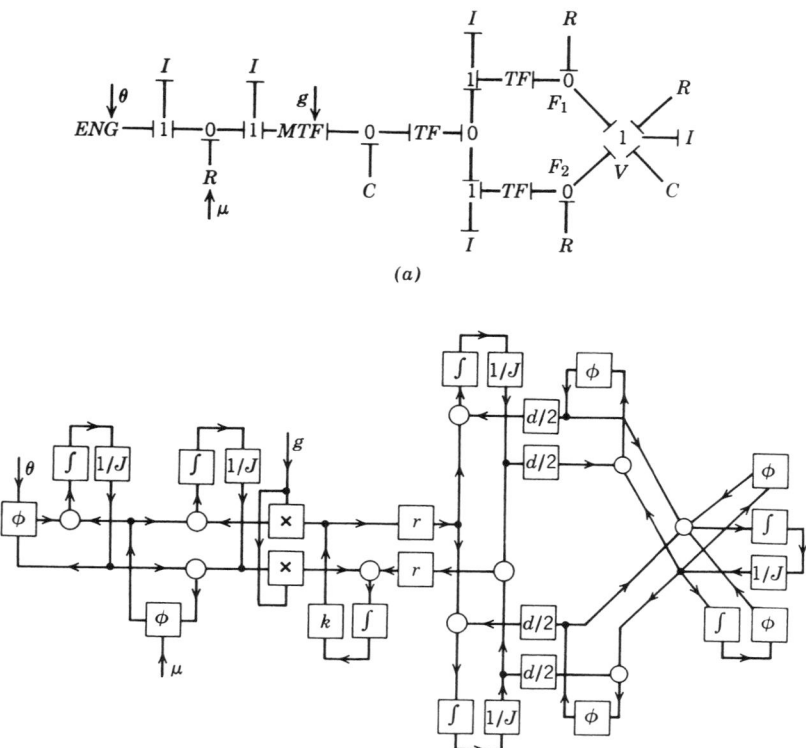

FIGURE 3.16. Interconnected drive train model. (*a*) Bond graph; (*b*) block diagram.

capacitors, and parallel and series connections (our 0- and 1-junctions) and sources analogous to voltage and current sources. There are no thermal inertias, however.

Since this analogy in which temperature is an effort and heat flow is a flow has proved useful, we present it here. There is one major hitch, however. The product of temperature and heat flow is *not* a power. Heat flow itself has the dimensions of power. We choose to call any bond graph in which *e* and *f* are not power variables a *pseudo–bond graph*. Such a graph cannot be coupled to a normal bond graph using power variables except by means of some ad hoc elements that do not obey the rules of normal bond graph elements. Bond graph techniques may be usefully applied to any pseudo–bond graph as long as the basic elements in the pseudo–bond graph correctly relate the *e*, *f*, *p*, and *q* variables. Later, it will be shown that a true bond graph results if temperature and entropy flow are used for effort and flow variables, but the thermodynamic arguments are more sophisticated than those necessary to establish the usefulness of the thermal pseudo–bond graph.

In Figure 3.17, two common situations that arise in the study of thermal systems are depicted. In both cases, we assume the temperature gradients and heat flow are present only in the *x* direction. The case of Figure 3.17*a* represents a pure resistance.

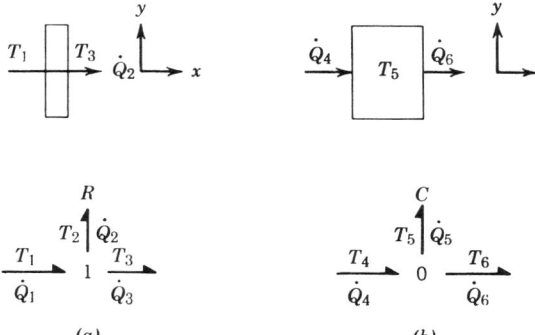

FIGURE 3.17. Basic elements for models of conduction heat-transfer systems. (*a*) Thermal resistor and 1-junction; (*b*) thermal capacitor and 0-junction.

If T_1 and T_3 are the temperatures (on any convenient scale) on the two sides of a slab of material of area A, we assume that the flow of heat through the slab, $\dot{Q}_2$, is a function of $T_2 = T_1 - T_3$. In the linear case, we say,

$$R\dot{Q}_2 = T_1 - T_3 = T_2, \tag{3.24}$$

where $\dot{Q}_2$ may be measured in Btu/s, cal/s, or any other power measure. (Because this is a pseudo–bond graph, the choice of units is not critical as long as all element parameters are defined in a consistent manner.) For a material with thermal conductivity k, thickness l, and area A,

$$R = \frac{l}{kA}.$$

Note that the thermal resistor implies a relation between T_2 and $\dot{Q}_2$, and the 1-junction implies $\dot{Q}_1 = \dot{Q}_2 = \dot{Q}_3$ and $T_1 - T_2 - T_3 = 0$. We have simply shown how the 1-port R and the 1-junction can be used together to constrain the common flow to be a function of the difference between two efforts.

Figure 3.17*b* represents a lump of material that changes temperature as a function of the net heat energy stored in it. That is, T_5 is a function of

$$Q_5 = \int^t \dot{Q}_5 \, dt = \int^t (\dot{Q}_4 - \dot{Q}_6) \, dt.$$

Since T_5 is an effort and Q_5, the integral of a flow, is a displacement, the element is a capacitor. Again, the bond graph of Figure 3.17*b* combines a C— with a 0-junction to indicate that $T_4 = T_5 = T_6$ and $\dot{Q}_5 = \dot{Q}_4 - \dot{Q}_6$. In the linear case a thermal capacitance C can be defined such that

$$T_5 = T_{50} + \frac{1}{C} \int_{t_0}^t \dot{Q}_5 \, dt, \tag{3.25}$$

where T_{50} is the temperature at $t = t_0$. The capacitance C can be found by assuming that the element does negligible work by expanding or contracting, so that changes in its internal energy are only the result of $\dot{Q}_5$. Then if c is a *specific heat*,

$$c = \frac{\partial u}{\partial T}, \tag{3.26}$$

where u is the internal energy per unit mass, then

$$C = mc,$$

where m is the mass of the substance. Strictly speaking, c is the specific heat *at constant volume*, but for most solids and liquids the work done by expansion is small, as compared with Q_5, so that c does not vary much even if the material is allowed to expand. (When gases do work as they are heated, we will find it much better to use the true bond graph representation described in Chapter 12.)

Figure 3.18 shows a typical use of the elements shown in Figure 3.17 and introduces a thermal effort source. A unit area of the wall of a pipe carrying hot fluid is modeled. A lumped-parameter model consisting of three resistors and two capacitors is used, and the inside and outside temperatures are assumed to be determined by effort (temperature) sources.

If T_1 changes because the temperature of the fluid flowing through the pipe changes, the bond graph model can give a prediction of the dynamic changes in temperature in the pipe wall. If the pipe is really made of a uniform material, then the

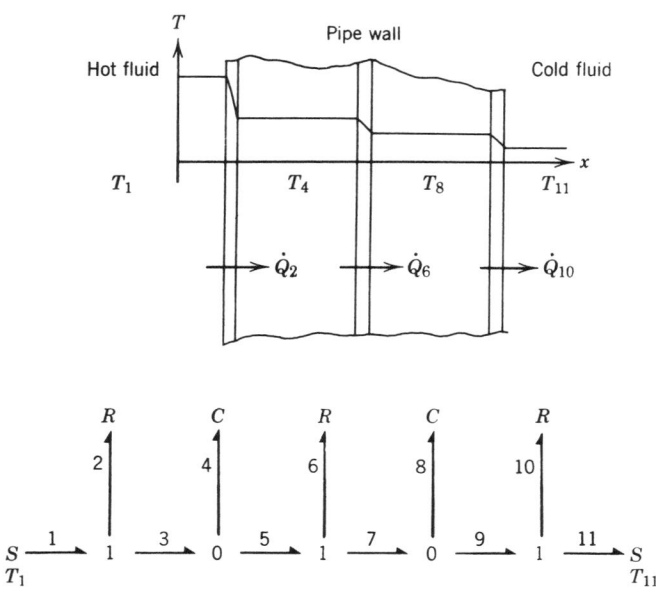

FIGURE 3.18. Heat transfer in pipe wall.

stepwise distribution of temperature with distance, x, will be a fairly crude approximation of the actual temperature distribution, which will be a continuous function of both time and x. A better approximation could be obtained by splitting the pipe into a large number of resistor and capacitor layers, but the model would be more complicated to use.

REFERENCE

[1] S. H. Crandall, D. C. Karnopp, E. F. Kurtz, and D. C. Pridmore-Brown, *Dynamics of Mechanical and Electromechanical Systems*, New York: McGraw-Hill, 1968.

PROBLEMS

3-1 A nonlinear dashpot has as its constitutive relation the "absquare law,"

$$F = AV|V|,$$

where F and V are the force and velocity across the dashpot and A is a constant. Plot this relation in a sketch, and indicate the bond graph sign convention implied if $A > 0$ and which causality the equation implies in the form given. Try to invert the constitutive law to yield the velocity as a function of the force.

3-2 A fluid of mass density ρ is pumped into an open-topped tank of area A. If P is the pressure at the tank bottom and Q the volume flow rate, the tank is approximately a $-C$ for slow changes in the volume of fluid stored. Is the $-C$ a linear element in this case, and if so, what is the capacitance? *Hint*: It is useful to compute the height of fluid, h, as a function of the total volume of fluid as an intermediate step.

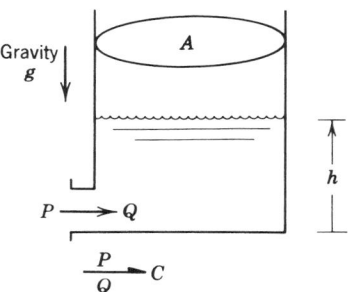

3-3 Reconsider Problem 3-2, but let the tank have sloping walls as shown. What does this do to the constitutive law for the device?

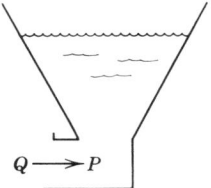

3-4 Consider a uniform cantilever beam of length L, elastic modulus E, and area moment of inertia I. If a force F is applied at the tip of the beam, it will deflect. If the beam is supposed to be massless, decide what type of 1-port it is and compute its constitutive law.

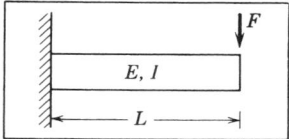

3-5 Consider a given mass of water as a thermal capacitor. Sketch the constitutive law for this element for a range of temperatures including the freezing point. Indicate the effect of the latent heat of freezing or melting.

3-6 Linear electrical inductors can be characterized by a law relating the voltage e to the rate of change of current:

$$L\frac{di}{dt} = e.$$

Convert this to a law relating the flow i to the momentum λ, which in this case is the time integral of e and is called the *flux linkage*. Plot a linear flow–momentum constitutive law, and show on the plot where the inductance L appears. Now sketch a nonlinear flow–momentum law. Convert the nonlinear law back to a relation between e and di/dt if this is possible.

3-7 A rigid pipe filled with incompressible fluid of mass density ρ has length L and cross-sectional area A.

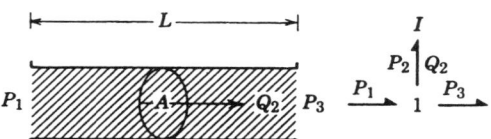

If P_1 and P_2 are the pressures at the ends of the pipe and the volume flow rate is Q_2, convince yourself that the bond graph shown correctly represents the pipe in the absence of friction. Show that the correct constitutive law relating pressure momentum and volume flow is

$$p_{P_2} = \int^t P_2 \, dt = \left(\rho \frac{L}{A} \right) Q_2$$

by writing Newton's law for the slug of fluid in the pipe. (It should come as a surprise that small-area tubes have a lot of inertia when P, Q variables are used.)

3-8 An accumulator consists of a heavy piston in a cylinder. If the pressure is determined primarily by the weight of the piston, sketch the constitutive law for this 1-port device.

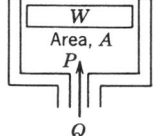

3-9 A heat transfer coefficient α for a surface is given in W/m^2-°C. If an area A of this surface is involved in a problem, show how the effect of the surface would be represented in a bond graph, and write the constitutive law for the element in terms of A and α.

3-10 A flywheel is a uniform disk of radius R and thickness t and is made of a material of mass density ρ. Write the constitutive law for the 1-port representing the flywheel in its flow–momentum form. Evaluate the inertance parameter for a steel disk 1 in. in thickness and 10 in. in diameter.

3-11 Assume any needed dimensions for the hydraulic ram of Figure 3.8e, and write the constitutive law for this 2-port.

3-12 Repeat Problem 3-11 for the devices of Figures 3.8b and c.

3-13 In Figure 3.9c assume that the rotor has moment of inertia J and spins at a high angular rate Ω. If the rotor is centered on an axle of length L, relate F_1, V_1, F_2, V_2, and thus demonstrate that the device is indeed approximately a —GY—.

3-14 Draw block diagrams for the following bond graphs, assuming all 1-ports are linear:

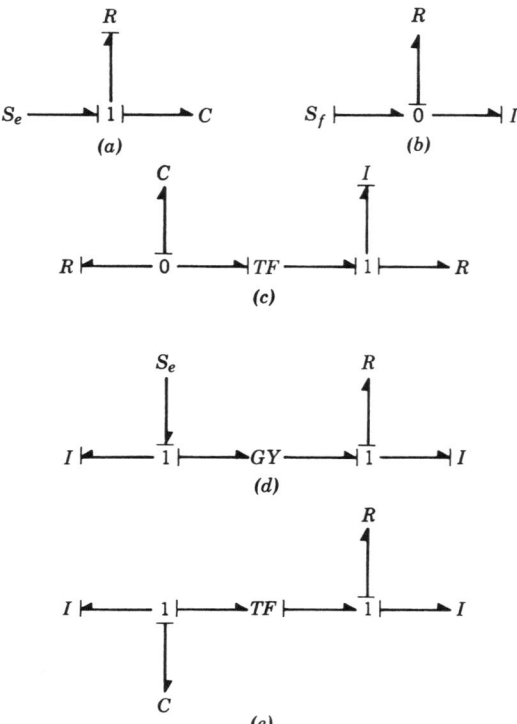

(a)   *(b)*

(c)

(d)

(e)

3-15 Draw a block diagram for the oscillator using the bond graph shown.

3-16 A block of material with total heat capacitance C is covered with insulating material with total thermal resistance R and surrounded by an atmosphere of temperature T_0. The bond graph is supposed to aid in estimating how fast the block will heat up if it starts at some temperature $T_i(t_0) < T_0$. Draw a block diagram from the bond graph.

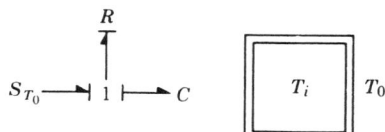

3-17 Consider an ideal rack and pinion with no friction losses:

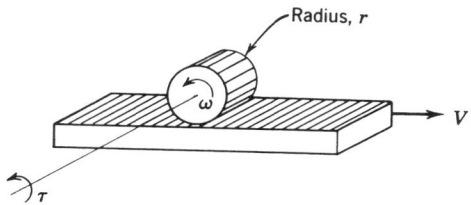

If the pinion has radius r and torque and speed τ and ω, and if the rack has velocity V and force F, what type of element would represent the device? Write the appropriate constitutive laws.

3-18 An electrodynamic loudspeaker is driven by a voice coil transducer described by Eq. (3.12). Show that if only the mass of the speaker cone is considered, then at the electrical terminals the device will act like a capacitor. The bond graph identity

$$\frac{e}{i}\overset{T}{G}\dot{Y}\frac{F}{V}I = \frac{e}{i}C$$

is to be verified using the element constitutive equations directly.

3-19 An air spring is idealized as a piston in a cylinder with no leakage and no heat transfer through the cylinder walls. The process the air undergoes is assumed to be isentropic, such that

$$PV^{\gamma} = P_0 V_0^{\gamma},$$

where V is the instantaneous volume,

$$V = V_0 - A_p x,$$

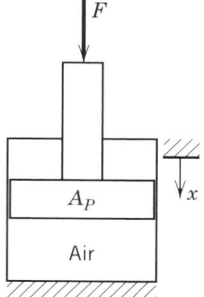

V_0 is the volume when $x = 0$, P_0 is atmospheric pressure, γ is the ratio of specific heats (≈ 1.4 for air), and P is the absolute cylinder pressure.

Derive the nonlinear constitutive relationship for $F = F(x)$.

3-20 Three springs are used in a parallel configuration as shown. Sketch the constitutive behavior as seen from the $F-v$ port. Show both compression and tension.

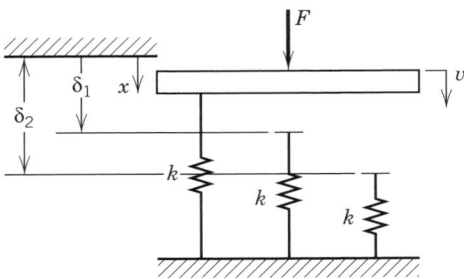

3-21 Friction is always dissipative and is therefore a resistance in a bond graph model. Its constitutive behavior is sketched in an ideal sense.

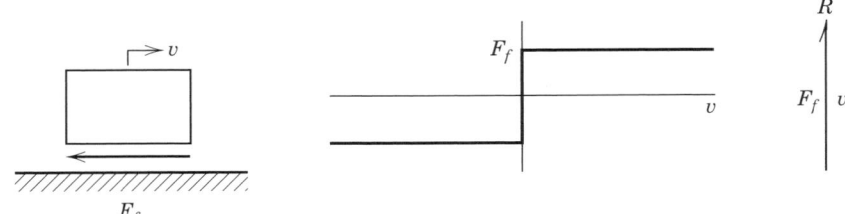

(a) Discuss the only possible causality for this element.

(b) Discuss the problem of using this device near $v = 0$. Propose a change to the constitutive behavior that will retain its fundamental character but avoid the problem near $v = 0$.

4

SYSTEM MODELS

You are now ready to model the world, having mastered the basic multiport elements. All that remains is to sally forth—armed with your bond graph arsenal of C, I, R, S_e, S_f, 0, 1, TF, and GY—ready to represent as a bond graph any physical system you may meet. This chapter is devoted to helping you do so. However, it is not true that every system you may encounter will be reducible to a simple bond graph. For electrical circuits, certain classes of mechanical systems, hydraulic circuits, and some transducer systems you will be entirely successful. Additional study and experience will enable you to extend your range, but at the end there will always be some problems that do not fall into your carefully spread snare. But be assured that the size of your snare will become amazingly large, with some effort on your part.

In this chapter we show how to represent any electrical circuit in bond graph form by using a direct, simple modeling procedure. Then we use a similar approach to problems involving mechanical translation. With slight extensions of the procedure, the next class of problems we treat contains fixed-axis rotation. We next couple translation and rotation and introduce plane motion dynamics. By generalizing our approach slightly, we are able to model hydraulic and acoustic circuits effectively, since they are similar in several respects to systems already treated.

All of the previously mentioned systems involve only one type of power within a single system. For that reason they are said to be *single-energy-domain* systems. There are many useful devices involving two (or occasionally more) types of power, and these have transducer elements (e.g., motors and pumps) coupling the different energy domains. We introduce some simple transducer system models as one more topic. Bond graphs are ideally suited to the study of multiple-energy-domain systems, so transducers are important devices worthy of your attention. The last section of this chapter is devoted to constructing bond graph models of multi-energy-domain systems. We will combine electrical, hydraulic, and mechanical devices to yield realistic overall physical systems that are ready for analysis and/or simulation.

In succeeding chapters, you will learn how to derive differential equations from a bond graph model, how to analyze certain types of equations, and how complex nonlinear systems can be handled in a computer. In this chapter, however, we want to emphasize that bond graphs are themselves precise mathematical models of systems that can be processed automatically by a computer program and that can, in many cases, yield time histories of interesting variables almost automatically, requiring only the specification of parameters, initial conditions, and forcing functions. A variety of computer programs are now available that will process bond graphs and put the implied equations into a form suitable for computational solution by a simulation program. The use of graphical displays of the results allows one to understand the dynamic responses of the system model and to optimize the system.

After studying the succeeding chapters, you will have an understanding of how automated simulation programs work, as well as the mathematical difficulties that may arise from some modeling decisions. For now, we wish only to emphasize that after a bond graph model has been devised, in most cases it is easy and convenient to perform computer experiments on the model and thus to size components or to change the system configuration if the predicted system behavior is not satisfactory.

4.1   ELECTRICAL SYSTEMS

We begin by observing that any electrical *circuit* can be modeled by a bond graph containing elements of the set $\{0, 1, C, I, R, S_e, S_f\}$. Notice that the elements TF and GY are not included. That is because these elements are properly used in representing electrical *networks*, a more general class than circuits. First we shall model circuits; then we shall extend the procedure to include network elements, arriving eventually at a complete result.

4.1.1   Electrical Circuits

In Chapter 3, the electrical resistance, inductance, and capacitance were shown to be bond graph $—R$, $—I$, and $—C$ elements. It remains to determine how to use the junction elements (0- and 1-junctions) to construct an overall bond graph model of an electrical circuit. Sometimes, for simple circuits, it is easy to recognize that some elements have the same current (flow) and others have the same voltage (effort). For these circuits, bond graph construction can be accomplished by inspection. For example, Figure 4.1a shows a simple circuit with positive voltage drops and current directions defined and node voltages labeled. For convenience, some node voltages are repeated to emphasize their association with specific elements. The circuit is grounded at the bottom, voltage labeled c, that is, $v_c = 0$.

To arrive at the bond graph in part b, we argue as follows: The elements C and R_1 have the same voltage ($v_a - v_c = v_a - 0 = v_a$) and thus are attached to the same 0-junction (common effort junction); the elements L and R_2 have the same current

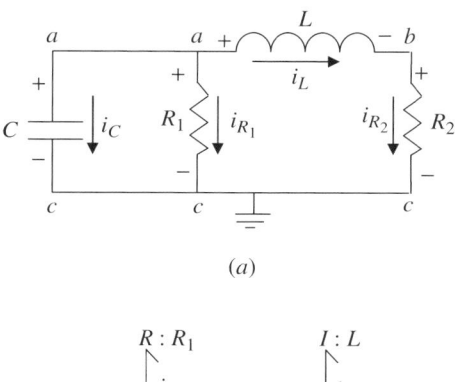

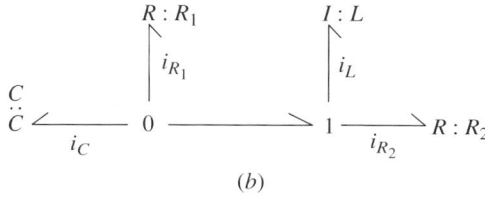

FIGURE 4.1. A simple electric circuit. Example 1.

$(i_L = i_{R_2})$, so they are attached to the same 1-junction (common flow junction). The bond joining the 0- and 1-junctions enforces the fact that the current through the inductor is the sum of the currents through the capacitor and the resistor, R_1 (actually, with the sign convention shown in Figure 4.1a, $i_L = -i_C - i_{R_1}$). Notice that all the 1-port R, C, and I elements have the power half-arrows defined such that whenever the voltage drop across the element is in the direction defined as positive in Figure 4.1a, and the current is simultaneously in the defined positive direction, the power is flowing into the element. We always define positive power directions for R, C, and I such that this is true.

Most of the time, electric circuits are too complex to model by inspection. There may be some parts that are obviously in series or parallel (common current or common voltage), but constructing the overall bond graph model is much easier if a procedure can be followed that ensures success regardless of the complexity of the circuit. Here we present a foolproof circuit construction procedure and develop the procedure along with an example shown in Figure 4.2. This is a voltage-excited circuit, grounded at the bottom, and open circuited at the right side exposing an output voltage, e_{out}.

Circuit Construction Procedure

1. **Assign a power convention to the circuit schematic.**
   This step must always be done regardless of the modeling procedure being used. If the ultimate goal of the model is to derive circuit dynamic equations or to simulate response, we must have a power convention. On the circuit, this is done by showing the positive voltage drop and current directions. For the

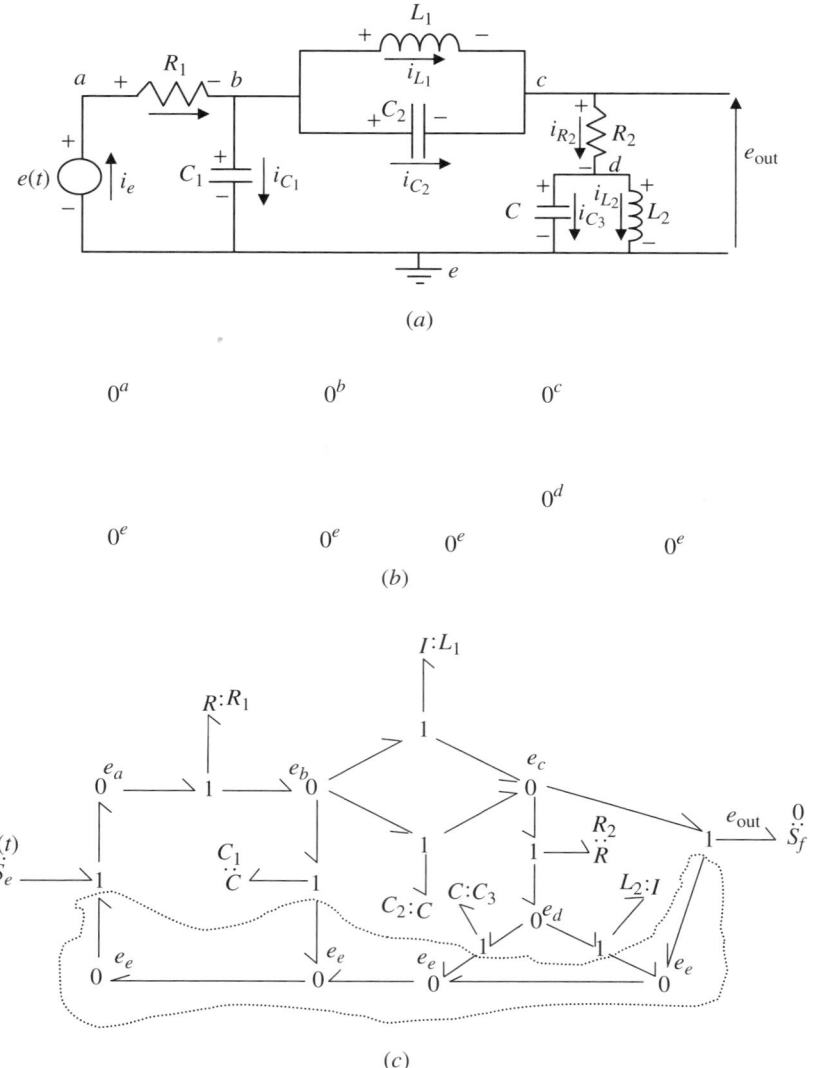

(a)

(b)

(c)

FIGURE 4.2. A bit more complicated electric circuit. Example 2.

$-I$, $-R$, and $-C$ elements, the positive voltage drop ($+$ to $-$) is shown in the same direction as the positive current. This ensures that power is directed inward on the corresponding bond graph element. For the source elements (S_f for current source and S_e for voltage source), it is not critical which directions are chosen for positive voltage drop and positive current. If positive current is defined such that the current moves "uphill" against the positive voltage as is done in Figure 4.2a, then positive power will come from the source into the rest of the circuit. If either the positive voltage direction or current direction is

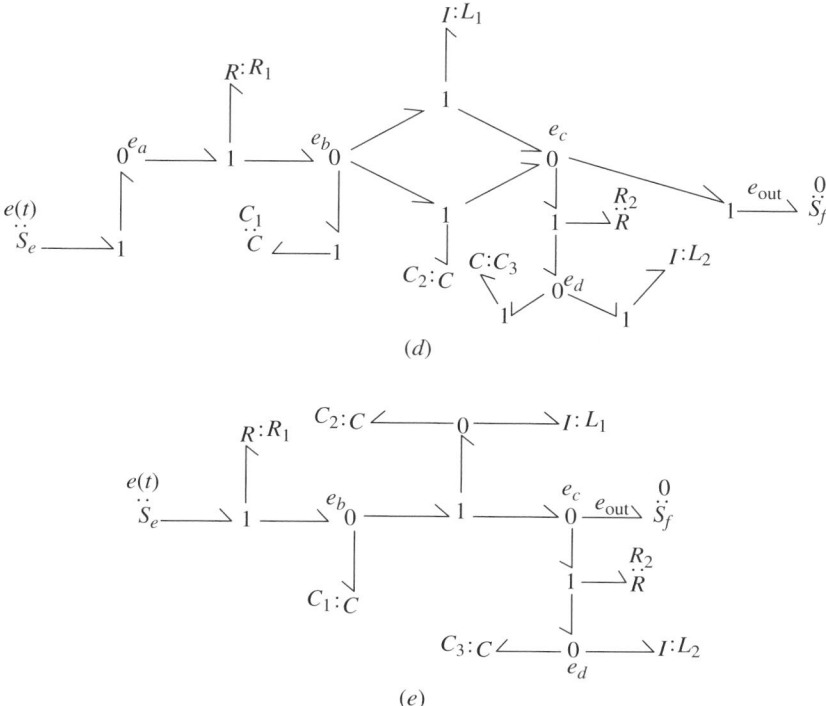

(d)

(e)

FIGURE 4.2. Continued

chosen in the opposite direction, then positive power will be absorbed by the
source. There is absolutely nothing wrong with this, and, in fact, real sources
may sometimes absorb power from the attached system and sometimes supply
power to the attached system.

2. **Label each node voltage on the circuit schematic and use a 0-junction to
   represent each node voltage as shown in Figure 4.2b.**
   A node voltage is the voltage above and below or to the left and right of each
   circuit element. In Figure 4.2a, the node voltages are labeled using letters. For
   convenience, the ground voltage, e, is repeated several times. Remember, every
   bond that touches a particular 0-junction has an identical voltage.

3. **Establish the positive voltage drops across the elements using 1-junctions.**
   Remember that 1-junctions add efforts (voltages) according to the power convention. By properly directing the half arrows on 1-junctions, the proper voltage drop can be established across each bond graph element. Figure 4.2c shows
   this construction. For example, for the —R element representing the resistor,
   R_1, the "effort" on the bond is $e_a - e_b$, which is the positive voltage drop
   defined in the circuit schematic. Notice that positive power is out of the voltage source element and the voltage on the source bond is $e_a - e_c$, as was

defined in the schematic. Also, the output voltage, e_{out}, is exposed using a flow source, S_f, of zero current. The voltage drop across this flow source is $e_{out} = e_c - e_e = e_c - 0$. The reader should check the other elements and ensure that all have their defined positive voltages.

4. **Remove all bonds that have zero power.**
   Before the bond graph can be used for equation derivation or simulation, the reference voltage must be established. Our reference is e_e, and it is zero since it is the ground voltage. Since every bond that touches a 0-junction has the identical voltage, all the bonds inside the curve of Figure 4.2c have zero voltage and each of those bonds carries no power. We can either append an effort source of zero voltage to one of the 0-junctions representing e_e, or we can simply erase all the bonds that carry no power, with the result shown in Figure 4.2d.

5. **Simplify the bond graph by using the bond graph identities defined in Chapter 3.**
   This is not an absolutely necessary step. By removing the 0- and 1-junctions with a through power convention, a much neater picture emerges, as shown in Figure 4.2e. Also, the loop structure from e_b to e_c has been reduced. We establish the voltage drop e_b–e_c once, and then attach the —I and —C elements associated with L_1 and C_2 to a 0-junction constrained to have this voltage drop. For equation derivation or automated simulation as described in Chapter 5, it is not necessary to reduce the bond graph to its simplest form, but it does make the final bond graph much nicer.

   Note that the flow source introduced to expose the output voltage, e_{out}, may also be erased since there is no power associated with the source bond. It is left in Figure 4.2e as a convenience to remind us that we are interested in that particular output voltage. Later we will see that any effort or flow on any bond can be simply established as an output, and we will not have to construct an artificial means to expose desired outputs.

   As a final example of circuit modeling using the procedure just presented, consider the Wheatstone bridge shown in Figure 4.3a. This circuit is typically used with strain gages as the resistive elements, R_1 through R_4, and the voltage across the load resistance, R_L, is the output that is indicative of any change in the bridge resistances. We are simply going to model this circuit as an exercise in using the bond graph circuit construction procedure. In Figure 4.3a, the positive voltage drop and current directions are shown along with labels for the node voltages. This is **Step 1** of the procedure.

   Step 2 of the procedure is done in Figure 4.3b, where the node voltages are represented by 0-junctions. **Step 3** is shown in Figure 4.3c where 1-junctions are used to appropriately add voltages such that the proper positive voltage drops are across each element. Also shown in c is the curve surrounding all the bonds at the ground voltage, $e_d = 0$. **Steps 4** and **5** are done in Figure 4.3d. The bonds with no power are erased, and the bond graph simplifications have been done where all 1-junctions with two bonds and a "through" power convention have been reduced to single bonds.

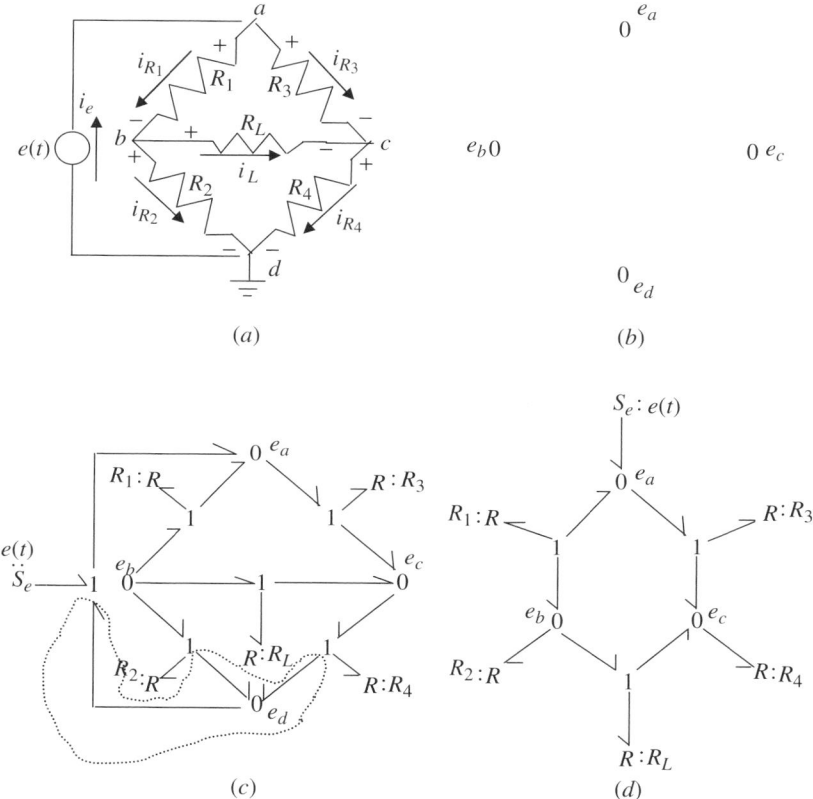

FIGURE 4.3. A Wheatstone bridge circuit. Example 3.

The resulting final bond graph reveals a beautiful structural symmetry quite similar to that of the hydrocarbon molecule called the "benzene ring." It was this similarity to chemical bond diagrams that prompted the "bond graph" name of our modeling approach. Quite nice, don't you think?

4.1.2   Electrical Networks

An electrical network is an extension of electrical circuits to include transformers and gyrators (see Reference [3]). An electrical transformer is a common electromagnetic device used to step voltages up or down while doing the opposite to the current. Electrical gyrators are exhibited in Hall effect transducers (see Reference [1]) where voltage across a semiconductor material is related to a current through the material perpendicular to the voltage drop direction. The basic rules for bond graph construction remain unchanged.

Figure 4.4a shows the electrical symbol for a transformer where N indicates the turns ratio across the device. Positive voltage drops and current directions are chosen

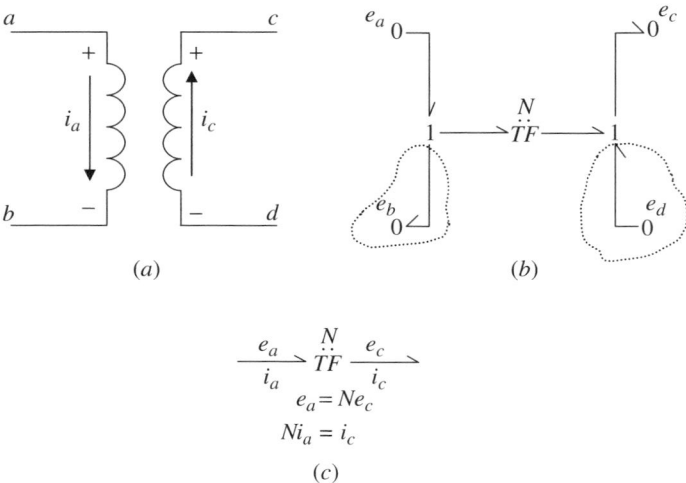

FIGURE 4.4. The ideal electrical transformer and its bond graph representation.

such that positive power is into the device on the left side and out of the device on the right side. Since transformers and gryrators are power conserving elements, it is logical to always define positive power such that it flows in on one side and out on the other. In Figure 4.4b, 1-junctions are used to establish the positive voltage drops across the input and output sides of the transformer. Notice that on the input side the voltage is $e_a - e_b$ and on the output side the voltage on the transformer output bond is $e_c - e_d$. These voltages are as defined in the schematic of Figure 4.4a. In most circuits, each side of a transformer would be referenced to the same ground voltage. If $e_b = e_d = 0$, then their associated bonds can be erased, some reductions can be performed, and the typical appearance of a transformer emerges as shown in Figure 4.4c.

An electrical network with an isolating transformer is shown in Figure 4.5a. Positive voltage drops and current directions are shown in this figure. Notice that positive power flows in on the left side of the transformer and out on the right side. Following the bond graph construction procedure of the previous section, Figure 4.5b uses 0-junctions to expose the node voltages labeled in part a, and Figure 4.5c uses 1-junctions to establish the proper voltage drops across all the elements. Also in part c are dotted lines enclosing all the bonds with zero power. These are erased in Figure 4.5d, and bond graph simplifications have been performed to yield the final result.

A final example of the using the construction procedure is the network shown in Figure 4.6a. Most of the network is composed of circuit elements that must be looking pretty routine by now, but the new component is the voltage modulated current source, or "controlled current source," labeled $I(t)$ in the figure. The meaning of this schematic symbol is that the voltage, e, from the left side of the circuit modulates, with virtually no power, the current source that is an input to the right side of the cir-

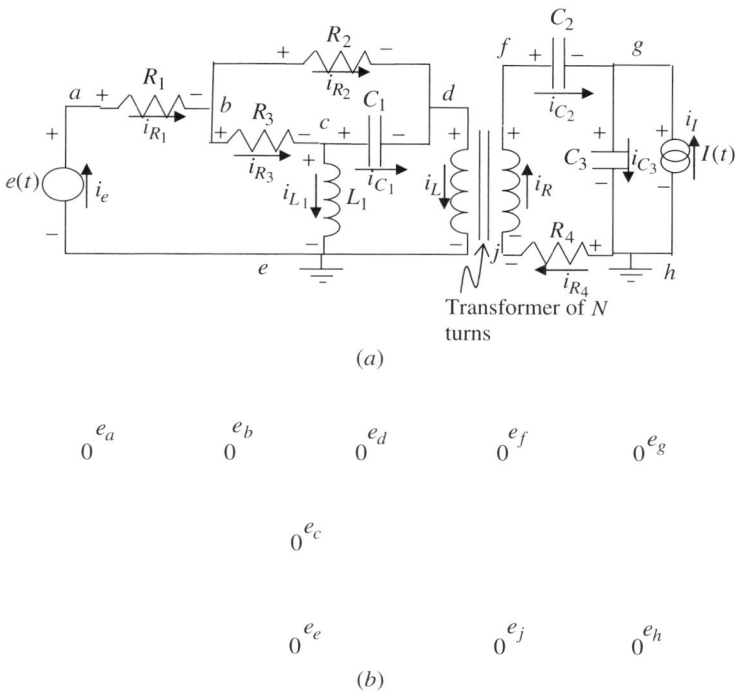

(a)

(b)

FIGURE 4.5. An electrical network with an isolating transformer.

cuit. Positive voltage and current directions are shown in the schematic. The labeled node voltages are shown using 0-junctions in Figure 4.6b, and 1-junctions are used in Figure 4.6c to establish voltage drops across the elements. Of particular interest is the establishment of the modulating voltage, e, and the use of the flow source, S_f, for the current source generating $I(t)$. Other than these elements, the circuit construction is identical to all the other examples presented.

An active bond is used to indicate the ideal modulation of the flow source. This means that the control voltage, $e(t)$, modulates the output current while requiring virtually no power from the system. The ground voltage is indicated by the closed curve, and the zero power bonds have been erased and simplifications performed to yield the final result of Figure 4.6d.

The network construction method presented here will allow you to model the most complicated electrical networks once you have acquired facility with the basic steps of node representation, element insertion, power definition, ground definition, and graph simplification. With practice you will begin to observe certain recurrent patterns, and your ability to do more by inspection will increase. Many interesting discoveries await you as you explore systems containing structures like *ladders*, *pi*'s, and *tees*.

Speaking of ladders, let us make use of our ability to model the topology of circuits by bond graph junction structures in order to examine the bond graph form of

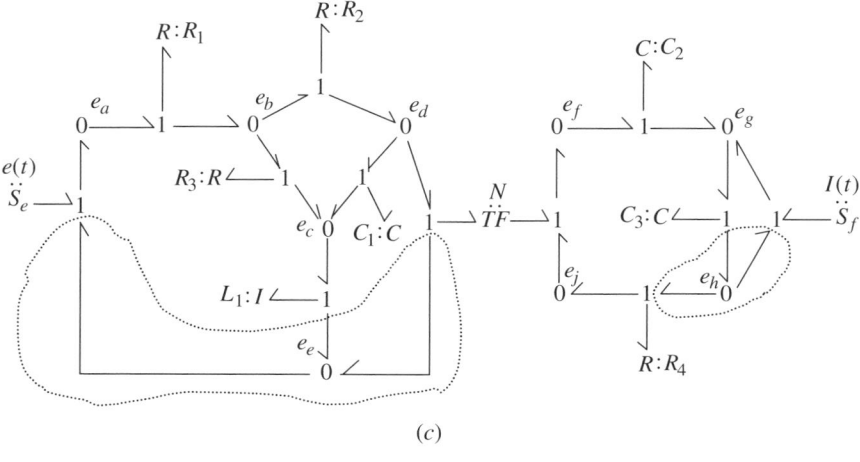

(c)

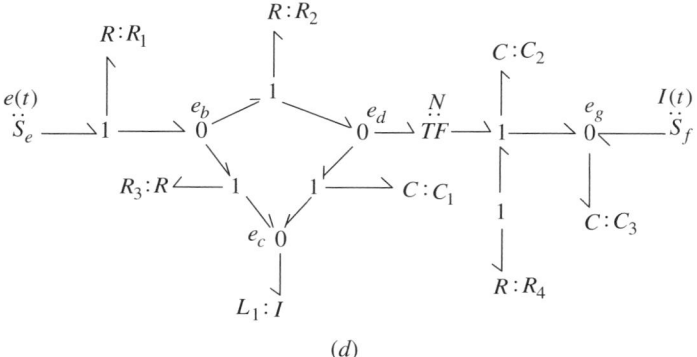

(d)

FIGURE 4.5. Continued

a ladder network. An example of a resistive ladder circuit is given in Figure 4.7a. We are not really concerned with the nature of the 1-ports (source and resistances), but rather with their interconnection patterns (or circuit topology). The bond graph of part b may be found by inspection, using arguments like "..., R_2 in parallel, R_3 in series, R_4 in parallel...." Or the formal construction procedure may be used. To display the system structure without regard to what is (or might be) connected to it, we use a bond graph like that of Figure 4.7c, which is a direct representation of the ladder structure. So ladders are nothing more than chains of alternating 3-port 0's and 1's.

While you are still recovering from the surprise of looking at the structure, or connection pattern, expressed as a bond graph, we shall use such a graph to explore the idea of dual topologies in circuits. A pi network is shown in Figure 4.8a in resistive form. The pi gets its name because of the Greek letter, pi-like appearance of the schematic. The structure is represented as parallel–series–parallel in part b by the

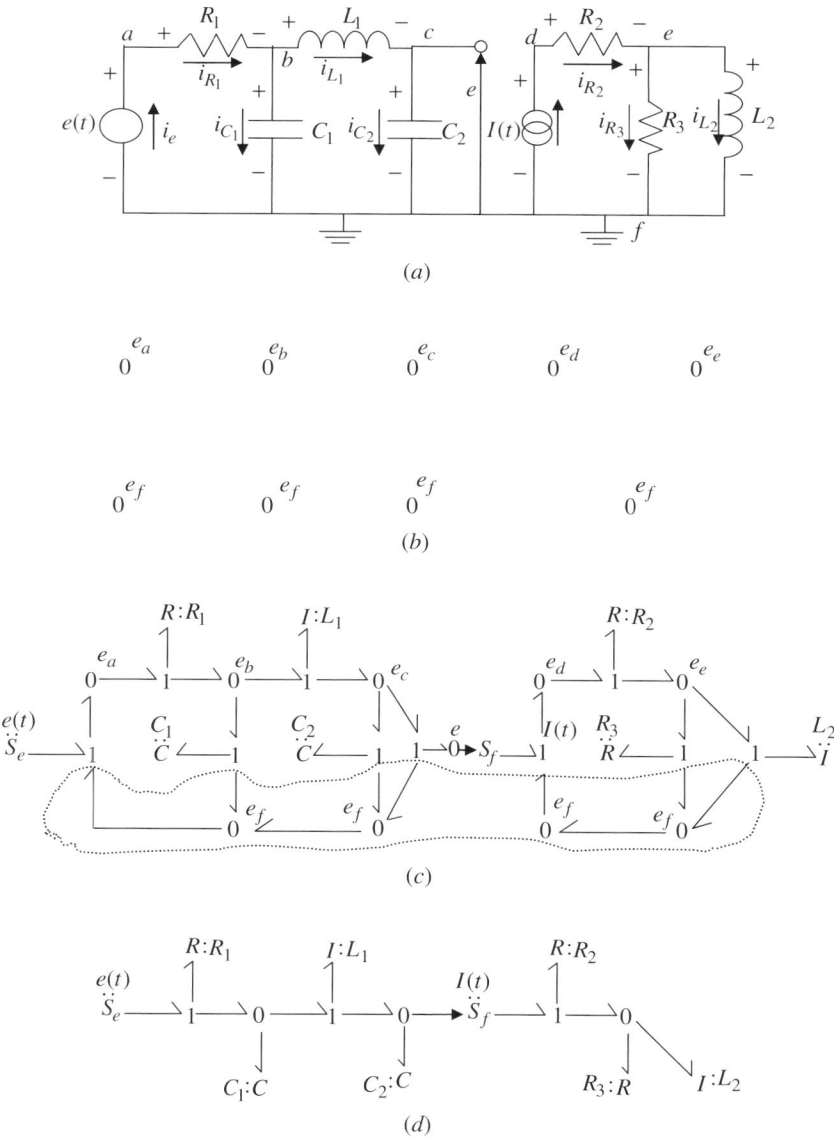

FIGURE 4.6. An electrical network with a controlled source.

bond graph. Resistances 1, 2, and 3 would go on the corresponding bonds to complete the model. In Figure 4.8c a tee network is shown, again in resistive form. The tee gets its name from the T-like appearance of the schematic. The bond graph in part d is a series–parallel–series type. We observe that part d can be obtained from part b by switching the roles of 0 and 1, and part b can be obtained from part d

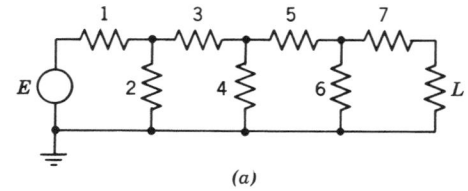

(a)

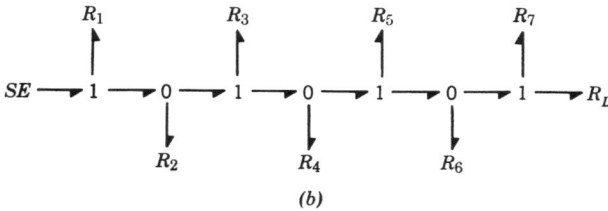

(b)

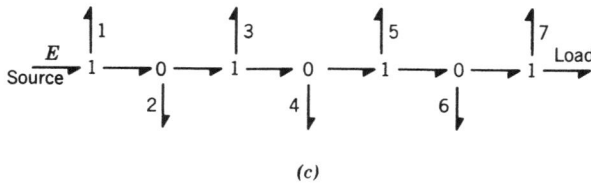

(c)

FIGURE 4.7. A ladder network example.

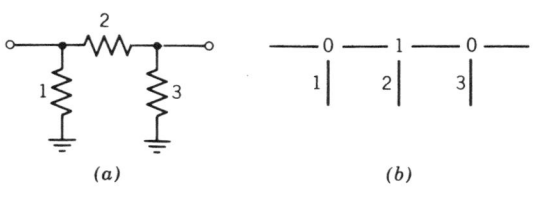

(a)   *(b)*

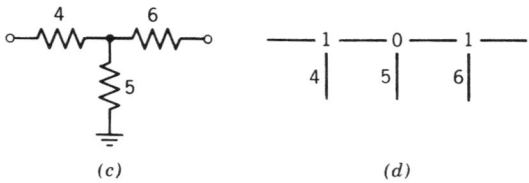

(c)   *(d)*

FIGURE 4.8. Truncations of the ladder structure—the pi and tee structures.

similarly. The formal idea behind the switching is that if the roles of voltage and current are interchanged, a *dual* network results. In terms of bond graph structure, this implies a switching of 0 and 1 elements. By this technique the topological dual of a complex circuit may be obtained from its bond graph in a very simple fashion.

The pi and tee structures are simplified versions of the ladder structure of Figure 4.7. We will see later in the text that transmission lines, both electrical and hydraulic, have dynamic behaviors that can be represented by ladder structures. The pi and tee structures become truncations of the ladder structure that allow for reasonable inclusion of transmission line dynamics without having the overall system become too large.

4.2 MECHANICAL SYSTEMS

Mechanical systems are those composed of such components as masses, springs, dampers, levers, flywheels, gears, shafts, and so forth. When dynamic systems are put together from these components, we must interconnect rotating and translating inertial elements with axial and rotational springs and dampers, and we must appropriately account for the kinematics of the system structure. Bond graphs are well suited for this task. In the next section we address the specifics of mechanical translation; that is, we restrict our attention to mechanical systems moving in a straight line. In Section 4.2.2 we focus only on mechanical systems that rotate, such as flywheels connected by a torsionally flexible shaft, or a gear pair in a transmission. Of ultimate importance is the dynamics of plane motion where we recognize that inertial elements have both mass and rotational inertia, and both often must be accounted for in a realistic system. Plane motion is covered in Section 4.2.3.

4.2.1 Mechanics of Translation

In the previous sections dealing with electrical systems, the construction procedure started with identifying the important node voltages and representing them with the common effort 0-junction. Then 1-junctions were used to create appropriate voltage drops across the elements. Finally, the reference voltage had to be identified and eliminated to yield the final model.

For mechanical systems, the effort variable is the force and the flow variable is the absolute velocity (velocity with respect to inertial space). It has been found from years of experience that constructing bond graph models for mechanical systems is easier if a procedure is followed that is dual to the procedure for electrical systems. For mechanical systems, we start by representing system velocities using 1-junctions (common flow, efforts add) and then we create the appropriate relative velocities across the springs and dampers using 0-junctions (common effort, flows add). This is best illustrated with a simple example shown in Figure 4.9. The system consists of a mass, m, suspended in a gravity field from a spring and damper, k, b, attached to an inertial frame.

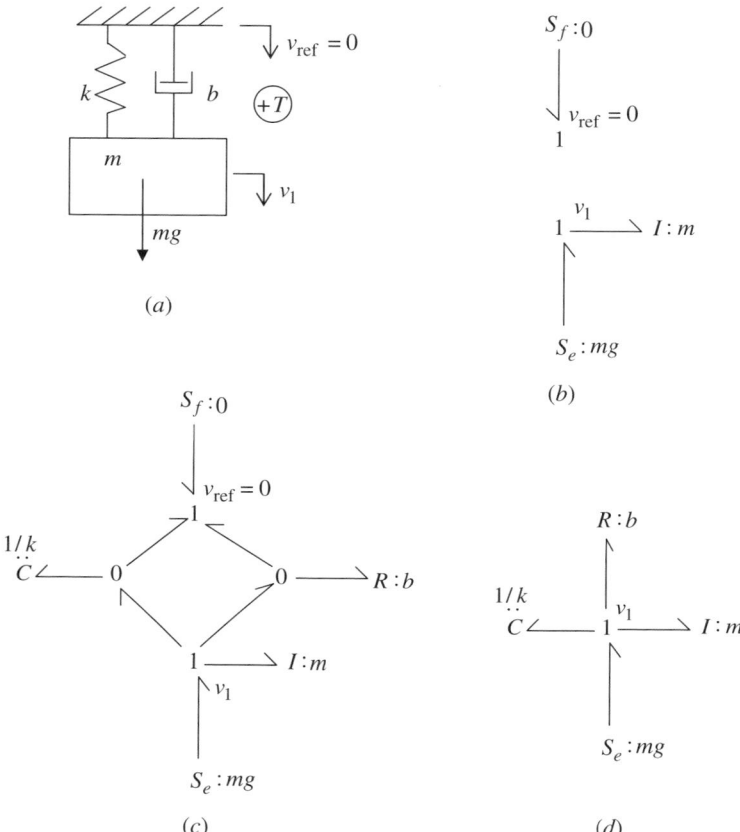

FIGURE 4.9. Mechanical translation. Example 1.

Just as with the electrical systems, a power convention must be established. This is a necessary step regardless of the modeling procedure being used. Figure 4.9a shows the positive velocity direction for the mass and states that the spring and damper are assumed to be positive in tension. Notice that the inertial reference is labeled as a velocity equal to zero. This is our reference velocity and ultimately will be erased, similarly to the reference voltage in electrical systems.

In Figure 4.9b, a 1-junction is shown such that any bonds that are connected to this junction will have the velocity v_1 and a second 1-junction is shown such that any bonds connected to it will have the reference velocity of zero. The mass is a 1-port inertia that has the absolute velocity, v_1, so the $-I$ element is attached to that 1-junction. The gravity force, mg, is modeled as an effort source in the bond graph (any force that is a known input to a system, whether time varying or constant, is modeled as an effort source in a bond graph). Since this force is moving at the velocity v_1, the effort source is attached to the 1-junction representing this velocity.

The power out convention on the effort source comes from the fact that if the velocity is in the positive, downward direction and the gravity force is acting downward, positive power is coming from the source into the system. A flow source equal to zero is attached to the reference 1-junction to enforce that the velocity is zero.

The spring and damper in general react to the relative velocity across them. The choice of how to properly add the velocity components at each end of these elements, a topic that causes much student frustration, will be dealt with in due time. For this first example, we simply state that for the elements positive in tension, with the positive velocity directions indicated, the relative velocity across both the spring and damper should be $v_1 - v_{\mathrm{ref}}$. This is accomplished using 0-junctions with the power convention indicated in Figure 4.9c. The reader should verify that the velocity on the C-element and the R-element bonds is the proper relative velocity. Finally, in Figure 4.9d, the bonds with zero power are removed and the final bond graph emerges after using some bond graph reductions. It makes sense that, since all the elements have the same velocity, they all end up on the same 1-junction. We will be able to do many mechanical system models by inspection after we gain some experience.

To formalize the bond graph construction procedure for mechanical translation, here are the steps to follow:

Construction Procedure for Mechanical Translation

1. On a schematic of the physical system, use arrows and symbols to indicate the positive direction of "distinct" absolute velocity components. These would include all individual mass elements, all prescribed input velocities, and the velocities of any other physical locations that may prove useful in establishing useful relative velocities. State whether force-generating elements, springs and dampers, are positive in compression or tension.

2. Use 1-junctions to represent each distinct velocity from 1. Label the 1-junction with the velocity symbol from the schematic. This will help to remind you which junction is associated with which velocity component. You might use a 1-junction to represent the reference of zero absolute velocity. This will later be eliminated, but it might be helpful for establishing relative velocities.

3. Attach to each 1-junction any element that relates to the absolute velocity represented by the junction. In general, distinct masses are inertias associated with distinct velocities represented by some of the 1-junctions. Remember, positive power is **always** directed into an element —I, —R, or —C element.

4. Use 0-junctions to establish proper relative velocities across the remaining elements so that the elements are positive in compression or tension as was assumed in the schematic.

5. Eliminate the bonds with zero velocity and reduce to the final model.

As an example of the use of the construction procedure, consider the quarter car model of Figure 4.10. This sort of represents one corner of an automobile, say the front right corner, where m_s is the sprung mass that is part of the vehicle body, m_{us} is the unsprung mass, which includes the tire, wheel, and some part of the brakes and

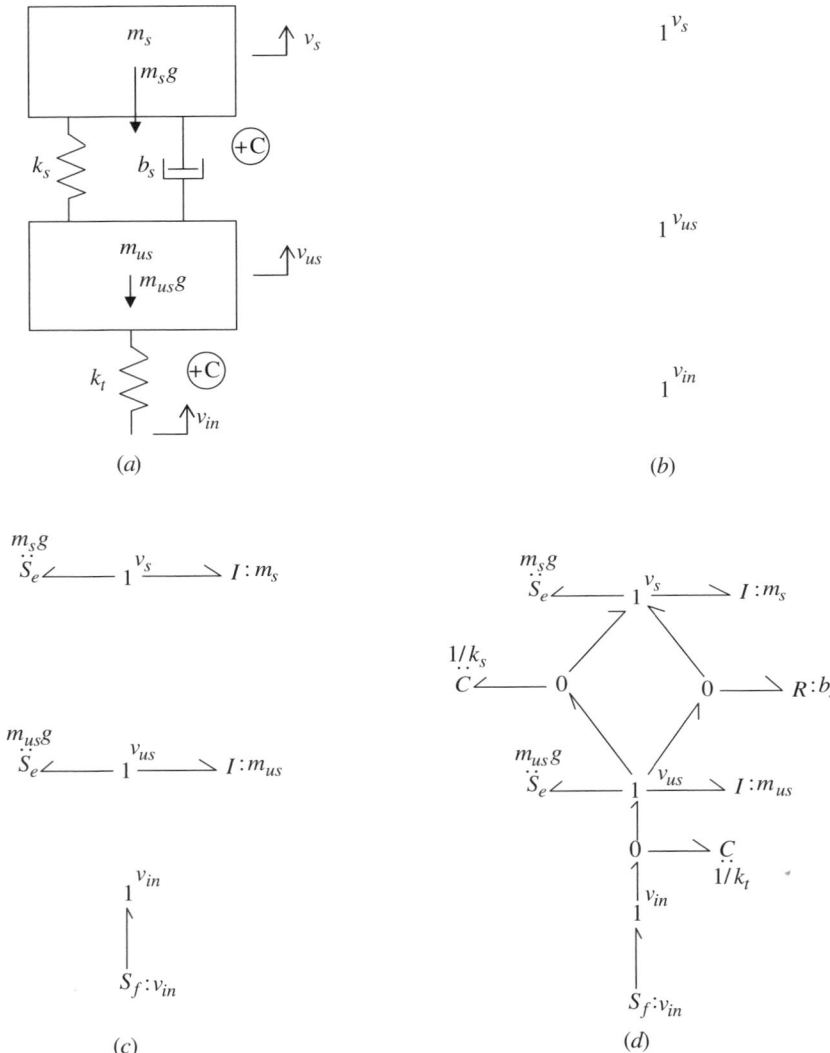

FIGURE 4.10. Quarter car model. Example 2.

suspension, k_s, b_s are the suspension spring and damper, and k_t is the tire stiffness. Quite some liberty has been taken in representing the suspension and tire as linear elements. The system is constrained to move only vertically, and the velocity input, $v_{in}(t)$, at the base is representing the roadway unevenness experienced as the vehicle would actually be moving forward. The vehicle is also under the influence of gravity acting vertically downward.

Construction rule 1 is used in the schematic of Figure 4.10a. The distinct velocities are labeled with arrows indicating positive directions, and, as indicated, the springs and dampers are all assumed positive in compression. In Figure 4.10b, Step 2 is performed where 1-junctions are used to represent each distinct velocity, and the junctions are labeled to remind us which velocity they represent.

Step 3 is accomplished in Figure 4.10c where the inertial —I elements, with inward power convention, are attached to the appropriate 1-junctions and labeled to indicate which mass they represent. Notice that the velocity input is represented by a flow source, S_f, attached to the 1-junction for v_{in}. The power arrow is out from the flow source because if the tire spring is in compression (positive by assumption) and the input velocity is up (positive by assumption), then power is flowing from the source into the system. The weight of each mass element, $m_s g$ and $m_{us} g$, is modeled as an effort source attached to the 1-junctions having the velocity of the associated mass. The power arrow is directed away from the 1-juntion and into the source because if the respective mass element is moving upward (positive by assumption) and the gravity force is acting downward (as it always must), then power is flowing from the system and into the source representing the gravity force.

For Step 4, shown in Figure 4.10d, 0-junctions are used to establish relative velocities across the remaining elements. For the positive velocity directions defined, and the positive in compression force directions defined, the proper relative velocity across the tire spring is $v_{\text{in}} - v_{us}$ and the proper relative velocity across the suspension spring and damper is $v_{us} - v_s$. These are the proper relative velocities because, if they are positive, then the respective elements will, in fact, be compressing. Notice that the relative velocities across the suspension spring and suspension damper are independently constructed, as dictated by the construction procedure. This is perfectly correct and can always be done; however, a simpler realization for these relative velocities will be shown below. There is one simplification that could be done for this model. The 1-junction at the bottom of Figure 4.10d could be removed, and the flow source could be attached directly to the 0-junction.

It should be recognized that we paid attention only to establishing the proper velocity components for this construction procedure, and we paid no attention to the forces. The beauty of using the power conservation properties of bond graphs is that we need only to constrain the velocities and the forces will automatically be balanced. For mechanical systems, we establish proper velocities and the forces are guaranteed to be correct. Consider the free body diagram for the quarter car example, shown in Figure 4.11. The masses are isolated and the forces from the springs, dampers, and gravity are exposed. The arrows for the forces are directed in the assumed positive directions, namely, in compression for this example, and the positive velocities are directed upward as was assumed positive previously. Notice that we must show all the forces as having equal but opposite effects, as required by Newton's laws. If we were to sum the forces on the sprung mass, positive upward, the resultant force, F_s, would be

$$F_s = F_{k_s} + F_{b_s} - m_s g, \tag{4.1}$$

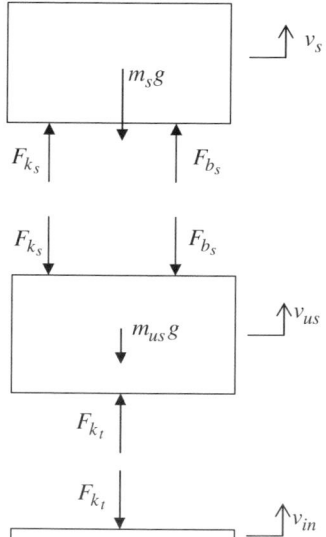

FIGURE 4.11. Free body diagram of the quarter car system.

and for the unsprung mass, the resultant force, F_{us}, would be

$$F_{us} = F_{k_1} - F_{k_s} - F_{b_s} - m_{us}g. \tag{4.2}$$

On the bond graph of Figure 4.10d, the forces on the respective mass elements are the efforts associated with the $-I$ elements emanating from the 1-junctions labeled v_s and v_{us}. The 1-junctions, in addition to being common flow or velocity, add efforts, or forces, according to the power convention. Amaze yourself by adding the forces on the $-I$ element bonds and seeing that the forces add exactly as required by Eqs. (4.1) and (4.2). The forces came for free after enforcing the velocity constraints. And we never had to show equal and opposite reaction forces.

The construction procedure used for the suspension spring and damper calls for the establishment of the relative velocity across each element using 0-junctions to appropriately add velocity components. Such adherence to the construction rules will always produce a correct result. The resulting bond graph structure, referred to as a "loop with a through power convention" or a "reducible loop," comes up a lot in bond graph modeling. This loop is isolated in Figure 4.12a, where it has been generalized using efforts and flows rather than forces and velocities. The relative velocities across the spring and damper are identical, and each is equal to $f_1 - f_2$. In Figure 4.12b, the relative velocity is established using a single 0-junction, and then a 1-junction is used to ensure that any bond attached to that 1-junction will have the relative velocity, $f_1 - f_2$. Since both the $-C$ and $-R$ elements have this relative velocity, they both get attached to the 1-junction as shown in Figure 4.12b. The reader can check to see that the forces still add up properly in this alternative representation. The result is

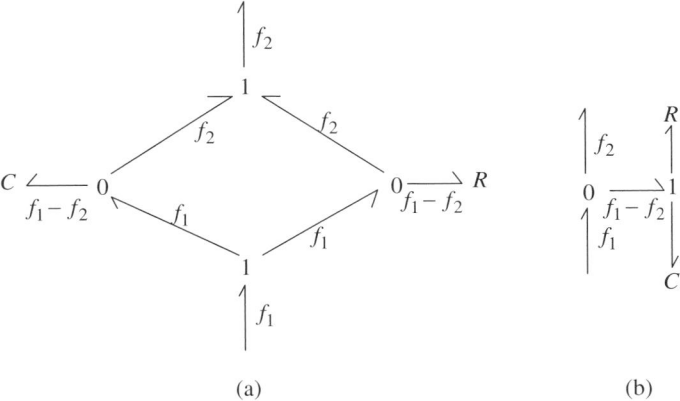

(a)   (b)

FIGURE 4.12. Loop with through power convention that can be reduced.

a slightly simpler bond graph representation when several elements have the same relative velocity or relative flow. It is not required to reduce the loop, but it will generally be done in this text just to produce nicer looking figures.

In addition to mechanical systems having one-directional translational motion, it requires only a modest extension of our procedure to deal with translational systems containing levers, pulleys, and other simple motion–force transforming devices. Consider the system shown in Figure 4.13a. A velocity input is prescribed in the horizontal direction on the end of spring, k_1. The other end of this spring goes around a

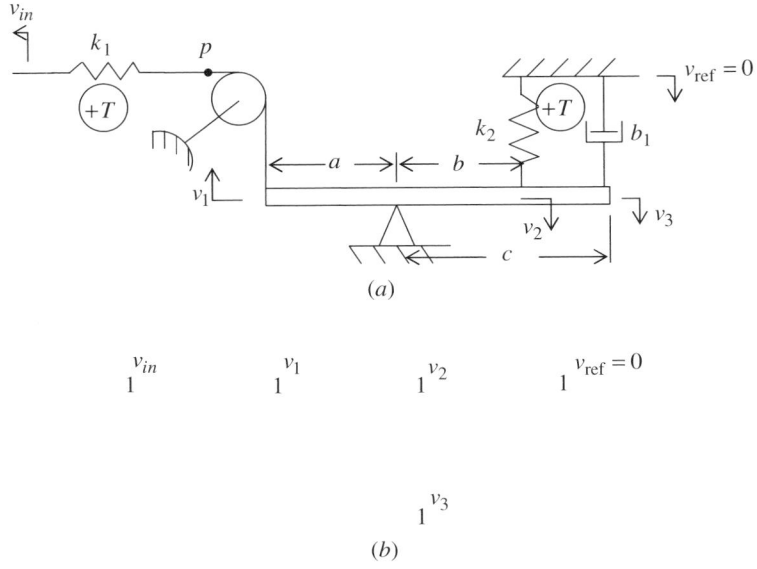

FIGURE 4.13. A pulley–lever mechanism.

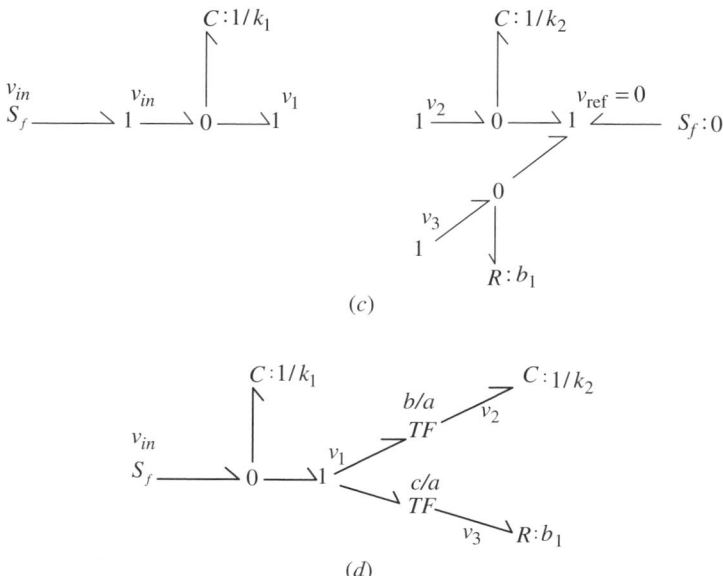

(c)

(d)

FIGURE 4.13. Continued

pulley and causes vertical motion at one end of the massless lever. A spring, k_2, and damper, b_1, are located on the right side of the lever with one end of each attached to inertial ground. The pivot for the lever is located a from the left end, while the spring and damper are located, respectively, b and c to the right of the pivot. Figure 4.13a in addition shows positive velocity directions and indicates that both springs and the damper are assumed positive in tension. Notice that v_1 is labeled only on the end of the lever, but it is the same as the horizontal velocity at the point P.

Following the construction procedure, we first use 1-junctions to represent the distinct velocities. A good question is why these particular velocities were chosen as distinct and worthy of representation with their own 1-junctions. As mentioned in Step 1 of the Construction Procedure, velocities are always assigned to distinct masses, as well as to other physical locations that might prove useful in establishing relative velocities across elements. You can never overspecify velocity components. If some 1-junctions turn out to not be needed, no bonds will end up attached to them, and they can be erased at the end of the model development. Looking carefully at Figure 4.13a, we see that v_1 (which is the velocity at point p) will be useful for establishing the relative velocity across the horizontal spring, and velocities v_2 and v_3 are needed for the relative velocities across the vertical spring and damper. Figure 4.13b shows the 1-junctions with the important velocities labeled. The only elements that have any of these specific absolute velocities are the source elements that establish the input velocity, $v_{in}(t)$, and the reference velocity, $v_{ref} = 0$. These are shown in Figure 4.13c where the relative velocities across the elements have been established using 0-junctions. Convince yourself that the arrows are correct for having all elements positive in tension. For example, with the velocities defined positive as they are, spring k_1 will be put into tension by the relative velocity, $v_{in} - v_1$, and this is how the flows add on the corresponding 0-junction. The other elements follow a similar argument. The rules of construction have been followed and the model is obviously not complete. There is no connection between the 1-junctions for v_1, v_2 and v_3.

From Chapter 3 we know that a massless lever is a transformer relating velocities and forces across it according to the constitutive relationships,

$$\frac{b}{a}v_1 = v_2 \quad \text{and} \quad \frac{c}{a}v_1 = v_3 \tag{4.3}$$

Figure 4.13d shows the final model with the transformers installed, the reference velocity eliminated, and some simplifications performed.

One of the many valuable features of bond graph modeling is the relative ease with which models can be modified. Let's say that a mass that we neglected turns out to be important in the system shown in Figure 4.13a. The modified schematic is shown in Figure 4.14a. No additional velocities need be labeled, since v_1 was needed in the previous version of the model. Looking at Figure 4.13c where v_1 is exposed, the new mass element, m_1, simply becomes an $—I$ element attached to the v_1 1-junction. Nothing else changes, and the final result is shown in Figure 4.14b.

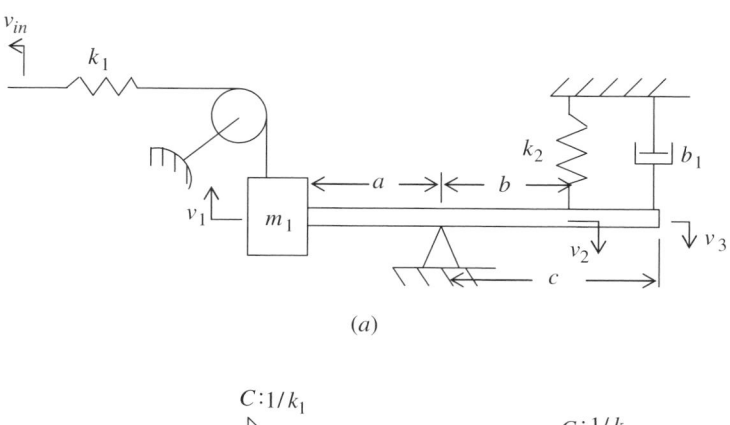

(a)

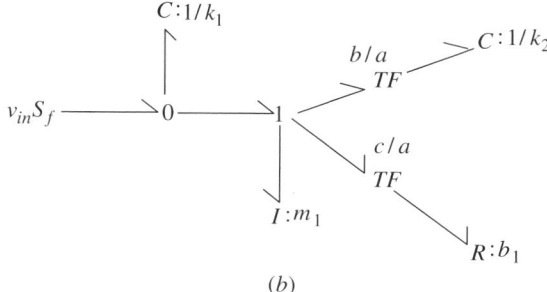

(b)

FIGURE 4.14. Modified pulley–lever mechanism.

A straightforward extension of translational mechanical systems is fixed-axis rotation, presented next.

4.2.2   Fixed-Axis Rotation

Fixed-axis rotation occurs in the mechanics of rotating machines such as electric motors and thermal engines. Gear boxes, transmissions, transfer cases, differentials, and drive shafts are all examples of systems that have some aspect of fixed-axis rotation. The procedure for modeling mechanical translation need only be modified slightly to handle systems with fixed-axis rotation. Because it is a straightforward extension of what we just did, this section starts with the procedure for bond graph construction of fixed-axis rotation.

Construction Procedure for Fixed-Axis Rotation

1. On a schematic of the physical system, use arrows and symbols to indicate the positive direction of "distinct" absolute angular velocity components. These would include all individual rotational inertial elements, all prescribed input angular velocities, and the angular velocity of any other physical locations that may prove useful in establishing useful relative angular velocities. State whether torque-generating elements, rotational springs and dampers, are positive when twisted clockwise or counterclockwise.

2. Use 1-junctions to represent each distinct angular velocity from 1. Label the 1-junction with the angular velocity symbol from the schematic. This will help to remind you which junction is associated with which angular velocity component. You might use a 1-junction to represent the reference of zero angular velocity. This will later be eliminated, but it might be helpful for establishing relative angular velocities.

3. Attach to each 1-junction any element that has that angular velocity. In general, distinct rotational inertias are $-I$ elements associated with distinct angular velocities represented by some of the 1-junctions. Remember, positive power is **always** directed into an $-I$, $-R$, or $-C$ element.

4. Use 0-junctions to establish proper relative angular velocities across the remaining elements so that the elements are positive when twisted one way or the other, as was assumed in the schematic.

5. Eliminate the bonds with zero angular velocity and reduce to the final model.

As the first example, consider a grinding wheel at the end of a torsionally flexible shaft, shown in Figure 4.15a. An electric motor is assumed to prescribe an input angular velocity, $\omega_{in}(t)$, at the left end of the shaft. The shaft is flexible and characterized by its torsional stiffness, k_τ, with units N-m/rad. The right end of the shaft has a disk of rotational inertia, J, with units kg-m^2. There is also rotational damping, b_τ, with units N-m/(rad/s), attached between the disk and ground. Figure 4.15b is a schematic of the physical system with angular velocities labeled and positive direc-

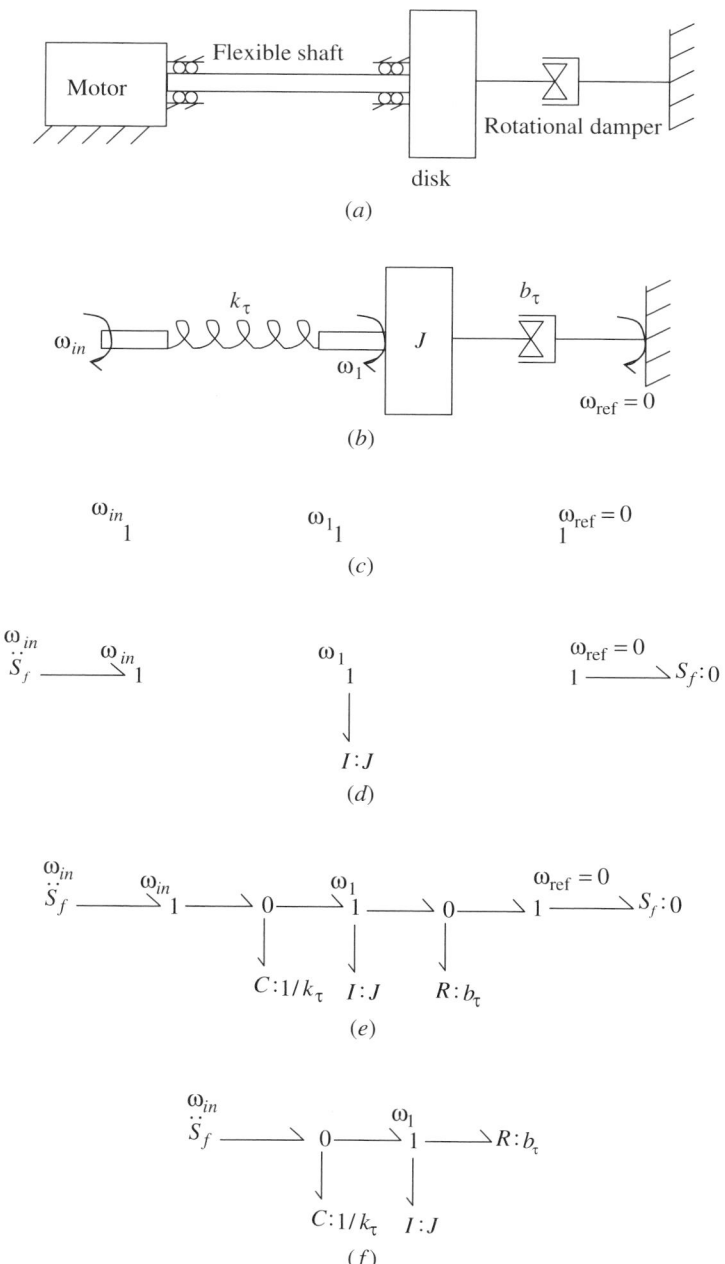

FIGURE 4.15. A grinding wheel model to demonstrate fixed-axis rotation. Example 1.

tions given. The twist in the shaft is assumed positive in the clockwise direction as viewed looking into the ends of the shaft, and the rotational damper torque is positive for clockwise rotation as viewed looking into ends of the damper. Figure 4.15c shows 1-junctions representing the distinct angular velocity components including the reference of zero angular velocity. In Figure 4.15d, the flow source that prescribes $\omega_{\text{in}}(t)$ is attached to the appropriate 1-junction, and the rotational inertia is attached to its appropriate 1-junction. In Figure 4.15e, 0-junctions are used to establish relative angular velocities across the torsional spring and torsional damper. In Figure 4.15f, the inertial reference is removed and some simplifications are carried out to reduce to the final model.

A more complex example of fixed-axis rotation is shown in Figure 4.16. Disks of rotational inertias J_1 and J_2 are rigidly attached to each end of a flexible shaft with torsional stiffness, k_{τ_1}. Disk 2 is attached at one end to a second torsional spring, k_{τ_2}, which is attached to a third disk, J_3, that is free to rotate on the shaft. This third disk is of radius R_3 and forms a gear set with the fourth disk, J_4, of radius R_4. Finally, disk 4 is attached through a torsional spring and damper, k_{τ_3} and b_τ, to a fifth disk of rotational inertia, J_5. Note that an intermediate angular velocity, ω', has been labeled in between the rotational spring and damper. This was done for convenience in anticipation of the need to establish the relative angular velocities across these two elements. We would ultimately like to predict the motion–time behavior of this system for a specified input torque, $\tau_{\text{in}}(t)$. We must wait to see how straightforward this is once we have a bond graph model. For now we must be satisfied with just constructing the model. On the physical schematic the positive angular velocity directions are indicated with arrows and labels, and the positive torque directions for the springs are indicated by the lines with full arrows using the right-hand-rule. The isolated shaft accompanying the schematic shows this notation.

In Figure 4.16b, 1-junctions have been used to represent all the distinct angular velocities from the schematic. Moreover, the bond graph elements that have these specific angular velocities have been attached directly to these 1-junctions. These include the rotational inertias and the input torque source that is moving at the angular velocity ω_1. In Figure 4.16c, 0-junctions have been used to establish relative angular velocities across the rotational spring and damper elements. The power convention for the 0-junctions was chosen to establish the positive torque directions from the schematic. Notice how useful the 1-junction for ω' is for establishing the relative angular velocities across k_{τ_3} and b_τ. Also notice that there is no connection between ω_3 and ω_4. The two disks of rotational inertias J_3 and J_4 are a gear set with the kinematic relationship

$$\frac{R_3}{R_4}\omega_3 = \omega_4. \tag{4.4}$$

This relationship is enforced by a transformer as shown in Figure 4.16d. In this figure, some final simplifications were done, including the removal of the unnecessary 1-junction representing ω'. After this 1-junction was removed, the two remaining 0-junctions were connected by a common bond enforcing that all connected bonds had the same torque (effort). All the bonds with a common effort can be connected

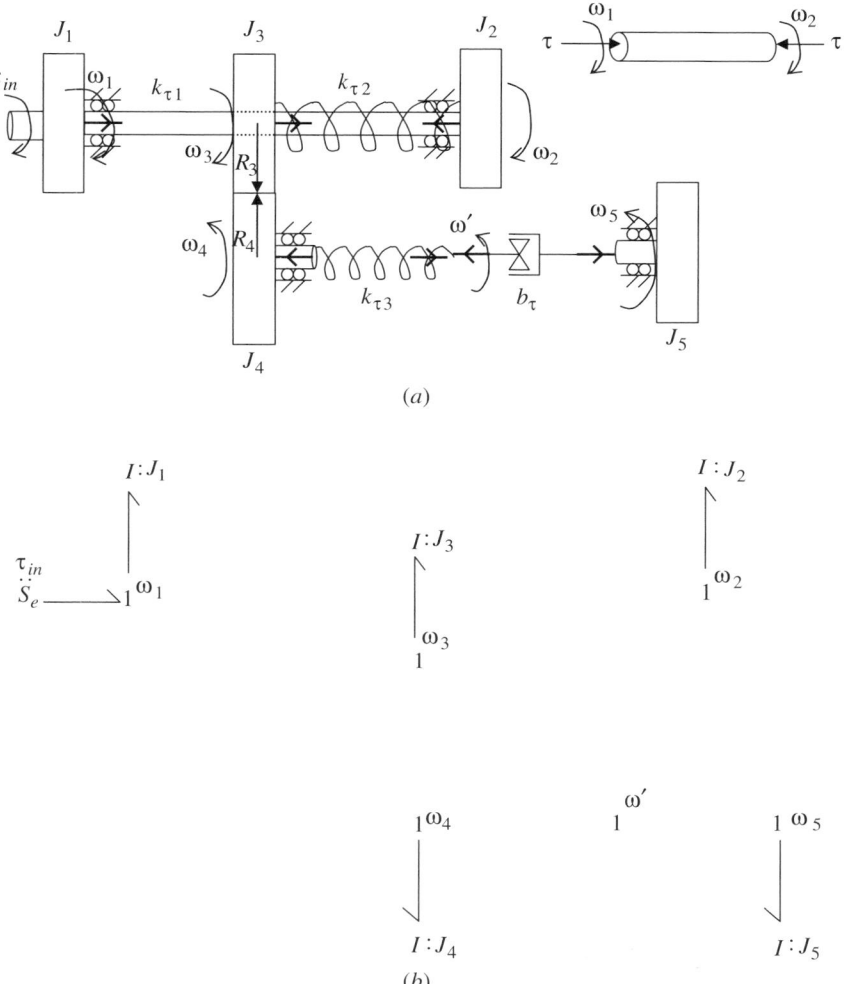

FIGURE 4.16. A system of shafts and gears. Example 2.

to a common 0-junction as was done in the final step, Figure 4.16*d*. This is a nice example of introducing 1-junctions for flows that might be convenient for model construction, and then having the modeling procedure dictate whether these 1-junctions remain in the final model. The modeler is free to define and introduce as many flows on 1-junctions (or efforts on 0-juntions) as desired. The modeling procedure will eliminate the unnecessary ones as bond graph simplifications are performed.

For most real engineering systems that have mechanical components, the mass elements have finite dimensions and must be represented by both their mass and rotational inertial properties. This is covered in the next section on the topic of dynamics of plane motion.

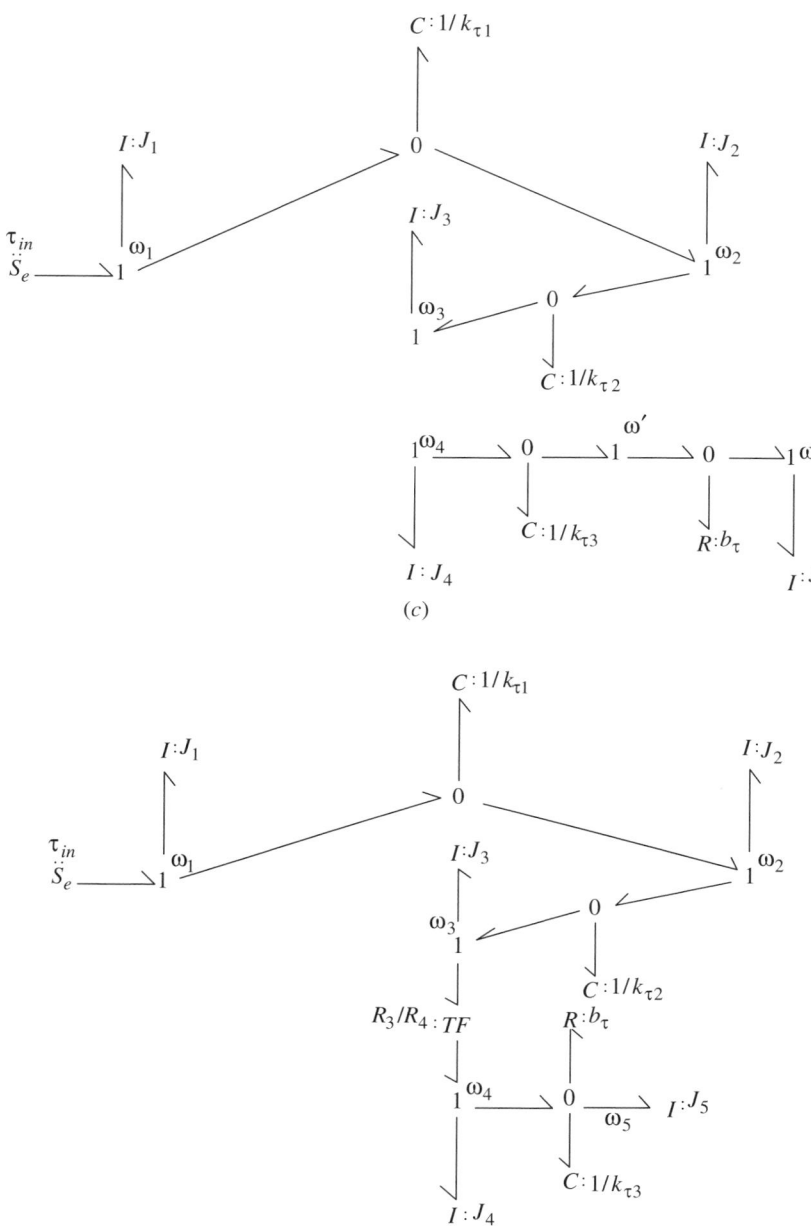

(c)

(d)

FIGURE 4.16. Continued

4.2.3 Plane Motion

For mechanical translation it was fine to imagine the mass elements as point masses, and for fixed-axis rotation it was fine to consider only the rotational inertia of a mass element. But, in general, mechanical pieces with finite mass both translate and rotate in a real application. Consider an automobile as it moves over roadway unevenness. The body of the vehicle moves forward, sideways, and vertically, and it rolls side to side, pitches front to rear, and rotates about a vertical axis as it turns. If we modeled the car body as rigid, we would have to characterize the body by its mass and the moments of inertia about three perpendicular axes. The motion of the vehicle would be quite complex, and certainly not able to be described by mechanical translation alone or fixed-axis rotation alone. Instead, the dynamics are governed by a simultaneous combination of translation and rotation. To describe this motion in three dimensions is quite complicated, but is greatly facilitated by using bond graphs. Chapter 9 discusses this complex topic. When our attention is restricted to plane motion, model construction is greatly simplified. Plane motion is the topic of this section.

Plane motion results when the inertial bodies of a physical system are constrained to translate in two dimensions and to rotate only about an axis perpendicular to the plane of motion. Figure 4.17a shows a rigid body of mass, m_c, and moment of inertia, J, about its center of mass or cg. The mass is translating in the X, Y plane with respect to the inertial X, Y axes, and it is rotating about the inertial Z axis, not shown, that is pointing out of the page according to the right-hand-rule. Any forces acting on the body from attached devices are not shown in the figure. The cg has an absolute velocity vector pointing in some direction in the plane, and this vector has been resolved into two mutually perpendicular components, v_X, v_Y, as shown on the figure. These components are aligned in the inertial X, Y directions. The kinetic energy of a rigid body is characterized by the velocity of the center of mass and the angular velocity of the body, such that for the body in Figure 4.17a, the kinetic energy, T, is

$$T = \frac{1}{2}mv_x^2 + \frac{1}{2}mv_Y^2 + \frac{1}{2}J\omega^2, \tag{4.5}$$

where m is the mass and J is the centroidal moment of inertia. Since bond graphs bookkeep energy, to account for all the energy of a rigid body in plane motion we must represent the translation of the body in two perpendicular directions, and we must also represent the rotation of the body. This is done in the bond graph fragment of Figure 4.17b. To account for the energy, 1-junctions are used to represent each center of mass velocity component and the angular velocity. Attached to the velocity 1-junctions is an $—I$ element with inertia parameter equal to the mass, m, and attached to the 1-junction for the angular velocity is an $—I$ element with the moment of inertia, J, as the parameter. This fragment characterizes all the energy of a body in plane motion. Any devices, such as springs and dampers, that interact with the rigid body will end up communicating with the 1-junctions of Figure 4.17b when incorporated into a bond graph model.

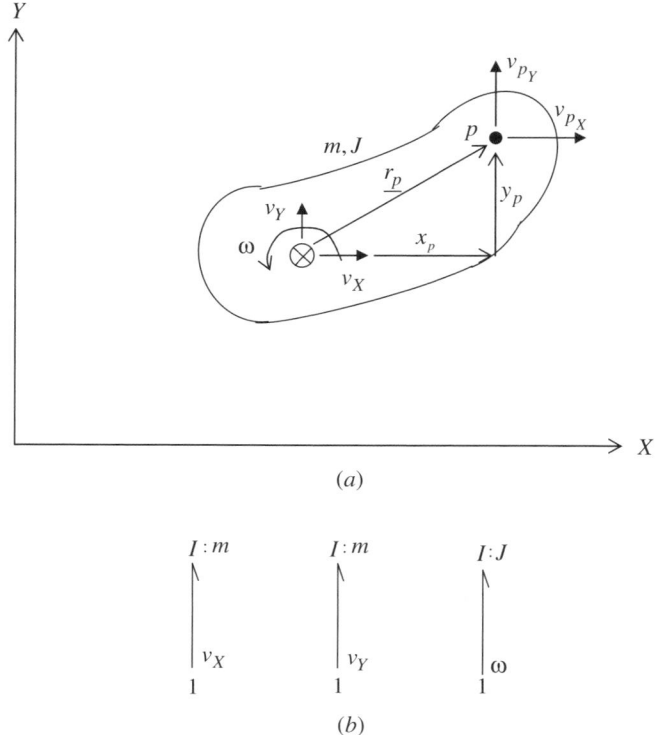

(a)

(b)

FIGURE 4.17. General plane motion of a rigid body.

In general, devices will be attached to rigid bodies at different places on the body. For example, the point p indicated in Figure 4.17a might be such a point. We will discover that we need the velocity components of the attachment points indicated as v_{P_x}, v_{P_y} in the figure. The attachment point is located with respect to the cg by the position vector, r_P, composed of the components x_P, y_P. The kinematic relationship (see Reference [2]) that allows determination of the velocity of any point on a rigid body with respect to the center of mass (or any other point on the rigid body) is

$$\vec{v_P} = \vec{v_{cg}} + \vec{\omega} \ x \ \vec{r_P}, \tag{4.6}$$

where the cross product term accounts for the contribution of rotation of the body to the velocity at points different from the cg. Eq. (4.6) is a vector equation where ω is the vector angular velocity with direction into or out of the plane of motion determined by the right-hand-rule. This relationship gets used over and over when modeling rigid bodies in plane motion. For plane motion, the angular velocity vector is always perpendicular to the position vector, thus making the use of Eq. (4.6) particularly simple. Its use will be demonstrated by example. For the geometry of Figure 4.17a, application of Eq. (4.6) yields

$$v_{P_x} = v_X - y_P \omega,$$

$$v_{P_Y} = v_Y + x_P \omega. \tag{4.7}$$

It should be noted that for large angular motions of the body, x_P and y_P will change as the body moves.

As a first example, consider the heave–pitch vehicle model shown in Figure 4.18a. This is an extension of the quarter car model done previously in Figure 4.10. This model is sometimes called the half car model. We are looking at the side of the vehicle with the front at the right and the rear at the left. The front and rear suspensions each consist of a tire spring, k_{t_f}, k_{t_r}, unsprung mass, m_{us_f}, m_{us_r}, and suspension spring and damper, k_{s_f}, b_{s_f}, and k_{s_r}, b_{s_r}. The inputs to the system are the prescribed input velocities at the front and rear, $v_{i_f}(t)$ and $v_{i_r}(t)$. The body is modeled as a rigid body with center of mass located a from the front suspension connection and b from the rear suspension connection. The body has mass, m, and centroidal moment of inertia, J. Gravity is acting vertically downward.

For this model the vehicle is not moving forward, but instead is constrained to move vertically and to rotate, or pitch. Thus, the rigid body is characterized by the vertical center of mass velocity, v_g, and the pitch angular velocity, ω. Other important velocities are oriented and labeled in the figure, including the vertical velocity components at the front and rear of the body, v_f and v_r. These components are introduced for convenience in anticipation of construction of the relative velocity across the front and rear suspension units. As indicated in Figure 4.18a, all spring and damper forces are assumed positive in compression.

The procedure for constructing a bond graph model of a system with rigid bodies in plane motion is identical to the procedure for mechanical translation and fixed-axis rotation, with the added complexity of dealing with kinematics described by Eq. (4.6). Figure 4.18b shows the use of 1-junctions to represent all the distinct velocities and angular velocities of the system. Attached to these 1-junctions are bond graph elements that have these specific velocity and angular velocity components. The front and rear suspension inertias and input velocities should look familiar because of the earlier quarter car example. The rigid body is characterized by the center of mass velocity and angular velocity, so —I elements for the mass, m, and moment of inertia, J, are attached to the appropriate 1-junctions. The weight of the body is an effort source attached to the center of mass vertical velocity. Positive power is directed into the source owing to the velocity convention (positive upward) adopted for this example. The same is true for the effort sources representing the weights of the unsprung masses, front and rear. In Figure 4.18c, 0-junctions have been used to establish the relative velocities across the springs and dampers. The power convention is such that all elements are positive in compression as indicated for this system. Some simplifications have been done. Notice how the 1-junctions for v_f and v_r were used to help in constructing the relative velocities across the front and rear suspensions. Also notice that for the suspension elements, the relative velocities were constructed only once, and then 1-junctions were used to enforce that the respective springs and dampers have the same relative velocity. From Figure 4.18c it is pretty

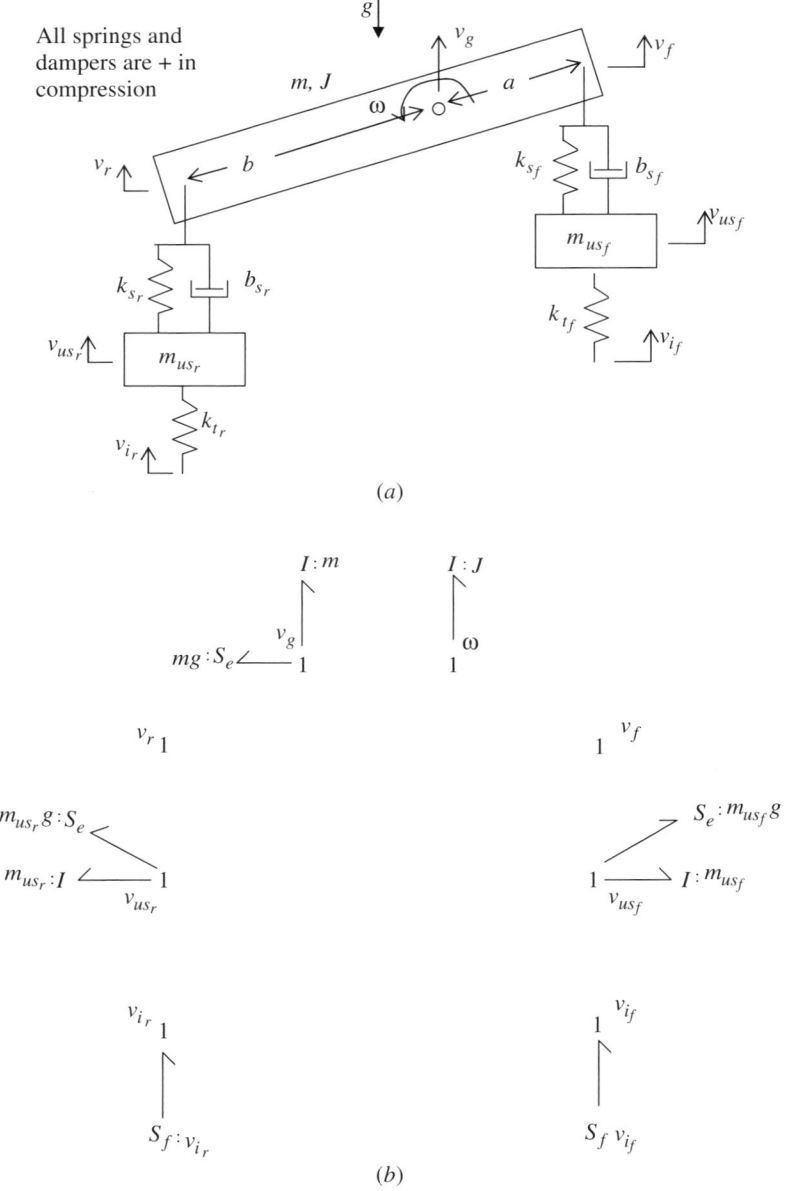

(a)

(b)

FIGURE 4.18. Heave-pitch vehicle model. Example 1 of plane motion.

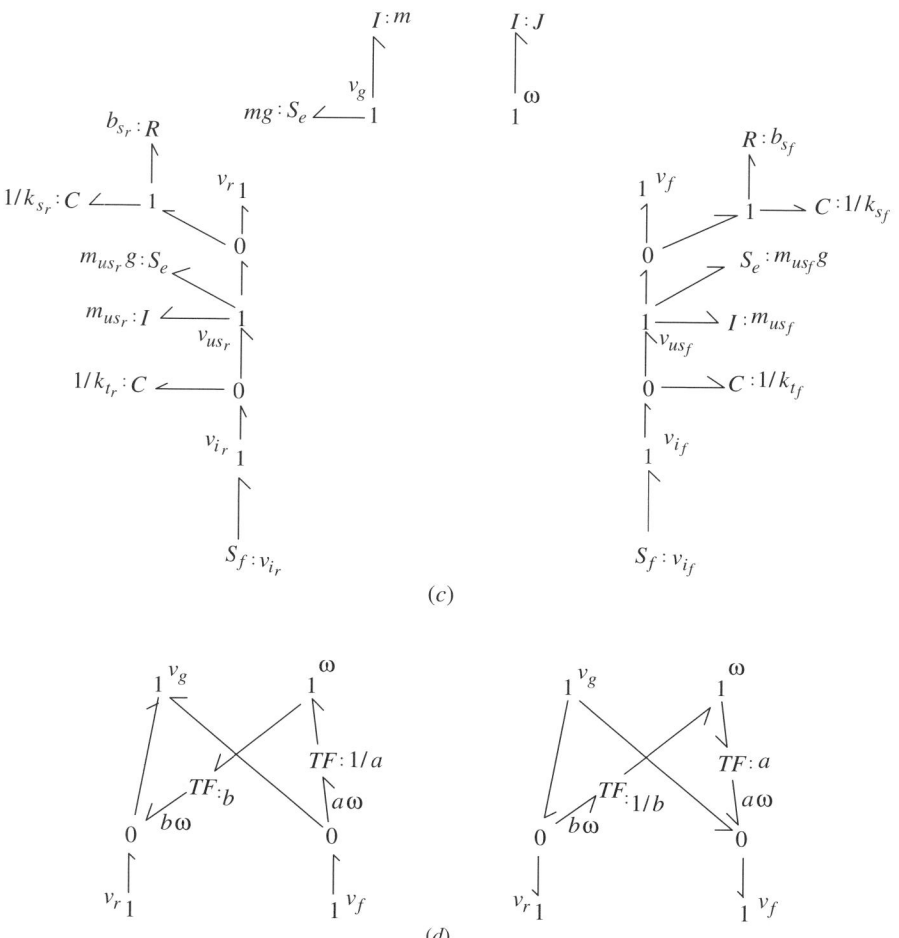

(c)

(d)

FIGURE 4.18. Continued

obvious that the model is not complete, and yet all elements have been incorporated. This is where the kinematics plays a major role.

For small angular motions, using Eq. (4.6) and our right hands, we can derive that

$$v_f = v_g + a\omega \quad \text{and} \quad v_r = v_g - b\omega. \tag{4.8}$$

These are called kinematic constraints, and they must be enforced correctly to end up with a correct model. Until now, it has been recommended that positive directions be shown on a schematic and then transferred to the bond graph model. If this is not done then the model may still be all right. But in a simulation of the system response, we might not know whether a positive velocity was up or down, or whether a positive

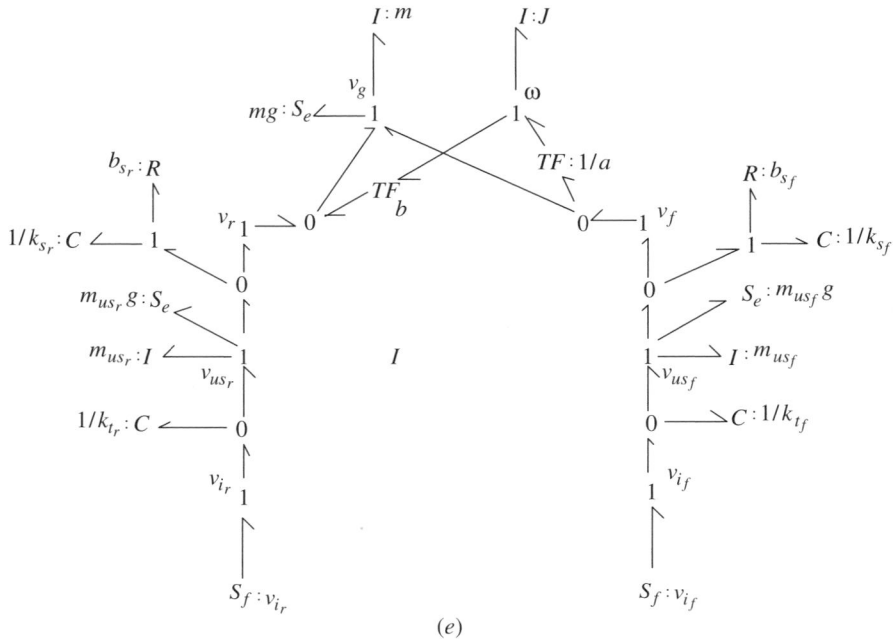

(e)

FIGURE 4.18. Continued

force was compressive or extensive. This is important information to have, so it is always a good idea to have a power convention in mind. However, if the kinematic constraints are not enforced correctly, then the model is just plain wrong.

Kinematic constraints are always relationships among flow variables. To add the flows according to the constraints we use 0-junctions. We just need to pay particular attention to the power convention when adding up the flows to enforce the constraints. In Figure 4.18d the constraints from Eqs. (4.8) are enforced two different ways using 0-junctions. The reader should make certain that the velocity components add correctly to produce Eqs. (4.8). The transformers are used to convert the angular velocity into velocity components, that is, ω into $a\omega$ or $b\omega$. On the transformers in the figure, the moduli are appended according to the definition of the transformer given in Chapter 3. Both versions in part d are correct, and either can be used for this example. If the first version is used, as is done in Figure 4.18e, then a bond graph simplification can be done at the 1-junctions for v_f and v_r.

A second example of plane motion is the system shown in Figure 4.19a. The system is a cart of mass, m, acted upon by an input force, $F(t)$. Attached to the cart through a spring and damper, k, b, is a cylinder that rolls without slip on the cart. The cylinder is a rigid body of mass, m_c, with center of mass in the center and moment of inertia about its center of mass, J_c, and is of radius R. We desire a model that, if solved or simulated, will predict the motion–time history of this system for any prescribed input force. The schematic shows positive velocity and angular velocity

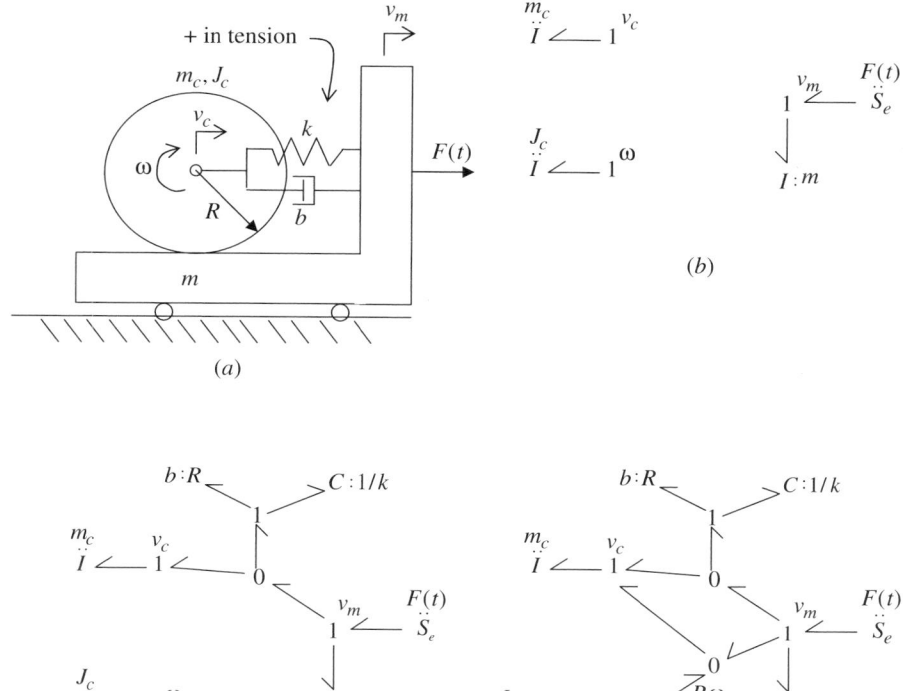

FIGURE 4.19. Cart with rolling cylinder. Example 2 of plane motion

directions, and the spring and damper are assumed positive in tension. Figure 4.19b uses 1-junctions to represent the distinct velocities and angular velocities, and the elements that have these distinct flows have been attached to the 1-junctions. The rigid body cylinder is a body in plane motion and is represented by its center of mass velocity, v_c, and its angular velocity, ω, and thus has the $-I$ elements for its mass and moment of inertia appropriately attached. In Figure 4.19c the relative velocity is established across the spring and damper. Perusal of this figure clearly indicates that the model is not complete, even though all energy has been accounted for.

Again, there is a kinematic relationship that must be derived before we can complete the model. Application of Eq. (4.6) to the velocity of the point at the bottom of the cylinder, v_p, positive to the right, yields

$$v_p = v_c - R\omega. \tag{4.9}$$

Since the cylinder rolls without slip, the velocity at the bottom of the cylinder must be the same as the velocity of the cart at the same point. Thus, the kinematic constraint is

$$v_m = v_c - R\omega. \tag{4.10}$$

In Figure 4.19d this constraint is enforced using a 0-junction and the model is complete.

As a final example of systems with plane motion dynamics, consider the system in Figure 4.20a. An input velocity, $v_{in}(t)$, is prescribed on one end of a spring, k_1, and the other end is attached to an inextensible rope that wraps around a floating cylinder of mass m_1, radius R_1, and moment of inertia J_1. The rope then passes over a second cylinder that is pinned at its center and has radius R_2 and moment of inertia J_2. The

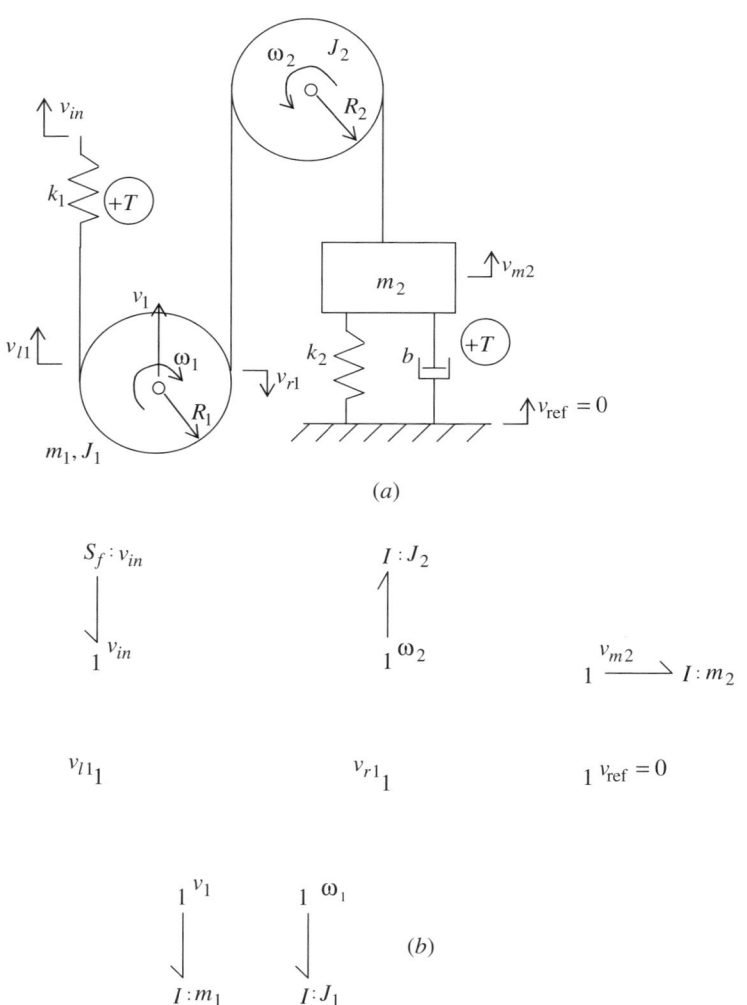

FIGURE 4.20. System with a floating pulley. Example 3 of plane motion.

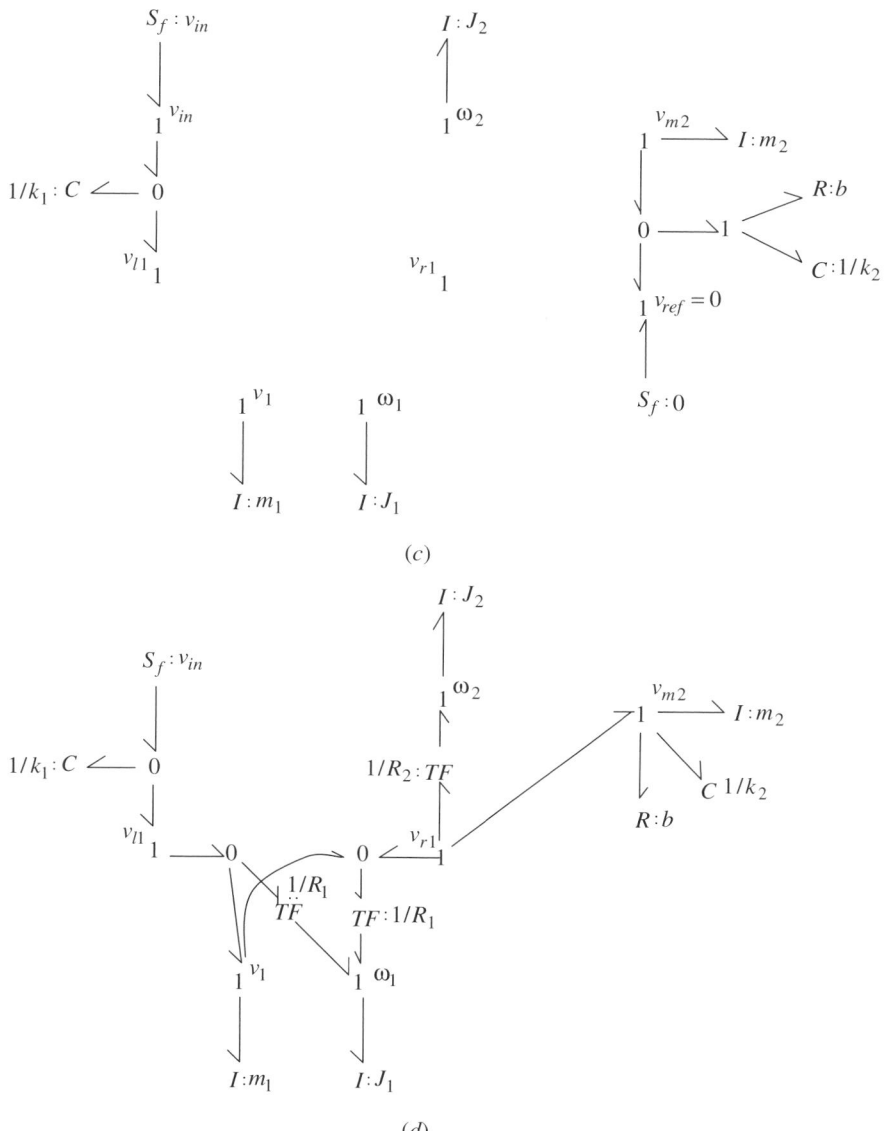

(c)

(d)

FIGURE 4.20. Continued

rope finally is attached to a mass, m_2, which is attached to ground through the spring
and damper, k_2 and b. The idea is to formulate a model that, if simulated, would
predict the motion–time history of the system to a prescribed input velocity. Positive
velocity and angular velocity directions are shown in the figure, and all springs and
dampers are assumed positive in extension.

In Figure 4.20b, 1-junctions are used to represent the distinct velocities and angular velocities. The floating cylinder requires a 1-junction for its center of mass vertical velocity and a 1-junction for its angular velocity. Notice that the velocities at the left and right of the floating cylinder have been defined. The one at the left, v_{l_1}, has been introduced as a convenience for establishing the relative velocity across the spring, k_1. The velocity at the right of the cylinder, v_{r_1}, will be useful for relating to the angular velocity, ω_2, of the pinned cylinder. Also, we can see that, since the rope is inextensible, v_{r_1} is equal to the mass velocity, v_{m_2}. Attached to these 1-junctions are the bond graph elements that have these distinct flows. The power arrow for the flow source, $-S_f$, is directed out of the source and into the system, because if the spring, k_1, is in tension, as defined as positive, and if the top end of the spring is moving upward, as defined as positive, then power will be provided by the source and power will be flowing into the system.

In Figure 4.20c, 0-junctions are used to establish the proper relative velocities across the springs and dampers so as to put these elements positive in tension. The spring and damper attached to m_2 have been modeled using the "reducible loop" simplification described earlier. The reference velocity, $v_{\text{ref}} = 0$, can be erased in the final bond graph. Clearly, the model in Figure 4.20c is not complete.

For cylinder 1, application of Eq. (4.6) yields the kinematic constraints

$$v_{l_1} = v_1 + R_1 \omega_1 \quad \text{and} \quad v_{r_1} = -v_1 + R_1 \omega_1. \tag{4.11}$$

Moreover, since the rope is inextensible, the velocity at the rim of cylinder 2 is the same as v_{r_1}, thus,

$$v_{r_1} = R_2 \omega_2. \tag{4.12}$$

Eqs. (4.11) are enforced using 0-junctions as shown in Figure 4.20d. The velocity v_{r_1} is related to ω_2, as required by Eq. (4.12). And v_{r_1} is set equal to v_{m_2} by simply connecting the 1-junctions with a bond. The model in Figure 4.20d is now complete. Some simplifications are possible, but this is left to the reader.

As has been demonstrated in all examples of mechanical systems with elements in plane motion, bond graph models are constructed by first establishing distinct velocities and angular velocities using 1-junctions that assign the same flow to each attached bond and add the efforts according to the power convention. This is followed by establishing the proper relative motions across compliances and resistances, using 0-junctions that add flows according to the power convention and assign the same effort to each attached bond. This part of the modeling procedure is straightforward and invariant. With some practice these steps will become automatic. The most difficult part of model construction is deriving and enforcing the kinematic constraints. There is no easy set of rules for recognizing and deriving these relationships. The constraints will always involve flow variables and will always involve application of the vector relationship, Eq. (4.6). Practice is the only way to become comfortable with this aspect of modeling.

Body-Fixed Coordinates* The concepts of plane motion were developed using the general motion of a rigid body, shown in Figure 4.17. In order to characterize all the energy of a rigid body in plane motion it is necessary to know the velocity of the center of mass and the angular velocity of the body. In Figure 4.17, the velocity vector of the center of mass was resolved into two mutually perpendicular components aligned in the inertial XY directions. When there are several places where a rigid body interacts with other parts of a system and when the rigid body can execute large angular motions, it is convenient to introduce the concept of "body-fixed coordinates." This is a coordinate frame that remains attached to the body at the center of mass and moves with the body as it translates and rotates under the action of whatever is attached to it. The virtue of such a frame is that the attachment points remain at fixed locations relative to this frame, and the inertial properties, that is, moments and products of inertia, remain invariant with respect to this frame. These properties are very important for three-dimensional rigid-body motion, which is discussed fully in Chapter 9. We introduce the concept here for plane motion.

The general rigid body of Figure 4.17 is shown again in Figure 4.21a with body-fixed coordinates. The xy-coordinate frame is attached to the body at the center of mass, and it is aligned in principal directions, although this is not essential. The instantaneous velocity vector that is pointing in some direction in the plane of motion is resolved into two mutually perpendicular directions aligned along the body-fixed axes. These components, v_x and v_y, are different from the inertial components in Figure 4.17 since they change direction as the body rotates. And this change in direction of velocity vector components results in very real acceleration components that must be accounted for. Also shown in Figure 4.21a is the attachment point, P, that is located with respect to the xy-coordinates by the fixed lengths x_p, y_p.

In Reference [2], it is shown that for any vector, say the velocity vector, $\vec{v}$, referred to a rotating frame, the absolute time rate of change of the vector, or acceleration in this case, is

$$\frac{d}{dt}\vec{v} = \frac{\partial}{\partial t}\vec{v}|_{\text{rel}} + \vec{\omega}x\vec{v}, \tag{4.13}$$

where the first term on the right is the acceleration as observed relative to the moving frame, and the second term accounts for the contribution of the frame rotation to the absolute acceleration. Applying this relationship to the rigid body of Figure 4.21a yields the cg absolute acceleration, a_x, a_y, as

$$a_x = \dot{v}_x - \omega v_y,$$
$$a_y = \dot{v}_y + \omega v_x, \tag{4.14}$$

where the components due to rotation are derived by carrying out the cross products using the right-hand-rule. For example, the ω-vector pointing out of the page, when crossed into the x-direction velocity component, v_x, yields a vector of length ωv_x pointing in the y-direction.

---

*This section can be omitted without any loss of generality.

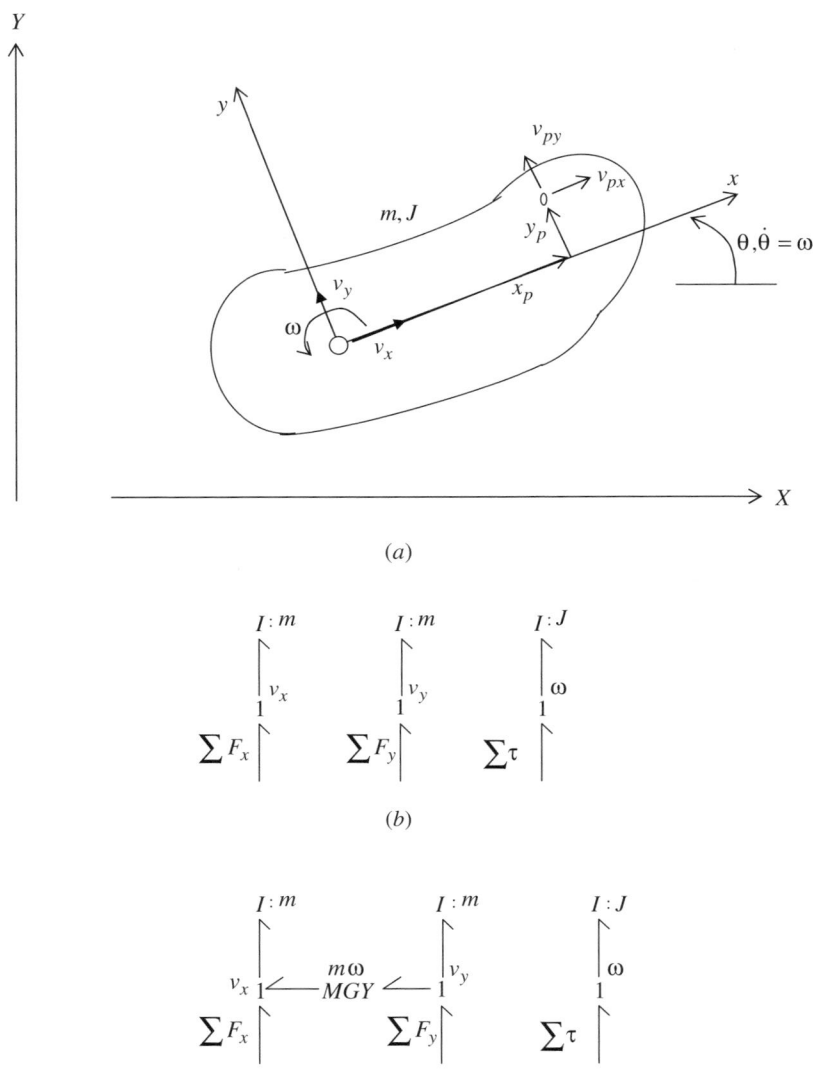

FIGURE 4.21. General plane motion with body-fixed coordinates.

Newton's laws apply to the absolute acceleration of the cg, and become

$$\sum F_x + m\omega v_y = m\dot{v}_x,$$
$$\sum F_y - m\omega v_x = m\dot{v}_y,$$ (4.15)

where the sum of forces in the x and y directions come from any attached systems to the body, and the rotational contributions to the acceleration have been brought to the force side of the equation after multiplying by the body mass, m.

Returning now to Figure 4.21b, it is tempting to represent the plane motion of the rigid body as shown, where the body-fixed velocity components, v_x, v_y are used rather than the inertial components, v_X, v_Y from Figure 4.17. The kinetic energy of the body is accounted for, but the absolute acceleration is not. In other words, if the forces from external elements were added using the 1-junctions of Figure 4.21b, then the resulting equations of motion would not be correct. The fix to this problem lies in Eqs. (4.15).

The cross product terms of Eqs. (4.15) have a remarkable symmetry. In the first equation, there is a force component in the x-direction that is equal to $m\omega$ times the velocity in the y-direction. And in the second equation there is a force component in the y-direction equal to $m\omega$ times the velocity in the x-direction. Since gyrators relate efforts to flows across them, consider Figure 4.21c where a modulated gyrator has been inserted between the 1-junctions representing v_x and v_y. A modulated gyrator is used because the modulus, $m\omega$, is not constant but varies as the body moves. Convince yourself that the forces on the $-I$ elements exactly duplicate the left-hand side of Eqs. (4.15). When body-fixed coordinates are used, the $-MGY-$ must be part of the bond graph.

For the attachment point in Figure 4.21a, the velocity components in the body-fixed directions are related to the cg velocity components using Eq. (4.6), with the result

$$v_{P_x} = v_x - y_P\omega \quad \text{and} \quad v_{P_y} = v_y + x_P\omega. \tag{4.16}$$

As an example of the use of body-fixed coordinates, consider the system in Figure 4.22a. The rigid body of mass, m, and centroidal moment of inertia, J, is attached to ground at points 1 and 2 through springs and dampers as shown. The ground side ends of the springs and dampers are on frictionless carts that ensure that horizontal elements remain horizontal regardless of the motion of the body. Horizontal spring 1 has a velocity input at its ground side end. The body is shown displaced from its starting orientation, and body-fixed coordinates are attached at the cg. The attachment points are located with respect to the body-fixed coordinate frame by the fixed distances x_1, y_1 for point 1 and x_2, y_2 for point 2 where x_2 is a negative quantity. In Figure 4.22b the body-fixed velocity components at the attachment points are indicated, and in Figure 4.22c 1-junctions are used to represent all distinct velocities. Attached to these 1-junctions are elements that have these absolute velocities. Notice the representation of the body when using body-fixed coordinates. The body-fixed velocity components at the attachment points are indicated using 1-junctions. These are defined as a convenience for ultimately determining the horizontal velocity at the attachment points for use as inputs to the horizontal spring–damper elements. As noted in Figure 4.22b, the body-fixed velocities at the attachments are

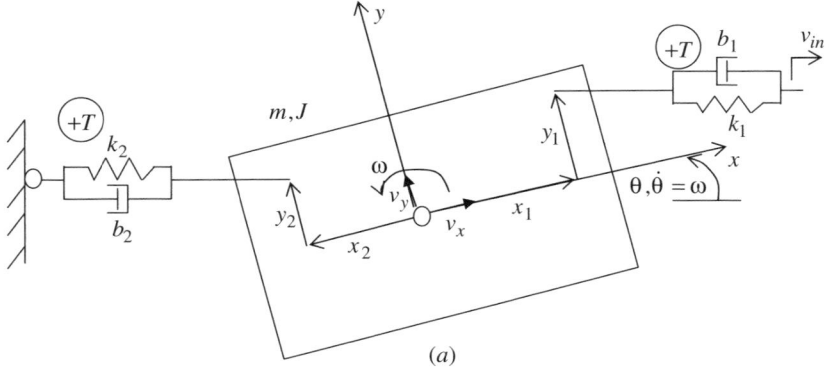

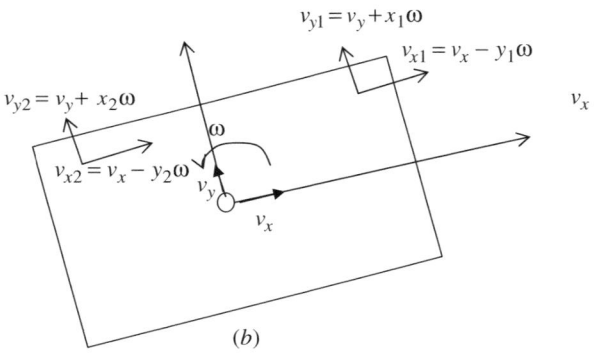

FIGURE 4.22. A rigid body attached to a frame undergoing large motions.

$$v_{x_1} = v_x - y_1\omega,$$

$$v_{y_1} = v_y + x_1\omega,$$

$$v_{x_2} = v_x - y_2\omega,$$

$$v_{y_2} = v_y + x_2\omega, \tag{4.17}$$

and the horizontal velocity at the attachments are

$$v_{h_1} = v_{x_1}\cos\theta - v_{y_1}\sin\theta,$$

$$v_{h_2} = v_{x_2}\cos\theta - v_{y_2}\sin\theta. \tag{4.18}$$

These kinematic relationships are enforced using 0-junctions as shown in Figure 4.22d. The transformers used to enforce constraints from Eqs. (4.18) are modulated transformers, —MTF—, because the moduli of these elements vary as the body moves. Figure 4.22d is the final bond graph for this example. The bond graph

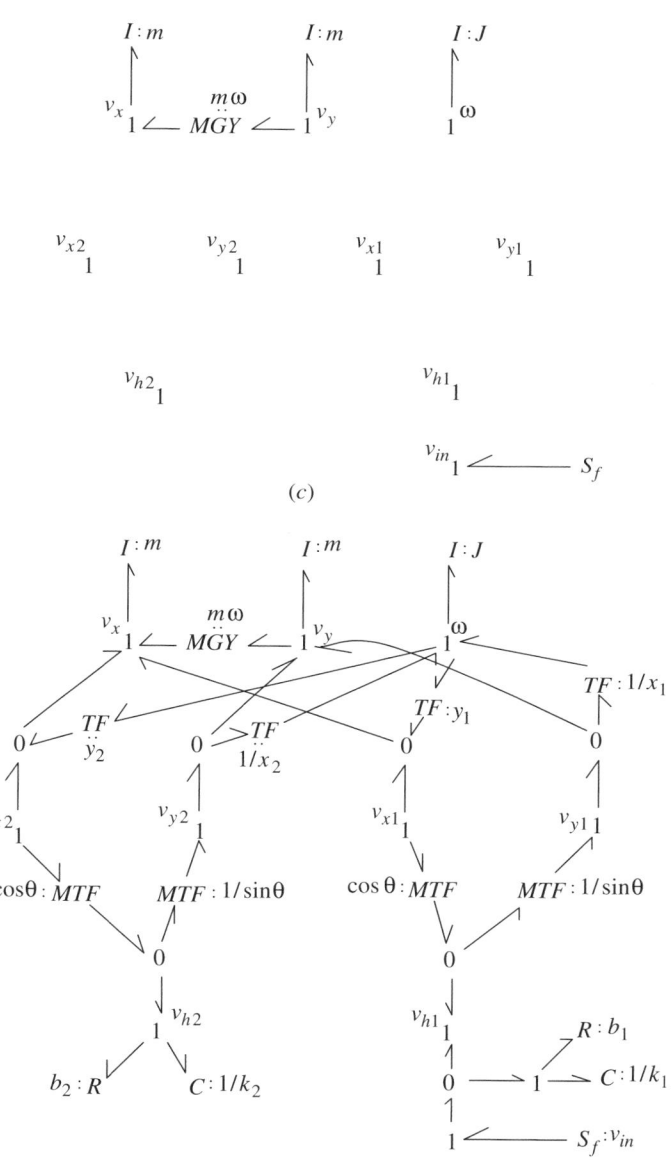

FIGURE 4.22. Continued

does appear a bit busy, but it contains a lot of information. In Chapter 9, some short-hand notation is introduced for complex systems that allows the bond graph to be constructed while showing fewer bonds.

4.3   HYDRAULIC AND ACOUSTIC CIRCUITS

In this section, special but important classes of fluid-flow systems will be modeled using bond graphs. The models to be used exhibit a close analogy to the mechanical and electrical systems studied in the previous section of this chapter. The variables to be used here have been discussed in Chapter 2 (Table 2.4), and the basic fluid system elements have been discussed in Chapter 3. Tables 3.1, 3.2, 3.3, and 3.4 and Figures 3.1, 3.2, and 3.4 show the type of 1-port fluid elements to be used in this section. Figure 3.8 shows a 2-port transformer with a fluid port, and Figure 3.11 shows 0- and 1-junction fluid elements. So far the discussion has been fairly general, but it is now time to be more specific about the modeling of fluid systems.

One category of systems that is important in engineering and can be modeled using the elements shown previously is commonly called *hydrostatic*. These are systems composed of pumps, motors, pipes, pistons, valves, filters, and accumulators that use nearly incompressible fluids such as water or hydraulic oil. Such systems are found in machine tools, earth-moving equipment, power transmissions, and aircraft control surface servomechanisms. Generally, these systems have high pressures and low fluid velocities so that the static pressure dominates the dynamic pressure. This is the origin of the name *hydrostatic*, even though the *dynamics* of such systems are typically of great interest.

A second category of fluid systems that can be treated with the elements to be used here involves compressible gases (such as air) but in which the pressure deviations are small enough that the *acoustic approximation* applies. Acoustic circuits can often be used to design some types of mufflers to reduce noise in air-conditioning systems, for example. However, pneumatic systems in which large pressure variations are expected need a more complicated modeling treatment since they are general thermofluid systems, as discussed in Chapter 12.

Most engineers are familiar with lumped parameter elements of mechanical and electrical systems such as rigid bodies, springs, dampers, resistors, capacitors, and inductors. The equivalent fluid elements are generally less familiar and less obvious. Thus, it is worthwhile to describe some 1-port elements that can be used to model hydraulic and acoustic circuits before showing how to assemble bond graph models of these systems.

4.3.1   Fluid Resistance

Figure 3.1 showed a 1-port fluid resistor that relates an effort, a pressure drop, P, to a flow, a volume flow rate, or volume velocity, Q. Figure 4.23 shows how resistors typically appear in a bond graph, together with example devices represented by a resistor. Part a of Figure 4.23 is a combination of a resistor and a 1-junction. This

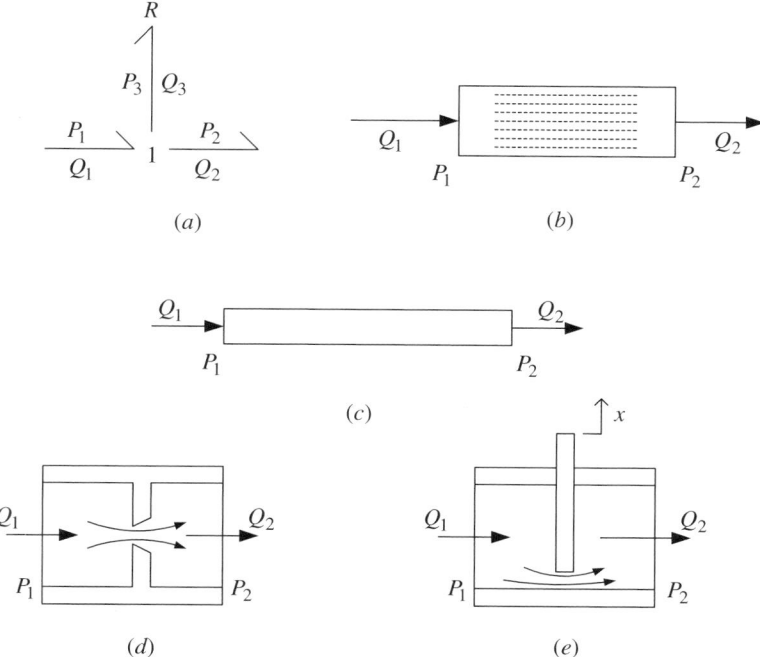

FIGURE 4.23. Fluid resistors: (*a*) Bond graph; (*b*) porous plug; (*c*) long pipe; (*d*) orifice; (*e*) valve with variable area, $A(x)$.

combination implies that the pressure drop, P_3, is related to the pressures P_1 and P_2 by the equation

$$P_3 = P_1 - P_2, \tag{4.19}$$

and that all volume flow rates are equal,

$$Q_1 = Q_2 = Q_3. \tag{4.20}$$

Following Table 3.1, the resistor implies that the pressure drop is related to the volume flow rate by a nonlinear function,

$$P_3 = \Phi_R(Q_3), \tag{4.21}$$

or, if a linear relation is assumed, by a resistance coefficient R_3

$$P_3 = R_3 Q_3. \tag{4.22}$$

It is an unfortunate fact that in many hydraulic and acoustic systems it is not easy to predict resistance functions or coefficients for parts of the system before it is constructed. However, experimentally it not difficult to measure pressure drops and flow rates under steady conditions and to characterize resistance effects. This means that there are at least guidelines in the engineering literature for suitable resistance law assumptions.

Part *b* of Figure 4.23 is intended to represent a porous plug in a pipe for which viscous forces on the fluid from the plug might be assumed to dominate. In this case it would be logical to use Eq. (4.22). In the absence of experimental data relating to the plug, one would have to experiment with values of the coefficient R_3 in studying a system model.

Another case in which a linear resistance can be assumed is shown in part *c* of Figure 4.23, which is supposed to represent a long, thin tube in which laminar flow of an incompressible fluid develops. In this case a theoretical value for the resistance can be given, (see Reference [4], Section 7.4):

$$R_3 = 128\mu l/\pi d^4, \tag{4.23}$$

where $\mu[Pa \cdot s]$ is the fluid viscosity coefficient, $l[m]$ is the length and $d[m]$ is the inside diameter. (Note that the Appendix lists typical property values, such as viscosity, for a number of materials. These are particularly useful for estimating 1-port element parameters and functions for the types of systems treated in this section.).

For incompressible flow in long pipes, it is useful to compute a Reynolds number, Re, defined as

$$\text{Re} = 4\rho Q/\pi d\mu, \tag{4.24}$$

where $\rho[kg/m^3]$ is the fluid mass density. When the Reynolds number is low, say about 200 or less, viscous forces predominate and Eqs. (4.22 and 4.23) can be used. At a higher Reynolds number, the flow becomes turbulent and the pressure–flow relationship under steady flow conditions is nonlinear. The transition to turbulent flow depends on the pipe dimensions l and d, the pipe surface roughness, and fluid properties.

For Reynolds numbers greater than about 5,000, the flow is likely to be turbulent, and a general function for the nonlinear relation from Eq. (4.21) is

$$P_3 = a_t Q_3 |Q_3|^{3/4}. \tag{4.25}$$

In this formula, the absolute value sign is necessary to make sure that P_3 is negative if Q_3 is negative. The constant a_t is often determined experimentally (see Reference [5]). Even when Eq. (4.25) is known to be valid for steady flow, its use in studying transient conditions is not necessarily valid. For oscillatory flows or other dynamic conditions, turbulence may not develop fully, so in these conditions, Eq. (4.25) should be regarded only as an approximation.

Parts *d* and *e* of Figure 4.23 show two important cases in which a pressure drop occurs over short lengths. The orifice is assumed to have a fixed area A_0 while the valve has a variable area $A(x)$, where x is a position coordinate. Although the valve is shown as if it were a gate valve, the basic equation for the pressure drop applies to a variety of other valve configurations. It is a standard exercise in fluid mechanics texts to derive the laws of an orifice using considerations of energy, momentum, and continuity. The main result is that the pressure drop is proportional to the square of the volume flow rate. One form of the law is (see Reference [6], Section 3.8)

$$P_3 = (\rho Q_3 |Q_3|)/2C_d^2 A_0^2, \tag{4.26}$$

where again the use of the absolute value sign corrects the sign of the pressure drop P_3 if Q_3 happens to be negative and C_d is a discharge coefficient. For a round, sharp-edged orifice, C_d can be predicted to have a value of 0.62, but for other shapes of holes, the values of C_d vary somewhat.

For valves, the area depends on a position coordinate and the discharge coefficient may also vary, so the relation equivalent to Eq. (4.26) can be written

$$P_3 = (\rho Q_3 |Q_3|)/2C_d^2(x) A^2(x). \tag{4.27}$$

In case it turns out that a causal analysis of the system model ultimately requires that the flow must be expressed in terms of pressure drop, the relationships of Eqs. (4.26 and 4.27) must be inverted. For example, the inverse version of Eq. (4.26) is

$$Q_3 = C_d A_0 (2|P_3|/\rho)^{1/2} \operatorname{sgn} P_3. \tag{4.28}$$

This form is slightly more complicated than the form usually given to allow the pressure drop and volume flow rate to take on both positive and negative values. (The *signum* function "sgn" is just $+1$ if P_3 is positive and -1 if it is negative.) In dynamic systems in which oscillating flows may exist, one must make sure that both plus and minus values of the variables are computed correctly.

Equation (4.28) is the inverse form of a nonlinear relation as indicated by Eq. (4.21). When a linear relation can be assumed as shown in Eq. (4.22), the inverse relation is much simpler:

$$Q_3 = P_3/R_3. \tag{4.29}$$

Chapter 5 will present methods for deciding which causal form of resistance laws are required for formulating the state equations of a system to be analyzed or simulated.

4.3.2 Fluid Capacitance

As shown in Table 3.2, a fluid capacitor imposes a relationship between a pressure (an effort variable) and the integral of the volume flow rate (a displacement variable). This integrated variable is a volume of fluid in the cases at hand. Figure 4.24 shows the common case in which capacitors are combined with common effort junctions (0-junctions). The nonlinear form of the capacitance constitutive relation for the bond graph of Figure 4.24*a* is

$$P_3 = \Phi_C^{-1}(V_3), \tag{4.30}$$

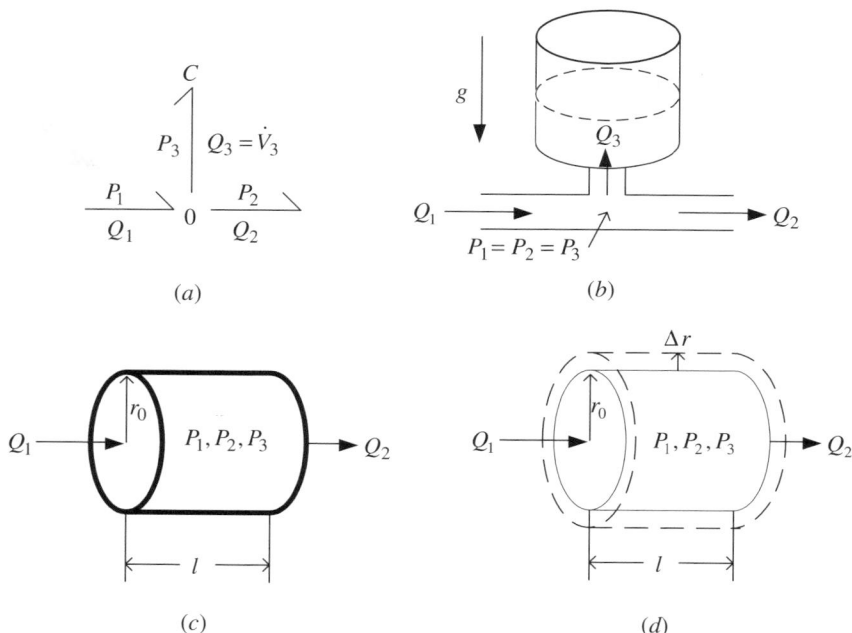

FIGURE 4.24. Fluid capacitors: (*a*) Bond graph; (*b*) water tank; (*c*) rigid pipe segment; (*d*) elastic pipe segment.

where

$$V_3 = \int_0^t Q_3 \, dt. \tag{4.31}$$

If the capacitor can be considered to be linear, a capacitance parameter, C_3, can be defined and the capacitor law of Eq. (4.30) simplifies to

$$P_3 = V_3/C_3. \tag{4.32}$$

The 0-junction laws are

$$P_1 = P_2 = P_3 \quad \text{and} \quad Q_3 = Q_1 - Q_2, \tag{4.33}$$

so V_3 in this case is the integral of the difference in the volume flow rates Q_1 and Q_2.

The first example, shown in Figure 4.24*b*, is a straight-walled water tank in a gravity field. In this case V_3 is the volume of water that has been stored in the tank. If h represents the height of the water above the horizontal connecting pipes, then the pressure at the bottom of the tank is just the hydrostatic value of $\rho g h$. (Strictly speaking, this is the *gage pressure*, counting atmospheric pressure to be zero. We also assume that the bottom of the tank is essentially at the location of the TEE junction connecting the pipes. The little stub shown entering the tank bottom is supposed to be very short.).

If the tank area is A_3, the volume of water in the tank, V_3, is $A_3 h$ and the pressure at the bottom of the tank is therefore

$$P_3 = \rho g V_3 / A_3. \tag{4.34}$$

Comparing Eqs. (4.34 and 4.32) we see that the capacitance of the tank is

$$C_3 = A_3 / \rho g. \tag{4.35}$$

A gravity tank with straight sides can be represented as a linear fluid capacitor.

The next example shown in Figure 4.24c is a segment of a rigid pipe that has a physical volume of

$$V_0 = \pi r^2 l. \tag{4.36}$$

For a truly incompressible fluid, the volume flow rate into the pipe, Q_1, and the flow rate coming out of the pipe, Q_2, would have to be equal, and thus the flow Q_3, according to Eq. (4.33), would be zero. On the other hand, if Q_1 and Q_2 were not exactly zero, Q_3 would represent the rate of compression of the fluid in the pipe. More important, V_3 according to Eq. (4.31), would represent the *decrease* in volume of the fluid in the pipe segment.

For liquids such as water or hydraulic oil that are only slightly compressible, a *bulk modulus B* is defined by the relation (see Reference [6])

$$dP = -B(dV|V). \tag{4.37}$$

The minus sign in Eq. (4.37) is due to the fact that when the volume of the fluid increases incrementally by an amount dV, the pressure decreases by the amount dP. For the capacitance model we will extend the relation of Eq. (4.37) to small but finite changes and note that ΔV_3 represents a small *decrease* in the volume of fluid in the pipe segment, so the minus sign in Eq. (4.37) disappears and the capacitance law for the pipe segment is

$$\Delta P_3 = (B/V_0)\Delta V_3, \tag{4.38}$$

where V_0 comes from Eq. (4.36) and the Δ may be eliminated if it is remembered that P_3 and V_3 in the bond graph of Figure 4.24a represent deviations from steady-state values in this case.

Comparing Eq. (4.32) with Eq. (4.38), we see that the capacitance for the rigid pipe segment is

$$C_3 = V_0 / B. \tag{4.39}$$

The Appendix gives values for the bulk modulus for some common liquids.

If the fluid is a gas and if the pressure and volume variations are small enough that the so-called *acoustic approximation* is valid, another relation can be given relating pressure change to volume change (Reference [7]):

$$\Delta P_3 = \rho_0 c^2 V_3 / V_0, \tag{4.40}$$

where ρ_0 is the density of the gas at the reference pressure and c is the speed of sound. This means that the capacitance for this case is

$$C_3 = V_0 / \rho_0 c^2. \tag{4.41}$$

Densities and speeds of sound for common gases can be found in the Appendix.

Figure 4.24d shows a case in which the capacitance in a fluid system is due to the elasticity of the vessel containing the fluid, rather than the compressibility of the fluid itself. The pipe segment has thin walls and is made of an elastic material. (It might represent a nonmetallic hydraulic brake line, for example.) An approximate analysis will be made in which ΔV_3 represents the *increase* in volume of the pipe segment due to an increase in pressure with the fluid being assumed incompressible. The stress and strain effects in the longitudinal direction will be neglected, and only the volume change due to the hoop stress in the pipe will be computed.

An elementary analysis yields the hoop stress in the pipe:

$$\sigma = r_0 P / t_w, \tag{4.42}$$

where r_0 is the nominal pipe radius and t_w is the (small) wall thickness. The circumferential strain is then

$$\epsilon = \sigma / E = r_0 P / t_w E = \Delta(2\pi r)/2\pi r_0 = \Delta r / r_0, \tag{4.43}$$

where E is the elastic modulus (see the Appendix). The change in volume is

$$\Delta V = \Delta(\pi r^2 l) = \pi l 2 r_0 \Delta r. \tag{4.44}$$

Combining Eqs. (4.43 and 4.44), the result is

$$\Delta V = (2\pi l r_0^3 / t_w E)\Delta P - (2r_0 V_0 / t_w E)\Delta P. \tag{4.45}$$

In terms relating to the bond graph of Figure 4.24a the result is

$$\Delta P_3 = (t_w E / 2 r_0 V_0)\Delta V_3.$$

This means that the capacitance due to pipe elasticity is

$$C_3 = 2 r_0 V_0 / t_w E. \tag{4.46}$$

Combining Capacitances It sometimes happens that both fluid compressibility and pipe elasticity contribute to a compliance effect. One possibility is to use two separate bond graphs like Figure 4.24a in cascade with parameters for the two cases of

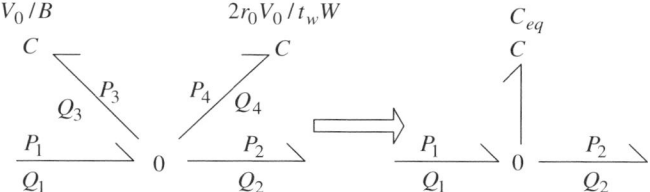

FIGURE 4.25. Single equivalent capacitance for liquid compressibility and pipe elasticity.

Figure 4.24c and d, but another possibility is to construct a single equivalent compliance. Figure 4.25 shows another way to think about the two sources of compliance. Bond 3 in the first bond graph has a C-element dealing with fluid compressibility with capacitance from Eq. (4.39), and bond 4 has a compliance dealing with pipe wall elasticity and using a capacitance from Eq. (4.46). The 4-port 0-junction implies that all bonds have the identical pressure and that the sum of the flows (considering the sign half arrows) add to zero. The bond graph implies that the difference between Q_1 and Q_2 is the sum of Q_3 and Q_4, that is, the difference between inlet and outlet flows is partly due to fluid compression and partly to pipe expansion. After integration, the volume variables are related as follows:

$$\Delta V_3 + \Delta V_4 = \int_0^t (Q_1 - Q_2)\, dt. \tag{4.47}$$

Using the capacitances from Eqs. (4.39 and 4.46) and calling the common pressure ΔP, Eq. (4.47) becomes

$$(V_0/B + 2r_0 V_0/t_w E)\Delta P = \int_0^t (Q_1 - Q_2)\, dt, \tag{4.48}$$

which implies that capacitances add on a 0-junction. The equivalent single capacitance shown in Figure 4.25 is just

$$C_3 = V_0(1/B + 2r_0/t_w E). \tag{4.49}$$

Note that this formula enables one to decide whether fluid compressibility or pipe wall elasticity is the more important factor or whether both effects should be considered together. Both effects have to do with the volume of the pipe segment, but other physical parameters and dimensions are also involved.

Nonlinear Capacitances   Although much of the dynamic modeling of fluid systems can be usefully accomplished using linearized component laws, there are circumstances in which nonlinear models are essential. This is the case for some types of fluid resistance and for some fluid capacitance devices. Consider, for example, the compressed gas *accumulator* shown in Figure 4.26. It consists of a pressure vessel

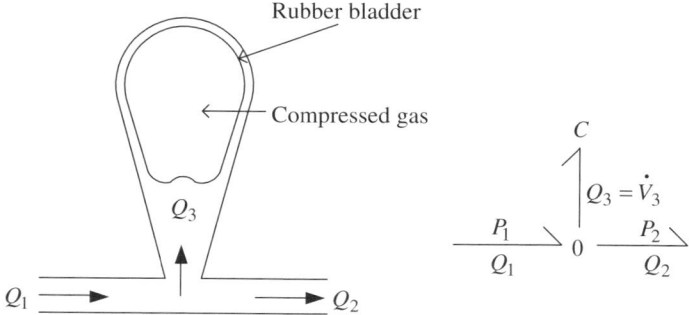

FIGURE 4.26. Compressed gas accumulator.

in which a flexible bladder contains a compressed gas. Such devices are often used to smooth out pressure and flow fluctuation in hydraulic systems or to provide short bursts of power under transient conditions.

The incompressible fluid flow, Q_3, compresses the gas in the bladder, and the volume of the compressed gas determines the pressure in the accumulator and at the 0-junction. Under dynamic conditions, the gas usually does not have time to exchange much heat with its surroundings. If this is the case, the isentropic pressure–volume law

$$PV^\gamma = P_0 V_0^\gamma = \text{constant} \tag{4.50}$$

is a good approximation. In this equation, γ is the ratio of specific heats at constant pressure and volume (see the Appendix), P and V are the instantaneous pressure and volume of the gas, and P_0 and V_0 are values at some initial time.

The bond graph of Figure 4.26 indicates that the gas volume is

$$V = V_0 - V_3 = V_0 - \int_0^t Q_3 \, dt = \int_0^t (Q_1 - Q_2) \, dt. \tag{4.51}$$

This means the nonlinear capacitance relation between P_3 and V_3 is

$$P_3 = P_0 V_0^\gamma / (V_0 - V_3)^\gamma, \tag{4.52}$$

upon use of Eq. (4.50).

4.3.3 Fluid Inertia

The final passive 1-port element that needs to be discussed for fluid circuits concerns the inertia associated with fluid in a pipe segment. Table 3.3 shows that fluid systems have inertia effects analogous to those of mechanical and electrical systems, but for most people the variables involved, *volume flow rate* and *pressure momentum*, are much less familiar than the corresponding mechanical or electrical variables. Also, as we will see, the inertia coefficient for a fluid circuit element is related to the parameters of a pipe segment in a counterintuitive way.

Figure 4.27 shows the simplest case of a straight pipe section of area A and length l filled with a fluid of density ρ. The bond graph shows the common situation in which the 1-port appears connected to a 1-junction. In this case, the volume flow rates at both ends of the pipe are equal and the pressure difference acting on the inertia element is

$$P_3 = P_1 - P_2. \tag{4.53}$$

The pressure momentum variable for the inertia element, as defined in Table 2.4, is

$$p_{P_3} \equiv \int^t P_3 \, dt - \int^t (P_1 - P_2) \, dt. \tag{4.54}$$

A linear inertia coefficient for this I-element is defined by a relation between the volume flow rate and the pressure momentum:

$$I_3 Q_3 = p_{P_3} \text{ or } Q_3 = p_{P_3}/I_3. \tag{4.55}$$

A simple derivation of the inertia coefficient for this case involves writing Newton's law for the fluid in the pipe. (A more thorough analysis is given in Section 12.3.1.) Assuming that the slug of fluid in the pipe moves as a rigid body, its velocity is Q_3/A, its acceleration is $\dot{Q}_3/A$, and its mass is $\rho A l$. The net force accelerating the slug involves the pressure times the area at the two ends, $P_1 A - P_2 A$. Then Newton's law becomes

$$(\rho A l)\dot{Q}_3/A = (P_1 - P_2)A. \tag{4.56}$$

Using Eqs. (4.53, 4.54, and 4.55), the results are

$$(\rho l/A)\dot{Q}_3 = P_3, \tag{4.57}$$

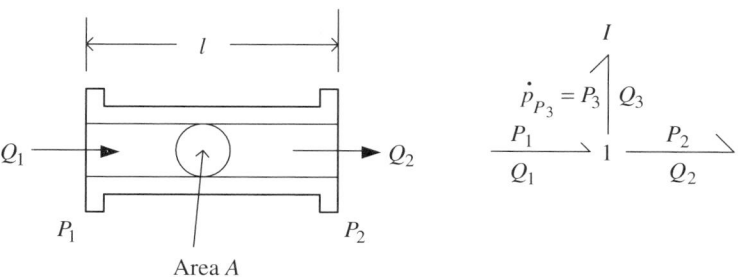

FIGURE 4.27. Fluid inertia for a pipe segment.

or

$$(\rho l/A)Q_3 = p_{P_3}, \tag{4.58}$$

which means that the inertia coefficient is

$$I_3 = \rho l/A. \tag{4.59}$$

This coefficient for the inertia of fluid in a pipe segment is analogous to the mass for a body in translation or the inductance in an electric circuit.

It is no surprise that the inertia coefficient is proportional to the mass density and length of the fluid slug, but that it is inversely proportionate to the area seems counterintuitive. The reason is that we are using pressure and volume flow rate as effort and flow variables instead of force and velocity. Many engineers have assumed that narrow tubes would exhibit little inertia effect since they contain little fluid mass, but, in fact, the narrower the tube, the more inertia the fluid in the tube will have when it is used in a fluid circuit. Oddly enough, fluid circuits with long, small-diameter pipes react sluggishly as would mechanical systems with large masses.

As discussed in Section 12.3.1, the inertia coefficient for a pipe whose area varies with distance s along the pipe, $A(s)$, is given by the formula

$$I_3 = \int_0^t \rho\, ds/A(s), \tag{4.60}$$

which shows that it is the parts of the pipe where the area is the smallest that contribute the most to the inertia coefficient.

It should be noted that although the inertia formula, Eq. (4.59), applies to both hydraulic systems and acoustic systems, there is a subtlety connected with the inertia coefficient for an open-ended pipe in an acoustic circuit. If the pipe is unflanged, the length in Eq. (4.59) should be corrected by adding about 0.6 times the radius to the physical length at both ends. This takes into account the radiation impedance associated with the connection to the atmosphere (see Reference [7], Section 9.2).

4.3.4 Fluid Circuit Construction

Hydraulic and acoustic circuits have a lot in common with electrical circuits. Remember that for electrical circuits, it was easy to establish a 0-junction for each location in a circuit at which an "absolute" voltage could be defined. The passive 1-port elements were then inserted between the appropriate 0-junctions using 1-junctions to establish the voltage differences acting on the elements. Finally, one of the 0-junctions was picked as representing a reference voltage and was eliminated together with all the bonds emanating from it. For fluid circuits, pressures act like voltages and volume flow rates act like electrical currents, so a short version of the circuit construction procedure is as follows:

1. For each distinct "absolute" pressure, establish a 0-junction. If necessary, include a 0-junction for atmospheric pressure.

2. Insert R, C, and I elements between appropriate 0-junctions using 1-junctions.

3. Assign power sign convention half arrows using a "through" scheme so that the elements react to pressure differences.

4. Attach any pressure or flow sources.

5. The 0-junction for atmospheric pressure can be eliminated if "gage" pressures are to be used.

6. The graph can be simplified if there are any 2-port 0- or 1-junctions with "through" sign convention half arrows.

Figure 4.28*a* shows a hydraulic pump–motor unit with a pressure-relief-valve bypass. We shall assume the pump to be a flow source, the relief valve and motor both to be resistances, and the reservoir to be at atmospheric pressure. For now we shall not deal with the fact that both the pump and the motor transduce hydraulic and mechanical powers. We also shall ignore the filter. In Figure 4.28*b* the two distinct pressures are identified. In part *c* the pump, valve, and motor are inserted. This is a correct bond graph model. Next we define the pressure at B with respect to at-

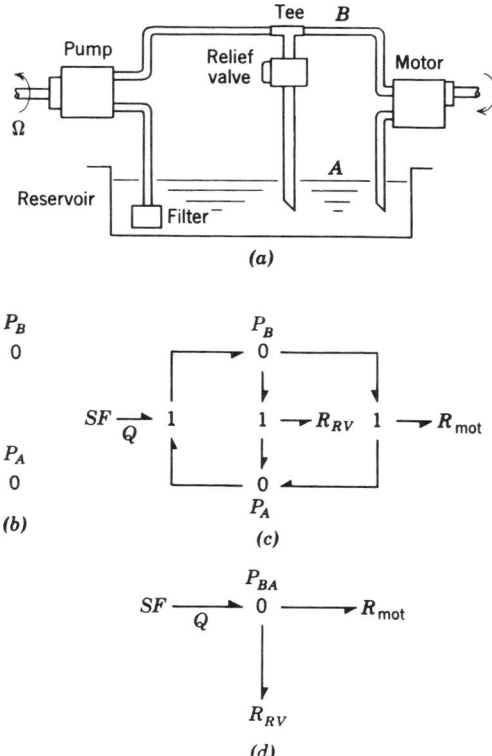

FIGURE 4.28. A hydraulic pump–motor unit.

mospheric pressure and eliminate the P_A junction. Simplification yields the system bond graph of Figure 4.28d. We are now dealing with "gage" pressures.

Frequently, hydraulic elements must be represented that do not show their pressure points quite so distinctly as in the previous example. Consider a length of flexible line connecting a pump to a load. If the line is long, if the pressure varies rapidly or widely, or if great accuracy in prediction is required, it may be necessary to include the effects of fluid inertia in the line, the flexibility of the line walls, and the internal line resistance to flow. In fact, all these effects are distributed throughout the line. Fortunately, it is often possible to get good predictive models by making a few simple assumptions. A long, flexible line is shown in Figure 4.29a. In part b a simple dynamic model indicating inertia, resistance, and compliance is shown; notice the pressure points defined in the process. The inertia acts to generate a flow (Q_{12}) in dynamic response to a pressure difference ($P_1 - P_2$). The resistance also generates a flow (Q_{23}) in response to a pressure difference ($P_2 - P_3$). Wall compliance generates a pressure (P_3) (relative to atmosphere) in dynamic response to

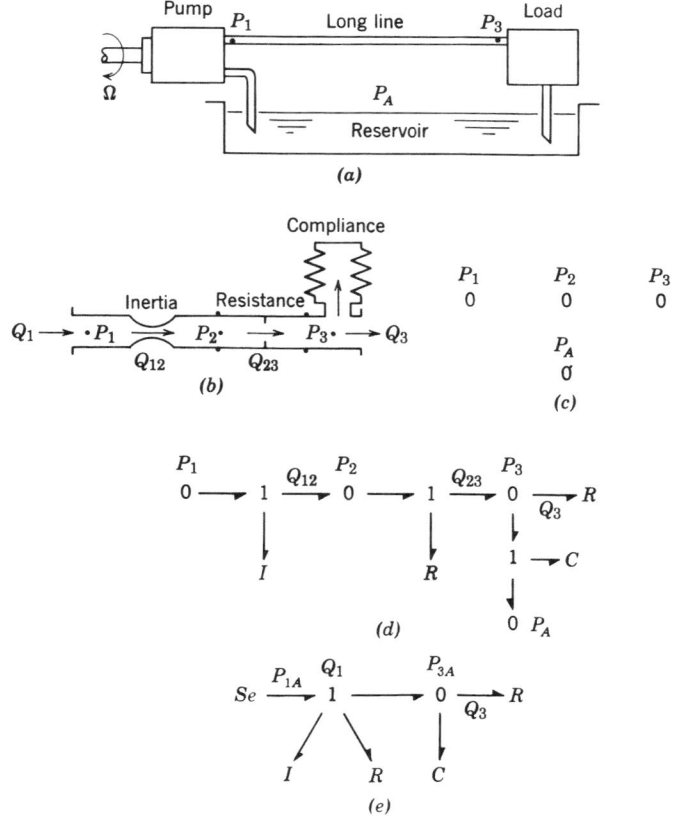

FIGURE 4.29. Hydraulic system with a long, flexible line.

a net flow. These pressures are displayed in Figure 4.29c. Elements are inserted as shown in part d. The inertia and resistance effects are inserted directly; the compliance goes between local pressure and atmospheric. Finally, the pump is added as a pressure source, the load as a resistance, and the graph is simplified. All pressures are measured relative to atmospheric.

Inspection of Figure 4.29e with regard to the line model itself shows that a simple 1–0 structure with I, R, and C attached is the basic 2-port line representation. By subdividing the line, making such a model for each unit, and then adjoining them port to port in cascade, a model of greater accuracy (albeit greater complexity) can be constructed. This approach applies equally well to many other types of devices that are distributed, such as rods, shafts, plates, and electrical lines. It is discussed again in Chapter 10.

4.3.5   An Acoustic Circuit Example

Figure 4.30 shows a reactive type of muffler that works primarily through the interaction of inertia and capacitive effects. (This is why it is called a reactive muffler.) Part a of the figure represents a cross section of a straight-through design consisting of a shell divided into two cavities, of volumes V_c and V_b, and by an internal plate and having two tubes allowing a steady flow to pass through. An imposed flow, having both steady and pulsating components, is imposed at the left side and a flow exits at

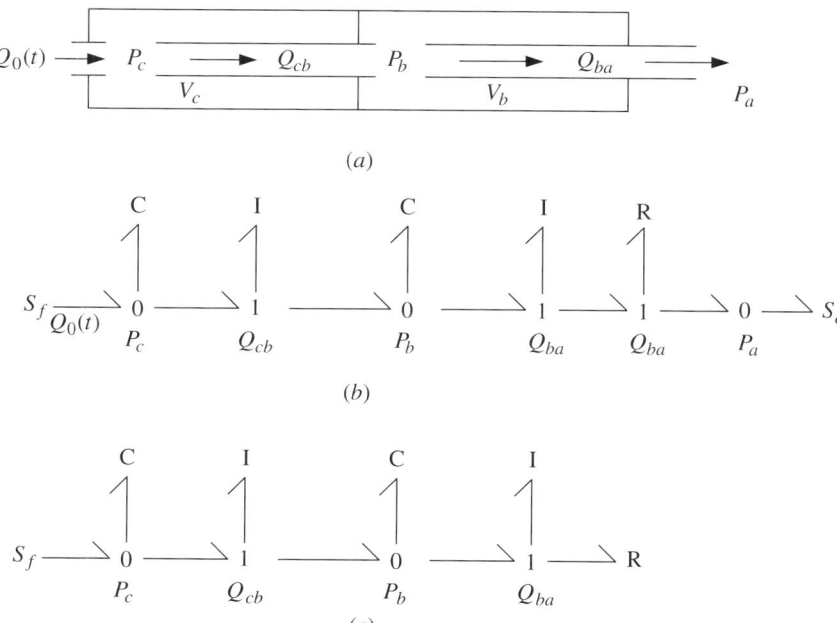

FIGURE 4.30. Low-frequency shell and tube reactive muffler model.

the right into the atmosphere. If the muffler is effective, the pulsating components of the flow at the exit should be reduced, as compared with the input fluctuations for some frequency range of interest.

For this example, we will assume that the frequencies of interest are sufficiently low that the tubes can be represented as a single I-element and the cavities can be represented by C-elements. For higher frequencies, a much more complex model might be required similar to the long line model shown in Figure 4.29. (It should also be noted that this model applies better to an air-conditioning duct muffler in which the acoustic approximation is valid than to an automobile muffler that has very large pressure fluctuations from the engine exhaust ports.).

The bond graph shown in Figure 4.30b can be constructed by following the procedure outlined above. First, we establish three 0-junctions for the pressures P_a, P_b, and P_c. The I-elements representing the tubes are inserted between the appropriate 0-junctions using 1-junctions, as in Figure 4.27. A minor complication at the exit is the R-element representing the part of the radiation impedance of the flow as it enters the infinite atmosphere. It is a standard exercise in acoustics to derive an expression for this impedance and to show that for low frequencies, the impedance can be represented as a resistance and an increased inertia as represented by the end correction discussed above. (For a more thorough discussion of these effects, see Reference [7], Sections 8.12 and 9.3.)

The flow through the radiation impedance R-element is the same as the flow through the last tube, so an extra 1-junction is used. Equivalently, a single 4-port 1-junction could be used for the I- and R-elements.

The atmospheric pressure is set by an effort source, and the internal pressures are set by capacitances, as in Figure 4.24. The input flow is determined by a flow source. The capacitance parameters are evaluated using Eq. (4.41), and the inertia parameters using Eq. (4.59).

A simplification is possible if all pressures are defined as deviations from atmospheric pressure. This means that $P_a \equiv 0$ and allows all bonds on which atmospheric pressure appears to be eliminated. When this is done and a 2-port 1-junction with a through sign convention to the R-element is eliminated, the bond graph of Figure 4.30c results.

Although the techniques and limitations of modeling hydraulic and acoustic circuits are quite complex in detail, it should be clear that there is a useful analogy between fluid and electrical circuit models.

4.4   TRANSDUCERS AND MULTI-ENERGY-DOMAIN MODELS

Transducers are devices that couple subsystems of distinct energy domains. For example, there are electromechanical devices such as motors, generators, and relays, there are hydraulic-mechanical devices such as pumps, motors, and rams, and there are even devices such as electrohydraulic valves that involve electrical, mechanical, and hydraulic power. Chapter 8 deals in some detail with various types of transducers, so here we will discuss only some relatively simple transducers based on 2-port transformers and gyrators. Some of these devices have already been shown schematically in Figures 3.8 and 3.9. These common transducers will allow the modeling of multi-energy-domain systems using bond graphs.

A major reason for studying bond graphs is that they provide a uniform and precise way to represent system models when several forms of power and energy are involved in a system to be modeled. Bond graphs practically force the modeler to consider how power and energy are conserved across an ideal transducer and how added R-elements modeling dissipative effects account for efficiencies of less than 100% for real devices. Finally, it is straightforward to add I- and C-elements to a bond graph to model the dynamics of transducers.

4.4.1   Transformer Transducers

Some transducers are quite easy to understand. Consider, for example, the hydraulic ram shown in Figure 3.8e and again in Figure 4.31.

From the schematic diagram in Figure 4.31a, it is clear that the pressure in the hydraulic fluid acts on the piston face of area A to create a force. It should also be clear that the velocity of the piston times the area equals the volume flow rate, Q. Thus, the ideal transducer is represented by the bond graph in part b, which implies the equations (3.10) given previously in Chapter 3:

$$F = AP, \quad AV = Q, \quad FV = PQ, \tag{4.61}$$

Note that the sign convention half arrows are consistent with the schematic diagram. If at some instant the force and velocity are positive in the directions indicated, then mechanical power is being supplied to the device and hydraulic power is being delivered to some other device. This "through" power convention for transformers ensures that the area A that appears twice in the equations has the same sign both times.

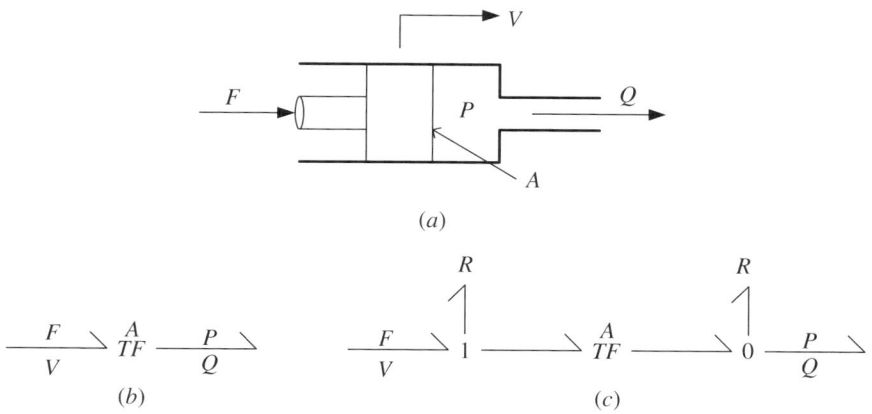

FIGURE 4.31. Hydraulic ram transducer: (a) Schematic diagram; (b) ideal transformer model; (c) friction and leakage resistors added.

(When the sign half arrows do not flow through the transformer, then $-A$ must appear in one of the equations and the mechanical power would always be the negative of the hydraulic power.)

Actually, the ideal transformer transducer really applies only to the effects at the face of the piston. All real devices contradict the idea that mechanical power can be converted with no loss to hydraulic power. As pointed out in Chapter 8, the modeling of a real transducer involves adding a number of elements to the ideal transducer to account for power losses and, possibly, for inertia and capacitance dynamic effects as well. Part c of Figure 4.31 shows how loss effects could be added to the simple ideal model.

On the mechanical side, the added 1-junction and resistor are used to include mechanical friction effects. Typical pistons have piston rings or packing that introduce frictional forces on the piston. This means that the net force on the piston rod is not just AP but also includes the frictional force. The 1-junction adds the forces and also enforces that the piston rod, the friction force element, and the piston all have the same velocity, V. Since the piston ring R-element can only absorb power, the hydraulic power will now be less than the mechanical power at the piston rod. On the hydraulic side, the combination of the 0-junction and resistance represent possible leakage past the piston. The pressure at the hydraulic outlet, on the piston face, and at the piston ring leakage location are assumed to be equal. The 0-junction makes the volume flow rate out of the ram to be AV minus the leakage flow through the leakage resistor. This hydraulic resistor also reduces the power efficiency of the transducer.

Positive displacement hydraulic pumps and motors consist of a number of pistons and a mechanism for moving them back and forth as a function of the angular position of a shaft. When there are a fairly large number of pistons, say seven or nine, the angular speed of the shaft ω is related to the volume flow rate Q by a nearly constant coefficient, T. (The coefficient actually varies a little with the position of the shaft. This makes the flow rate have a little ripple on top of a steady flow when the shaft rotates steadily.) Considering that the hydraulic ram can be represented by a transformer, it is not surprising that positive displacement machines can also be modeled using transformers, as shown in Figure 4.32.

In part a, the schematic diagram shows the torque and the angular speed of the shaft and the pressures at the inlet and outlet ports P_A and P_B. When the leakage is

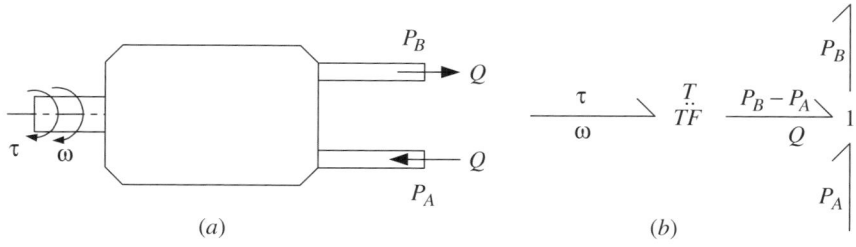

FIGURE 4.32. Hydraulic pump or motor modeled using a transformer.

negligible, the same flow Q enters and leaves the device. The hydraulic power has to do with the flow Q and the pressure difference $P_B - P_A$.

The ideal model of the pump or motor is shown in bond graph form in Figure 4.32b. If the transduction coefficient relating angular speed and volume flow rate is defined by the relation

$$T Q = \omega, \tag{4.62}$$

then it follows that the torque–pressure relation must be

$$P_B - P_A = T \tau, \tag{4.63}$$

since the power relation for the ideal case must be

$$(P_B - P_A) Q = \tau \omega. \tag{4.64}$$

The choice of signs on the transformer may seem to indicate that the model represents a pump, but even with these sign choices, the mechanical and hydraulic powers could both be negative, indicating that the device model could as well represent a hydraulic motor.

Although the bond graph in its present form is ideal with no losses and no dynamic behavior, it should be clear that the model could be augmented by adding resistors to model losses, as was done in Figure 4.31, and one could add an I-element on a 1-junction on the mechanical side to represent the moment of inertia of the rotating parts. Chapter 8 discusses the modeling procedure based on starting with an ideal model and then adding loss and dynamic elements as appropriate.

4.4.2   Gyrator Transducers

While many types of transducers that can be represented as transformers seem to be fairly easily understood, those represented by gyrators often seem harder to understand. The importance of electromechanical transducers, such as rotary and linear motors or voice coils in electrodynamic loudspeakers, justifies a short discussion of why gyrators are used to describe a number of useful devices.

Figure 4.33 shows the basis of the gyrator models. A current carrying conductor is moving in a magnetic field under the action of an applied force that is equal but opposite to a magnetic force. Because of the motion of the conductor, a voltage is induced in the conductor.

Because all the various vectors are at right angles to each other, the laws governing this situation take particularly simple forms. Faraday's law of magnetic induction relates the induced voltage to the velocity:

$$e = B l V. \tag{4.65}$$

The Lorentz force law relates the applied force to the current:

$$B l i = F. \tag{4.66}$$

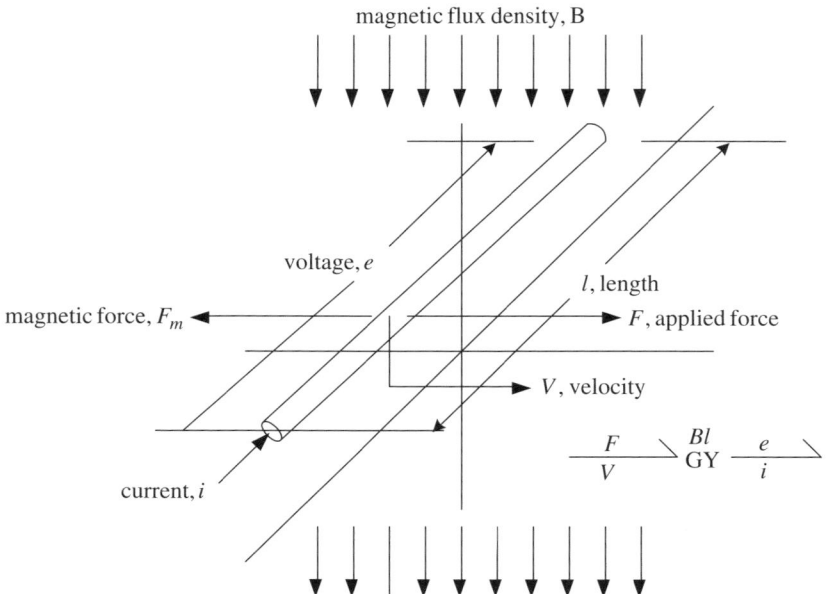

FIGURE 4.33. Gyrator model of a current-carrying conductor moving in a magnetic field.

Note that the electrical and mechanical powers are always equal,

$$ei = FV, \tag{4.67}$$

and in both Eqs. (4.65 and 4.66) an effort is related to a flow. Thus, the power-conserving 2-port transducer of Figure 4.33 is properly represented as a gyrator.

Just as the simple case of the hydraulic ram leads to the representation of hydraulic pumps and motors as transformers, so too does the case of a conductor in a magnetic field lead to the representation of electric motors and generators as gyrators. Figure 4.34 shows several representations of d-c motors or generators. Internally, most motors have complex arrangements of coils of wires carrying currents moving in magnetic fields. In the case shown, it is assumed that the field is due to permanent magnets and a commutator or other means switches the coils of wire in such a way that the voltage at the terminals induced by the rotary motion is essentially proportional to the angular speed of the rotor. In addition, the torque produced is proportional to the current at the terminals of the device. (More detailed models of motors are able to predict minor ripples in the torque of actual d-c motors as well as small voltage fluctuations in d-c generators.)

Using the symbol T as a transduction coefficient, the equations for the ideal model of Figure 4.34*b* are

$$e = T\omega, \quad Ti = \tau, \quad \text{and} \quad ei = \tau\omega. \tag{4.68}$$

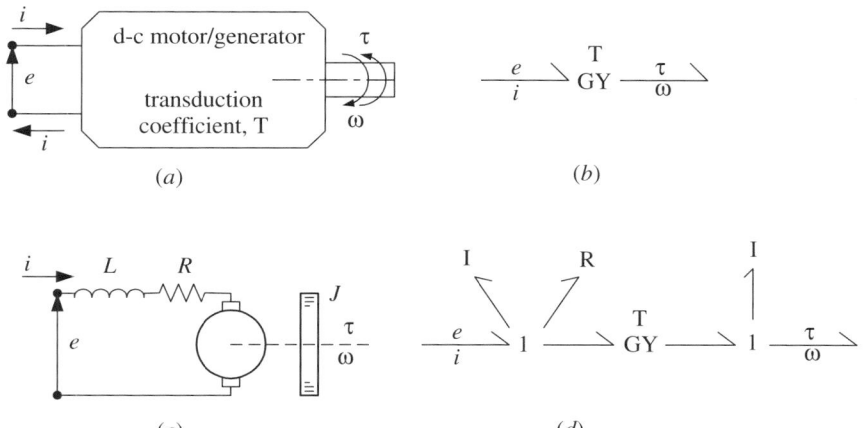

FIGURE 4.34. Motor/generator models: (*a*) Sketch of device; (*b*) bond graph of ideal model; (*c*) schematic diagram of extended model; (*d*) bond graph of extended model.

The gyrator correctly represents the essence of the transduction process, but in many cases it needs to be supplemented to model real devices that are never completely ideal. The schematic diagram of Figure 4.34*c* indicates that the windings of the motor have inductance and resistance and that the rotor has a moment of inertia, J. Part *d* of the figure shows an extended bond graph including these effects. The 1-junction on the electrical side enforces the fact that the same current flows from the terminals through the inductance and the resistance and participates in the transduction.

On the mechanical side, the 1-junction indicates that there is a single angular velocity for the output shaft, the rotary inertia, and the transduction process. In this model, the electrical power is not instantaneously equal to the mechanical power, as it was for the ideal gyrator model. Not only is there now a loss element, R, but also the I-elements can store and release energy during transient conditions. The idea of starting with ideal transducers and adding realistic effects is developed further in Chapter 8.

4.4.3 Multi-Energy-Domain Models

Figure 4.35*a* shows a system for positioning a machine tool table bed. It includes an electric d-c motor connected to a power supply driving a hydraulic pump. The power supply is modeled as a voltage source. The motor and pump may run continuously, because the combination valve (with handle position variable, x) can recycle some of the hydraulic fluid to the sump and can divert some of the flow to the hydraulic motor. The hydraulic motor in turn drives a rack and pinion arrangement that moves the table bed.

There are a number of transducer devices in this system. The d-c motor changes electrical power to mechanical power, and the hydraulic pump and motor exchange

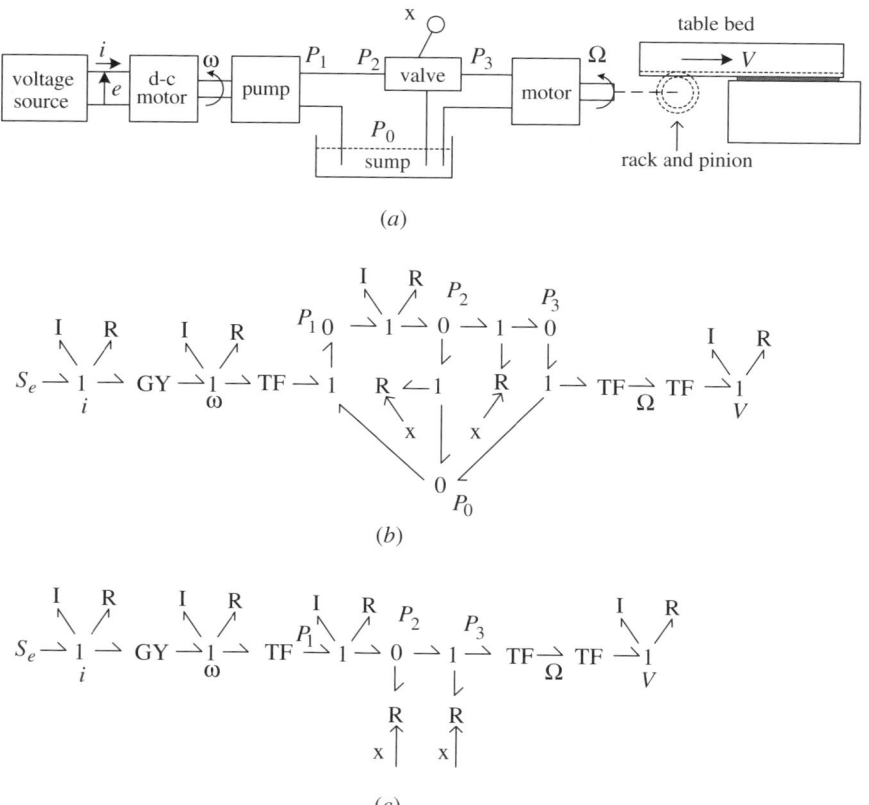

FIGURE 4.35. An electrical-hydraulic-mechanical system: (*a*) Schematic diagram; (*b*) bond graph including atmospheric pressure; (*c*) bond graph using gage pressures.

mechanical and hydraulic power. One could even consider the rack and pinion as a transducer between rotary and translational mechanical power. (A transformer represents the rack and pinion, because the contact radius of the pinion relates the angular velocity of the pinion to the linear velocity of the table, as well as the force on the table to the torque on the pinion.) This system is intended to demonstrate that bond graph models are capable of representing multi-energy-domain systems in a uniform and precise manner.

The schematic diagram of the system shown in part *a* of Figure 4.35 shows a number of important system variables such as pressures and angular velocities, but it does not make clear just how the various devices are to be modeled. The first stage in the construction of the bond graph for the system shown in part *b* of Figure 4.35 begins to make clear the assumptions that are being made to model the various devices.

The left side of the bond graph in part *b* represents the voltage source and the d-c motor. The motor model of Figure 4.34 has been used with some modifications.

The first I- and R-elements clearly represent the inductance and resistance of the armature windings, but the next $I-R$ combination attached to the 1-junction, with ω as the common flow, requires some explanation. The I-element represents not only the rotary inertia of the motor, as in Figure 4.34, but also the inertia of the pump, since both devices rotate with the same angular speed ω. The R-element represents frictional losses associated with both the electric motor and the pump.

Transformers represent the hydraulic pump and motor, as in Figure 4.32. The hydraulic pressures P_0, P_1, P_2 and P_3 are represented as existing on 0-junctions, as in the hydraulic circuit construction method described in Section 4.3.4. A study of the sign half arrows will reveal that the pressure difference acting on the pump is $P_1 - P_0$ and the pressure difference acting on the motor is $P_3 - P_0$.

The line between the pump outlet and the combination valve has been modeled as an $I-R$ combination that includes fluid inertia, as in Figure 4.27, and fluid resistance, as in Figure 4.23. Evidently, no fluid or pipe compliance effects are included in the model. This is a modeling decision that may or may not be justified in a particular case. In any case, the bond graph is absolutely clear about which possible effects are included and which are not.

The combination valve has a variable resistance reacting to the pressure difference $P_2 - P_0$ and another variable resistance reacting to the pressure difference $P_2 - P_3$. Both these valve resistances change with the handle position, x. The double arrow active bond carrying the variable x indicates this. It is thus assumed that moving the handle does not transfer any significant power to the system, although the handle position does change the valve resistance laws relating pressure and volume flow rate.

Finally, the rack and pinion transformer connects the hydraulic motor to the table bed model, which is represented by an I-element for the table mass and an R-element for the table friction.

The last step to arrive at the bond graph of Figure 4.35c is to consider that the atmospheric pressure P_0 is zero. This allows three bonds to be removed and several 2-port 0- and 1-junctions with "through" sign half arrows to be eliminated. This procedure is similar to the procedure for electrical circuits, but in this case the result is that the efforts representing pressures are to be considered gage pressures.

As this example should make clear, there is an art, as well as a science, to making a useful mathematical model. There is a scientific basis for the ideal models that involve power or energy conservation, and these are elegantly incorporated in bond graph elements. However, there are a number of effects that can be added to the ideal elements to account for effects that occur in real systems. These extra loss and dynamic elements must be added with restraint, since an unduly complicated model that is hard to understand is often just as bad as an oversimplified model. In every case the model should be adjusted to provide guidance for the specific problems at hand. The best model is generally the simplest model capable of demonstrating the behavior that is desired for the system.

Good modelers are always ready to modify a preliminary model by simplifying it, by removing elements, or by adding elements to bring in new effects. Often the changes are suggested based on the results of analysis or simulation of the model, particularly if there is a lack of agreement between the model predictions and experimental results.

REFERENCES

[1] D. K. Schroder, *Semiconductor Material and Device Characterization*, New York: John Wiley & Sons, Inc., 1998.

[2] J. L. Merriam and L. G. Kraige, *Engineering Mechanics: Statics and Dynamics*, 4th Ed., New York: John Wiley & Sons, Inc., 1997.

[3] A. G. Bose and K. N. Stevens, *Introduction to Network Theory*, New York: Harper & Row, 1965.

[4] R. H. Sabersky, A. J. Acosta, and E. G. Hauptmann, *Fluid Flow: A First Course in Fluid Mechanics*, 2nd Ed., New York: Macmillan, 1971.

[5] J. F. Blackburn, G. Reethhof, and J. L. Shearer, *Fluid Power Control*, New York: John Wiley & Sons, Inc., 1960.

[6] J. Thoma, *Modern Hydraulic Engineering*, London: Trade and Technical Press, 1970.

[7] L. E. Kinsler, A. R. Frey, A. B. Coppens and J. V. Sanders, *Fundamentals of Acoustics*, 3rd Ed., New York: John Wiley & Sons, Inc., 1982.

PROBLEMS

4-1   Make a bond graph model for each of the following electrical circuits. Use the inspection method whenever possible. If the circuit has voltage and current orientations, make the bond graph have equivalent power directions.

Note: The E-meter may be modeled as an open-circuit pair, and the auditor may be modeled as a resistance, R_A.

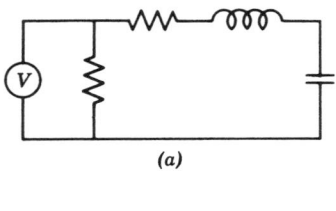

(a)

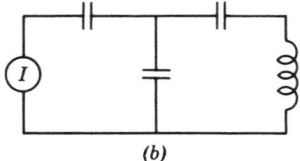

(b)

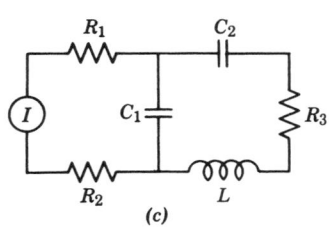

(c)

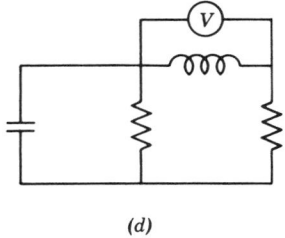

(d)

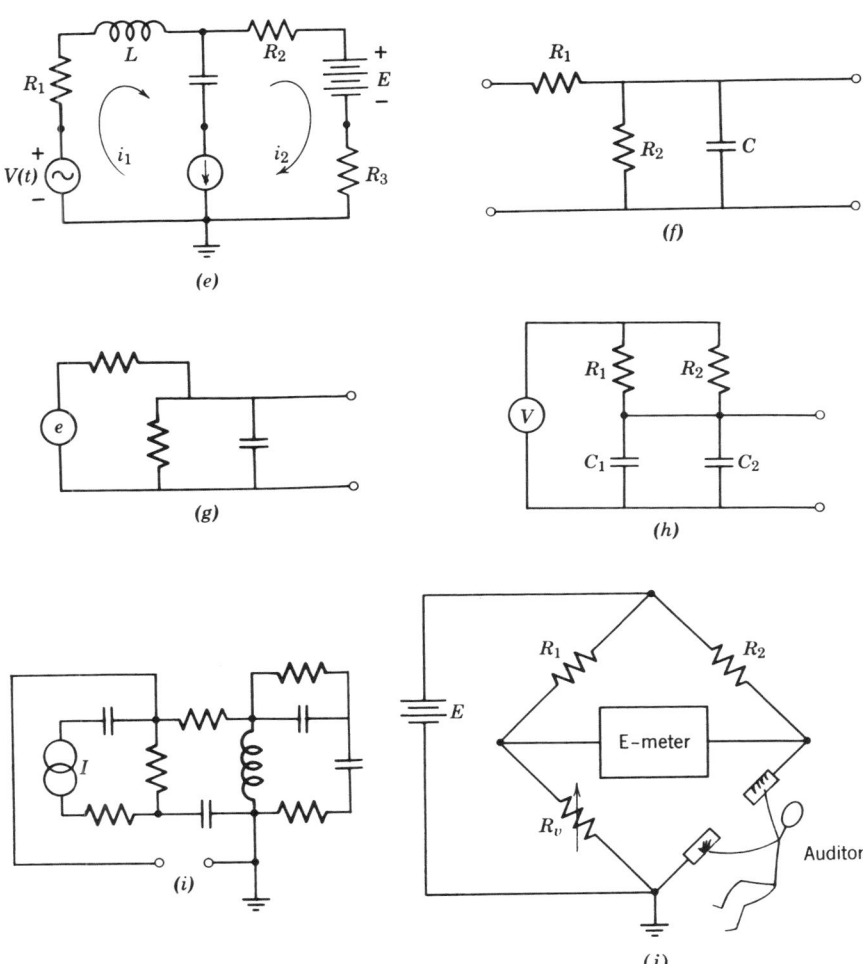

(e)

(f)

(g)

(h)

(i)

(j)

4-2   Make a bond graph model of each of the electrical networks shown below.

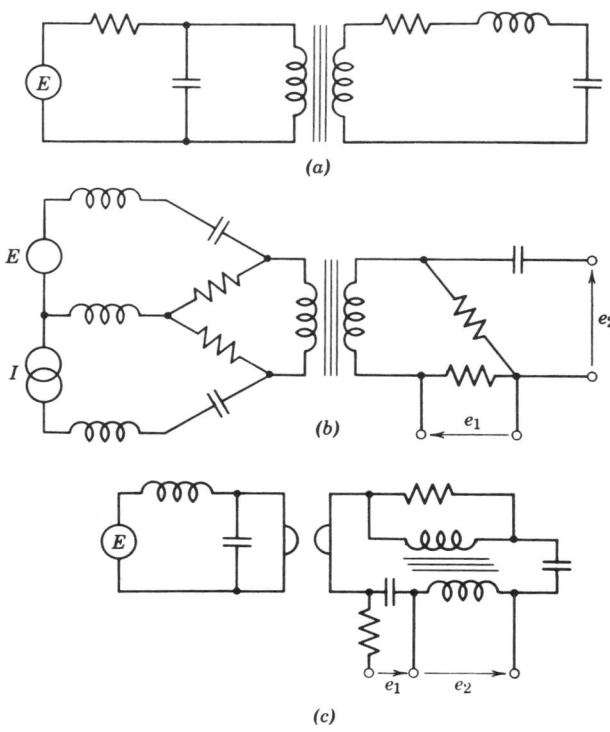

4-3   Make a bond graph model of each of the translational mechanical systems shown below.

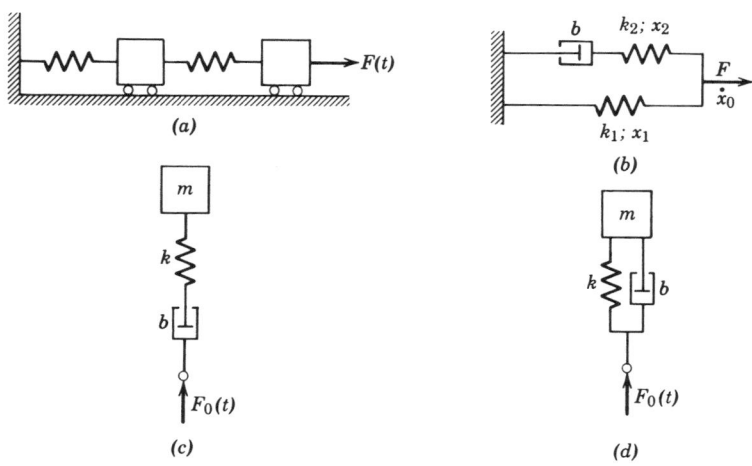

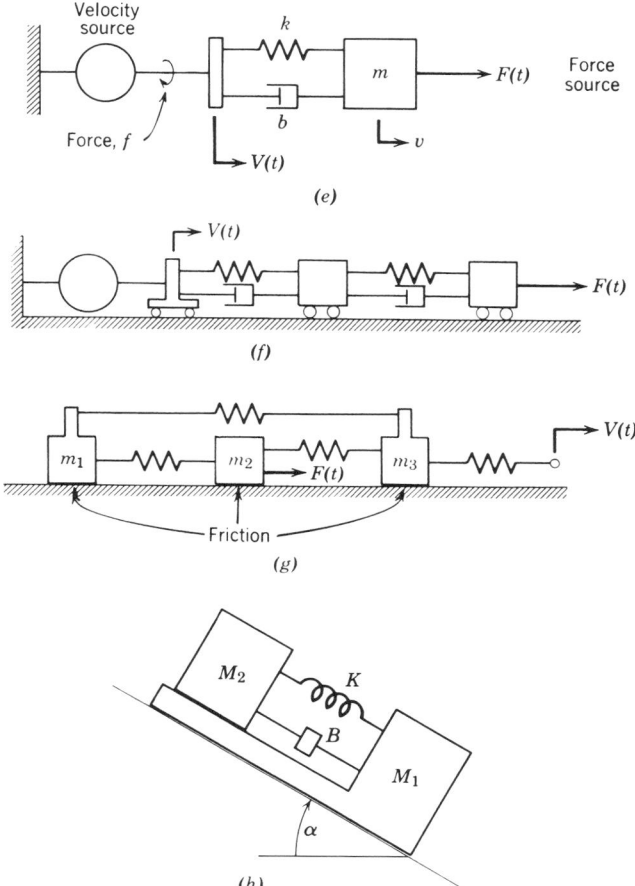

(e)

(f)

(g)

(h)

4-4   Find a bond graph model for the following mechanical networks.

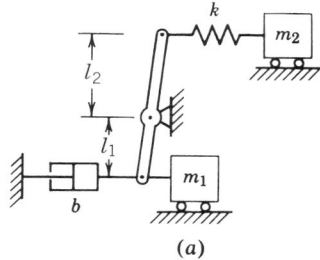

(a)

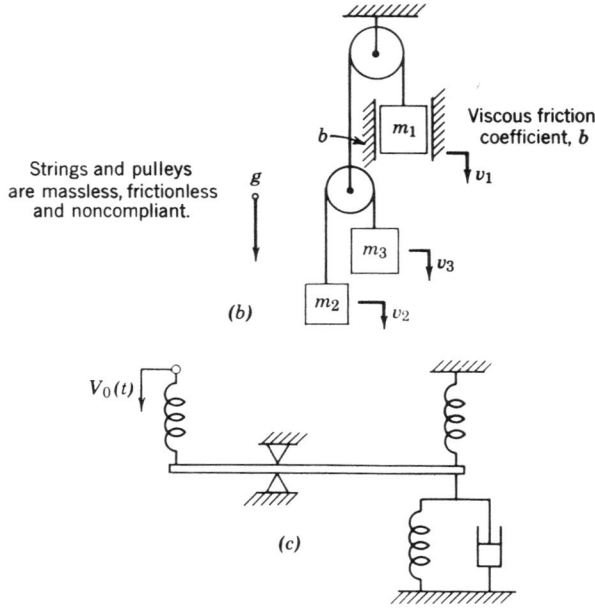

Viscous friction coefficient, **b**

Strings and pulleys
are massless, frictionless
and noncompliant.

(b)

(c)

4-5   For the following mechanical systems involving fixed-axis rotation, find a
bond graph model in each case. In the rack-and-pinion system shown, several
frictionless guides and bearings are omitted for clarity.

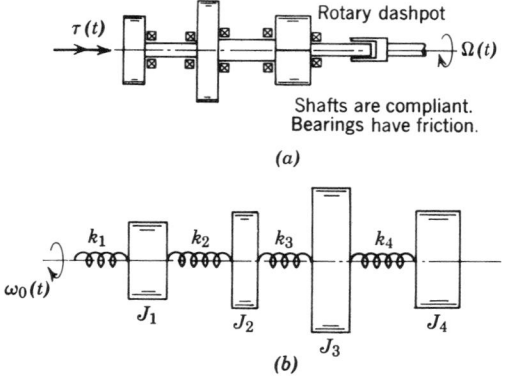

Rotary dashpot

$\Omega(t)$

Shafts are compliant.
Bearings have friction.

(a)

$\omega_0(t)$

(b)

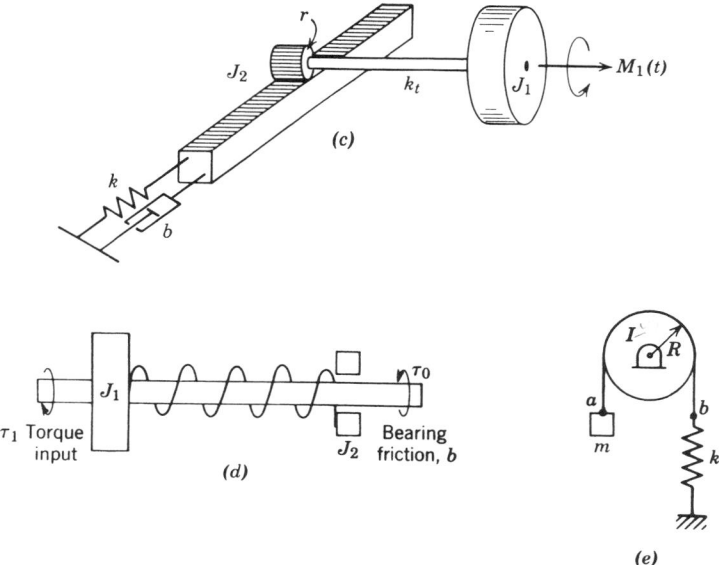

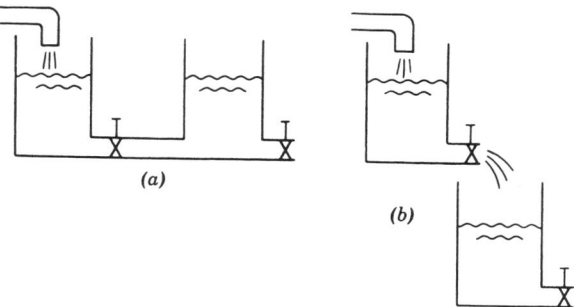

4-6 Two hydraulic systems that are closely related but not identical are shown below. Make a bond graph model for each.

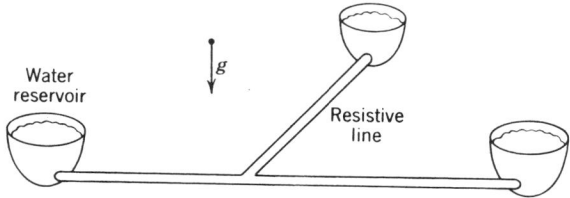

(a)

(b)

4-7 A simplified model of a water storage system is shown below. Assume the three tanks behave as (nonlinear) capacitances. Assume the three conduits have resistance only.

(a)   Make a bond graph model.

(b)   Introduce inertance effects for each of the conduits into your model for (a).

4-8   In the hydraulic system shown below, the pipes containing flows Q_4 and Q_5 have both inertia and friction effects present. The outflow line (Q_6) has only friction effects. The control pump maintains a desired flow, Q_C, independent of pressure. Make a bond graph model of the system.

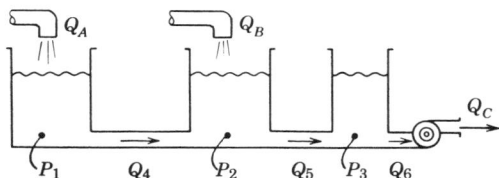

4-9   In the positive-displacement pump shown below, the piston moves back and forth and the check valves act like unsymmetrical resistances, allowing relatively free forward flow while impeding the back flow greatly. Inertia effects are important in the outlet line, while the nozzle is a resistive restriction.

(a)   Make a bond graph model that can be used to relate the mechanical port variables (F, V) to the fluid port variables (P, Q).

(b)   Noting that an electrical diode is analogous to a fluid check valve, use the bond graph of part (a) to find an equivalent electrical network to the pump system.

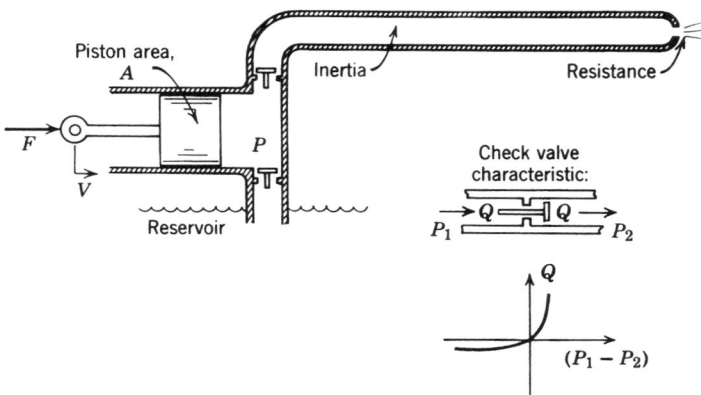

4-10   Consider a permanent-magnet generator of the type often used to power bicycle lights:

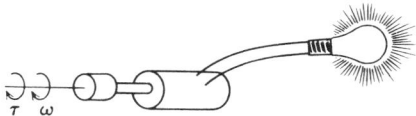

(a)   Suppose the generator to be an ideal gyrator in its transducer action. Make a bond graph model of the system.

(b)   Suppose the lamp shortcircuits. Is your model from (a) still useful? If not, modify it so that it can handle this situation.

4-11   Two related hydraulic devices are shown below, a hydraulic jack and a ram.

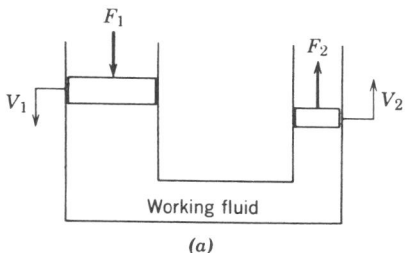

(a)

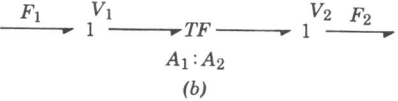

(b)

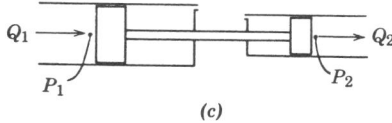

(c)

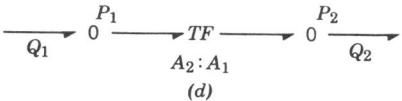

(d)

(a)   If the inertia of the sliding parts and friction at the seals are important, modify the models given in the figure for the devices to include these effects.

(b) If compliance of the working fluid for the jack and of the shaft for the ram is important, modify your models from (a) above to include these effects.

4-12 A schematic diagram of a basic d-c machine is shown below. The principal electrical effects are armature resistance and inductance and field resistance and inductance. The coupling between armature and shaft powers is modulated by the strength of the magnetic field, set by i_f. A simple assumption for coupling is

$$\tau = T(i_f)i_a = (Ki_f)i_a$$

and

$$e_a = T(i_f)\omega = (Ki_f)\omega$$

where $T(i_f)$ is the modulating value, assumed linear with i_f.

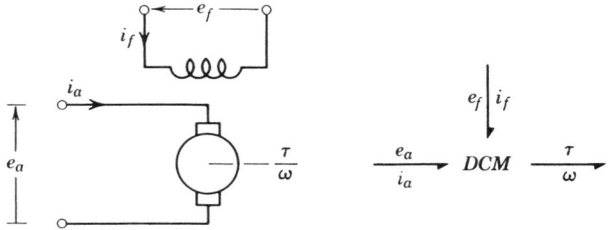

(a) Make a bond graph model of the basic d-c machine using a modulated gyrator (MGY) as the heart of the coupling.

(b) Introduce mechanical rotational inertia and dissipation effects into your model from (a) above.

(c) Operate your model from (b) in the generator mode. What changes must be made?

(d) Calculate the power efficiency for the d-c machine operating as a motor and as a generator.

4-13 The network shown below includes a voltage-controlled current source. Make a bond graph model of the network using an active 2-port as the coupling element.

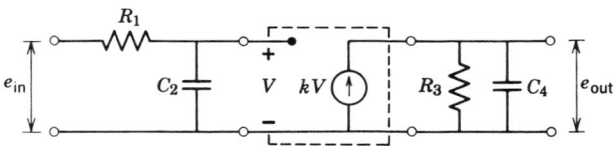

4-14   A load of mass M is suddenly dropped onto piston 1 from height L.

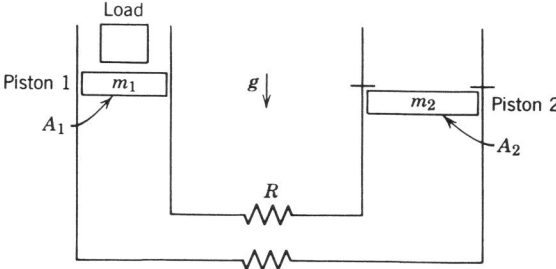

(a)   Make a bond graph model of the system, assuming piston 2 is locked in place by pins. Also assume the load stays joined to piston 1.

(b)   If piston 2 is not locked in place, but pistons 1 and 2 were in equilibrium before the load was dropped, modify the model in (a).

4-15   Model the spring-loaded accumulator device shown below. Include inertia and dissipation effects.

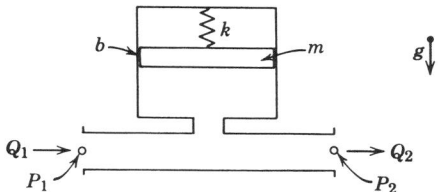

4-16   A slab is shown below being quenched in a cooling bath. Assume the bath temperature is maintained constant at T_b. You may neglect heat transfer from the ends and edges.

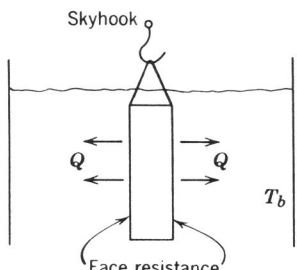

(a) Make a bond graph model of the cooling process.

(b) Suppose the water bath is in a tank of volume V_b and the temperature is not maintained constant. Modify the model to cover this case.

4-17 There is a simple analogy between lumped-parameter pseudo–bond graph models of conduction heat transfer systems and electrical circuits. Reinterpret the bond graph of Figure 3.18 as an electric circuit and point out the analogous thermal and electrical effort and flow variables.

4-18 For the device shown, construct a bond graph model. For the positive directions indicated, assign a power convention.

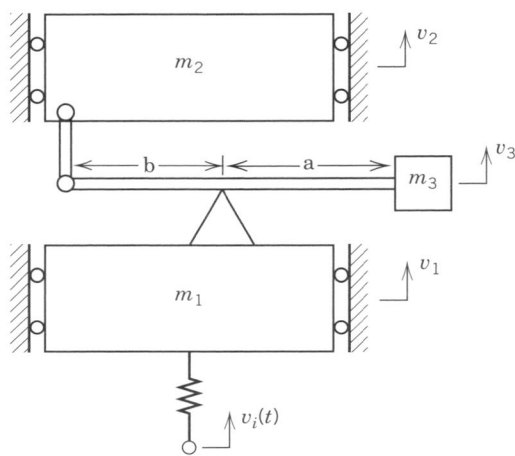

4-19 The coupled mechanical and hydraulic system is used to isolate the mass, m, from the ground motion, $v_i(t)$. Construct a bond graph model, show your assumed positive e and f directions, and assign an appropriate power convention.

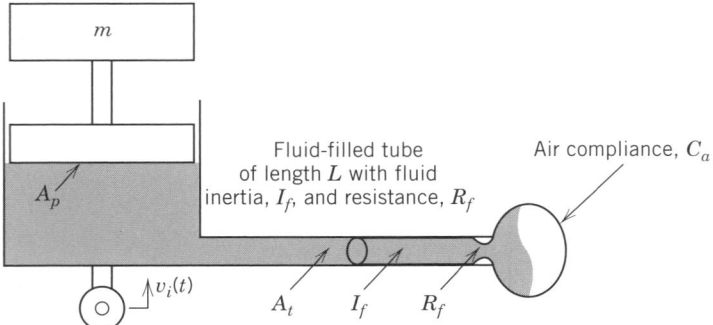

4-20 A quarter car model of a vehicle is shown with an idealized actuator capable of generating any control force, F_c.

  (a) Construct a bond graph model, identify positive directions, and assign a power convention.

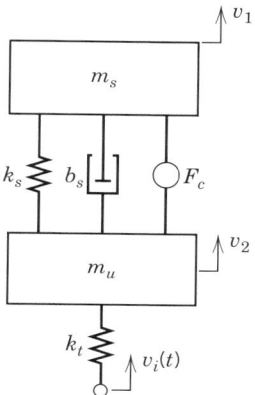

  (b) A voice coil actuator with winding resistance, R_w, is to be installed as the force actuator to replace the idealized control force. The actuator will be voltage, e_c, controlled. Include the force actuator in the quarter car model and construct the overall bond graph model.

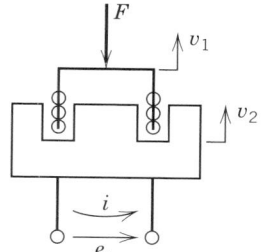

winding resistance, R_w
coil length, l
permanent magnetic with flux density B
transduction constant $T = BL$
force, $F = Ti$

4-21 The hydraulic piston cylinder device is a fundamental component of many motion control systems. Hydraulics can generate large forces over large distances, which is necessary for big motion platforms such as flight simulators. The ideal form of the device shown is a transformer, as indicated in the bond graph.

An actual device has inertia, friction, leakage past the piston, and compliance of the volumes of oil on both sides of the piston. For the realistic effects shown, construct a bond graph model and assign a power convention.

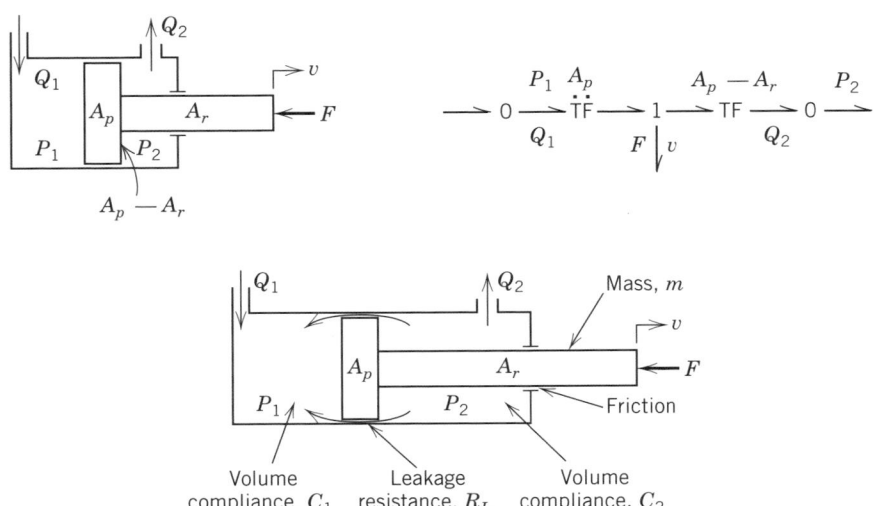

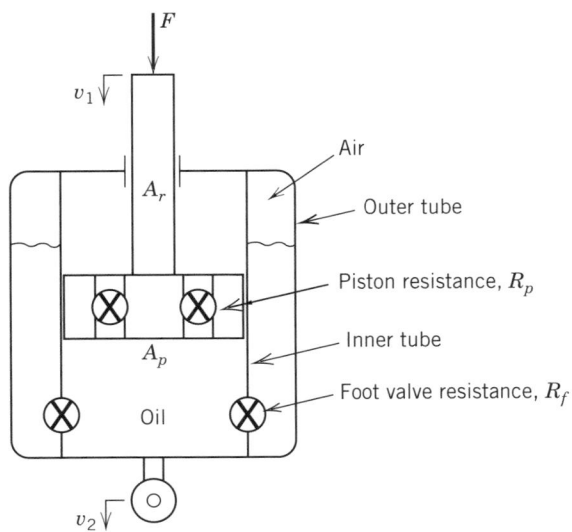

4-22 Shock absorbers for automobiles are very sophisticated devices. The piston head has several levels of check valves to generate a force–velocity characteristic quite different from what might be expected from simply forcing oil through an orifice. Also, in order to provide volume for the piston rod as the shock is compressed, most shock absorbers are of a twin-tube construction, as shown below. The valve that connects the inner and outer tubes is called the foot valve.

(a) Construct a bond graph model incorporating the dynamic elements shown and assign a power convention.

(b) State your assumptions about more realistic dynamic effects and include them in your model.

4-23 Install the shock absorber model from Problem 4-22 into the quarter car model of Problem 4-20. Replace the damper with the shock model.

4-24 The system is similar to the example system of Figure 4.14. For this problem, the lever has mass, m, and cg moment of inertia, J_l. In addition, the pivot point has prescribed vertical motion set by the specified velocity, $v_{i2}(t)$. The spring at the left end has prescribed vertical velocity at the top, and there is a mass with spring and damper attached at the right end of the lever. The lever executes plane motion since it simultaneously translates and rotates.

(a) Derive the kinematic constraints for this system by relating v_l, v_{i2}, and v_m to the cg velocity and angular velocity, v_g and ω.

(b) State whether springs and dampers are positive in compression or tension and construct a bond graph model for this system.

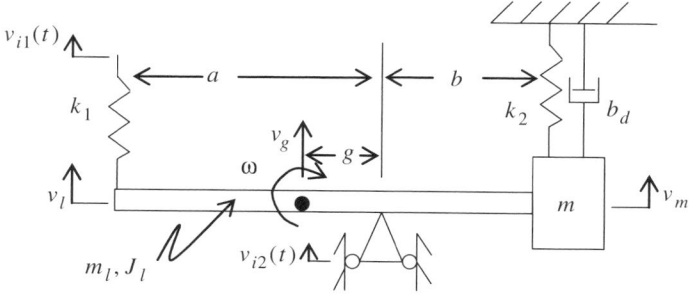

4-25 Masses m_1 and m_2 move only horizontally. Mass m_1 has an input force prescribed, and m_2 has an attached spring, k. The two masses are connected by

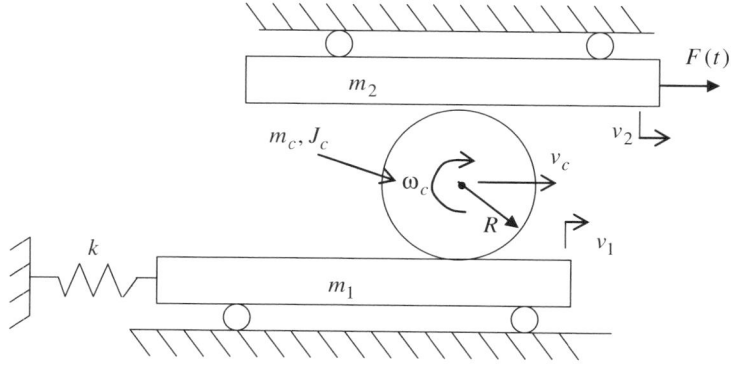

a rolling element of mass, m_c, cg moment of inertia, J_c, and radius, R. The rolling element rolls without slip at the contact point with each mass.

(a)   Derive the kinematic constraints relating the mass velocities v_1 and v_2 to the velocity and angular velocity of the rolling element.

(b)   Construct a bond graph model of this system and show an appropriate power convention.

4-26   A gantry robot is a large manufacturing machine that consists of a very stiff table-like structure, atop which sits a platform that can be driven in the X, Y plane. Suspended below the platform is the robot arm that will have a cutting tool of some kind attached at its lower end. The robot arm can be moved vertically, and the cutting tool can be moved in several directions. The result is a multi-axis cutting tool that can create large 3-D shapes. Although gantry robots are made as stiff and rigid as possible, sometimes small motions at the cutting tool end can limit the speed or precision of the machine.

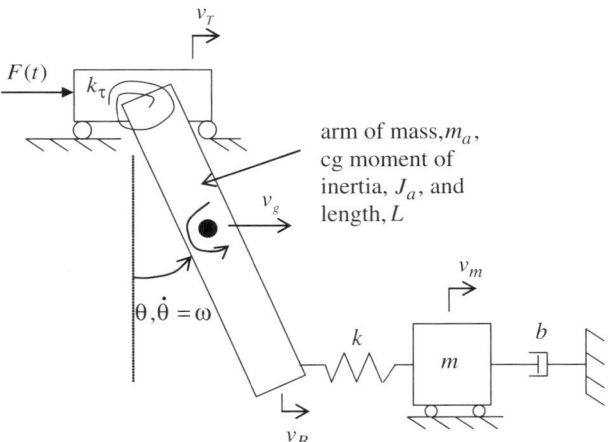

A simple model of the gantry robot is shown in the figure. An input force, $F(t)$, is used instead of the actual drive motors, and the top of the arm is constrained to move horizontally. The arm is rigid and has mass, m_a, cg moment of inertia, J_a, and length, L. The cg is located in the middle of the arm. The flexibility in the frame is represented by the torsional spring, k_τ, and the load being cut is represented by the spring, mass, damper, k, m, b, attached to the lower end of the arm. We desire a model that will allow the prediction of the motion–time history of the system for small angular deflections. Perhaps we can come up with some control strategy for the drive motors that will limit the vibratory motion of the arm end. For now,

(a)   Derive the kinematic constraints relating v_T and v_B to the arm motion, v_g, and ω.

(b) Construct a bond graph model of this system and show an appropriate power convention.

4-27 Consider a simple model of a so-called Helmholtz oscillator formed by attaching a circular tube of diameter d and length l to a larger circular tube of diameter D and length L that is sealed at the end to form an acoustic volume.

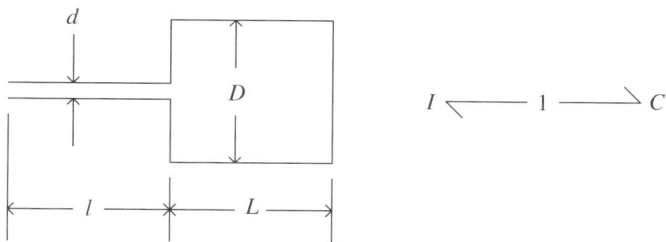

Using the elementary bond graph shown, which neglects any resistance effects, the natural frequency of the oscillator is given by the formula

$$\omega_n = (1/IC)^{1/2} \text{ [rad/s] or } f_n = (1/IC)^{1/2}/2\pi \text{ [cycles/s] or [Hz]}.$$

(a) Using the formulas for acoustic inertia and capacitance given in the text, find an expression for ω_n and f_n in terms of the dimensions of the device.

(b) What would be the natural frequency in air at standard conditions if $d = 0.01 \text{ m}$ and $D = l = L = 0.1 \text{ m}$? Do you think you could hear this frequency if you blew across the end of the open tube as you would across an empty bottle?

4-28 Consider a length of hydraulic hose to be used to operate the ram of a high-speed actuator. There is a concern that the compliance in the system may slow down the system response. You are to decide whether the hydraulic compliance due to the compressibility of the hydraulic fluid or the flexibility of the hose walls is most important. Use these dimensions: nominal radius of the hose 15 mm, wall thickness 5 mm. Assume that the modulus of elasticity for the hose is about the same as for hard rubber and the bulk modulus for hydraulic oil has the value given in the Appendix.

4-29 Consider a system that is a combination of Figure 4.28 and Figure 4.29. That is, a hydraulic pump is connected to a hydraulic motor by one line to a relief valve and by another line from the valve to the motor. In the new system, both lines are long and to be represented by a combination of inertia, resistance, and capacitance, as in Figure 4.29. In addition, both the pump and the hydraulic motor are positive displacement machines, represented as in Figure 4.32. The hydraulic motor is driven by a d-c electrical motor as represented by the gyrator in Figure 4.34.

The load that the hydraulic motor drives can be represented by a combination of a rotary mechanical inertia and a rotary resistance. Show a complete bond graph for the system after setting the atmospheric pressure to zero (using gage pressures).

4-30   The figure shows an accumulator similar to the one in Figure 4.26. In this case, there is a distinction between the fluid pressure P_3 and the pressure of the gas in the rubber bladder under some conditions. Before the fluid system is pressurized, when P_3 is zero, the pressure in the gas is P_0 and the volume of the gas is V_0. The bladder has a rigid section at the base that prevents it from expanding into the inlet pipe. Until the fluid pressure exceeds P_0, no fluid enters the accumulator; $Q_3 = 0$. However, after some fluid has entered the accumulator, the gas is compressed and the gas pressure and the fluid pressure are essentially identical.

Consider the volume of fluid that has entered the accumulator, $V_3 = \int_0^t Q_3 \, dt$, to be zero at $t = 0$ when the system is unpressurized. Make a sketch of the nonlinear capacitance relationship between P_3 and V_3 using the isentropic assumption discussed in the text. Note that the P_3 can rise to P_0 even when $V_0 = 0$ and that $P_3 \to \infty$ when $V_3 \to V_0$.

4-31   The figure shows a pipe of area a and length l, containing hydraulic fluid, connected to a hydraulic ram of area A. This combination will eventually be used to model a hydraulic system, and the concern is whether the pipe hydraulic inertia, $\rho l / a$, is significant in comparison with the mass, m, of the ram and its connected load mass.

The bond graph on the right shows the two inertia elements separately, and the one on the left shows the two inertias combined into a single equivalent inertia associated with the velocity of the ram, V. One way to compute the equivalent inertia is to express the kinetic energy $T = mV^2/2 + IQ^2/2$ (where $I = \rho l / a$) in terms of the mass velocity by recognizing that the mass velocity and the volume flow rate are related by the transformer law $Q = AV$.

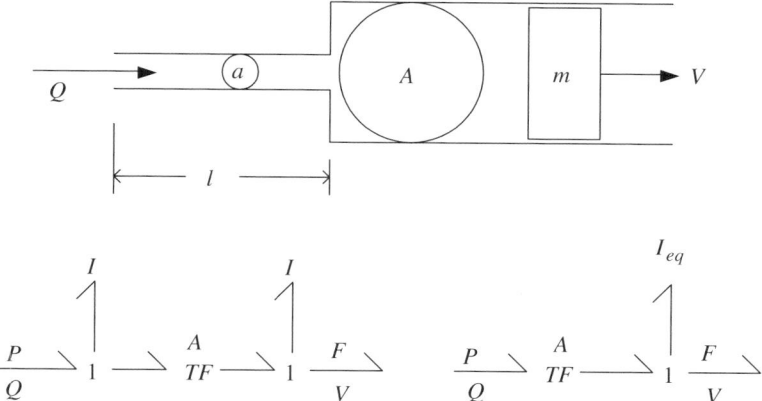

(a) Give an expression for the equivalent inertia (or equivalent mass) of the ram and pipe combination.

(b) Perform a sample calculation to find out what radius the pipe should have to make the fluid inertia of the pipe have the same effect on the effective mass of the system as the mass of the ram and load mass. Use the following parameters:

$$\rho = 900[\text{kg/m}^3], \quad l = 0.5[\text{m}], \quad m = 40[\text{kg}],$$
$$A = \pi R^2, \quad R = 50[\text{mm}], \quad a = \pi r^2$$

Answers:

(a) $I_{eq} = (m + \rho l A^2/a)$,

(b) $a = \rho l A^2/m, r = 15[\text{mm}]$.

4-32 The figure shows the physical part of a load positioning system without the sensors and the controller electronics that control the voltage source. You are to make a bond graph model of the system, including the following effects:

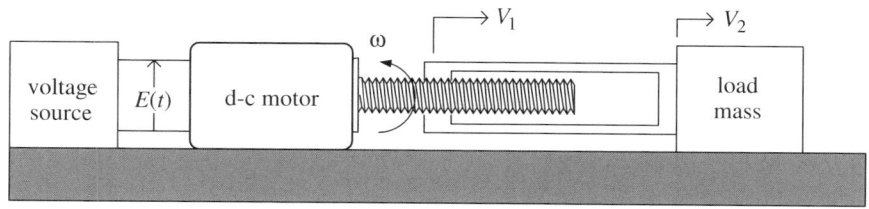

(a) The voltage is a controlled effort source. Indicate this by including a signal or active bond with a double arrow impinging on an effort source.

(b) The d-c motor has inductance and resistance in the armature coils.

(c)   Include a rotary inertia for the motor armature and the screw drive.

(d)   Use a resistor to model the friction moment associated with the motor and the screw drive.

(e)   A constant S relates the angular velocity of the screw with the nut: $V_1 = S\omega$.

(f)   The tube connecting the nut with the load mass is elastic with a spring constant k.

(g)   The load mass has significant friction with the ground.

5

STATE-SPACE EQUATIONS AND AUTOMATED SIMULATION

In the last chapter, bond graph models were developed for many physical engineering systems. We first treated each energy domain separately, but our goal was always to connect the energy domains into overall system models using the same symbols and structure for the entire system. This was done in Section 4.4 of the previous chapter and will continue to be done for the rest of this text. The nine basic bond graph elements are capable of representing a very large cross section of engineering systems.

The reason for constructing bond graph models will be very apparent in this chapter. It will be demonstrated just how straightforward it is to derive system equations and to obtain computer solutions to these equations. In some cases we can go from a bond graph directly to a simulation using a computer program without first obtaining equations of motion manually. In other cases we will derive the equations of motion and use them to obtain analytical information about the system behavior. The real virtue of bond graph modeling becomes overwhelmingly apparent when one considers that, from a bond graph point of view, all systems appear the same. Thus, only one formulation and solution procedure is needed for any of the systems we model.

To enforce the virtue of having a formulation procedure that works for any bond graph, consider the physical system shown schematically in Figure 5.1a for part of an electric power steering system. A d-c motor with winding resistance, R_w, rotor inertia, J_m, and rotary damping, b_τ, drives a flexible output shaft of rotary stiffness, k_τ. The shaft is connected to a rack and pinion setup where the gear has radius, R, and the rack has mass, m. The rack is attached to a spring and damper, k, b. We desire to predict the motion of this system for a voltage input to the motor.

Without a procedure for deriving equations of motion, one might start by drawing a free body diagram of the entire system. This is done in Figure 5.1b. The electric circuit voltage, e_{mf}, is the back emf of the motor. The electrical torque, τ_e, is the

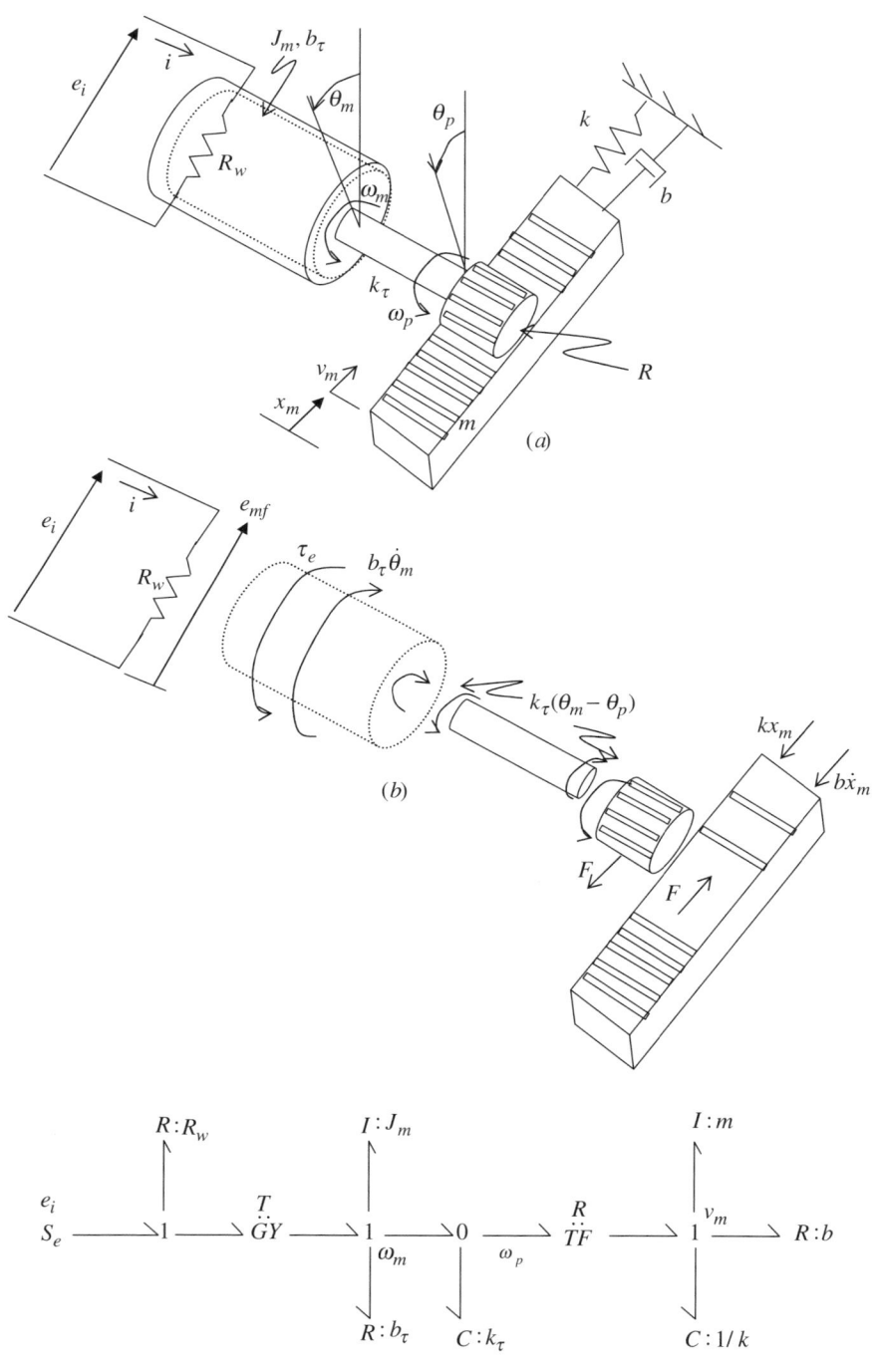

FIGURE 5.1. Part of an electric power steering system.

torque on the rotor due to the electromechanical coupling of the motor. The torque in the output shaft is shown acting on the rotor and the gear. And the internal force, F, is shown acting on the gear and the rack. With bond graphs, we always define efforts and flows, regardless of the energy domain involved. This is done in the system schematic. The more conventional approach to formulation often starts with displacement variables, and these are also indicated in the schematic.

The next step is to use Kirchoff's law for the circuit, with the result

$$e_i - R_w i - e_{mf} = 0. \tag{5.1}$$

Then use Newton's laws for translation and rotation, with the result

$$\tau_e - b_\tau \dot{\theta}_m - k_\tau(\theta_m - \theta_p) = J_m \ddot{\theta}_m, \tag{5.2}$$

$$F - kx - b\dot{x} = m\ddot{x}. \tag{5.3}$$

We must recognize the kinematic, no-slip constraint between the pinion gear and the rack,

$$\dot{x} = R\dot{\theta}_p, \tag{5.4}$$

and the moment equilibrium for the inertia-free pinion gear,

$$k_\tau(\theta_m - \theta_p) = FR. \tag{5.5}$$

At this point it might be useful to count the unknowns and see how we are doing. The unknowns are i, e_{mf}, τ_e, θ_m, θ_p, x, and F. Thus we have seven unknowns and five equations. We are not done yet. The missing information is the electromechanical physics relating the electrical torque to the current and the back emf to the motor speed. These relationships are

$$\tau_e = Ti,$$

$$e_{mf} = T\dot{\theta}_m, \tag{5.6}$$

where T is the coupling constant for the motor. We now have equal numbers of equations and unknowns and can theoretically proceed to a solution.

If we were to pursue this example further, we would next have to figure out substitutions to make to eliminate unwanted variables while retaining desired variables. There is no set procedure for this substitution, we just use our intuition and start substituting. If we are successful at coming up with a computable set of equations, the act of doing so will not benefit us much the next time we are confronted with a new physical system.

With bond graphs there is a procedure for choosing the proper variables. There is a procedure for deriving a computable set of equations. It is the same procedure each time equations are derived. If one struggles the first time, it will be a little easier the second time, and easier each successive time. Figure 5.1c is a bond graph of the example system. All the dynamics in the schematic are included in the bond graph.

Problem 5-16 revisits this system. You will amaze yourself as to how easy it is to derive the governing equations as compared with what was required by the conventional approach illustrated above. The formulation procedure is presented next.

5.1   STANDARD FORM FOR SYSTEM EQUATIONS

One of the most remarkable features of bond graphs is that a study of equation formulation may be carried out prior to writing any equations. To understand how this can be so, we first consider some particular forms of equations that are used to represent a system, and select one form—the state-space type—as our goal.

Briefly, we may state that there are two limit forms for systems of differential equations, plus a wide range of possibilities within these limits. An nth-order system may be represented by

1. a single nth-order equation in terms of one unknown variable;
2. n first-order coupled equations in terms of n unknown variables; or
3. various combinations of unknowns and equations of appropriate orders (not necessarily equal).

Many important mathematical problems, methods, and results are organized in terms of the first form. At first, almost all engineering mathematics was cast in that form. The second form has certain advantages that have recommended it to mathematicians* for development of theory, but, even more important from our point of view, it is a convenient form for use by engineering system analysts, control engineers, and people conducting digital and analog computer studies. An interesting example of the third form is to be found in the sets of second-order equations generated by the Lagrangian approach to system analysis.

To illustrate each of the forms, and to give an appreciation of some of the differences, let us consider the mechanical double-oscillator example shown in Figure 5.2. There are several choices of unknowns available. Let us choose x_2 and express the system behavior in terms of that single displacement. The system equation in terms of x_2 turns out to be

$$\dddot{x}_2 + \left[\frac{k_2}{m_2} + k_1 \left(\frac{1}{m_1} + \frac{1}{m_1} \right) \right] \ddot{x}_2 + \frac{k_1 k_2}{m_1 m_2} x_2 = k_1 \left(\frac{1}{m_1} + \frac{1}{m_2} \right) g. \qquad (5.7)$$

If a Lagrangian approach is used, with x_3 and x_4 (the spring extensions) shown in Figure 5.1b as the generalized coordinates, the following pair of coupled second-order equations may be developed:

$$\ddot{x}_3 + k_1 \left(\frac{1}{m_2} + \frac{1}{m_2} \right) x_3 - \frac{k_2}{m_2} x_4 = 0,$$

$$\ddot{x}_4 - \frac{k_1}{m_2} x_3 + \frac{k_2}{m_2} x_4 = g. \qquad (5.8)$$

*See, for example, G. D. Birkhoff, *Dynamical Systems*, Providence: Amer. Math. Soc., 1966.

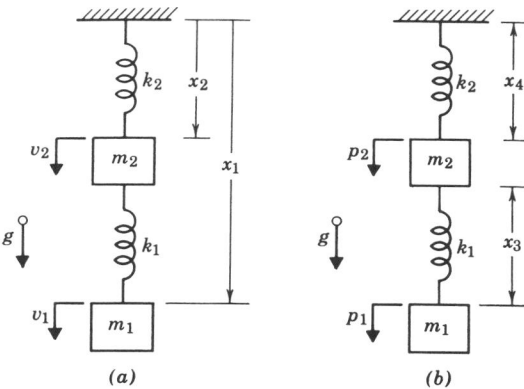

FIGURE 5.2. Mechanical double-oscillator example.

Finally, if one concentrates closely upon the energy in the system, associating a momentum or displacement variable with each distinct energy element, a set of four coupled first-order equations in terms of p_1, p_2, x_3, and x_4 may be found, as shown in Eq. (5.9). Clearly, this set could also be converted to a geometric-variable form by replacing the momentum variables by the velocities to which they are related:

$$\dot{p}_1 = -k_1 x_3 + m_1 g, \qquad \dot{p}_2 = k_1 x_3 - k_2 x_4 + m_2 g,$$

$$\dot{x}_3 = \frac{p_1}{m_1} - \frac{p_2}{m_2}, \qquad \dot{x}_4 = \frac{p_2}{m_2}. \tag{5.9}$$

It is possible, in principle, to transform from one of the given forms to any other, whether the system is linear or nonlinear. However, for nonlinear systems the desired transformations may be very difficult to discover. Furthermore, if one starts with the form (5.7), for example, choosing additional unknowns is a rather haphazard process unless one has considerable insight into the system being studied. Yet if a physically meaningful set of variables is already available [e.g., as displayed by (5.9)], elimination of unwanted variables can be carried out with considerable insight.

In studying engineering systems using bond graphs, there is an ideal opportunity to start the formulation in terms of significant physical variables and to generate simultaneous sets of first-order equations from the bond graphs. This is the approach we shall follow. When the system being studied is nonlinear, the form we seek is given by

$$\dot{x}_1(t) = \phi_1(x_1, x_2, \ldots, x_n; \quad u_1, u_2, \ldots, u_r),$$

$$\dot{x}_2(t) = \phi_2(x_1, x_2, \ldots, x_n; \quad u_1, u_2, \ldots, u_r),$$

$$\vdots$$

$$\dot{x}_n(t) = \phi_n(x_1, x_2, \ldots, x_n; \quad u_1, u_2, \ldots, u_r), \tag{5.10}$$

where the x_i are the *state* variables, the $\dot{x}_i$ are the time derivatives of the x_i, the u_i are inputs to the system, and the ϕ_i are a set of static (or algebraic) functions.*

If the system is linear, Eqs. (5.10) take on a simpler form, as shown below:

$$\dot{x}_1(t) = a_{11}x_1 + a_{12}x_2 + \cdots + a_{1n}x_n + b_{11}u_1 + b_{12}u_2 + \cdots + b_{1r}u_r,$$

$$\dot{x}_2(t) = a_{21}x_1 + a_{22}x_2 + \cdots + a_{2n}x_n + b_{21}u_1 + b_{22}u_2 + \cdots + b_{2r}u_r,$$

$$\vdots$$

$$\dot{x}_n(t) = a_{n1}x_1 + a_{n2}x_2 + \cdots + a_{nn}x_n + b_{n1}u_1 + b_{n2}u_2 + \cdots + b_{nr}u_r, \qquad (5.11)$$

where the a_{ij} and b_{ij} are constants in most cases. For a linear time-varying system the a_{ij} and b_{ij} may vary with time, but they must not depend on the x-variables.

For linear systems, equations of the form (5.11) can be put into a standard matrix formulation,

$$\dot{X} = AX + BU, \qquad (5.11a)$$

where X is the vector of state variables,

$$X = \begin{bmatrix} x_1 \\ x_2 \\ \bullet \\ \bullet \\ \bullet \\ x_n \end{bmatrix},$$

U is the vector of inputs,

$$U = \begin{bmatrix} u_1 \\ u_2 \\ \bullet \\ \bullet \\ \bullet \\ u_r \end{bmatrix},$$

The A-matrix is composed of the coefficients, a_{ij}, where

$$A = \begin{bmatrix} a_{11} & a_{12} & \bullet & \bullet & a_{1n} \\ a_{21} & \bullet & \bullet & \bullet & a_{2n} \\ \bullet & \bullet & \bullet & \bullet & \bullet \\ \bullet & \bullet & \bullet & \bullet & \bullet \\ a_{n1} & a_{n2} & \bullet & \bullet & a_{nn} \end{bmatrix},$$

and the B-matrix contains the coefficients, b_{ij}, where

---

*This simply means that, given the values of the arguments on the right-hand side of Eq. (5.4), a set of values for the derivatives may be found by algebraic means.

$$B = \begin{bmatrix} b_{11} & b_{12} & \bullet & \bullet & b_{1r} \\ b_{21} & \bullet & \bullet & \bullet & b_{2r} \\ \bullet & \bullet & \bullet & \bullet & \bullet \\ \bullet & \bullet & \bullet & \bullet & \bullet \\ b_{n1} & b_{n2} & \bullet & \bullet & b_{nr} \end{bmatrix}.$$

We will discover how useful this formulation is when the system being studied can be modeled as a linear one.

Our task for the rest of this chapter is to learn how to select significant system variables from among the many effort and flow variables contained in a bond graph and to organize them into relations of the form (5.10) or the matrix form (5.11a) for linear models.

5.2   AUGMENTING THE BOND GRAPH

In the previous chapters we have developed an orderly procedure for representing all kinds of interacting physical systems using bond graphs. At this point we assume that a bond graph has been constructed and that an appropriate power convention has been assigned. Before writing any equations, it is desirable to prepare the bond graph with additional information that will make writing of equations take on a very orderly pattern. Providing the additional information consists of

1. Numbering all the bonds in consecutive order
2. Assigning to each bond a causal sense for the e and f variables

By numbering the bonds, every element and variable can be referred to unambiguously. Thus $I4$ is the inertia on bond 4, e_6 is the effort on bond 6, q_{11} is the displacement on bond 11, and so forth. For beginners, it is suggested that after the physical system has been modeled and a bond graph exists with all the physical labels, a second, identical bond graph be drawn without the physical labels. It is this bond graph that will receive the numbered bonds. We call this bond graph the "computational bond graph." Since the bond graphs for all physical systems are so similar, it is also suggested that the physical bond graphs be kept nearby so that the user can keep track of the physical systems involved.

The second step in the augmentation process is the *assignment of causality*. Basic considerations of causality for the various elements were introduced earlier (Section 3.4). It is now appropriate to apply such information to the entire system in an orderly fashion.

In a causal sense there are two distinct types of bond graph elements. Source elements (S_e and S_f) and junction elements (0, 1, TF, and GY) must meet certain causal conditions or their basic definitions are no longer valid. As an obvious example, if a source of force does not have a causality showing that it defines a force on the system to which it is connected, then it has no meaning. We generalize this observation and assert that *every source element must have its appropriate causal form assigned to it*.

For each of the 2-port junction elements, *TF* and *GY*, there are two possible causal forms that preserve the basic definition of the element. If neither of these forms can be assigned, the concept of input and output associated with the element is not valid. Consequently, we assert that *every TF and GY must have one of its two allowable causal forms assigned to it.* The choice of form will generally be indicated by the adjoining system on the basis of other considerations to be discussed.

In a similar vein, we argue with respect to the ideal multiport junctions 0 and 1 that *each 0 and 1 must have one of its appropriate causal forms assigned to it* or the basic definition of the particular element will not be valid. The selection of a particular causal form will be motivated by other system considerations in general.

If a system cannot meet the causal conditions outlined above, the basic physical model on which the bond graph is based must be restudied. The indications are that an impossible situation has been created that is not capable of sensible mathematical resolution. Two such examples are shown in Figure 5.3. In part *a*, source element 2 is defined invalidly. Inspection of an electrical circuit interpretation indicates the nature of the difficulty; obviously, a modeling error has been made in putting two supposedly independent current sources in series. In part *c* of Figure 5.3, a *TF* is found to have invalid causality. If this bond graph were derived from a fluid circuit, as shown in part *d*, the interpretation would be that two independent pressure sources had been joined by an ideal transformer of pressure (the *TF*), leading to a physically incompatible situation. The next step is up to the system modeler, who must correct the model in an appropriate way.

Continuing our discussion of assigning causality, we come to the energy storage elements, $-I$ and $-C$. Bond graphs "bookkeep" energy, and the instantaneous energy of the system is indicated by the energy variables (p's on $-I$'s and q's on $-C$'s) associated with the energy storage elements. The junction structure elements,

$$-0-, -1-, -TF-, -GY-,$$

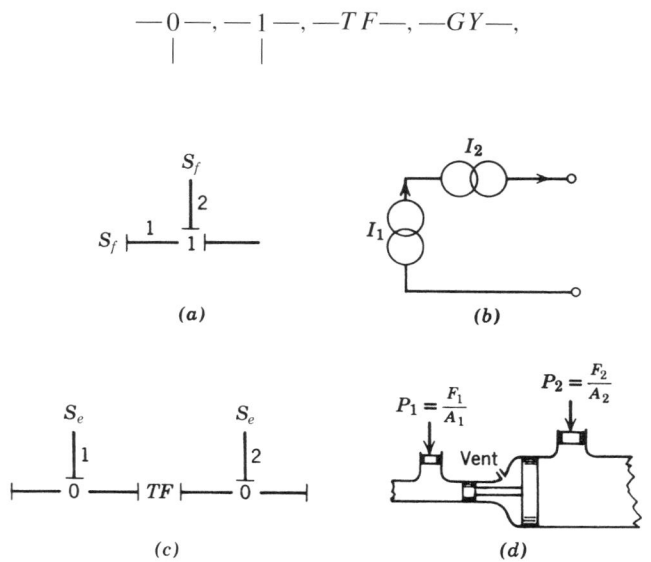

FIGURE 5.3. Two examples of invalid causality and physical interpretations.

merely shuttle the power from and to the source elements, the energy storage elements, and the dissipative elements. We choose to use the energy variables as the state variables for the systems we model. Knowledge of the energy variables dictates the energy state of the system at each instant of time. As we shall see, an energy storage element is independent if it can accept integral causality. Thus, in our causal assignments, we make every effort to assign integral causality to each energy storage element. We will see that this is not always possible. If an energy storage element is forced to accept derivative causality, then that element is not independent, and its energy variable is algebraically related to the other energy variables in the system. The energy storage element in derivative causality still stores energy, but its contribution to the system energy can be calculated algebraically from knowledge of the other energy variables, and, if the energy variable of the derivative element should happen to enter the equation formulation, it can be eliminated from the final state equations. Thus, for the formulation procedure proposed here, *the state variables are the p-variables on I-elements and the q-variables on C-elements that are in integral causality.*

At the system level the causality associated with *R*-elements is largely a matter of indifference. The major exception is in the case of a nonlinear constitutive law that is not bi-unique (e.g., coulomb friction). Then the causality associated with a unique input–output relation for the element should be used. Otherwise, *R*-elements accept whatever causality they are assigned.

The basic causality assignment procedure is summarized as follows:

Sequential Causal Assignment Procedure

1. Choose any source (S_e, S_f), and assign its required causality. Immediately extend the causal implications through the graph as far as possible, using the constraint elements (0, 1, *GY*, *TF*).

2. Repeat step 1 until all sources have been assigned.

3. Choose any storage element (*C* or *I*), and assign its preferred (integration) causality. Immediately extend the causal implications through the graph as far as possible, using the constraint elements (0, 1, *GY*, *TF*).

4. Repeat step 3 until all storage elements have been assigned a causality. In many practical cases all bonds will be causally oriented after this stage. In some cases, however, certain bonds will not yet have been assigned. We then complete the causal assignment as follows:

5. Choose any unassigned *R*-element and assign a causality to it (basically arbitrary). Immediately extend the causal implications through the graph as far as possible, using the constraint elements (0, 1, *GY*, *TF*).

6. Repeat step 5 until all *R*-elements have been used.

7. Choose any remaining unassigned bond (joined to two constraint elements), and assign a causality to it arbitrarily. Immediately extend the causal implications through the graph as far as possible, using the constraint elements (0, 1, *GY*, *TF*).

8. Repeat step 7 until all remaining bonds have been assigned.

The procedure is straightforward and orderly. Some practice on examples will convince you of the ease and rapidity with which causality can be assigned. It is important to recognize that the constraint elements represent the physical structure in the system (e.g., Kirchoff's voltage and current laws; Newton's law and geometric compatibility), and assigning causality to them means that they will be used correctly in a particular input–output fashion. Further discussion of the use and interpretation of causality in bond graphs is given in Chapter 7.

There are several situations that can arise when applying causality according to the procedure given:

1. All storage elements have integral causality, and the graph is complete after step 4. This simple, common case is discussed in Section 5.3.
2. Causality is completed by using R-elements or bonds, as indicated in steps 5–8. This situation is discussed in Section 5.4.
3. Some storage elements are forced into differentiation causality at step 3. This case is discussed in Section 5.5.

Now let us consider several examples of using the procedure.

Assignment of causality is carried out step by step in the example in Figure 5.4. Part a shows a numbered bond graph without causality or power convention. We would normally have assigned a power convention by the time we arrived at causality assignment. However, to enforce upon the reader that power convention and causality

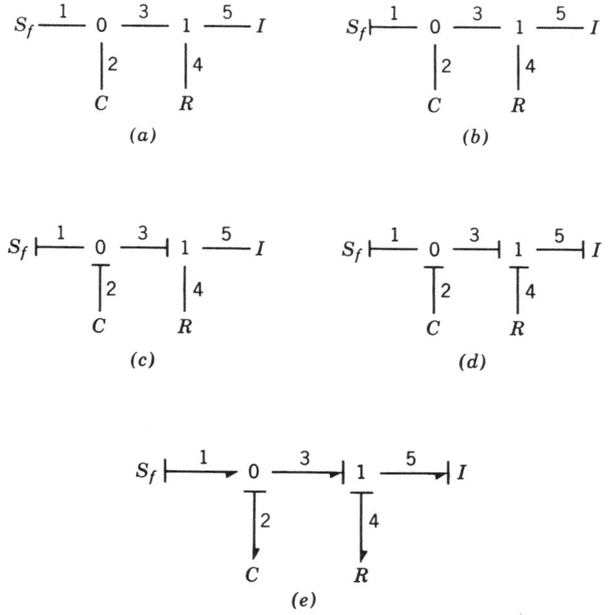

FIGURE 5.4. Causality assignment and complete augmentation of a bond graph. Example 1.

assignment are totally unrelated, we work a few examples in which the reader is not distracted by a power sign convention.

A graph without causality is sometimes said to be *acausal*. In part *b* bond 1 is directed according to the meaning of the source element (a source of flow). Since the 0-junction has only one flow variable determined, the other bonds cannot yet be assigned a causality. There are no more sources, so we turn to step 3. In part *c* bond 2 is causally directed to produce integral causality for the C-element. Immediately, bond 3 may be directed because of the 0-junction, which can have only one effort input. However, bonds 4 and 5 are as yet undirected. Bond 5 is directed, as shown in Figure 5.4*d*, to give integral causality to the I-element. Immediately, bond 4 can (nay, must) be causally directed as shown, due to the 1-junction, which can have only one flow input. In the grand finale of part *e* we have added a set of power directions to the graph; the result is a completely augmented bond graph. That is, the bonds are labeled, power directions have been chosen, and causality has been assigned. Such a graph will yield its state equations to us with very little resistance, as we shall show in the next section. However, let us first augment another graph or two to gain some experience.

Figure 5.5 shows a bond graph derived from a fluid example involving pipes and reservoirs. There are three pressure sources (S_{e1}, S_{e2}, and S_{e3}) feeding through three pipes (roughly, the 1-junction complexes) into a tank (the 0-junction and C_{10}). In part *a*, the graph is labeled. In part *b*, bonds 1, 2, and 3 associated with the sources have been assigned causality. In each case no extension of causality is possible. In

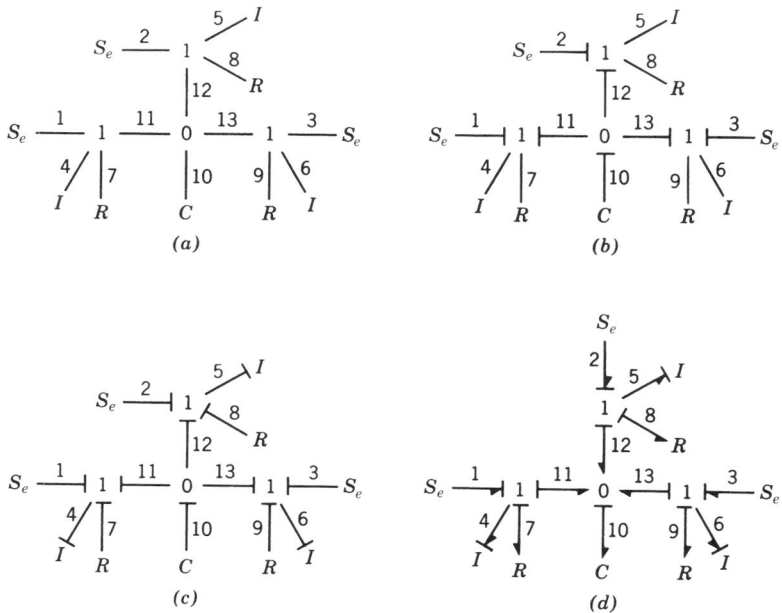

FIGURE 5.5. Augmentation of a bond graph. Example 2.

addition, bond 10 has been directed according to integral causality for the element C_{10}, and causality has been extended to bonds 11, 12, and 13 by the 0-junction, using the effort identity condition. In part c, bond 4 has been directed; consequently, so has bond 7 due to the 1-junction, and so on for bonds 5 and 8 and bonds 6 and 9. Finally, power directions have been added to produce the fully augmented graph shown in part d. In the next section, we shall show how this graph yields four first-order equations with three inputs to the system.

As a final example in this section, consider the acausal bond graph in Figure 5.6a. The model is derived from a study of a pressure-controlled valve and includes both mechanical and fluid mechanical power. The elements S_{e1} and S_{e2} represent sources of pressure and force, respectively, and the element TF couples the two power domains. In part b, bonds 1 and 2 are causally assigned, one at a time. The causal information cannot be extended using constraint elements at this point. Bonds 3 and 4 are assigned next, and causality is extended as shown in part c. Bond 4 causality does not extend to other bonds, but bond 3 has implications for bonds 8 and 9, as well as for bonds 10 and 11. Inspection of Figure 5.6d reveals that assigning causality to bond 5 (associated with the element I_5) determines the causality on bonds 6 and 7. With powers assigned as shown in part e, the bond graph is completely augmented.

Two further points may be made with respect to augmentation. The first is that, in assigning causality, the results do not depend on the order of bonds chosen except in special circumstances. These will be discussed in Section 5.4. The second point is that assignment of causality and assignment of power directions are two entirely

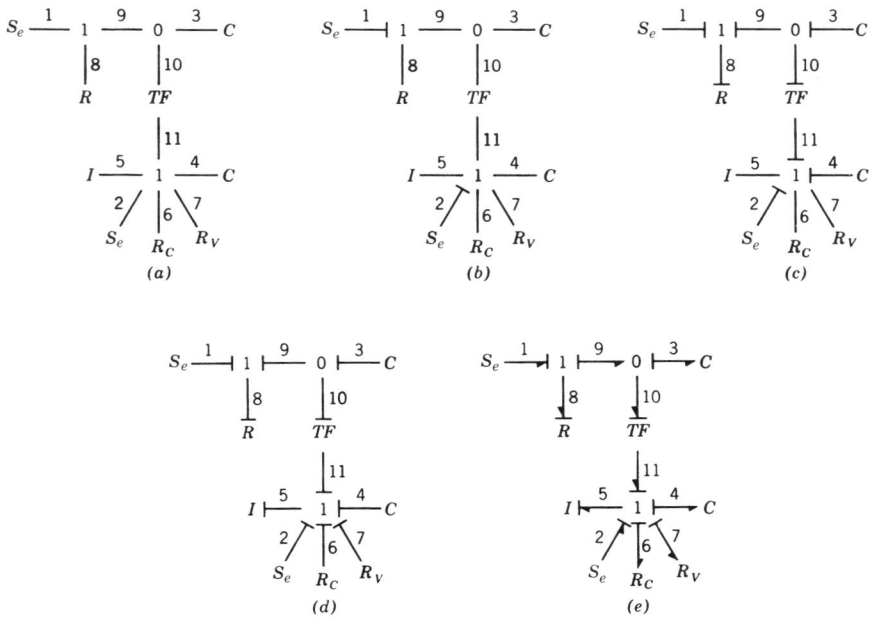

FIGURE 5.6. Augmentation of a bond graph. Example 3.

independent operations. Either one may be performed first. Typically, power directions will be first, but on occasion one may not bother to assign powers in studying aspects of system equation formulation.

5.3   BASIC FORMULATION AND REDUCTION

Once a fully augmented bond graph model is available, the equations for the system can be developed in a very orderly fashion. Frequently, when the system is small or uncomplicated in structure, state-space equations can be written down directly. However, as system size and complexity grow, the need for an organized procedure for equation generation becomes apparent.

Very general and powerful procedures are available for producing sets of system equations. In this section we shall concentrate on a basic pattern that is applicable in a large majority of cases encountered in engineering practice. There are three simple steps to be followed:

1. select input and energy state variables;
2. formulate the initial set of system equations; and
3. reduce the initial equations to state-space form.

Selection of *inputs* is straightforward. For each source element write on the graph the input variable to the system. These variables will appear in the final state-space equations if they have any effect on system behavior. The list of input variables will be called U.

As mentioned earlier, the state variables are the p-variables on I-elements and the q-variables on C-elements that are in integral causality. For many system models, we will discover that after assigning causality sequentially by the rules of the previous section, all energy storage elements are in integral causality and there are no unassigned bonds. This is the simplest occurrence with regard to equation derivation, and it is this case that is discussed in this section on basic formulation procedures. When this causal pattern emerges, then explicit first-order differential equations of the type indicated in Eqs. (5.10) for nonlinear systems or Eqs. (5.11a) for linear systems result, and equation derivation is straightforward.

The state variables can be placed in a vector called the state vector, X. The derivative of the state variables will always be either an effort or a flow variable, since $\dot{p} = e$ and $\dot{q} = f$. On the computational bond graph it is convenient to show $\dot{p}$'s as the efforts on the I-elements in integral causality and $\dot{q}$'s as the flows on C-elements in integral causality. In addition, it will prove useful to indicate the *co-energy variables* on the energy storage elements. These are the flow variables on the I-elements and the effort variables on the C-elements. We will discover that the co-energy variables will appear in the initial formulation and then will be eliminated during the reduction process.

As a first example of basic formulation, consider the augmented bond graph in Figure 5.7*a*, which is repeated from Figure 5.4*e*. The causality has been assigned,

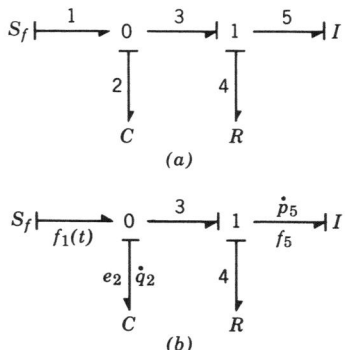

FIGURE 5.7. Equation formulation. Example 1.

and all energy storage elements are in integral causality, all bonds have a causal assignment, and there are no causal conflicts. The state variables are q_2 and p_5. The formulation always starts by writing the rate of change of a state variable equals an effort or a flow. We then use causality to track the efforts and flow through the bond graph. Causality will show us the substitutions to make to eliminate unwanted variables while retaining appropriate ones, as the procedure leads us to explicit first-order equations. It makes no difference which state variable we start with. We will later learn that computer programs exist that can derive the equations following the same rules we use. These programs number the bonds consecutively and derive equations starting with the state variable with the lowest bond number. Not to be outdone by a computer, we will do the same here. Thus, starting with q_2, we write,

$$\dot{q}_2 = f_2, \tag{5.12}$$

and follow the causality of the model to direct us through the bond graph and lead us directly to the state equations. From Figure 5.7 we see that f_2 is an *input to* the C-element and an *output from* the 0-junction. This causal output is caused by the inputs, f_1 and f_3, and according to the power convention,

$$\dot{q}_2 = f_1 - f_3. \tag{5.13}$$

The flow f_1 is the input flow from the flow source, and this variable is known and remains in the formulation. The flow, f_3, is not wanted in the formulation, so we must look further.

We just used f_3 as an input to the 0-junction, but it is also an output from the 1-junction. It is true that f_3 equals both f_4 and f_5, but it is *caused by* f_5 (look at the causality on the 1-junction). Thus, we write,

$$\dot{q}_2 = f_1 - f_5. \tag{5.14}$$

The flow, f_5, is a co-energy variable and directly related to the state variable, p_5, through the constitutive relationship for the element. If I_5 is a nonlinear element, then,

$$f_5 = \Phi_I^{-1}(p_5), \tag{5.15}$$

and, if I_5 is a linear element, as it is for this example, then

$$f_5 = \frac{p_5}{I_5}. \tag{5.16}$$

With this substitution we obtain the first state equation as

$$\dot{q}_2 = f_1 - \frac{p_5}{I_5}. \tag{5.17}$$

The second state equation starts with

$$\dot{p}_5 = e_5, \tag{5.18}$$

and we need to follow the causal path and track down e_5. From Figure 5.7 we see that e_5 is an *input to* the I-element and an *output from* the 1-junction. The 1-junction indicates that e_5 is *caused by* e_3 and e_4, and according to the power convention,

$$e_5 = e_3 - e_4, \tag{5.19}$$

thus,

$$\dot{p}_5 = e_3 - e_4. \tag{5.20}$$

Neither of these efforts is wanted in the final formulation, so we must continue to use causality to determine what substitutions to make. The effort e_3 is an output from the 0-junction and is *caused by* e_2. But e_2 is a co-energy variable and is directly related to the state variable, q_2, by the constitutive relationship for the C-element,

$$e_2 = \Phi_C^{-1}(q_2), \tag{5.21}$$

or, for the linear element used in this example,

$$e_2 = \frac{q_2}{C_2}. \tag{5.22}$$

This substitution will be made in (5.20), but first let's track down e_4. The effort, e_4, is an output from a resistance element where, in this causality,

$$e_4 = \Phi_R(f_4), \tag{5.23}$$

or, for the linear assumption being used for this example,

$$e_4 = R_4 f_4. \tag{5.24}$$

Eq. (5.24) is used in (5.20) to eliminate e_4, but we introduce, f_4. This flow is a *causal output* from the 1-junction where the causal flow input is the co-energy variable, f_5, which is directly related to the state variable p_5, as indicated in (5.16). Using the substitutions from (5.22), (5.24), and (5.16) in (5.20), we obtain the second state equation as

$$\dot{p}_5 = \frac{q_2}{C_2} - R_4 \frac{p_5}{I_5}. \tag{5.25}$$

Eqs. (5.17) and (5.25) are the two state equations for this system. Since this example is linear, the final step is to place these equations into the standard matrix form of (5.11),

$$\frac{d}{dt}\begin{bmatrix} q_2 \\ p_5 \end{bmatrix} = \begin{bmatrix} 0 & -\dfrac{1}{I_5} \\ \dfrac{1}{C_2} & -\dfrac{R_4}{I_5} \end{bmatrix}\begin{bmatrix} q_2 \\ p_5 \end{bmatrix} + \begin{bmatrix} 1 \\ 0 \end{bmatrix} f_1. \tag{5.26}$$

This is the perfect starting point for analysis if the system is linear. One need only follow the causality, and the substitutions needed to eliminate unwanted variables while retaining state and source variables automatically occur. The reader may not believe this yet, so we present a second example.

Figure 5.8 shows an augmented bond graph that originally came from an electrical circuit. When causality was assigned, all energy storage elements ended up in integral causality, and all bonds had a causal assignment. This is always an indication that formulation of equations is straightforward. The state variables for this example are the momentum variable on bond 2, p_2, and the displacement variable on bond 5, q_5. If the physical bond graph were nearby we would realize that the momentum variable is the flux linkage for the inductor of L_2 henrys and the displacement variable is the charge on the capacitor of C_5 farads. The derivatives of the state variables are indicated on the energy storage elements, as are the associated co-energy variables, f_2 and e_5.

The formulation starts by writing

$$\dot{p}_2 = e_2 \tag{5.27}$$

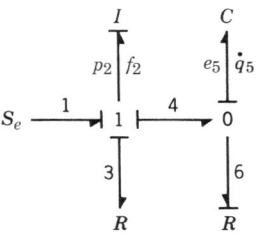

FIGURE 5.8. Equation formulation. Example 2.

and then pursuing e_2 by using the causality of the bond graph. The effort, e_2, is an output from the 1-junction and is caused by e_1, e_3, and e_4, according to the power convention; thus,

$$\dot{p}_2 = e_1 - e_3 - e_4. \tag{5.28}$$

The effort, e_1, is a source variable and remains in the final equations. The effort e_4 is an output from the 0-junction and is caused by the causal input, e_5. But e_5 is a co-energy variable directly related to a state variable by

$$e_5 = \frac{q_5}{C_5}. \tag{5.29}$$

Before substituting into (5.28), notice that e_3 is an output from a resistance with causality dictating

$$e_3 = R_3 f_3, \tag{5.30}$$

and f_3 is caused by the co-energy variable, f_2, and f_2 is related directly to a state variable by

$$f_2 = \frac{p_2}{I_2}. \tag{5.31}$$

Using (5.29), (5.30), and (5.31) in (5.28) yields the state equation

$$\dot{p}_2 = e_1 - R_3 \frac{p_2}{I_2} - \frac{q_5}{C_5}. \tag{5.32}$$

The second state equation starts with

$$\dot{q}_5 = f_5, \tag{5.33}$$

followed by using causality to track the path for f_5. The flow, f_5, is the input to C_5 and the output from the 0-junction. This output is caused by the causal inputs, f_4 and f_6, according to the power convention; thus,

$$\dot{q}_5 = f_4 - f_6. \tag{5.34}$$

The flow f_4 is the output from the 1-junction, caused by the co-energy variable, f_2, where

$$f_2 = \frac{p_2}{I_2}. \tag{5.35}$$

The flow f_6 is the causal output from a resistance, where causality indicates that

$$f_6 = \Phi_R^{-1}(e_6) = \frac{1}{R_6} e_6 \tag{5.36}$$

for this linear example. But e_6 is caused by the co-energy variable, e_5, which is directly related to a state variable according to (5.29). Using (5.35) and (5.36) in (5.34) yields the final state equation as

$$\dot{q}_5 = \frac{p_2}{I_2} - \frac{1}{R_6} \frac{q_5}{C_5}. \tag{5.37}$$

The final step for linear systems is to put state equations (5.32) and (5.37) into the standard matrix format,

$$\frac{d}{dt} \begin{bmatrix} p_2 \\ q_5 \end{bmatrix} = \begin{bmatrix} -\dfrac{R_3}{I_2} & -\dfrac{1}{C_5} \\ \dfrac{1}{I_2} & -\dfrac{1}{R_6 C_5} \end{bmatrix} \begin{bmatrix} p_2 \\ q_5 \end{bmatrix} + \begin{bmatrix} 1 \\ 0 \end{bmatrix} e_1. \tag{5.38}$$

Perhaps the reader is starting to see the power of using causality and the equation formulation procedure presented in this section. Any approach to equation derivation always requires elimination of some variables in favor of others in order to end up with a computable set of equations. Bond graphs and causality simply make the substitutions obvious so that appropriate variables are eliminated while others are retained, without having to guess the appropriate substitutions to make. One more example should be sufficient to demonstrate the consistency of our approach to the formulation of state equations.

Figure 5.9 is the augmented bond graph originally shown in Figure 5.6 for development of causality. The bond graph represents a physical system with hydraulic and mechanical parts. To keep track of what parts are hydraulic and what parts are mechanical, we really need to have the physical bond graph. For equation formulation, we need only have the augmented bond graph since equation formulation follows the same procedure regardless of the physical system. This example has a transformer that couples the hydraulic and mechanical sides of the model. The modulus of the

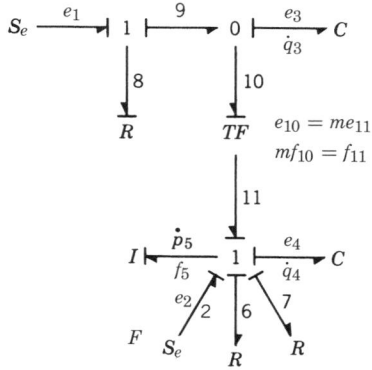

FIGURE 5.9. Equation formulation. Example 3.

transformer is physically an area, A. In Figure 5.9, the modulus, m, is defined by noting the constitutive relationships for the transformer. We will derive the equations in terms of the generic definition of the transformer shown in the figure, but before we can actually use the equations to predict system response, we will have to determine how the modulus, m, relates to the area, A. (A physical variable bond graph would tell us whether m as defined is actually A or $1/A$.)

The causal assignment once again yielded all energy storage elements in integral causality and no bonds were left unassigned. The state variables are q_3, the displacement variable on C_3; q_4, the displacement variable on C_4; and p_5, the momentum variable on I_5. Also noted on the bond graph are the associated co-energy variables. Equation derivation starts with

$$\dot{q}_3 = f_3, \tag{5.39}$$

but

$$f_3 = f_9 - f_{10},$$

and

$$f_9 = f_8 = \frac{1}{R_8} e_8 = \frac{1}{R_8}(e_1 - e_9) = \frac{1}{R_8}(e_1 - e_3) = \frac{1}{R_8}\left(e_1 - \frac{q_3}{C_3}\right). \tag{5.40}$$

The flow f_{10} is the output flow from the transformer, caused by the input flow, f_{11}, according to,

$$f_{10} = \frac{1}{m} f_{11} = \frac{1}{m} f_5 = \frac{1}{m} \frac{p_5}{I_5}. \tag{5.41}$$

Substituting (5.40) and (5.41) into (5.39) yields the first state equation as

$$\dot{q}_3 = \frac{1}{R_8}\left(e_1 - \frac{q_3}{C_3}\right) - \frac{1}{m} \frac{p_5}{I_5}. \tag{5.42}$$

The second state equation is particularly simple,

$$\dot{q}_4 = f_4 = f_5 = \frac{p_5}{I_5}. \tag{5.43}$$

And the final state equation is

$$\dot{p}_5 = e_5 = e_2 - e_4 - e_6 - e_7 + e_{11}. \tag{5.44}$$

The effort e_2 is a source variable and remains in the final formulation. The effort e_4 is directly related to a state variable,

$$e_4 = \frac{q_4}{C_4}. \tag{5.45}$$

The efforts e_6 and e_7 are outputs from resistances, where

$$e_6 = R_6 f_6 = R_6 f_5 = R_6 \frac{p_5}{I_5}$$

$$e_7 = R_7 f_7 = R_7 f_5 = R_7 \frac{p_5}{I_5}, \tag{5.46}$$

and e_{11} is the output effort from the transformer,

$$e_{11} = \frac{1}{m} e_{10} = \frac{1}{m} e_3 = \frac{1}{m} \frac{q_3}{C_3}. \tag{5.47}$$

Substituting (5.45), (5.46), and (5.47) into (5.44) yields the final state equation as

$$\dot{p}_5 = e_2 - \frac{q_4}{C_4} - (R_6 + R_7) \frac{p_5}{I_5} + \frac{1}{m} \frac{q_3}{C_3}. \tag{5.48}$$

Equations (5.42), (5.43), and (5.48) are a complete set of linear state equations and can be cast into the standard matrix format,

$$\frac{d}{dt} \begin{bmatrix} q_3 \\ q_4 \\ p_5 \end{bmatrix} = \begin{bmatrix} -\dfrac{1}{R_8 C_3} & 0 & -\dfrac{1}{m I_5} \\ 0 & 0 & \dfrac{1}{I_5} \\ \dfrac{1}{m C_3} & -\dfrac{1}{C_4} & -\dfrac{(R_6 + R_7)}{I_5} \end{bmatrix} \begin{bmatrix} q_3 \\ q_4 \\ p_5 \end{bmatrix} + \begin{bmatrix} \dfrac{1}{R_8} & 0 \\ 0 & 0 \\ 0 & 1 \end{bmatrix} \begin{bmatrix} e_1 \\ e_2 \end{bmatrix}. \tag{5.49}$$

We hope that by now the reader is convinced that by choosing the state variables to be the p's on I's and the q's on C's in integral causality, and by following the input/output causal paths, the state equations are straightforwardly derived. If the system is linear, then the resulting equations can be put into a standard matrix format that is the perfect starting point for linear analysis, which is covered in the next chapter. If the system is nonlinear, equations are still derived easily by using bond graphs and causality, but there is no nice matrix representation from which to launch an analysis. Instead, simulation is required, which is covered later in this chapter and at an advanced level in Chapter 13.

Unfortunately, models do not always lend themselves to straightforward formulation. Sometimes modeling assumptions lead to perfectly good models that have mathematical formulation difficulties. Such problems are the topic of the next two sections.

5.4   EXTENDED FORMULATION METHODS—ALGEBRAIC LOOPS

Constructing models of physical systems always requires modeling assumptions. Any physical system can be infinitely complicated. It is we, as engineers, who must construct models that are simple enough to work with and yet retain sufficient dynamics to answer the questions being asked. Along the way we decide whether the mass of some component is important, whether a part is sufficiently stiff to be modeled as rigid, whether components can be represented as linear over some operating range, and so forth. The result of these assumptions can sometimes lead to mathematical formulation problems that have nothing to do with the quality of the model, yet they must be addressed in order to make use of the model. One of these formulation problems is caused by "algebraic loops." This topic is discussed at an advanced level in Section 13.2 of Chapter 13. Here, the basic problem is introduced.

Consider the physical system in Figure 5.10a. It consists of a mass, m, acted upon by the input force, $F(t)$. The mass is attached to some dampers and a spring, as shown, and the spring has an input velocity at its right end, $v_i(t)$. The mass velocity is labeled v_m, and the attachment point of the dampers and spring is indicated as v'. The dampers and spring are assumed positive in tension, and the attachment point of these elements is assumed massless. This model is perfectly reasonable, and yet it has a serious formulation problem that is vividly exposed with the use of bond graph causality.

Figure 5.10b shows the bond graph for this system constructed using the procedure for mechanical translation. Some liberty has been taken in that physical labels are on the bond graph along with numbering the bonds. Normally, we would have a physical bond graph and a separate computational bond graph. Also in Figure 5.10b, causality has been assigned using the sequential assignment procedure presented in Section 5.2.

From the sequential assignment of causality, the inertial element, I_2, has set the flow input to the 1-junction, constraining bond 3 to have flow output from the 1-junction. Since f_3 is a causal flow input to a 0-junction, there is no constraint to assign further causality. The compliance element, C_8, is in integral causality, and e_8 sets the effort on the 0-junction. This constrains the effort, e_7, to be an output from the 0-junction and an input to the 1-junction with bonds 5 and 6 unassigned. At this point there are no further causal assignments that we are constrained to make. Both of the energy storage elements are in integral causality, but there remain causally unassigned bonds.

Whenever causality has been assigned according to the procedure of Section 5.2 and there are causally unassigned bonds remaining, the system has formulation problems and state equations of the form (5.10) or (5.11) cannot be derived in the straightforward manner of the previous section. When this situation arises, it means that there is an "algebraic loop" where some effort or flow variables are algebraically related to themselves as well as to other source variables and state variables. We will demonstrate what happens if this condition is ignored.

The procedure for handling algebraic loops is to first make an arbitrary causal assignment on one of the causally unassigned bonds and then propagate the causal

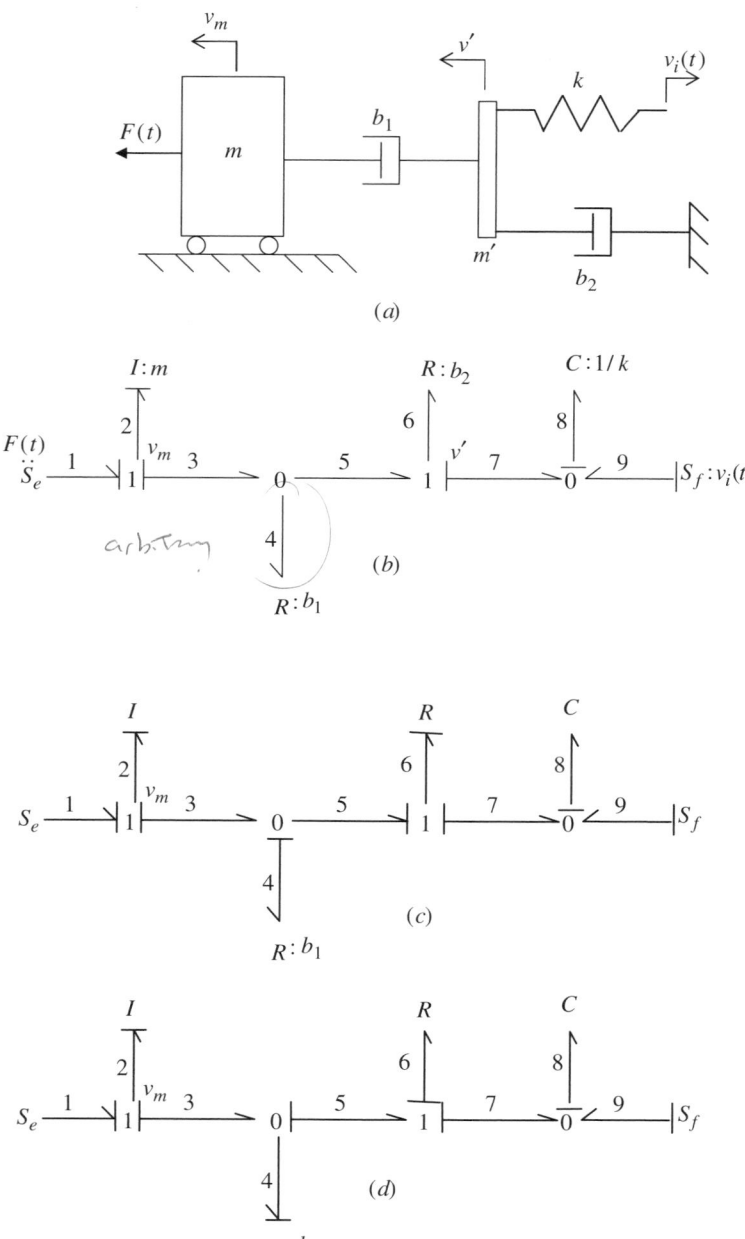

FIGURE 5.10. Extended formulation—algebraic loop. Example 1.

information as far as possible. If the causality completes after one arbitrary assignment, then there is one algebraic loop. If a second arbitrary assignment is required, then a second algebraic loop is present, and so on. The more loops involved, the more difficult the final formulation becomes. In Figure 5.10c an arbitrary assignment has been made on R_4, which remained causally unassigned after the sources and energy storage elements had been assigned according to the procedure. The effort, e_4, has been assigned as an output from the R-element. This requires e_5 to be an output from the 0-junction, and now with e_5 and e_7 as inputs to the 1-junction, we are constrained to have R_6 in the causality indicated. Thus, making one arbitrary assignment has yielded a causally complete bond graph, and there is only one algebraic loop for this example.

From the appearance of Figure 5.10c, one would not realize that there is a problem. The bond graph causality is complete, all energy storage elements are in integral causality, and there are no causal conflicts. Let's see what happens if we attempt to derive the state equations using the procedure of the previous section. The state variables are p_2 and q_8, and the co-energy variables are f_2 and e_8. Then,

$$\dot{p}_2 = e_1 - e_3 = e_1 - e_4 = e_1 - R_4 f_4 = e_1 - R_4 \left(\frac{p_2}{I_2} - f_6 \right). \tag{5.50}$$

So far, all looks fine. Then,

$$f_6 = \frac{1}{R_6} e_6 = \frac{1}{R_6} (e_5 - e_7) = \frac{1}{R_6} \left(e_4 - \frac{q_8}{C_8} \right), \tag{5.51}$$

but,

$$e_4 = R_4 f_4 = R_4 (f_3 - f_5) = R_4 \left(\frac{p_2}{I_2} - f_6 \right) = \dots \tag{5.52}$$

The flow, f_6, comes from Eq. (5.51), which leads back to e_4 again and we find ourselves in an endless loop of substitutions. This loop is not the fault of bond graphs. It is a direct result of the modeling assumptions that went into the schematic of Figure 5.10a. However, bond graphs and causality let you know there is a problem before any equations are derived.

The procedure for handling algebraic loops is to first recognize the problem from the resulting incomplete causality. This is followed by making an arbitrary causal assignment and extending the information according to the sequential assignment rules. This continues until all bonds are assigned. While it makes no difference which bond is selected for the arbitrary assignment, it is recommended from step 5 of the *sequential causality assignment procedure* that when a resistance element is involved in the algebraic loop (i.e., it has a causally unassigned bond), the resistance element be chosen for the arbitrary assignment. This was done in Figure 5.10c where the element R_4 was chosen for the arbitrary assignment.

As mentioned earlier, an algebraic loop is an indicator that some effort or flow variable involved in the loop is algebraically related to itself and other source and state variables. The next step in the procedure is to derive this algebraic relationship. When a resistance element is involved in the loop, it is customary to choose the variable that was arbitrarily selected to be the output from the R-element and the input to the rest of the system. For the example here, since R_4 was assigned arbitrarily, we select e_4 as the variable to be related to itself. Thus, we write e_4 on the left-hand side of an equation and then use causality in the normal way to derive the desired relationship, thus,

$$e_4 = R_4 f_4 = R_4 (f_3 - f_5) = R_4 \left(\frac{p_2}{I_2} - f_6 \right) = R_4 \left(\frac{p_2}{I_2} - \frac{1}{R_6} e_6 \right)$$

$$= R_4 \left(\frac{p_2}{I_2} - \frac{1}{R_6} \left(e_4 - \frac{q_8}{C_8} \right) \right). \tag{5.53}$$

Eq. (5.53) is an algebraic relationship that relates e_4 to source variables, state variables, and to itself.

The next step is to solve the resulting relationship for the algebraic variable, e_4 in this case,

$$e_4 = \frac{1}{1 + \frac{R_4}{R_6}} R_4 \frac{p_2}{I_2} + \frac{1}{1 + \frac{R_4}{R_6}} \frac{R_4}{R_6} \frac{q_8}{C_8}. \tag{5.54}$$

The final step is to derive the state equations following the standard procedure, only stop the formulation when the algebraic variable enters the formulation. For our example,

$$\dot{p}_2 = e_1 - e_3 = e_1 - e_4,$$

$$\dot{q}_8 = f_9 + f_7 = f_9 + f_6 = f_9 + \frac{1}{R_6} \left(e_4 - \frac{q_8}{C_8} \right). \tag{5.55}$$

We now substitute (5.54) into (5.55) and end up with the state equations in standard explicit form,

$$\dot{p}_2 = e_1 - \frac{1}{1 + \frac{R_4}{R_6}} R_4 \frac{p_2}{I_2} - \frac{1}{1 + \frac{R_4}{R_6}} \frac{R_4}{R_6} \frac{q_8}{C_8},$$

$$\dot{q}_8 = f_9 - \frac{1}{R_6} \frac{q_8}{C_8} + \frac{1}{R_6} \left[\frac{1}{1 + \frac{R_4}{R_6}} R_4 \frac{p_2}{I_2} + \frac{1}{1 + \frac{R_4}{R_6}} \frac{R_4}{R_6} \frac{q_8}{C_8} \right]. \tag{5.56}$$

To complete this discussion of algebraic loops, Figure 5.10d shows the same example system with a different arbitrary causal assignment. This time f_4 is assigned as the output from R_4, and the causality completes as shown. It makes no difference what bond is assigned arbitrarily. This *system* has an algebraic problem consisting of one algebraic loop. Choosing f_4 as the variable to relate to itself, we write,

$$f_4 = \frac{1}{R_4}e_4 = \frac{1}{R_4}e_5 = \frac{1}{R_4}(e_6 + e_7) = \frac{1}{R_4}\left(R_6 f_6 + \frac{q_8}{C_8}\right)$$

$$= \frac{1}{R_4}\left(R_6(f_3 - f_4) + \frac{q_8}{C_8}\right) = \frac{1}{R_4}\left(R_6\left(\frac{p_2}{I_2} - f_4\right) + \frac{q_8}{C_8}\right). \quad (5.57)$$

Solving for f_4 yields

$$f_4 = \frac{1}{1 + \frac{R_6}{R_4}}\frac{R_6}{R_4}\frac{p_2}{I_2} + \frac{1}{1 + \frac{R_6}{R_4}}\frac{1}{R_4}\frac{q_8}{C_8}. \quad (5.58)$$

Equation derivation would follow the standard procedure except that we would stop when f_4 enters the formulation,

$$\dot{p}_2 = e_1 - R_4 f_4,$$

$$\dot{q}_8 = f_9 + f_5 = f_9 + \frac{p_2}{I_2} - f_4. \quad (5.59)$$

The formulation is completed when (5.58) is substituted into (5.59). It is left to the reader to show that the resulting state equations are identical to (5.56).

So far in this section we have determined that some modeling assumptions can lead to algebraic formulation problems, and that bond graphs identify such a problem by the presence of incomplete causality after using the sequential causal assignment procedure. A procedure has been presented through an example of a linear system that shows how to solve the algebraic problem and end up with state equations in standard form. For linear systems, this procedure will work for multiple loops involving several algebraic variables, although solving by hand becomes formidable rather quickly. For nonlinear systems, it may not be possible to invert some of the algebraic relationships. For example, friction is a very common resistance effect in mechanical systems and is represented by an R-element in a bond graph. For the most common friction representation (see Section 8.1 of Chapter 8), the velocity (or flow) must be an input and the friction force (or effort) must be an output from the element. If the friction element was involved in an algebraic loop that required its inversion (calculation of velocity from knowledge of the force), it could not be done.

We probably should look into the modeling assumptions that created the problem in the first place. In Figure 5.10a, the connection of the spring and damper was *assumed* to be massless. The connection is not actually massless, it was just assumed to be and the assumption was probably justifiable during the modeling process. Consider Figure 5.11a, where a small mass, m', has been associated with the connection point. Figure 5.11b shows the bond graph for the system with causality assigned. There is no arbitrary assignment required and there is no algebraic loop. Equation derivation is straightforward for this system. If this had been the first model constructed, we would never have known that a potential problem exists. In Chapter 13, the concept of algebraic loops is dealt with more formally, and the idea of revisiting modeling assumptions when formulation issues arise is developed in much more de-

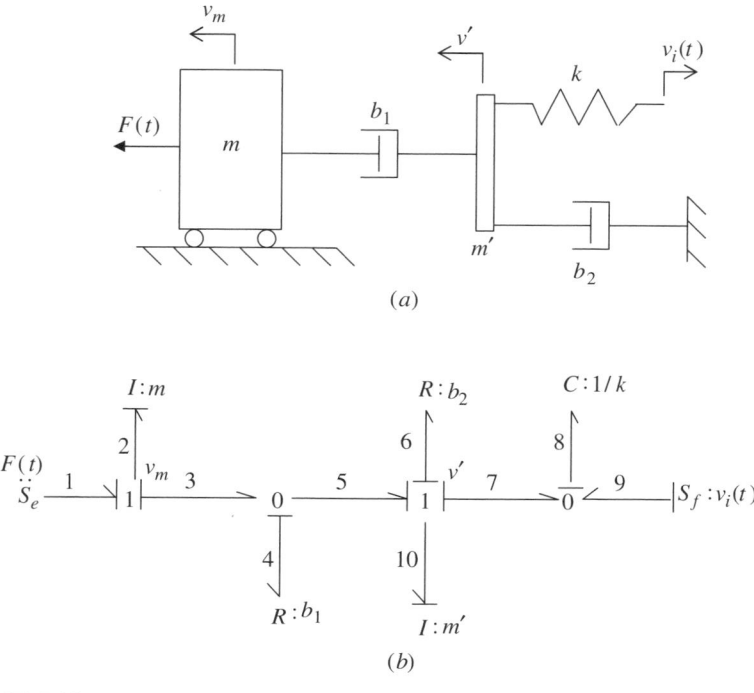

FIGURE 5.11. Eliminating an algebraic loop by revisiting modeling assumptions in Example 1.

tail. Reference [1] describes a formal vector-field method for generating state space equations.

5.4.1   Extended Formulation Methods—Derivative Causality

In this section we study systems where, after sequential causal assignment, one or more energy storage elements end up in derivative causality. As with algebraic loops, the occurrence of derivative causality is due to the modeling assumptions that went into the model construction. Such energy storage elements are not dynamically independent, and they do not contribute a state variable to the system. The state variables remain the p's on I's and the q's on C's in *integral* causality. These derivative elements do store energy, and this energy is accounted for in the bond graph. By being dynamically dependent, their associated energy variable, p or q, is algebraically related to the state variables. So, knowledge of the state variables allows algebraic determination of the energy variables on the derivative elements, and thus algebraic calculation of their contribution to the system energy. The algebraic relationships, relating the energy variables associated with the derivative causal elements to the energy variables that are the state variables, must be derived before it is possible to have a state representation in standard form. This is demonstrated through example.

Consider the physical system shown in Figure 5.12a. It consists of a permanent magnet d-c motor driving a rotational load, J, through a flexible shaft of torsional stiffness, k_τ. There is also some rotational damping, b_τ. On the electrical side of the motor, the winding resistance, R_w, and the winding inductance, L, are important dynamic effects. It has been decided, for this application, that the rotational inertia of the motor is negligible and it is not included. Figure 5.12b shows a bond graph

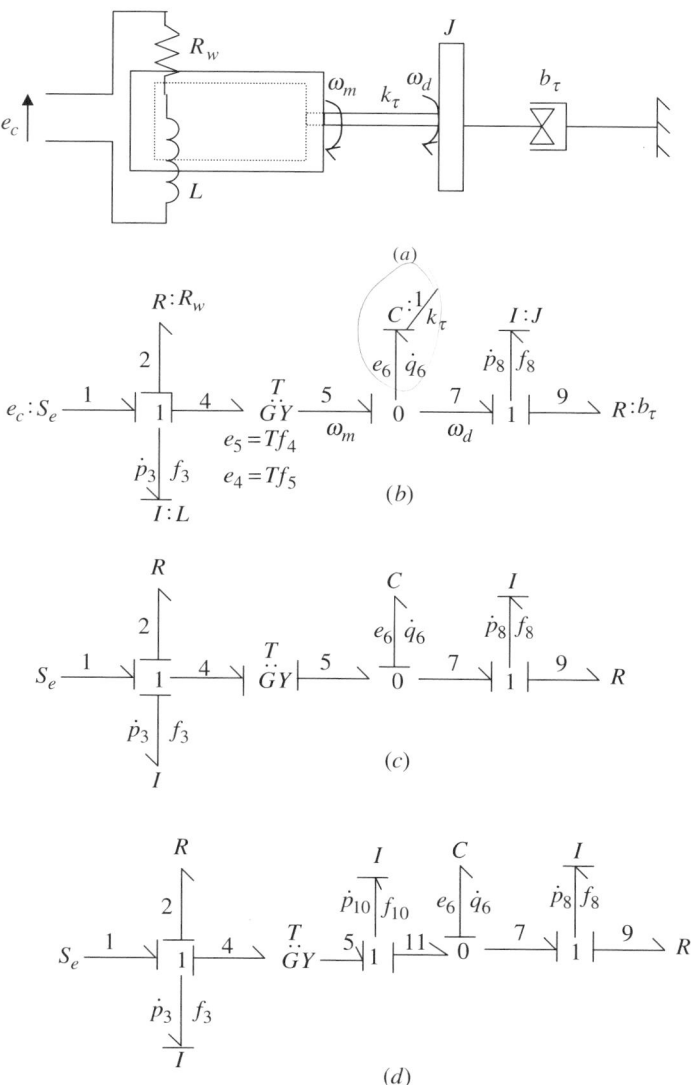

FIGURE 5.12. Derivative causality in a bond graph. Example 1.

for this system where the motor has been characterized as a gyrator with coupling constant, T N $\cdot$ m/A (see Section 4.5.4 to remind you how to model d-c motors).

In the figure, the bonds have been numbered and causality has been assigned using the sequential assignment procedure. The inductance, I_3, has been put into integral causality. As a result, the compliance, C_6, was constrained to be in derivative causality. The state variables for this system are p_3 and p_8, and only two state equations need to be derived. The energy variable, q_6, is not a state variable, but is algebraically dependent on the state variables.

We cannot simply ignore the presence of the derivative element. If we try to derive the equations without its consideration, then we would write,

$$\dot{p}_3 = e_1 - e_2 - e_4 = e_1 - R_2 \frac{p_3}{I_3} - T f_5 = e_1 - R_2 \frac{p_3}{I_3} - T \left(f_6 + \frac{p_8}{I_8} \right). \quad (5.60)$$

The appearance of f_6 is where the derivative element enters the formulation, and f_6 must be dealt with before final formulation can be done.

We next recognize that

$$f_6 = \dot{q}_6,$$

but there is no state equation for q_6, and more work must be done.

The next step is to derive the algebraic relationship relating q_6 to the state variables and source variables. When we do this, we start by writing the constitutive law "backwards," that is,

$$q_6 = C_6 e_6. \quad (5.61)$$

(For an I-element in derivative causality, we would write $p_i = I_i f_i$. Also, for nonlinear elements, we would start with $q_6 = \Phi_C(e_6)$ for compliance elements in derivative causality, and $p_i = \Phi_I(f_i)$ for inertial elements in derivative causality.)

Starting with Eq. (5.61), we follow causality in the standard fashion and find

$$e_6 = e_5 = T f_4 = T f_3 = T \frac{p_3}{I_3}. \quad (5.62)$$

Thus,

$$q_6 = \frac{C_6 T}{I_3} p_3. \quad (5.63)$$

This is the algebraic relationship that allows the energy variable, q_6, to be algebraically determined from the state variable, p_3. Taking the derivative of (5.63) yields

$$\dot{q}_6 = \frac{C_6 T}{I_3} \dot{p}_3. \quad (5.64)$$

We can now substitute (5.64) into (5.60), with the result

$$\dot{p}_3 = e_1 - R_2 \frac{p_3}{I_3} - T \left(\frac{C_6 T}{I_3} \dot{p}_3 + \frac{p_8}{I_8} \right). \tag{5.65}$$

When dealing with derivative causality, we end up with derivatives of state variables on both sides of the equation. Eq. (5.65) is particularly simple to reduce to the standard form because q_6 depends on only one state variable. We can simply bring $\dot{p}_3$ from the right side to the left, with the result

$$\dot{p}_3 = \frac{e_1}{1 + \frac{C_6 T^2}{I_3}} - \frac{R_2 \frac{p_3}{I_3}}{1 + \frac{C_6 T^2}{I_3}} - \frac{T \frac{p_8}{I_8}}{1 + \frac{C_6 T^2}{I_3}}. \tag{5.66}$$

The second and last state equation starts with

$$\dot{p}_8 = e_7 - R_9 \frac{p_8}{I_8}.$$

But e_7 is not *caused* by the compliance element, but instead,

$$e_7 = e_5 = T f_4 = T \frac{p_3}{I_3}, \tag{5.67}$$

with the final result

$$\dot{p}_8 = T \frac{p_3}{I_3} - R_9 \frac{p_8}{I_8}. \tag{5.68}$$

Eqs. (5.66) and (5.68) can be put into the standard matrix format for linear systems,

$$\frac{d}{dt} \begin{bmatrix} p_3 \\ p_8 \end{bmatrix} = \begin{bmatrix} -\dfrac{R_2/I_3}{Q} & -\dfrac{T/I_8}{Q} \\ \dfrac{T}{I_3} & -\dfrac{R_9}{I_8} \end{bmatrix} \begin{bmatrix} p_3 \\ p_8 \end{bmatrix} + \begin{bmatrix} 1/Q \\ 0 \end{bmatrix} e_1 \tag{5.69}$$

where

$$Q = 1 + \frac{C_6 T^2}{I_3}.$$

When this example was presented, the I-element on bond 3 in Figure 5.12b was put into integral causality and this resulted in the C_6 element being in derivative causality. In Figure 5.12c the bond graph is repeated, but this time the C_6 element is put into integral causality, with the result that the I_3 element ends up in derivative causality. We have some flexibility in choosing which elements are in integral causality, but, once chosen, some elements are constrained to be in derivative causality. The derivative causality resulted from the original modeling assumptions, and without changing these assumptions, the model will have derivative causality regardless of causal assignment choices.

When deriving equations for Figure 5.12c, the effort $e_3 = \dot{p}_3$ will enter the formulation. Thus, we will need to write the constitutive law "backwards" for the I_3 element and determine how the energy variable, p_3, is algebraically related to the state variables. Thus,

$$p_3 = I_3 f_3 = I_3 f_4 = I_3 \frac{1}{T} e_5 = I_3 \frac{1}{T} \frac{q_6}{C_6} \tag{5.70}$$

and

$$e_3 = \dot{p}_3 = \frac{I_3}{TC_6} \dot{q}_6. \tag{5.71}$$

Now we can derive the equations using standard procedures, starting with

$$\dot{q}_6 = f_5 - f_7 = \frac{1}{T} e_4 - \frac{p_8}{I_8} = \frac{1}{T}(e_1 - e_2 - e_3) - \frac{p_8}{I_8}. \tag{5.72}$$

We see that e_3 entered the formulation, and we will eliminate it using (5.71); we should note that $e_2 = R_2 f_2$ and f_2 is *not caused* by I_3, but must be tracked through the causal information. Continuing,

$$\dot{q}_6 = \frac{1}{T}\left(e_1 - R_2 \frac{1}{T}\frac{q_6}{C_6} - \frac{I_3}{TC_6}\dot{q}_6\right) - \frac{p_8}{I_8}. \tag{5.73}$$

Again we see that a derivative of a state variable is on both sides of the equation. And again, this equation is very simple to get into the standard form by simply bringing $\dot{q}_6$ from the right-hand side to the left, resulting in

$$\dot{q}_6 = \frac{\frac{1}{T}}{1 + \frac{I_3}{T^2 C_6}} e_1 - \frac{\frac{R_2}{T^2}}{1 + \frac{I_3}{T^2 C_6}} \frac{q_6}{C_6} - \frac{1}{1 + \frac{I_3}{T^2 C_6}} \frac{p_8}{I_8}. \tag{5.74}$$

The second state equation is

$$\dot{p}_8 = \frac{q_6}{C_6} - R_9 \frac{p_8}{I_8}. \tag{5.75}$$

Comparing (5.66) and (5.68) with (5.74) and (5.75), we see that the two state representations of the same system have different appearances. This is not surprising since we are using different state variables for the two representations. Rest assured that if both sets of equations were solved, the predicted system dynamics would be identical from both state representations.

It has been emphasized that modeling assumptions are the cause of algebraic problems in formulations. For the example system we have just shown it was assumed that the rotational inertia of the electric motor was negligible and was not included in the model. Figure 5.12d shows the bond graph for the system when the rotational inertia of the motor, J_m, is included. The angular velocity of the motor, ω_m, is on bond 5 in the bond graphs in parts b and c. In part d, a 1-junction has been inserted, exposing ω_m, and the I-element representing the motor inertia has been attached (it is bond 10 in Figure 5.12d). Causality has been assigned, and there is no derivative causality. The new modeling assumption has eliminated any formulation problems. If these modeling assumptions had been made originally, we would have never known that a potential formulation problem exists. It should be noted that the model in Figure 5.12d has four state variables and four state equations, whereas the model in part b has only two state variables and two state equations. Having included the motor inertia has increased the number of state variables by two, one being the new I-element in integral causality and the other coming from the derivative causal element that is now in integral causality. If we were going to do pencil and paper analysis, then two additional equations is a significant further complication. If our intention is computer solution, then the change is insignificant. The concept of revisiting modeling assumptions to alleviate formulation problems is discussed in detail in Chapter 13.

One more example involving derivative causality may serve to sufficiently illustrate the pattern of formulation and reduction. Figure 5.13a shows masses m_1 and m_2 on the left and right ends of a massless lever. A force, $F(t)$, acts on m_1, and a spring, k, is attached to m_2. Positive force and velocity directions are indicated. Part b of the figure shows a bond graph for this system, with bonds numbered and the transformer modulus defined by the relationships shown for the transformer. In part c causality has been assigned, putting I_2 into integral causality, with the result that I_1 is in derivative causality. The state variables are p_2 and q_3, and the energy variable p_1 is not a state variable but will enter the equation formulation. Thus, we write the constitutive law backwards for this element and derive the algebraic relationship relating p_1 to the other state variables. Thus,

$$p_1 = I_1 f_1 = I_1 \frac{a}{b} f_6 = I_1 \frac{a}{b} \frac{p_2}{I_2} \tag{5.76}$$

and

$$e_1 = \dot{p}_1 = \frac{I_1}{I_2} \frac{a}{b} \dot{p}_2. \tag{5.77}$$

Equation derivation starts with

$$\dot{p}_2 = e_6 - e_3 = \frac{a}{b} e_5 - \frac{q_3}{C_3} = \frac{a}{b}(e_4 - \dot{p}_1) - \frac{q_3}{C_3}. \tag{5.78}$$

We substitute from (5.77), with the result

$$\dot{p}_2 = \frac{a}{b}\left(e_4 - \frac{I_1}{I_2}\frac{a}{b}\dot{p}_2\right) - \frac{q_3}{C_3}, \tag{5.79}$$

where the state variable derivative appears on both sides of the equation. The final result is

$$\dot{p}_2 = \frac{\frac{a}{b}}{1 + \left(\frac{a}{b}\right)^2 \frac{I_1}{I_2}} e_4 - \frac{1}{1 + \left(\frac{a}{b}\right)^2 \frac{I_1}{I_2}} \frac{q_3}{C_3}, \qquad (5.80)$$

and the second state equation is

$$\dot{q}_3 = \frac{p_2}{I_2}. \qquad (5.81)$$

The reader must appreciate that the two examples shown for derivative causality are quite simple in that (1) only one energy storage element was in derivative causality and (2) the derivative element, energy variable was algebraically related to only one state variable. For more complex systems the algebra required to reduce a set of equations to a standard equation format can be formidable at best, and, for nonlinear systems, impossible at worst.

It is always possible to revisit the modeling assumptions that resulted in the formulation problem and change these assumptions to yield a model that is virtually the same as the one originally intended but without the formulation problems. There are some computational penalties that accompany the addition of very small inertias or very stiff compliances in a model. But for some systems these penalties are a very small price to pay for eliminating formulation problems. As mentioned, this topic is discussed in detail in Chapter 13.

For the system in Figure 5.13, it is interesting to see how derivative causality could have been avoided. Figure 5.14a shows a physical system where the lever has been divided into two massless pieces, with the right end of one and the left end of

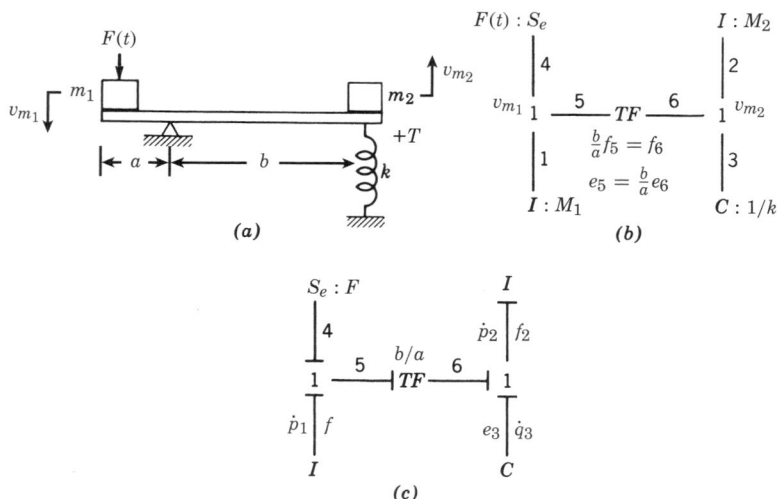

FIGURE 5.13. Derivative causality in a bond graph. Example 2.

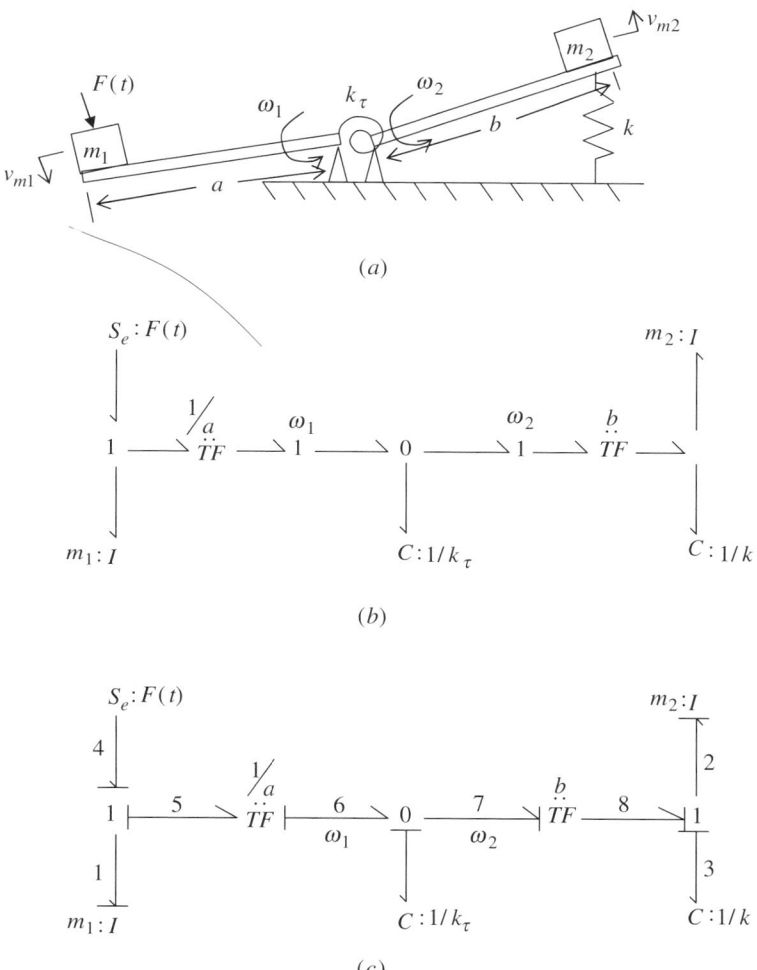

FIGURE 5.14. Avoiding derivative causality by changing the modeling assumptions for Example 2.

the other pivoted at the same point. The two pieces are coupled by a torsional spring of stiffness, k_τ, which represents the fact that the lever is not actually rigid, but has some flexibility. The system is otherwise identical to the system of Figure 5.13a. The angular velocity of the left side lever is ω_1 and that of the right side lever is ω_2. The torsional spring "sees" the difference between these angular velocities, and the bond graph for the system is shown in Figure 5.14b.

Figure 5.14c shows the bond graph reduced and with causality assigned. There is no derivative causality, and equation formulation would be straightforward. If the torsional spring is made stiff enough, then there will be very little difference between ω_1 and ω_2, and a close approximation to the original system will result.

5.5   OUTPUT VARIABLE FORMULATION

We have seen that it is very convenient to derive state equations for all types of systems using the energy variables, p's and q's, as the state variables. More often than not, however, the system outputs that we want to observe and plot are not the p and q variables. Can you imagine interpreting the motion of a mechanical system from a plot of a mass momentum, or the behavior of a circuit from a plot of the flux linkage in a coil? You might want to look at these responses, but you also might want to look at mass velocities and displacements, circuit voltages and currents, hydraulic pressures, and so on. It turns out that any effort or flow on any bond in a bond graph can be related to the state variables. Thus, we solve systems of equations in terms of the state variables, and then derive output equations that relate the desired outputs to the state variables. If the system solution is done computationally, then it is a simple matter to append output equations at the end of the program so that desired outputs can be viewed.

Consider the system in Figure 5.9, which was used earlier to demonstrate equation formulation. If we desire the flow on bond 1 as an output, then we simply follow the causal path and relate f_1 to the state and source variables. Thus,

$$f_1 = f_8 = \frac{1}{R_8} e_8 = \frac{1}{R_8}(e_1 - e_9) = \frac{1}{R_8}\left(e_1 - \frac{q_3}{C_3}\right). \tag{5.82}$$

By following causality we have related the desired output to the known source variable, e_1, and the state variable, q_3. If we wanted the effort on bond 5, then

$$e_5 = e_2 - e_6 - e_7 - e_4 + e_{11} = e_2 - (R_6 + R_7)\frac{p_5}{I_5} - \frac{q_4}{C_4} + \frac{1}{m}\frac{q_3}{C_3}. \tag{5.83}$$

These two output equations were derived assuming that the system was linear, which it was when this system was used for equation derivation. When the system is linear, the output equations can be put into a matrix format,

$$Y = CX + DU, \tag{5.84}$$

where Y is a vector of desired outputs,

$$Y = \begin{bmatrix} y_1 \\ y_2 \\ \bullet \\ \bullet \\ y_k \end{bmatrix}, \tag{5.85}$$

U is the vector of inputs,

$$U = \begin{bmatrix} u_1 \\ u_2 \\ \bullet \\ \bullet \\ u_r \end{bmatrix} \tag{5.86}$$

and C is a $k \times n$ matrix of coefficients and D is a $k \times r$ matrix of coefficients where the coefficients are made up of system parameters. For the output Eqs. (5.82) and (5.83), the matrix form would be

$$\begin{bmatrix} f_1 \\ e_5 \end{bmatrix} = \begin{bmatrix} -\dfrac{1}{R_8 C_3} & 0 & 0 \\ \dfrac{1}{mC_3} & -\dfrac{1}{C_4} & -\dfrac{(R_6 + R_7)}{I_5} \end{bmatrix} \begin{bmatrix} q_3 \\ q_4 \\ p_5 \end{bmatrix} + \begin{bmatrix} \dfrac{1}{R_8} & 0 \\ 0 & 1 \end{bmatrix} \begin{bmatrix} e_1 \\ e_2 \end{bmatrix}, \tag{5.87}$$

If the system is nonlinear, the desired outputs can still be expressed in terms of the state variables and source variables, but the nice matrix notation of Eq. (5.84) is not available. For example, in Figure 5.9, if the resistance element R_8 and the compliance element C_3 are nonlinear such that f_8 is a function of e_8,

$$f_8 = \Phi_R^{-1}(e_8), \tag{5.88}$$

and e_3 is a function of q_3,

$$e_3 = \Phi_C^{-1}(q_3), \tag{5.89}$$

then the output, f_1, would be derived as,

$$f_1 = f_8 = \Phi_R^{-1}(e_8) = \Phi_R^{-1}(e_1 - e_3) = \Phi_R^{-1}\left(e_1 - \Phi_C^{-1}(q_3)\right). \tag{5.90}$$

Causality still shows us the path from the output of interest to the known variables, but the relationships are not linear. From a computational point of view it makes absolutely no difference.

If the output we desire is the power on some bond, then the output equation will be nonlinear whether or not the system is linear. Power is always composed of the product of an effort and a flow. In Figure 5.9, if we desire the power on bond 1, P_1, as an output variable, then

$$P_1 = e_1 f_1 = e_1 \frac{1}{R_8}\left(e_1 - \frac{q_3}{C_3}\right), \tag{5.91}$$

where f_1 is the linear version from Eq. (5.82). Since we are most often dealing with computational solutions of system equations, it really does not matter as to the complexity or linearity of the output equations, other then getting them typed and formatted correctly for the program being used.

5.6   AUTOMATED AND NONLINEAR SIMULATION

A primary reason for constructing a physical engineering model of a dynamic system is to predict the response of the system prior to actually building it. A good model can go a long way toward avoiding mistakes in a first prototype of a complex system. Sometimes it is justified to assume that a system behaves linearly, thus allowing the model to be constructed from linear inertance, compliance, and resistance elements and restricted to small motions with no nonlinear kinematics. When this is justified, linear analysis can be performed, sometimes analytically but most often computationally. Some of the more important linear concepts for engineers are developed in the next chapter.

More often than not, at some point in the system development process, a more realistic nonlinear model must be developed in order to really understand the system, or to test a control strategy that was developed using linear thinking, or to see what happens when displacements hit hard boundaries or power amplifiers saturate, or the like. For nonlinear systems there is really only one choice for predicting the system response, and that is computer simulation. This topic is discussed thoroughly in Chapter 13. Here we discuss how to organize system equations for computer simulation and demonstrate with an example. We also discuss the concept of automated simulation where bond graph models are inputted graphically to the computer and causality and state equations are automatically generated, even for nonlinear systems.

5.6.1   Nonlinear Simulation

Figure 5.15a is the quarter car model repeated from Figure 4.10, where it was used to demonstrate bond graph construction procedures. This time, the suspension spring

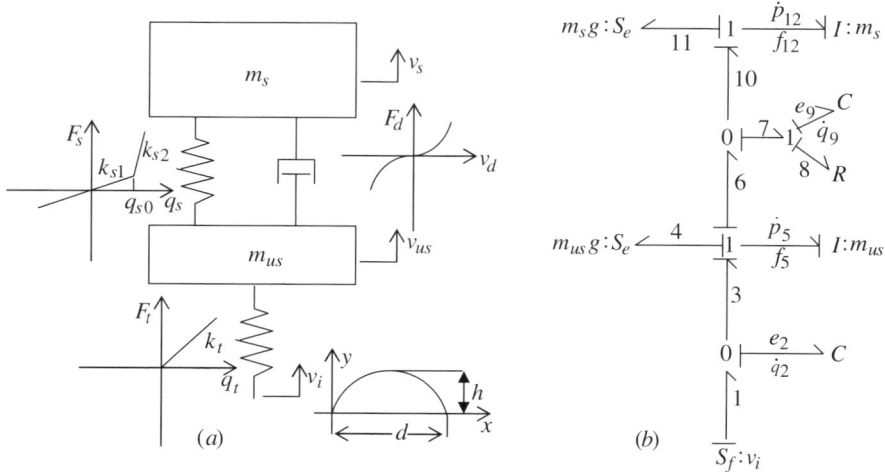

FIGURE 5.15. Quarter car simulation example with nonlinear elements.

and damper and tire spring are all nonlinear elements. A sketch of the constitutive relationships for each element is in the schematic. Also shown next to $v_i(t)$ is a roadway bump the vehicle is going to hit. This bump is shown as a height versus distance and has the appearance of a speed bump such as found in parking lots. We will characterize the bump as $\frac{1}{2}$ of a sine wave, and interpret the bump as an equivalent velocity as required for bond graph flow inputs. For the spatially distributed input,

$$y = h \sin \pi \frac{x}{d} \quad \text{for } 0 < \frac{x}{d} \leq 1$$

and

$$y = 0 \quad \text{for } \frac{x}{d} > 1. \tag{5.92}$$

If we imagine the vehicle traveling at constant horizontal velocity, U, toward the right, then $x = Ut$ and (5.92) becomes

$$y = h \sin \pi \frac{U}{d} t \quad \text{for } 0 < \frac{U}{d} t \leq 1. \tag{5.93}$$

A reasonable characterization of the vertical input velocity is to multiply the forward velocity times the slope of the roadway, dy/dx, thus,

$$v_i(t) = \frac{h}{d} \pi U \cos \pi \frac{U}{d} t \quad \text{for } 0 < \frac{U}{d} t \leq 1,$$

$$v_i(t) = 0 \text{ otherwise.} \tag{5.94}$$

We will characterize the suspension spring as a linear spring with slope, k_{s1}, up until a compressive displacement of q_{s0}, at which point the slope changes to a stiffer spring, k_{s2}. Thus,

$$F_s = k_{s1} q_s \quad \text{for } q_s \leq q_{s0},$$

$$F_s = k_{s1} q_{s0} + k_{s2}(q_s - q_{s0}) \quad \text{for } q_s > q_{s0}. \tag{5.95}$$

The damper will be characterized as

$$F_d = B v_d^3, \tag{5.96}$$

and, finally, the tire spring will be characterized as linear in compression while generating no force in extension, that is, no force when the tire is off the ground. Thus,

$$F_t = k_t q_t \quad \text{for } q_t \geq 0,$$

$$F_t = 0 \quad \text{for } q_t < 0. \tag{5.97}$$

The bond graph for the example system is shown in Figure 5.15b. Causality has been assigned and bonds have been numbered. Notice that there are no parameters listed with the suspension spring and damper or with the tire spring. These are nonlinear elements and this is indicated on the bond graph by the absence of an identifying parameter. The state variables are q_2, p_5, q_9, and p_{12}.

Eqs. (5.10) are the functional form of standard nonlinear state space equations. These are called explicit first-order differential equations and can be expressed in shorthand form as

$$\dot{x} = f(x, U), \tag{5.98}$$

where x is the vector of state variables,

$$x = \begin{bmatrix} x_1 \\ x_2 \\ \bullet \\ \bullet \\ x_n \end{bmatrix},$$

U is the vector of inputs,

$$U = \begin{bmatrix} u_1 \\ u_2 \\ \bullet \\ \bullet \\ u_r \end{bmatrix},$$

and f is a vector of functions of states and inputs,

$$f(\bullet) = \begin{bmatrix} f_1(x, U) \\ f_2(x, U) \\ \bullet \\ \bullet \\ f_n(x, U) \end{bmatrix}.$$

It turns out that equations that can be derived in this condensed form are particularly easy to solve numerically (see Chapter 13). However, it is enough to know that equations could be put into this form, but it is not necessary to do so in order to carry out numerical simulation. In fact, it is rare to derive equations in the explicit form of (5.98) when the intention is numerical simulation. This is demonstrated with our example.

Equation formulation for the bond graph of Figure 5.15b begins with

$$\dot{q}_2 = f_1 - \frac{p_5}{I_5}, \tag{5.99}$$

where f_1 is the input velocity, $v_i(t)$, from (5.94). This state equation turns out to be linear even though the system is nonlinear. The second equation is

$$\dot{p}_5 = -e_4 + e_2 - e_7. \tag{5.100}$$

The effort $e_4 = m_{us}g$, the effort, e_2 comes from the nonlinear tire spring characterized by (5.97), and

$$e_7 = e_8 + e_9. \tag{5.101}$$

The effort e_8 is the damper force from (5.96), and the effort e_9 is from the nonlinear suspension spring from Eq. (5.95).

The next state equation is

$$\dot{q}_9 = \frac{p_5}{I_5} - \frac{p_{12}}{I_{12}}, \tag{5.102}$$

which is another linear state equation.

The final state equation is

$$\dot{p}_{12} = -e_{11} + e_7, \tag{5.103}$$

where $e_{11} = m_s g$ and e_7 has already been discussed.

The equation formulation is now complete. Bond graph causality told us that explicit equations would result, but there is no need to reduce the equations to this final form. It is perfectly all right to leave the equations in uncondensed form as long as all the variables are accounted for. Eqs. (5.94) through (5.97) with (5.99) through (5.103) are a complete, computable set of equations ready to simulate the quarter car system. We need only specify initial conditions for the state variables, and the simulation can proceed.

In addition, many commercial equation solvers have a "sort" feature. This means that the user need not pay any attention to the order in which the equations appear in the program. The program determines the order in which the equations must be used and properly "sorts" the equations for the user. For example, Eq. (5.100) requires e_7, but e_7 appears below (5.100) in (5.101). If this ordering were done in a nonsorting program, an error would occur, but with a sorting program it presents no problem. Some simulation results for this example are presented below, but first we discuss automated simulation.

5.6.2 Automated Simulation

It is hoped that the reader recalls how straightforward it is to assign causality to a bond graph and thus know, before deriving any equations, if any formulation problems exist. With causality assigned, state variables are automatically known. Computers can follow the same rules as humans and can assign sequential causality to a bond graph, choose state variables, and derive the first-order state equations. There are several commercial programs that start from a graphical description of a bond graph and automatically deliver state equations ready for simulation. The user need only input the parameters, define any nonlinearities, specify the inputs, and specify the initial conditions. The program will then simulate the system and plot any desired outputs, including any state variable, any effort or flow variable, or any output defined by the user as a combination of state variables and input variables. These programs are so straightforward to use that a user unfamiliar with bond graph modeling could carry out a simulation if a fully augmented bond graph were provided. In Chapter 1 some references are given that describe some of the more well known commercial bond graph processors.

For our example system, a bond graph processor program would follow the causality of Figure 5.15b and deliver the following uncondensed equations:

$$e_1 = e_2 \qquad\qquad e_7 = e_8 + e_9$$

$$e_3 = e_2 \qquad\qquad \dot{q}_9 = f_9$$

$$f_2 = f_1 - f_3 \qquad\qquad e_8 = R_8 f_8??$$

$$\dot{q}_2 = f_2 \qquad\qquad f_{10} = f_{12}$$

$$f_3 = f_5 \qquad\qquad f_{11} = f_{12}$$

$$f_4 = f_5 \qquad\qquad e_{12} = e_{10} - e_{11}$$

$$f_6 = f_5 \qquad\qquad \dot{p}_{12} = e_{12}$$

$$e_5 = e_3 - e_4 - e_6 \qquad\qquad e_2 = \frac{q_2}{C_2}??$$

$$\dot{p}_5 = e_5 \qquad\qquad f_5 = \frac{p_5}{I_5}??$$

$$e_6 = e_7 \qquad\qquad e_9 = \frac{q_9}{C_9}??$$

$$e_{10} = e_7 \qquad\qquad f_{12} = \frac{p_{12}}{I_{12}}??$$

$$f_7 = f_6 - f_{10}$$

$$f_8 = f_7$$

$$f_9 = f_7 \qquad\qquad\qquad\qquad\qquad\qquad (5.104)$$

If the reader follows the causality of Figure 5.15b it will be discovered that every input/output causal implication is reflected in Eqs. (5.104). Of particular interest are the equations with question marks next to them. When an automated program derives equations, it has no idea which components are linear or nonlinear nor how to characterize any specific nonlinearities. The program does recognize which elements might be nonlinear, and it highlights these with a question mark or some other indicator. For this example, the resistance element, R_8, is nonlinear and the user would have to modify the linear equation in (5.104) with

$$e_8 = B f_8^3 \qquad\qquad\qquad\qquad\qquad (5.105)$$

from (5.96) above. Neither of the inertial elements is nonlinear, so the user would only need to provide the inertia parameter for each. Both compliance elements are

TABLE 5.1. Parameters for the Example Simulation

$m_s = 320$ Kg

$\dfrac{m_{us}}{m_s} = \dfrac{1}{6}$

f_s, suspension frequency, $= 1.0$ Hz

$\omega_s = 2\pi f_s$, rad/s

For the suspension spring, $k_{s1} = m_s \omega_s^2$, N/m

$k_{s2} = 10k_{s1}$, N/m

For the tire spring, $k_t = 10k_{s1}$, N/m

Initial conditions for springs, $q_{s-ini} = \dfrac{m_s g}{k_{s1}} \quad q_{us-ini} = \dfrac{(m_s + m_{us})g}{k_t}$

Breakpoint for suspension spring, $q_{s0} = 1.3q_{s-ini}$

B, damper parameter, $= 1500.0$N/(m/s)^3

For the input bump, $h = 0.25$ m

$d = 1.0$ m

Forward velocity, $U = 2$ and 30 mph $= 0.9$ and 13.5 m/s

---

nonlinear, so the equation for e_2 in (5.104) would have to be modified using (5.97), and the equation for e_8 would have to be modified according to (5.95). Finally, the input, f_1, would be defined according to (5.94), and e_4 and e_{11} must be defined as the respective weights of the unsprung and sprung vehicle masses. Along with initial conditions, the simulation is ready to be run as long as a sorting equation solver is used.

For the example system, the parameters used in the simulation are shown in Table 5.1.

The system was simulated using a commercial equation solver for two different forward speeds, U, listed in Table 5.1. Figure 5.16 shows the velocity input from (5.94) for the lower vehicle forward speed of 2 mph. Figure 5.17 shows the suspension spring and damper forces along with the tire spring force. The vehicle starts from static conditions where the masses are not moving vertically and the springs are compressed to their initial values. As the bump is traversed, the springs become more compressed and the damper force becomes nonzero as a relative velocity is established across the suspension. The sharp change in slope of the suspension spring force at about 1.2 s is due to the spring displacement becoming more compressed than the breakpoint, q_{s0}, and thus the spring becomes much stiffer. The tire displacement never becomes negative in this simulation, and thus the tire force follows a linear law as indicated in (5.97).

When the vehicle forward speed is increased to 30 mph, the force results are as shown in Figure 5.18. This time the tire force becomes zero just after the vehicle strikes the bump. The tire spring first compresses and then uncompresses until the tire deflection becomes negative, the wheel is off the ground, and the tire force becomes

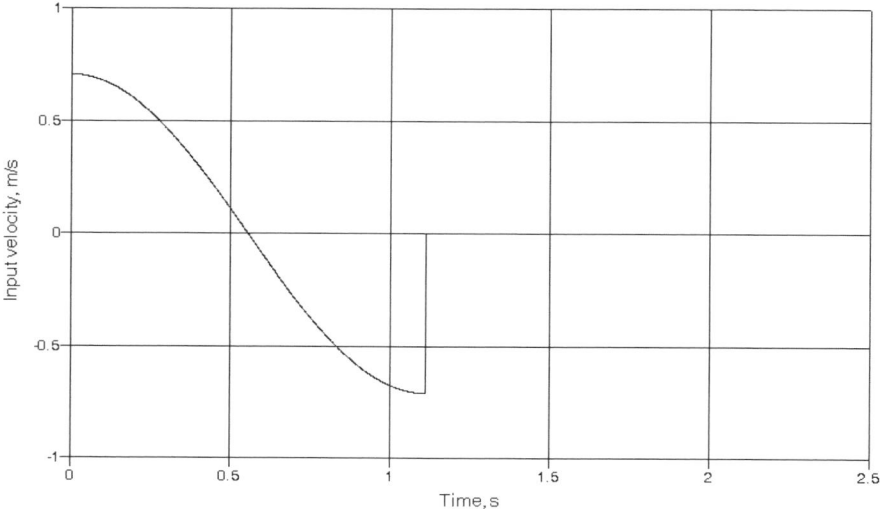

FIGURE 5.16. Input velocity for the simulation example for a vehicle speed of 2 mph.

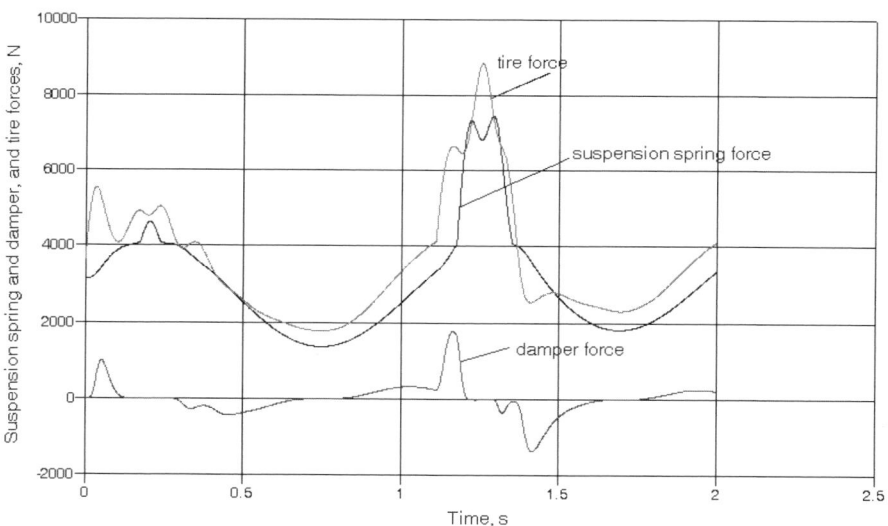

FIGURE 5.17. Suspension spring and damper forces and tire forces for a vehicle speed of 2 mph.

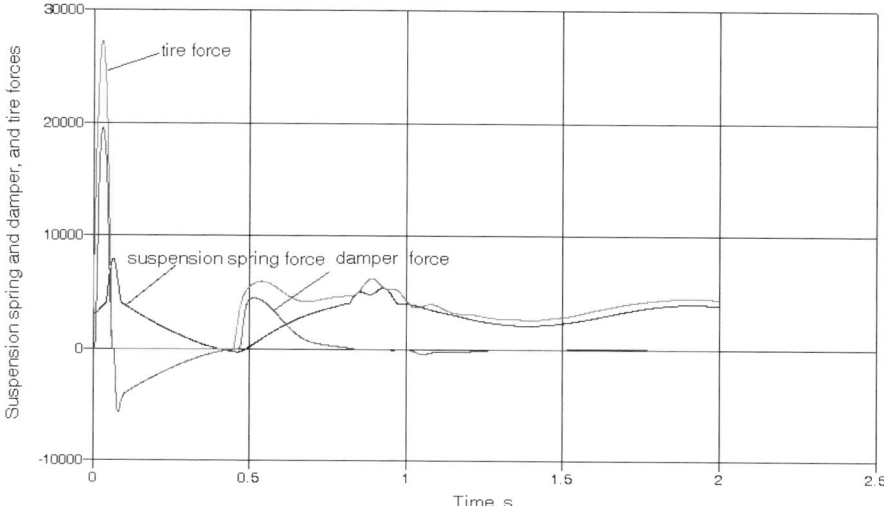

FIGURE 5.18. Suspension spring and damper forces and tire forces for a vehicle speed of 30 mph.

zero. When the wheel returns to the ground, the tire spring becomes compressed once again. It is interesting to plot the damper force versus the relative velocity across the suspension as shown in Figure 5.19. The cubic constitutive law is very obvious in this view. Figure 5.20 shows the suspension spring force versus the suspension deflection,

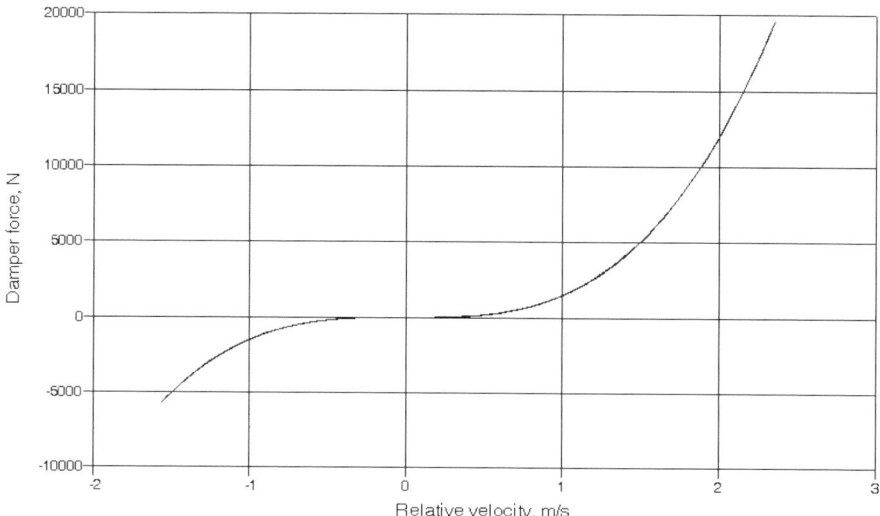

FIGURE 5.19. Damper force versus relative velocity across the suspension.

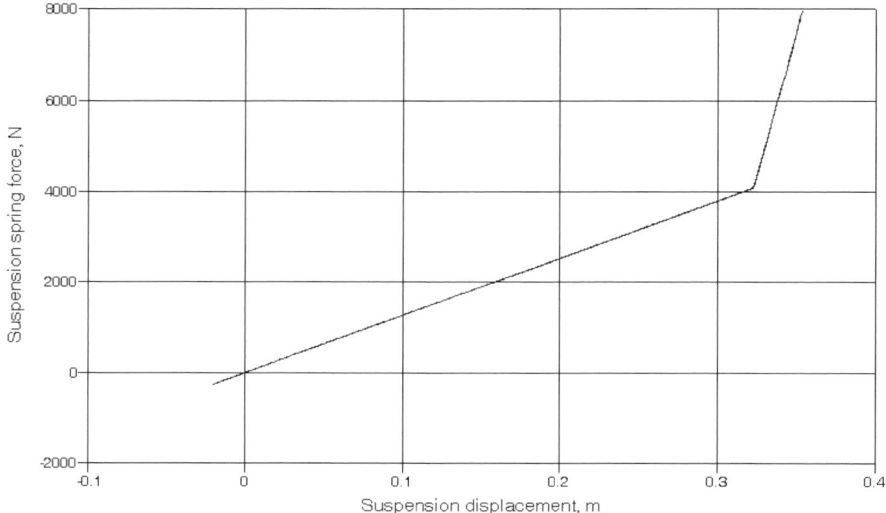

FIGURE 5.20. Suspension spring force versus suspension deflection.

and again the nonlinear constitutive behavior is very apparent. Finally, Figure 5.21 shows the tire force plotted against the tire deflection, and once again the nonlinear behavior is exposed.

It is hoped that the reader has gained some appreciation for the virtues of bond graph modeling for obtaining simulation results for nonlinear systems. Chapter 13

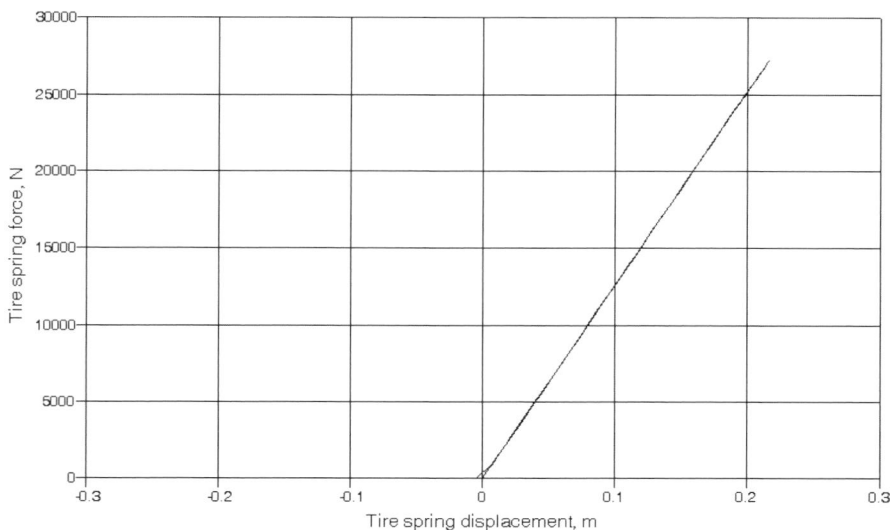

FIGURE 5.21. Tire spring force versus tire deflection.

covers this topic more thoroughly. In the previous chapters, we felt it necessary to introduce the modeling concepts using linear systems, since there are a lot of new concepts to learn, including system construction rules, power convention, causality, and equation derivation. While linear systems play a very large part in understanding the basic behavior of systems, the fact is that most systems are nonlinear, and, at some point in the development of an engineering system, the nonlinear behavior will have to be included. And the nonlinear system will probably be simulated rather than solved analytically.

In the following chapter, some of the more important linear analysis tools are presented; however, the later chapters are devoted to constructing realistic overall system models that include nonlinear physics when appropriate.

REFERENCE

[1] R. C. Rosenberg, "State Space Formulation for Bond Graph Models of Multiport Systems," *Trans. ASME J. Dyn. Syst. Meas. Control*, **93**, Ser. G, No. 1, 35–40 (Mar. 1971).

PROBLEMS

5-1 For each of the following bond graphs, assign causality, predict the number of state variables, and write a set of state equations. All elements may be assumed to be linear with constant coefficients.

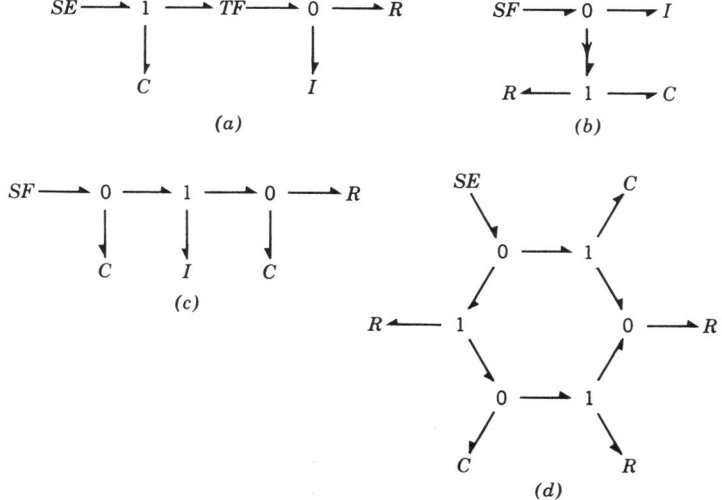

5-2   For each of the problems below, you are asked to find an equivalent expression for the system given. Try it first for constant-coefficient elements and then for nonlinear characteristics using the first case as a guide.

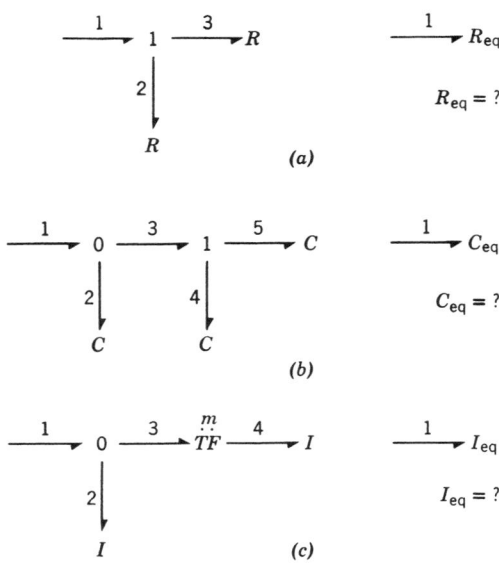

(a)

(b)

(c)

5-3   For the problems listed below, in each case make a bond graph model of the circuit, schematic, or network diagram. Then augment the graph and write state equations. Interpret the equations physically. You may assume the elements are linear.

(a)   Problem 4-1(e), electrical circuit

(b)   Problem 4-3(h), mechanical translation

(c)   Problem 4-5(a), mechanical rotation

(d)   Problem 4-6(a) and (b), hydraulic

(e)   Problem 4-5(c), mechanical transduction

(f)   Problem 4-5(e), mechanical transduction

5-4   For the electrical circuit shown below, verify the bond graph, write state equations, and develop an output equation for the voltage across the load resistor (R_L) in terms of state variables and input variables. Assume all elements have constant coefficients.

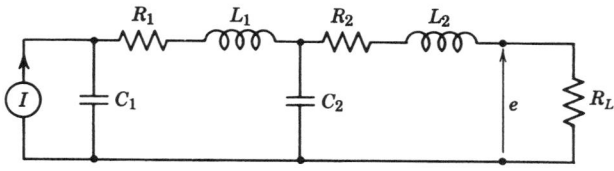

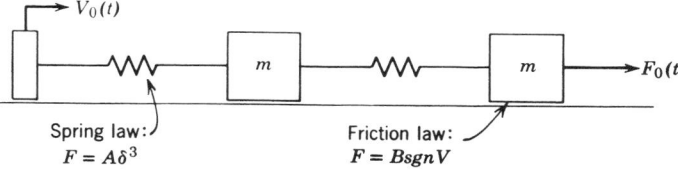

5-5   The mechanical system shown below has two nonlinear springs with constitutive laws, as shown. Note that δ is a deflection. The friction relations are also nonlinear, with a signum (sign) characteristic as shown. Make a bond graph and write state equations.

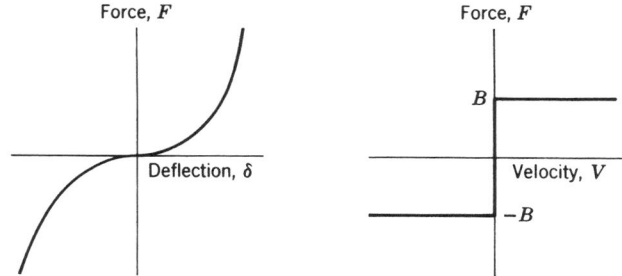

5-6   Make a bond graph model of the pulley system shown in part b of Problem 4-4. Augment the graph and predict the number of state variables. Write a suitable set of state equations. Modify the graph to include pulley inertias and pin-joint friction. Write state equations for the resulting graph.

5-7   A ball is suspended at the end of a spring in a container of fluid, as shown below. The extension of the spring measured from free length is x_s, and the

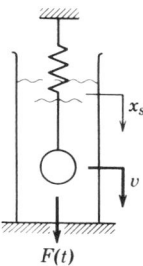

$$F(t)$$

velocity of the mass is v, measured downward. The spring is nonlinear, characterized by the relation

$$F_S = \phi_S(x_S).$$

The damping effect of the fluid is proportional to the square of the velocity, corrected for sign:

$$F_D = b|v|v.$$

(a) Formulate state equations for the system in terms of x_S and p (mass momentum).

(b) Transform the equations from x_S, p variables to x_S, v variables.

5-8 Consider the simple model of a vehicle shown below, where the suspended mass is M; the main suspension stiffness is K; the damper coefficient is B; the tire mass is m; the tire spring stiffness is k; and the velocity input from the roadway is $V(t)$.

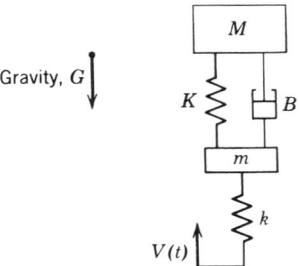

(a) Make a bond graph model, augment the graph, and write state equations.

(b) The tire mass is assumed to be negligible: let $m \rightarrow 0$, but retain all other values. Make a bond graph and derive state equations.

5-9 A commercial type of air spring isolator is shown. The mass is supported by
air pressure contained by means of a virtually frictionless bellows arrangement (not shown). Damping is provided by adjusting the area of the orifice
between the two chambers. The orifice resistance law is generally nonlinear,
as is the capacitance relation, since the pressure–volume relation for the gas
is approximately $PV^n = \text{const.}$

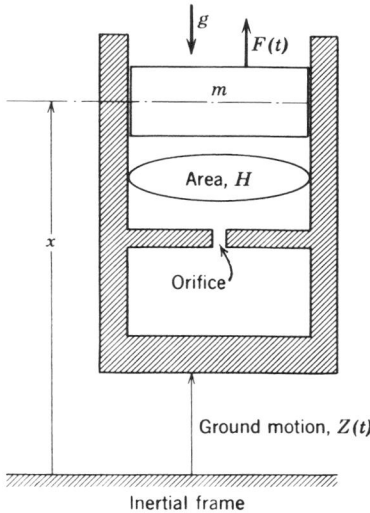

(a) Make a bond graph for the system that allows both ground motion $Z(t)$
and force $F(t)$ as inputs.

(b) Write state equations, using general functions for resistance and capacitance relations.

(c) Using the bond graph, show an analogous all-mechanical system.

5-10 For the water storage system of Problem 4-7, including the inertia effects
introduced in part b, augment the graph, predict the number of state variables,
and write state equations. (If you encounter formulation difficulties, indicate
clearly the steps to be taken to achieve state equations in standard form.)

5-11 On the diagram, physical-system variables and parameters are identified,
where

$$\theta_0 = \text{output position angle;}$$

$$v_{in} = \text{input voltage;}$$

$$v_a = \text{output voltage of linear amplifier;}$$

$$i_a = \text{motor armature current;}$$

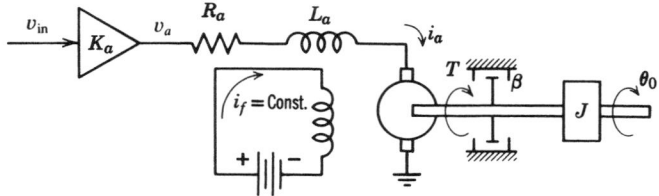

i_f = motor field current, assumed constant;

K_a = gain of linear amplifier, assumed to have no significant time constants;

R_a = resistance of armature winding;

L_a = inductance of armature winding;

J = inertial load;

b = viscous-damping constant;

K_T = torque constant of motor;

K = back-emf constant of motor.

The differential equations that govern the dynamics of the system are

$$J\ddot{\theta}_0 + \beta\dot{\theta}_0 = K_T i_a, \qquad L_a\dot{i}_a + R_a i_a = V_a - K_v\dot{\theta}_0.$$

(a)   Construct a bond graph for the system.

(b)   Write state-space equations, and verify that they are equivalent to those listed above.

(c)   Compare the two methods for analyzing this system. For example, is the system third or second order? Are K_T and K_v related in any way?

5-12   Consider the seismometer sketched below:

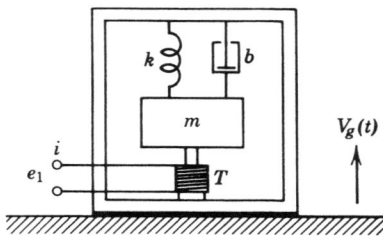

The input is ground motion $V_g(t)$; an electrical transducer using a perma-nent magnet moving in a coil reacts to the relative motion between the case and the seismic mass m.

(a) Construct a bond graph model of the device, leaving the electrical port as a free bond.

(b) Assume the device is connected to a voltage amplifier, so that $i = 0$. Find an expression relating $V_g(t)$ and e_1.

(c) Sometimes it is preferable to use current rather than voltage as output, due to noise considerations. Suppose a current amplifier is used, so that $e_1 = 0$. Relate output current i to input signal $V_g(t)$ for the case when coil resistance and inductance are neglected.

(d) Reconsider the problem of (c) with coil resistance included.

5-13 The three systems shown below each possess a feature that makes the writ-ing of state-space equations interesting. In each case assign causality sequen-tially and predict any difficulties. Where possible, write state-space equations assuming linear element characteristics.

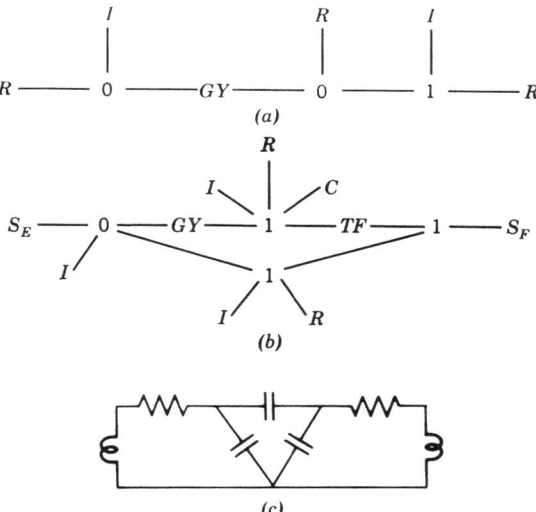

5-14 Consider the mechanical system shown below.

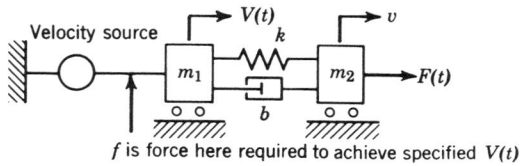

f is force here required to achieve specified $V(t)$

The mass, m, plays a somewhat unusual role in the system.

(a) Show that a state space for the system can be found that does not even depend on m_1.

(b) Show that most system output variables depend statically on the state variables and the inputs, F and V, but because of m_1 this is *not* true of f (defined above).

5-15 The resistive circuit shown below offers an opportunity to exploit the use of causality and helping (or auxiliary) variables.

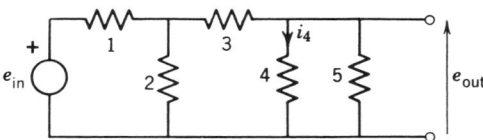

(a) Make a bond graph model of the circuit, and assign causality.

(b) Find e_{out} in terms of e_{in}.

(c) Find i_4 in terms of e_{in}.

(d) If a load resistance R_L is put on at the e_{out} port, modify the graph and your solution to part (b) above.

5-16 The d-c motor shown has winding resistance R_w, rotary inertia J_m, and output shaft compliance k_T. The pinion gear of radius R drives a rack of mass m with attached spring k and damper b. The bond graph for this system is shown below, where T is the motor constant such that the motor torque τ is related to the current i by $\tau = Ti$. Derive a state-space representation of this system and put into matrix form, $\dot{x} = Ax + be$:

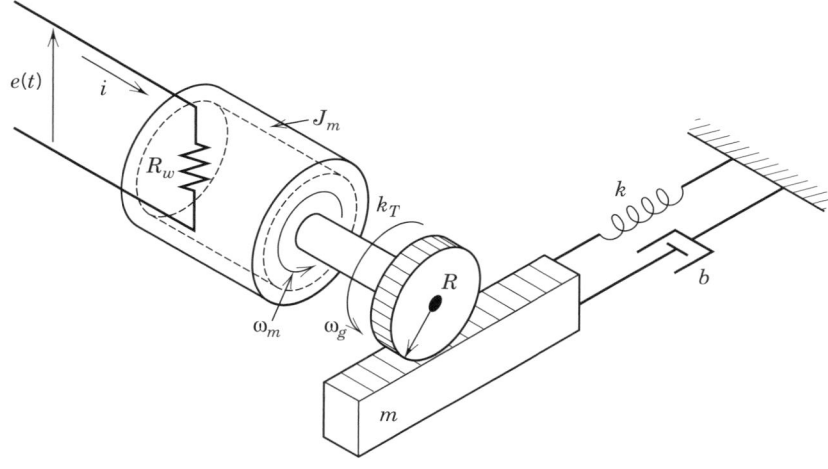

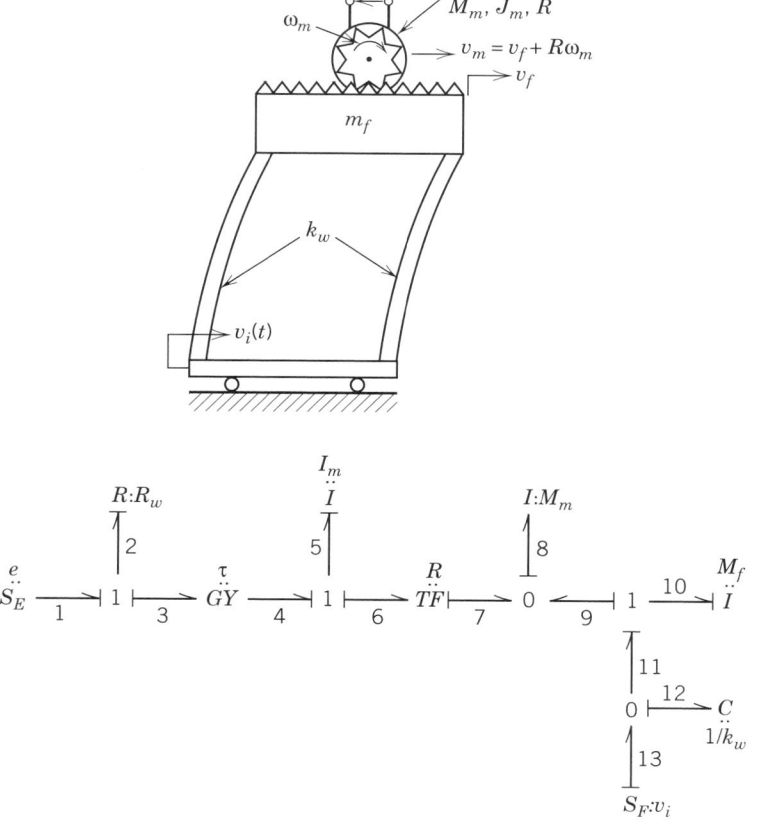

$R{:}R_w$

$I{:}J_m$ $C{:}1/k_T$

$I{:}m$

$$\begin{array}{c}2\end{array}$$

$$5$$ $$7$$

$$10$$

$S_e \xrightarrow{\ \ 1\ } 1 \xmapsto{\quad} \overset{T}{\underset{3}{\overset{\cdot\cdot}{GY}}} \xrightarrow{\quad 4\quad} 1 \xrightarrow{\ 6\ } 0 \xrightarrow{\ 8\ } \overset{R}{\overset{\cdot\cdot}{TF}} \xrightarrow{\ 9\ } 1 \xmapsto{\quad 11\quad} R{:}b$

$$12$$

$C{:}1/k$

$$e_4 = T\, f_3$$
$$e_3 = T\, f_4$$

$$e_9 = \frac{1}{R}e_8$$
$$f_8 = \frac{1}{R}f_9$$

Write output equations for e_4, f_7, and e_2.

5-17 A one-story building model is shown with a d-c motor on the "roof." The motor can be accelerated such that a reaction force is generated. The idea would be to control the motor voltage so that the reaction force helps calm the building in the event of an earthquake, represented by input velocity $v_i(t)$.

The bond graph for this system has derivative causality. Using procedures from this chapter, derive the state equations for this system. The algebra may become tedious, but describe with equations and words how you would come up with final state equations.

5-18   For Problem 5-17, propose the addition of a physical energy storage element that eliminates derivative causality. Redraw the bond graph and describe what the added element represents. Derive the state equations.

5-19   The system shown has an algebraic loop.

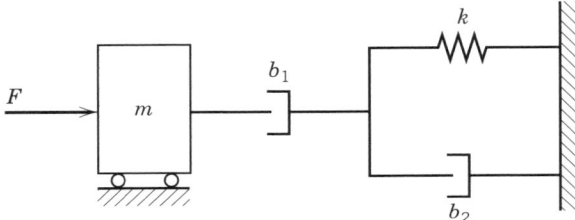

(a)   Construct a bond graph model, assign causality, and expose the algebraic loop.

(b)   Make an arbitrary causal assignment and perform the algebra necessary to allow derivation of state equations. Derive the state equations.

(c)   Add a physical energy storage element that will eliminate the algebraic loop, and redraw the bond graph. Derive the system equations.

6

ANALYSIS OF LINEAR SYSTEMS

6.1 INTRODUCTION

In the past, a major reason for studying linear models of systems was convenience, both in the model formulation and in the solution of the resulting differential equations. But in this book, we have taken pains to point out that with the use of bond graph methods, nonlinear models are not much more difficult to formulate than linear ones. This is particularly true when only the 1-port elements have nonlinear characteristics. Chapters 7, 9, and 12 demonstrate that some types of physical systems have even more complicated types of nonlinearities, but even in these cases bond graph methods result in equation sets that can be handled by computer simulation, as discussed in Chapter 13.

An important reason for the study of linear system models is that complete analytical solutions to linear differential equations exist, and this allows one to develop an intuitive feeling for system dynamics in general that is hard to acquire if only computer solutions are available. Everyone dealing with system dynamics is expected to be familiar with concepts relating to eigenvalues, free response, and frequency response, even though they do not strictly apply to real (nonlinear) systems.

In addition, there are many fields of engineering such as acoustics, electric circuit design, structural vibration, and the design of actuators of various types in which linear models have proved to be particularly effective. Moreover, the theory and practice of automatic control have traditionally been based on linear system theory. So there are good reasons for taking time in this chapter to review linear system theory.

Since linear ordinary differential equations are treated in virtually all engineering and scientific curricula, we will not attempt to present a mathematically rigorous and complete account here. Our purpose is rather to point out some features of the theory of linear ordinary differential equations that are significant for the understanding and design of dynamic systems.

Although a system model can, in principle, be described by a single high-order differential equation, as mentioned in Chapter 5, in this chapter we will concentrate on equivalent sets of first-order equations. This *state equation* form is produced naturally by bond graph models, and it is particularly suitable for machine computations. When the model is linear, it is convenient to use a vector matrix form as a way of representing the equations. In fact, matrices will be used mainly as a convenient notation to describe system models of any order. Only minimal knowledge of matrix operations will be assumed.

It is still quite common for books on system dynamics or automatic control to introduce Laplace transforms for solving constant coefficient linear differential equations (see Reference [1]). This technique changes a differential equation problem to an algebraic problem involving the Laplace variable s instead of time, t. At one time, the use of Laplace transform techniques was one of the convenient ways to solve linear differential equations analytically. But with the advent of computer simulation, it has become easy to solve linear equations numerically, so now this aspect of Laplace transform methods is of diminished utility.

In the present exposition an alternative approach will be used based on the general assumption that solutions are of the form e^{st} for special values of the variable s. The results derived below are the same as if Laplace transforms had been used, but the approach seems more fundamental and avoids the necessity of discussing the transform techniques in detail.

6.2 SOLUTION TECHNIQUES FOR ORDINARY DIFFERENTIAL EQUATIONS

Before discussing the characteristics of linear state equations, it is useful to note that it is not difficult in principle to find one or more solutions to general state equations, using a computer if necessary. In most of the examples encountered so far, the physical system model yielded state equations of the form

$$\dot{x}_1 = f_1(x_1, x_2, \ldots, x_n; u_1, u_2, \ldots, u_r),$$

$$\dot{x}_2 = f_2(x_1, x_2, \ldots, x_n; u_1, u_2, \ldots, u_r),$$

$$\vdots$$

$$\dot{x}_n = f_n(x_1, x_2, \ldots, x_n; u_1, u_2, \ldots, u_r), \tag{6.1}$$

where $x_1, \ldots, x_n$ are *state variables* (typically p's or q's in bond graph terms) and $u_1, \ldots, u_r$ are input variables (typically e's or f's from sources).

In addition, there may be *output variables* $y_1, \ldots, y_s$ statically related to the state and input variables:

$$y_1 = g_1(x_1, x_2, \ldots, x_n; u_1, u_2, \ldots, u_r),$$

$$y_2 = g_2(x_1, x_2, \ldots, x_n; u_1, u_2, \ldots, u_r),$$

$$\vdots$$

$$y_s = g_s(x_1, x_2, \ldots, x_n; u_1, u_2, \ldots, u_r). \tag{6.2}$$

In general, the f and g functions are nonlinear, but we will mainly consider linear functions. The linear versions of Eqs. (6.1) and (6.2) are

$$\dot{x}_1 = a_{11}x_1 + a_{12}x_2 + \cdots + a_{1n}x_n + b_{11}u_1 + b_{12}u_2 + \cdots + b_{1r}u_r,$$

$$\dot{x}_2 = a_{21}x_1 + a_{22}x_2 + \cdots + a_{2n}x_n + b_{21}u_1 + b_{22}u_2 + \cdots + b_{2r}u_r,$$

$$\vdots$$

$$\dot{x}_n = a_{n1}x_1 + a_{n2}x_2 + \cdots + a_{nn}x_n + b_{n1}u_1 + b_{n2}u_2 + \cdots + b_{nr}u_r, \tag{6.1a}$$

and

$$y_1 = c_{11}x_1 + c_{12}x_2 + \cdots + c_{1n}x_n + d_{11}u_1 + d_{12}u_2 + \cdots + d_{1r}u_r,$$

$$y_2 = c_{21}x_1 + c_{22}x_2 + \cdots + c_{2n}x_n + d_{21}u_1 + d_{22}u_2 + \cdots + d_{2r}u_r,$$

$$\vdots$$

$$y_s = c_{s1}x_1 + c_{s2}x_2 + \cdots + c_{sn}x_n + d_{s1}u_1 + d_{s2}u_2 + \cdots + d_{sr}u_r. \tag{6.2a}$$

Instead of writing out all the a, b, c, and d coefficients for linear systems, as in Eqs. (6.1a) and (6.2a), one can write the equations in a condensed vector matrix form as follows:

$$[\dot{x}] = [A][x] + [B][u], \tag{6.1b}$$

$$[y] = [C][x] + [D][u]. \tag{6.2b}$$

A comparison of this symbolic form with the explicit forms, Eqs. (6.1a) and (6.2a), should make it clear that $[x]$ is an n-dimensional column vector of the state variables, $[\dot{x}]$ is a vector of the derivatives of the state variables, $[u]$ is an r-dimensional column vector of the input variables, $[A]$ is an $n \times n$ matrix of the a_{ij} coefficients, $[B]$ is an $n \times r$ matrix of the b_{ij} coefficients, $[C]$ is an $s \times r$ matrix of the c_{ij} coefficients, and $[D]$ is an $s \times r$ matrix of the d_{ij} coefficients. All that one needs to know at this point is that *row by column* multiplication of matrices and column vectors in Eqs. (6.1b) and (6.2b) yields the explicit forms of Eqs. (6.1a) and (6.2a).

In this chapter, we will be concerned with the special case of system equations that are not only linear but also have constant coefficients. Occasionally, the situation arises in which linear equations have coefficients that vary in time. In matrix form this could be represented as

$$[\dot{x}] = [A(t)][x] + [B(t)][u]. \tag{6.3}$$

The theory of systems in which some of the coefficients vary in time is quite complicated, and since such equations do not arise often, we will not deal with them here.

Equations (6.1), (6.1a), (6.1b), (6.2), (6.2a), and (6.2b) describe *state determined systems*. To completely specify a problem to be solved, more information than just the equations is needed. Typically, the input variables are specified as functions of time:

$$u_1 = u_1(t),$$

$$u_2 = u_2(t),$$

$$\vdots$$

$$u_r = u_r(t). \tag{6.4}$$

In addition, a particular trajectory in state space must be singled out for attention. Our main concern will be with *initial-condition problems* in which the state at some initial time, t_0, is known:

$$x_1(t_0) = x_{10},$$

$$x_2(t_0) = x_{20},$$

$$\vdots$$

$$x_n(t_0) = x_{n0}, \tag{6.5}$$

where x_{10} to x_{n0} are the initial states. Note that given the information in Eqs. (6.4) and (6.5), the initial output variables are determined; for example,

$$y_1(t_0) = g_1\left[x_{10}, x_{20}, \ldots, x_{n0}; \ u_1(t_0), u_2(t_0), \ldots, u_r(t_0)\right].$$

One may also determine $\dot{x}_i(t_0)$ to $\dot{x}_n(t_0)$ using Eq. (6.1); for example,

$$\dot{x}_1(t_0) = f_1\left[x_{10}, x_{20}, \ldots, x_{n0}; \ u_1(t_0), u_2(t_0), \ldots, u_r(t_0)\right].$$

In principle, it is easy to compute how the system changes in a short interval of time, Δt, using the concept of a derivative directly. For example,

$$x_1(t_0 + \Delta t) \cong x_1(t_0) + \dot{x}_1(t_0)\Delta t = x_{10} + \dot{x}_1(t_0)\Delta t. \tag{6.6}$$

This equation is really just a rearrangement of the definition of a derivative,

$$\dot{x}_1 \equiv \frac{dx_1}{dt} = \lim_{\Delta t \to 0}\left[\frac{x_1(t_0 + \Delta t) - x_1(t_0)}{\Delta t}\right],$$

in which Δt is small but finite. Since $\dot{x}_1(t_0)$ depends on the initial state and the u's at t_0 that are known, one can use Eq. (6.6) to find the state at $t_0 + \Delta t$. This is sometimes called *Euler's formula* for integrating equations, and it may be applied to all the state equations at $t = t_0$. It can be reapplied at $t = t_0 + \Delta t$, since the u's are known at any time, and after applying Eq. (6.6) at t_0, the state at $t_0 + \Delta t$ is then known. The equation may then be applied recursively to march the solution along in time as far as desired. The process described above is readily adapted to automatic digital computation, but there is a snag. Equation (6.6) is not exactly correct except in the limit as $\Delta t \to 0$. As Δt is made very small, the approximation generally becomes increasingly accurate, but it takes more steps and more evaluations of the functions to cover any given time span of interest. The problem of choosing suitable time increments Δt and recursion formulas analogous to Eq. (6.6) to optimize digital integration accuracy falls in the domain of *numerical methods*, and there is a wealth of literature on this subject. Here it suffices to note that methods to integrate state equations exist in profusion and that, despite its obviousness, Euler's formula often turns out to be quite inefficient compared to other techniques.

The integration of state equations is straightforward, but when the equations are linear, one may say a great deal about the totality of possible systems behavior without the necessity of finding many solutions for various initial conditions and forcing functions. As will be demonstrated, linear systems obey the *principle of superposition*, which states that any scaled sum of solutions to linear state equations is also a solution. The superposition property allows one to consider a particular solution to the state equations to be the sum of a *response due to initial conditions*, or *free response* (in which all input variables vanish) and a *response due to inputs* or *forced response*. In addition, since the free-response component of the total system response has to do only with the system equations, this response may be used to characterize the system in a useful manner. We may discover natural frequencies, time constants, and stability properties that, in contrast to the general case, have nothing to do with initial conditions or forcing terms and help in understanding how the system would behave in a variety of situations.

6.3 FREE RESPONSE AND EIGENVALUES

Consider now the linear case represented in the form of Eqs. (6.1a) and (6.2a) or in the vector matrix form of Eqs. (6.1b) and (6.2b) and assuming that the input variables are all zero. When a solution for this case is found, it will be called the *free response* and it will have to do only with initial conditions, not input variable forcing. We consider a first-order system at the onset and then move on to an example with multiple state variables.

6.3.1 A First-Order Example

In Figure 6.1, a block of hot steel is shown after it has been lowered into a large oil bath. We can predict the time history of the temperature of the steel using a simple lumped-parameter thermal system model if we make several assumptions:

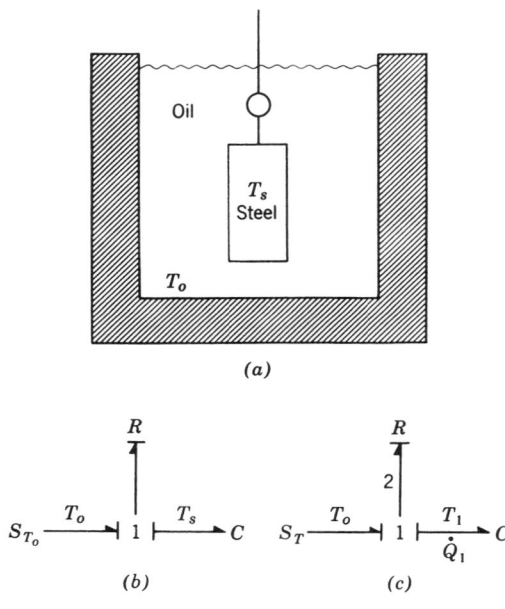

FIGURE 6.1. First-order example system. (*a*) Schematic diagram; (*b*) bond graph showing temperatures, T_o for oil and T_s for steel; (*c*) bond graph with bonds numbered for convenience in writing equations.

1. We assume that the temperature gradients within the steel block are small as compared with the gradient between the cool oil and the hot steel. We then define a single average temperature for the steel, T_s.

2. We assume that the oil temperature does not change much as the steel cools, so that the temperature, T_o, is nearly constant.

3. We assume that the heat flow rate $\dot{Q}$ from the steel to the oil is proportional to $T_o - T_s$.

4. We assume that the change in T_s is proportional to the total heat exchanged, Q.

Clearly, this simplifying set of assumptions is physically reasonable for some situations, but many details of the thermal process have been glossed over. It would be easy to complicate the model in the style of Figure 3.18 to allow more detailed prediction of temperatures, but our purpose here is to construct a first-order system. Two pseudo–bond graphs using temperature as an effort and heat flow rate as a flow are shown in Figures 6.1b and c. The constitutive laws for R and C can be found as follows. Using assumption 3, we write

$$\dot{Q} = \frac{1}{R}(T_s - T_o),$$ (6.7)

where the resistance R has the units °C/W, for example. Often, R is estimated from a heat transfer coefficient n and the total area of the steel block, A:

$$\frac{1}{R} = hA. \tag{6.8}$$

In this case, h has the units

$$\frac{W}{°C \cdot m^2}$$

Using assumption 4, we write

$$T_s - T_{s0} = \frac{Q}{C}, \tag{6.9}$$

where T_{s0} is the initial temperature of the steel when no heat has been transferred. The capacitance C, with units J/°C, may be found from

$$C = mc, \tag{6.10}$$

where the specific heat per unit mass, c, has units J/°C-kg.

Using the methods of Chapter 5, we may easily write the state equations for the bond graph of Figure 6.1c, which embodies the relations discussed above:

$$\dot{Q}_1 = \dot{Q}_2 = \frac{1}{R}(T_o - T_1) = \frac{1}{R}\left(T_o - \frac{Q_1}{C} - T_{10}\right),$$

$$\dot{Q}_1 = -\frac{Q_1}{RC} + \frac{1}{R}(T_o - T_{10}), \tag{6.11}$$

where $T_1 = T_s$ and $T_{10} = T_{s0}$ in Eq. (6.9). Another version of the state equation, which is found by substituting Eq. (6.9) into Eq. (6.11), is

$$RC\dot{T}_s = -T_s + T_o, \tag{6.12}$$

where we have returned to T_s in place of T_1. It is sometimes convenient to measure temperatures from some datum so that some constants are absorbed in the definitions. For example, if we define

$$T = T_s - T_o, \tag{6.13}$$

then Eq. (6.12) becomes

$$RC\dot{T} = -T. \tag{6.14}$$

Also, if we modify Eq. (6.9) to read

$$T = T_s - T_o = \frac{Q_1}{C} = T_1 - T_o, \tag{6.15}$$

then Eq. (6.11) becomes

$$\dot{Q} = \frac{1}{R}(T_0 - T_1) = \frac{1}{R}\left(T_0 - \frac{Q_1}{C} - T_0\right),$$

or

$$\dot{Q}_1 = \frac{-1}{RC}Q_1. \tag{6.16}$$

In Eq. (6.16), when $Q_1 = 0$, we have $T_s = T_o$ and the initial value of Q_1 is related to the initial value of T_s through Eq. (6.15).

For simplicity, we will use Eq. (6.16) for our discussion, but the presence of extra constants in state equations such as those in Eqs. (6.11) and (6.12) will cause no problems, since the constants simply act as constant forcing functions for the system. The solution to a first-order equation such as Eq. (6.16) or Eq. (6.14) is well known. First we note that one solution to Eq. (6.16) is obviously $Q_1 = 0$. This solution is called the *trivial solution*, since it is of little use. In particular, if $Q_1(0)$, the initial value of Q_1, is nonzero, then the trivial solution is of no use in fitting the initial condition. A general assumption that proves useful for linear systems is that the solution is exponential, for example,

$$Q_1(t) = Ae^{st}, \tag{6.17}$$

where the constants A and s are to be adjusted, if possible, to make sure that $Q_1(t)$ fits the differential equation and the initial conditions.

Substituting Eq. (6.17) into Eq. (6.16), we find

$$Ase^{st} = \frac{-1}{RC}Ae^{st},$$

or

$$\left(s + \frac{1}{RC}\right)Ae^{st} = 0. \tag{6.18}$$

Now, either

$$Ae^{st} = 0,$$

in which case we return to the trivial solution, or

$$s = \frac{-1}{RC}, \tag{6.19}$$

in which case the solution is

$$Q_1(t) = Ae^{-t/RC} = Ae^{-t/\tau}, \tag{6.20}$$

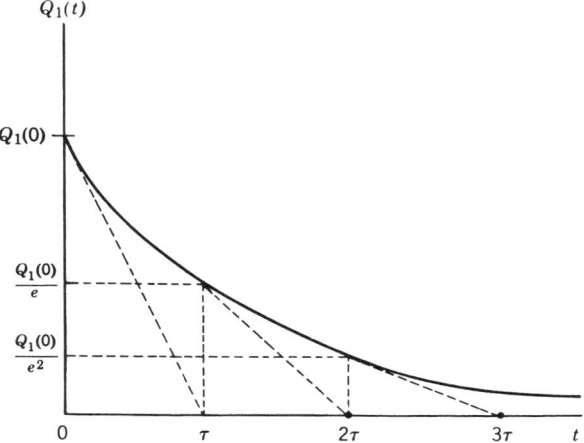

FIGURE 6.2. Free response of system of Figure 6.1.

where the *time constant*, τ, has been identified with RC. To complete the solution, the constant A can be found from the initial value $Q_1(0)$ of Q_1:

$$Q_1(0) = Ae^{-0} = A, \qquad (6.21)$$

where $Q_1(0)$ can be related to the initial temperature through Eq. (6.15). The solution is then

$$Q_1(t) = Q_1(0)e^{-t/\tau}, \qquad t \geq 0. \qquad (6.22)$$

This solution is shown in Figure 6.2. For this simple exponential solution, it is readily shown that, at any point on the curve, if one extends a straight line along the local slope of the curve to the steady-state value of Q_1 (zero in this case), it will intersect after τ time units have elapsed. Thus, τ sets the basic time scale of the response. If a system response is measured experimentally, one can easily estimate τ as shown in Figure 6.2 and one can check to see how well a linear, first-order constant-parameter system would reproduce the measured response by computing τ at several points on the experimental curve.

6.3.2 Second-Order Systems

Although a knowledge of first-order system response will prove very useful, it must be obvious that more complex systems respond in ways much more complicated than the exponential of Figure 6.2. Surprisingly, the pattern of analysis, somewhat generalized, does carry over for arbitrarily high-order systems. In fact, by studying a second-order system, the pattern will become evident, and the extension to nth-order systems is very straightforward.

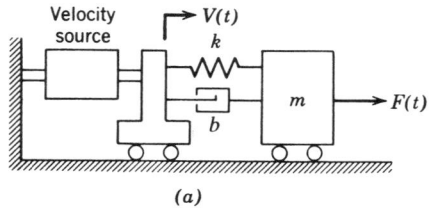

$$\dot{x} = 0\,x - \frac{1}{m}p + 1[V(t)] + 0[F(t)]$$

$$\dot{p} = k\,x - \frac{b}{m}p + b[V(t)] + 1[F(t)]$$

(c)

FIGURE 6.3. Example system. (a) Schematic diagram; (b) bond graph; (c) state equations.

Consider, then, the simple second-order example shown in Figure 6.3. To make the correspondence between the system of the figure and the general equations (6.1a) and (6.2a), we write the state equations in a standard form:

$$\dot{x} = 0x - \frac{1}{m}p + 1V + 0F,$$

$$\dot{p} = kx - \frac{b}{m}p + bV + 1F. \tag{6.23}$$

The force f is just one of several possible output variables:

$$f = kx - \frac{b}{m}p + bV + 0F. \tag{6.24}$$

The system has two state variables, $x_1 = x$, $x_2 = p$; two input variables, $u_1 = V$, $u_2 = F$, and one output variable, $y_1 = f$. The matrices for the forms of Eqs. (6.1b) and (6.2b) are then related to the system parameters as follows:

$$[A] = \begin{bmatrix} 0 & -1/m \\ k & -b/m \end{bmatrix}, \quad [B] = \begin{bmatrix} 1 & 0 \\ b & 1 \end{bmatrix}, \quad [C] = [k \quad -b/m], \quad [D] = [b \quad 0].$$

(6.25)

Note that in Figure 6.3c the notations $V(t)$ and $F(t)$ are used as a reminder that these input quantities must be specified functions of time before the state equations can be solved.

For the free response we consider that $V(t) = F(t) = 0$, so that

$$\begin{bmatrix} \dot{x} \\ \dot{p} \end{bmatrix} = \begin{bmatrix} 0 & -1/m \\ k & -b/m \end{bmatrix} \begin{bmatrix} x \\ p \end{bmatrix}. \tag{6.26}$$

In general terms, this equation could be written simply as $[\dot{x}] = [A][x]$, which is Eq. (6.1b) with $[u] = [0]$.

One trivial solution to Eq. (6.26) is evident. It is

$$x \equiv 0, \quad p \equiv 0. \tag{6.27}$$

This solution corresponds to the case in which the spring is unstretched and the mass has no momentum (or velocity). This solution is, again, not very useful and is of no help in fitting given initial conditions. Suppose a solution satisfying

$$x(t_0) = x_0, \quad p(t_0) = p_0 \tag{6.28}$$

is desired, where x_0 is the initial stretch in the spring and $p_0 = mv_0$ is the initial momentum of the mass.

To find the solution, assume that both energy state variables may be represented by an amplitude and an exponential time function. In the example, we assume

$$x(t) = Xe^{st}, \quad p(t) = Pe^{st}, \tag{6.29}$$

where X and P are constant amplitudes and s is a complex or real number with the dimension of inverse time. If the assumption of Eq. (6.29) is substituted into Eq. (6.26), the result is

$$\begin{bmatrix} sX \\ sP \end{bmatrix} e^{st} = \begin{bmatrix} 0 & -1/m \\ k & -b/m \end{bmatrix} \begin{bmatrix} X \\ P \end{bmatrix} \tag{6.30a}$$

The left side of Eq. (6.30a) can be written in a form similar to the right side by using the 2×2 unit matrix

$$[I] = \begin{bmatrix} 1 & 0 \\ 0 & 1 \end{bmatrix}$$

so that

$$\begin{bmatrix} sX \\ sP \end{bmatrix} e^{st} = \begin{bmatrix} 1 & 0 \\ 0 & 1 \end{bmatrix} \begin{bmatrix} X \\ P \end{bmatrix} se^{st} = \begin{bmatrix} s & 0 \\ 0 & s \end{bmatrix} \begin{bmatrix} X \\ P \end{bmatrix} e^{st}$$

With this idea, the two sides of Eq. (6.30a) can be combined. A single matrix is then involved.

$$\begin{bmatrix} s & 1/m \\ -k & s+b/m \end{bmatrix} \begin{bmatrix} X \\ P \end{bmatrix} e^{st} = \begin{bmatrix} 0 \\ 0 \end{bmatrix}. \tag{6.30b}$$

We can see now that the single matrix on the left side of Eq. (6.30b) is just $[sI - A]$. This expression will arise for systems of any order when analyzing the free response where $[I]$ is the general $n \times n$ unit matrix with 1's on the main diagonal and 0's elsewhere, and $[A]$ is the system matrix from Eq. (6.1b).

Now Eq. (6.30b) shows that the assumption of Eq. (6.29) has reduced the differential equation problem to an algebraic problem. (The Laplace transformation also achieves this reduction.) One way to solve Eq. (6.30b) would be to have e^{st} vanish. Not only is this not really possible for any finite value of s, but the result leads only to the trivial solution of Eq. (6.27). Rejecting this possibility, one can divide out the e^{st} term and concentrate on solving for X and P that satisfy

$$\begin{bmatrix} s & 1/m \\ -k & s+b/m \end{bmatrix} \begin{bmatrix} X \\ P \end{bmatrix} = \begin{bmatrix} 0 \\ 0 \end{bmatrix}. \tag{6.30c}$$

Once again, the trivial solution appears, since clearly $X = 0$, $P = 0$ is a solution to Eq. (6.30c). In fact, the elementary theory of linear algebra contains the result that there is only one solution to a set of linear simultaneous equations if the determinant of the coefficients does not vanish. Cramer's rule furnishes a recipe for expressing the solution in terms of ratios of determinants, the denominator of each being the determinant of the equation coefficients. Since we have one trivial solution by inspection, our assumed solution will prove useless unless the solution to Eq. (6.30a) turns out *not* to be unique.

This will occur if

$$\begin{vmatrix} s & \dfrac{1}{m} \\ -k & s+\dfrac{b}{m} \end{vmatrix} = 0 \tag{6.31a}$$

or

$$s^2 + \frac{b}{m}s + \frac{k}{m} = 0. \tag{6.31b}$$

Note that the determinant that must vanish in order for there to be a nontrivial free response is just the determinant of the matrix $[sI - A]$.

The requirement that the determinant vanish leads to the requirement that a polynomial in s must vanish. This equation is called the *characteristic equation* and leads to a set of *characteristic values* for the parameters. Equation (6.19) was the characteristic equation for our first example. (In German, a characteristic value is called an *eigenwert*, and in English the mongrel term *eigenvalue* is commonly used. Why "characteristic wert" never caught on remains a mystery.) In what follows, we will generally use the term "eigenvalue" in place of the words "characteristic value."

Solving Eq. (6.31), the result is

$$s_1 = \frac{-b/m + [(b/m)^2 - 4k/m]^{1/2}}{2},$$

$$s_2 = \frac{-b/m - [(b/m)^2 - 4k/m]^{1/2}}{2}. \tag{6.32}$$

Note that s_1 and s_2 will be complex for small values of b. For each value of s that is determined from an equation such as Eq. (6.31), a solution satisfying Eq. (6.30c) may be found. For an nth-order system, there will generally be n special values of s; in our second-order example, there are two. For high-order systems it will generally not be possible to solve for the characteristic values without resorting to numerical methods, but the existence of n roots of an nth-order characteristic equation is ensured by the theory of linear algebra.

Much useful information is contained in the eigenvalues of Eq. (6.32) themselves, but before discussing what can be attained by simply solving the characteristic equation, let us outline the program of finding a complete free response satisfying given initial conditions. For each value of s, s_1, and s_2, let us attempt to determine X and P values from Eq. (6.30c). For s_1 we attempt to find X_1 and P_1:

$$s_1 X_1 + \frac{1}{m} P_1 = 0, \qquad -k X_1 + \left(s_1 + \frac{b}{m}\right) P_1 = 0,$$

or

$$\frac{-b/m + [(b/m)^2 - 4k/m]^{1/2}}{2} X_1 + \frac{1}{m} P_1 = 0, \qquad (6.33)$$

$$-k X_1 + \left(\frac{-b/m + [(b/m)^2 - 4k/m]^{1/2}}{2} + \frac{b}{m}\right) P_1 = 0. \qquad (6.34)$$

Of course, one should not expect to be able to solve uniquely for X_1 and P_1 from Eqs. (6.33) and (6.34), since s_1 was chosen to make the determinant of the coefficients vanish. In fact, after some algebraic manipulation, it can be shown that both the equations are equivalent to the single equation

$$X_1 - \frac{b + (b^2 - 4km)^{1/2}}{2km} P_1 = 0, \qquad (6.35)$$

Whenever a characteristic value is substituted into equations for amplitudes such as Eq. (6.30c), the equations will be found *not* to be independent, which means that the ratio of X_1 to P_1 in (6.35) is determined, but not the magnitude of either one. The most one can expect to solve for is the ratios of $n - 1$ amplitudes in terms of one arbitrary amplitude.

Substituting s_2 from Eq. (6.32) into Eq. (6.30c) and solving for amplitudes X_2 and P_2, one again finds that both equations are equivalent to the single equation

$$X_2 + \frac{-b + (b^2 - 4km)^{1/2}}{2mk} P_2 = 0. \qquad (6.36)$$

Now whenever X_1 and P_1 are related according to Eq. (6.35), Eq. (6.30c) will be satisfied, and the original differential equation Eq. (6.36) will be satisfied by

$$x(t) = X_1 e^{s_1 t}, \qquad P(t) = P_1 e^{s_1 t}.$$

Similarly, another solution is

$$x(t) = X_2 e^{s_2 t}, \qquad p(t) = P_2 e^{s_2 t}$$

whenever X_2 and P_2 satisfy Eq. (6.36). It is also easy to verify that another valid solution is obtained by adding the two solutions together:

$$x(t) = X_1 e^{s_1 t} + X_2 e^{s_2 t}, \qquad p(t) = P_1 e^{s_1 t} + P_2 e^{s_2 t}. \tag{6.37}$$

This is an example of the principle of superposition for linear systems and is readily verified by simply substituting Eq. (6.37) into Eq. (6.26) and noting that as long as the components of the sum satisfy the differential equations, so does the sum.

Although s_1 and s_2 are completely determined by Eq. (6.32), the four quantities X_1, X_2, P_1, and P_2 are so far related only by the two relations (6.35) and (6.36). Two more conditions are necessary to specify the free response completely, and these are provided by the two initial conditions of Eq. (6.28):

$$x_0 = x(t_0) = X_1 e^{s_1 t_0} + X_2 e^{s_2 t_0}, \tag{6.38}$$

$$p_0 = p(t_0) = P_1 e^{s_1 t_0} + P_2 e^{s_2 t_0}. \tag{6.39}$$

Thus, the four equations (6.35), (6.36), (6.38), and (6.39) completely specify the free response of the system, satisfying the given initial conditions. Although in this simple case one could solve these equations completely, the solution is too complex for easy understanding, so we shall discuss some useful special cases.

6.3.3   Example: The Undamped Oscillator

If the dashpot parameter b is set to zero, or if the dashpot is removed from the system, the equations of the system are somewhat simplified. Equations (6.23) become

$$\begin{bmatrix} \dot{x} \\ \dot{p} \end{bmatrix} = \begin{bmatrix} 0 & -1/m \\ k & 0 \end{bmatrix} \begin{bmatrix} x \\ p \end{bmatrix} + \begin{bmatrix} 1 & 0 \\ 0 & 1 \end{bmatrix} \begin{bmatrix} V \\ F \end{bmatrix} \tag{6.40a}$$

Equation (6.24) becomes

$$f = kx + 0p + 0V + 0F.$$

For the free response we need study only a simple version of Eq. (6.26):

$$\dot{x} = \frac{-p}{m}, \qquad \dot{p} = kx.$$

Following the development leading to Eq. (6.31), the following characteristic equation is obtained:

$$s^2 + \frac{k}{m} = 0, \tag{6.40b}$$

which yields the eigenvalues

$$s_1 = +j \left(\frac{k}{m} \right)^{1/2} = j\omega_n, \qquad s_2 = -j \left(\frac{k}{m} \right)^{1/2} = -j\omega_n, \qquad (6.41)$$

where $j \equiv (-1)^{1/2}$ and the *undamped natural frequency* ω_n is defined as $(k/m)^{1/2}$. In distinction to the first-order example, we now have encountered purely complex eigenvalues.

In attempting to solve for X_1 and P_1, it is found that only the single relation (6.35) is available:

$$X_1 - \frac{jP_1}{(km)^{1/2}} = 0. \qquad (6.42)$$

In solving for X_2 and P_2, one encounters a simplified version of Eq. (6.36):

$$X_2 + \frac{jP_2}{(km)^{1/2}} = 0. \qquad (6.43)$$

Finally, we must satisfy the initial conditions in Eqs. (6.38) and (6.39).

Using Eqs. (6.42) and (6.43) to eliminate P_1 and P_2 in Eqs. (6.38) and (6.39), the result is

$$X_1 e^{s_1 t_0} + X_2 e^{s_2 t_0} = x_0, \qquad (6.44)$$

$$-j(km)^{1/2} X_1 e^{s_1 t_0} + j(km)^{1/2} X_2 e^{s_2 t_0} = p_0. \qquad (6.45)$$

These equations are readily solved, the result being

$$X_1 = \frac{1}{2} \left(x_0 + \frac{jp_0}{(mk)^{1/2}} \right) e^{-j\omega_n t_0}, \qquad (6.46)$$

$$X_2 = \frac{1}{2} \left(x_0 + \frac{jp_0}{(mk)^{1/2}} \right) e^{+j\omega_n t_0}, \qquad (6.47)$$

in which Eq. (6.42) has been used. For completeness, P_1 and P_2 may also be found:

$$P_1 = -j(mk)^{1/2} X_1 = \frac{1}{2} \left[-j(mk)^{1/2} x_0 + p_0 \right] e^{-j\omega_n t_0}, \qquad (6.48)$$

$$P_2 = +j(mk)^{1/2} X_2 = \frac{1}{2} \left[-j(mk)^{1/2} x_0 + p_0 \right] e^{+j\omega_n t_0}. \qquad (6.49)$$

We have now achieved our objective. The free response is given by Eq. (6.37), in which we now know s_1 and s_2, and we can determine X_1, X_2, P_1, and P_2 from arbitrary initial conditions. It remains only to interpret the solution.

The solution just determined is of the form

$$x(t) = X_1 e^{j\omega_n t} + X_2 e^{-j\omega_n t}, \qquad (6.50)$$

$$p(t) = P_1 e^{j\omega_n t} + P_2 e^{-j\omega_n t} \qquad (6.51)$$

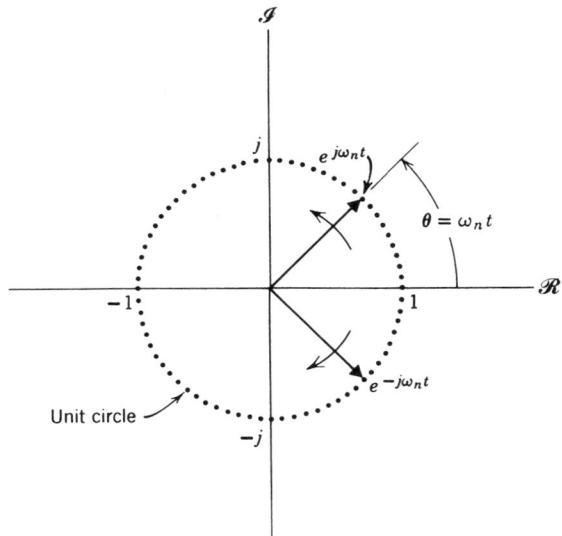

FIGURE 6.4. Complex plane representation of $e^{+j\omega_n t}$ and $e^{-j\omega_n t}$.

and is to be interpreted as valid for $t \geq t_0$. The time-dependent parts, $e^{j\omega_n t}$ and $e^{-j\omega_n t}$, may be represented in the complex plane as rotating vectors or, as they are sometimes called, *phasors* (see Figure 6.4). The vectors rotate at constant angular rate, ω_n, around the unit circle. The sketch of Figure 6.4 provides an easy way to remember the relations

$$e^{+j\omega_n t} = \cos \omega_n t + j \sin \omega_n t \qquad (6.52)$$

and

$$e^{-j\omega_n t} = \cos \omega_n t - j \sin \omega_n t. \qquad (6.53)$$

It is also easy to see from the sketch, or algebraically from Eqs. (6.52) and (6.53), that the following relations are true:

$$\cos \omega_n t = \frac{e^{j\omega_n t} + e^{-j\omega_n t}}{2} \qquad (6.54)$$

and

$$\sin \omega_n t = \frac{e^{j\omega_n t} - e^{-j\omega_n t}}{2j}. \qquad (6.55)$$

From the form of Eq. (6.50) one can see that if X_1 were real and equal to X_2, then $x(t)$ would be a cosine wave with amplitude $2X_1$. This would be the case if, in Eqs. (6.46) and (6.47), $t_0 = p_0 = 0$. Then Eq. (6.50) would reduce to $x(t) = x_0 \cos \omega_n t$.

In the general case, X_1 and X_2 are complex numbers that are easily sketched as the product of the complex numbers

$$\frac{x_0}{2} \pm j\frac{p_0}{2}(mk)^{1/2}$$

and the complex exponentials $e^{\pm j\omega_n t_0}$. See Figure 6.5, and recall that in multiplying two complex numbers, the magnitudes multiply and the angles add. It is no accident that X_1 and X_2 (and P_1 and P_2) turn out to be complex conjugates. According to Eq. (6.50), the real quantity $x(t)$ must be the sum of X_1 and X_2 multiplied by the complex conjugate quantities $e^{j\omega_n t}$ and $e^{-j\omega_n t}$, respectively. Figure 6.6 shows how the sum of complex quantities adds up to a real variable, $x(t)$.

Another way to represent Eq. (6.50) is as a cosine wave with a phase angle φ and an amplitude A,

$$x(t) = A\cos(\omega_n t + \varphi). \tag{6.56}$$

From Figure 6.6 it is easy to verify that

$$A = 2|X_1| = 2|X_2| = 2\left(\frac{x_0^2}{4} + \frac{p_0^2}{4mk}\right)^{1/2} \tag{6.57}$$

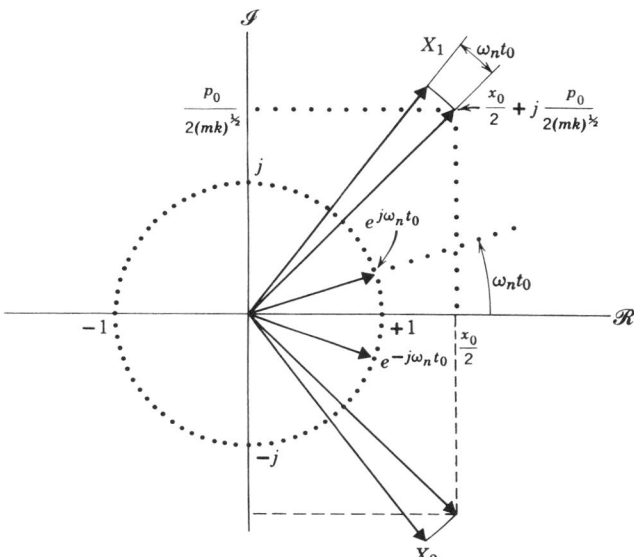

FIGURE 6.5. Representation of X_1 and X_2 in the complex plane. See Eqs. (6.46) and (6.47).

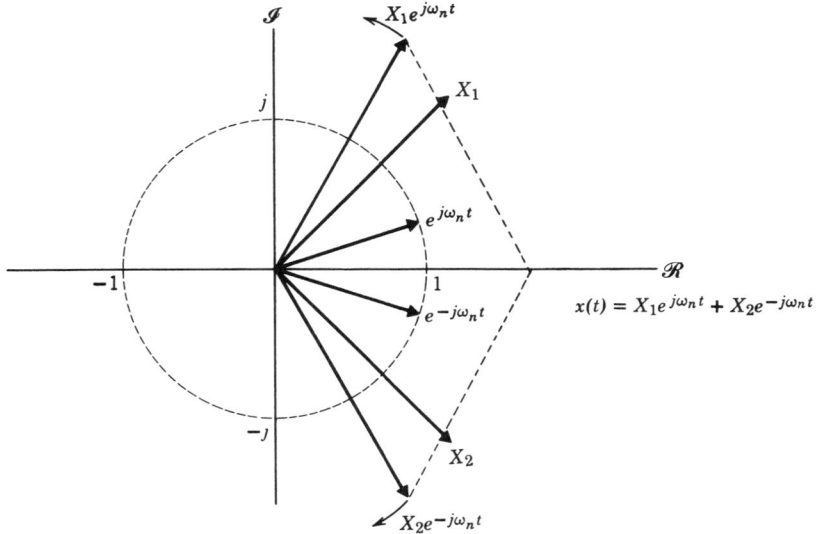

FIGURE 6.6. Sketch of Eq. (6.50) in the complex plane.

and

$$\varphi = \angle X_1 = \left[\tan^{-1}\frac{p_0}{x_0(mk)^{1/2}}\right] + \omega_n t_0, \qquad (6.58)$$

where $|\ |$ stands for "magnitude of" and $\angle$ stands for "angle of."

Another commonly used notation involves the "real part of" operator $\mathrm{Re}(\ )$ and the "imaginary part of," operator $\mathrm{Im}(\ )$. Sine and cosine functions are readily expressed in terms of these function as follows:

$$\cos \omega_n t = \mathrm{Re}(e^{j\omega_n t}), \qquad (6.59)$$

$$\sin \omega_n t = \mathrm{Im}(e^{j\omega_n t}). \qquad (6.60)$$

Then $x(t)$ may be expressed in yet another way as

$$x(t) = A\mathrm{Re}(e^{j\omega_n t + \varphi}) = \mathrm{Re}(2X_1 e^{j\omega_n t}). \qquad (6.61)$$

Equation (6.61) is sometimes simpler to use than some other equivalent representations, since it deals with only one of the pair X_1, X_2 and one of the pair $e^{+j\omega_n t}$, $e^{-j\omega_n t}$. Both pairs must be complex conjugates in order for x to be real; therefore, knowledge of one member of each pair is sufficient.

Finally, it should be noted that the discussion of $p(t)$ in Eq. (6.51) exactly parallels that of $x(t)$ just given. We may note from Eqs. (6.48) and (6.49) that P_1 and P_2 are proportional to X_1 and X_2 but are 90° away due to the j-factors. (Remember that multiplication of a complex number by j results in a number with identical magnitude, but with an angle advanced by $\pi/2$ rad, or 90°.) In fact, if one were to sketch $P_1 e^{j\omega_n t}$ and $P_2 e^{-j\omega_n t}$ as rotating vectors, as was done for the components of x in Figure 6.6, the vectors for P_1 and P_2 would lag the vectors for X_1 and X_2 by 90°. This corresponds with the fact that $p(t)$ is proportional to either the negative of the derivative of $x(t)$ or the integral of $x(t)$, Eq. (6.40a). (The reader should verify that the time derivative of $X_1 e^{+j\omega_n t}$ is a phasor advanced by 90° and the integral of $X_1 e^{+j\omega_n t}$ is a phasor retarded by 90° and that similar statements can be made for the counter-rotating vector $X_2 e^{-j\omega_n t}$.)

The 90° phase difference between x and p means that if x is a cosine wave, p is a sine wave. Then, in the state space, x, p, the trajectories are closed ellipses that can be made to be circles upon proper scaling. The direction that the representative point travels on the circles depends on the sign convention used in setting up the original differential equations, but, in any case, the representative point moves around the center of the circular trajectory at an angular rate of ω_n radians per second.

6.3.4   Example: The Damped Oscillator

When the dashpot coefficient b is positive and not too large, the free response of the system will be oscillatory but will eventually damp out. The eigenvalues of Eq. (6.32) clearly change character, depending on whether the expression under the radical is positive or negative. If

$$\left(\frac{b}{m}\right)^2 < 4\frac{k}{m}, \tag{6.62}$$

the system is said to be *underdamped*; if

$$\left(\frac{b}{m}\right)^2 > 4\frac{k}{m}, \tag{6.63}$$

the system is said to be *overdamped*; and if

$$\left(\frac{b}{m}\right)^2 = 4\frac{k}{m}, \tag{6.64}$$

the system is said to be *critically damped*.

Often the characteristic equation (6.31) is written as follows:

$$s^2 + 2\zeta\omega_n s + \omega_n^2 = 0, \tag{6.65}$$

in which the undamped natural frequency ω_n is as defined by Eq. (6.41), and the *damping ratio* ζ is given as

$$\zeta \equiv \frac{b}{2(mk)^{1/2}}. \tag{6.66}$$

Then the system is underdamped, overdamped, or critically damped, depending on whether $\zeta < 1, \zeta > 1$, or $\zeta = 1$. With these definitions, the eigenvalues for the underdamped case of Eq. (6.32) may be written

$$s_1 = -\zeta\omega_n + j\omega_n(1 - \zeta^2)^{1/2},$$
$$s_2 = -\zeta\omega_n - j\omega_n(1 - \zeta^2)^{1/2}. \tag{6.67}$$

Equations (6.35) and (6.36) become

$$X_1 - \frac{\zeta\omega_n + j\omega_n(1 - \zeta^2)^{1/2}}{k} P_1 = 0 \tag{6.68}$$

and

$$X_2 + \frac{-\zeta\omega_n + j\omega_n(1 - \zeta^2)^{1/2}}{k} P_2 = 0. \tag{6.69}$$

The free response solution, Eq. (6.37), becomes a bit more complicated than in the undamped case because the eigenvalues are now complex numbers rather than pure imaginary numbers. A typical term is

$$X_1 e^{s_1 t} = X_1 e^{-\zeta\omega_n t + j\omega_n(1-\zeta^2)^{1/2} t} = X_1 e^{-\zeta\omega_n t} e^{j\omega_n(1-\zeta^2)^{1/2} t}. \tag{6.70}$$

As before, X_1 may be a complex number determined partly by Eqs. (6.68) and (6.69) and partly by the necessity to fit initial conditions (6.38) and (6.39).

The term $e^{-\zeta\omega_n t}$ is real and decreases exponentially with time. (In the previous example, ζ was equal to zero, and this factor remained unity.) The last factor in Eq. (6.70) is a complex exponential form that represents a sinusoidal wave of frequency ω_d, the *damped natural frequency*,

$$\omega_d \equiv \omega_n(1 - \zeta^2)^{1/2}, \tag{6.71}$$

or

$$\left(\frac{\omega_d}{\omega_n}\right)^2 + \zeta^2 = 1.$$

When there is damping, the sinusoidal part of the system response has a frequency somewhat different from the undamped natural frequency, and, as Eq. (6.71) shows, the variation of ω_d with ζ may be plotted as a segment of a circle. When ζ is small as compared with unity, $\omega_d \cong \omega_n$, but $\omega_d \to 0$ as $\zeta \to 1$ (see Figure 6.7).

Although it is possible to solve for X_1, X_2, P_1, P_2 in terms of general initial conditions x_0, p_0, the result is not very useful, since it is usually easier to solve the equations for some particular initial conditions rather than to specialize the general result. What is important is to understand the nature of the solution. Figure 6.8 is the equivalent of Figure 6.6 for the underdamped case. The major difference between the two figures is that both components of the x-response in Figure 6.8 rotate at an

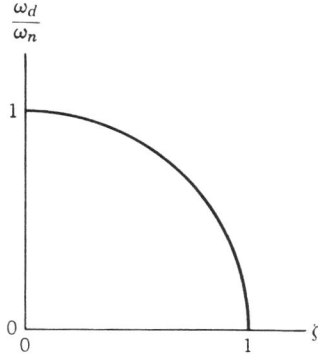

FIGURE 6.7. Variation of damped natural frequency with damping ratio.

angular speed ω_d instead of ω_n and that there is an attenuation factor, $e^{-\zeta\omega_n t}$, that reduces the amplitudes exponentially, so that the complex components spiral toward the origin of the complex plane. The corresponding response in time is sketched in Figure 6.9. Note that the response passes through zero whenever the sinusoidal component does; thus, the zeros occur periodically. The envelope of the time response is the exponentially decreasing amplitude shown. When the sinusoidal component is unity, the waveform $x(t)$ is locally tangent to the exponential. As a result, the actual maxima of the wave are not quite periodic.

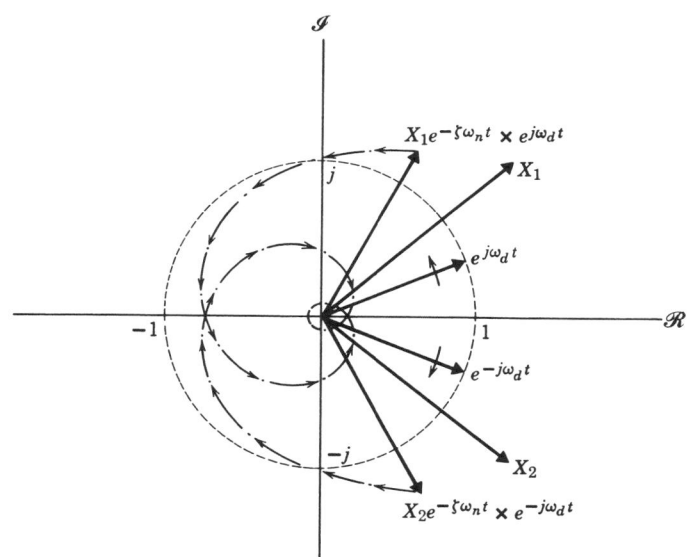

FIGURE 6.8. Complex representation of the response of an underdamped system.

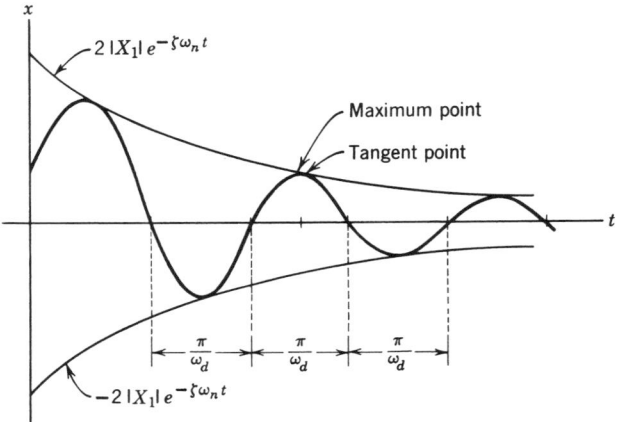

FIGURE 6.9. Time response for underdamped system.

Since so much understanding of the general nature of the free response is contained in the roots of the characteristic equation, it is useful to show how the damping ratio and the damped and undamped natural frequencies are correlated with the locations in the imaginary plane of the eigenvalues s_1 and s_2. This is shown in Figure 6.10. Note that for any system in which the state variables are real quantities, any complex eigenvalues will occur in complex conjugate pairs, and it is easy to read off natural frequencies and the damping ratio corresponding to each pair.

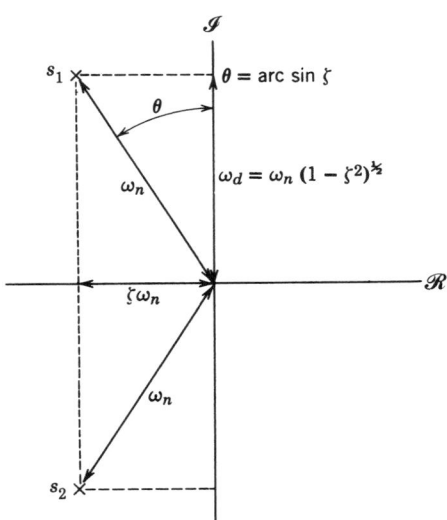

FIGURE 6.10. Correlation of natural frequencies and damping ratio with location of eigenvalues in the complex plane for underdamped case.

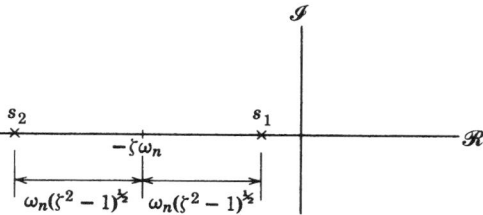

FIGURE 6.11. Eigenvalues for overdamped case.

Stable systems result in eigenvalues that lie in the *left* half of the complex plane; that is, for stable systems the real component of e^{st} is a *decreasing* exponential. *Unstable* systems correspond to s-values in the *right* half plane, and the intermediate case of the undamped oscillator corresponds to purely imaginary values of s on the j-axis.

For the overdamped oscillator, the character of the eigenvalues of Eq. (6.32) changes, and it is convenient to rewrite Eqs. (6.67) as

$$s_1 = -\zeta\omega_n + \omega_n(\zeta^2 - 1)^{1/2}, \qquad s_2 = -\zeta\omega_n - \omega_n(\zeta^2 - 1)^{1/2}. \qquad (6.72)$$

These roots are plotted in Figure 6.11. Note that for $1 < \zeta < \infty$ the roots s_1 and s_2 lie in the range $-\infty < s_1, s_2 < 0$; that is, the eigenvalues are negative and real. In such a case, terms of the form e^{st} are conveniently written in terms of a time constant τ, as was done for the first-order case. Suppose, for example, s_1 were equal to $-a$. Then

$$e^{s_1 t} = e^{-at} = e^{-t/\tau}, \qquad (6.73)$$

where

$$\tau = \frac{1}{a} = \frac{-1}{s_1}.$$

The factor $e^{-t/\tau}$ in the second-order system solution has the same effect as was shown in Figure 6.2 for the first-order system. Thus, in Figure 6.11, s-values far out on the negative real axis correspond to quickly decreasing exponentials. Also, it is clear that the real part of s_1 and s_2 in Figure 6.10, $\zeta\omega_n$, corresponds to a time constant of $1/\zeta\omega_n$. Eigenvalues on the positive real axis also possess time constants but represent exponential solutions that increase in time. A value of $s = 0$ implies a time response component of $e^{0t} = 1$, or a constant component: an infinite time constant.

6.3.5 The General Case

Although only first- and second-order systems have been studied above, the characteristics of the free response of all linear systems are not much more complicated.

The procedure for finding the free response is summarized below.

1. Neglect all forcing terms. In the general equations of (6.1a) or (6.1b), let $u_1, u_2, \ldots, u_r$ all vanish.

2. Assume each state variable x_i will possess solutions of the form $X_i e^{st}$. After canceling out the e^{st} terms, Eq. (6.1a) becomes a purely algebraic equation:

$$(s - a_{11})X_1 - a_{12}X_2 - \cdots - a_{1n}X_n = 0,$$

$$-a_{21}X_1 + (s - a_{22})X_2 - \cdots - a_{2n}X_n = 0,$$

$$\vdots$$

$$-a_{n1}X_1 - a_{n2}X_2 - \cdots + (s - a_{nn})X_n = 0. \qquad (6.74)$$

Note that this equation is the general version of Eq. (6.30b) and, in condensed form, can be written $[sI - A][X] = [0]$. This set of equations always possesses a trivial solution, $X_1 = X_2 = \cdots = X_n = 0$, which is the only solution unless

$$\begin{vmatrix} s - a_{11} & -a_{12} & \cdots & -a_{1n} \\ -a_{21} & s - a_{22} & \cdots & -a_{2n} \\ \vdots & \vdots & & \vdots \\ -a_{n1} & -a_{n2} & \cdots & s - a_{nn} \end{vmatrix} = 0. \qquad (6.75)$$

This is the general version of the determinant shown in Eq. (6.31a) and, in condensed form, could be written $\det[sI - A] = 0$.

When the determinant is expanded out into an nth-order polynomial, the characteristic equation results.

3. Solve the characteristic equation for the n eigenvalues $s_1, s_2, \ldots, s_n$. The s-values will either be real or, if complex, occur in complex conjugate pairs. The location of the eigenvalues in the complex plane indicates the type of component of the free response that is associated with the eigenvalues. Figure 6.12 indicates how root location may be correlated with time response. Stable systems possess eigenvalues with negative real parts.

4. For each value of s corresponding to an eigenvalue, Eq. (6.74) yields a partial solution. When the eigenvalues are distinct, that is, when no two of the n eigenvalues are equal, then $n - 1$ of the n equations in Eq. (6.74) will be found to be independent when an eigenvalue is substituted for s.* One could then solve for $n - 1$ of the X's in terms of the last X, for example. The total solution for the free response is

---

*The special case when roots of the characteristic equation are repeated is treated in all books on differential equations. It is rarely important in practice, so we will not treat it here.

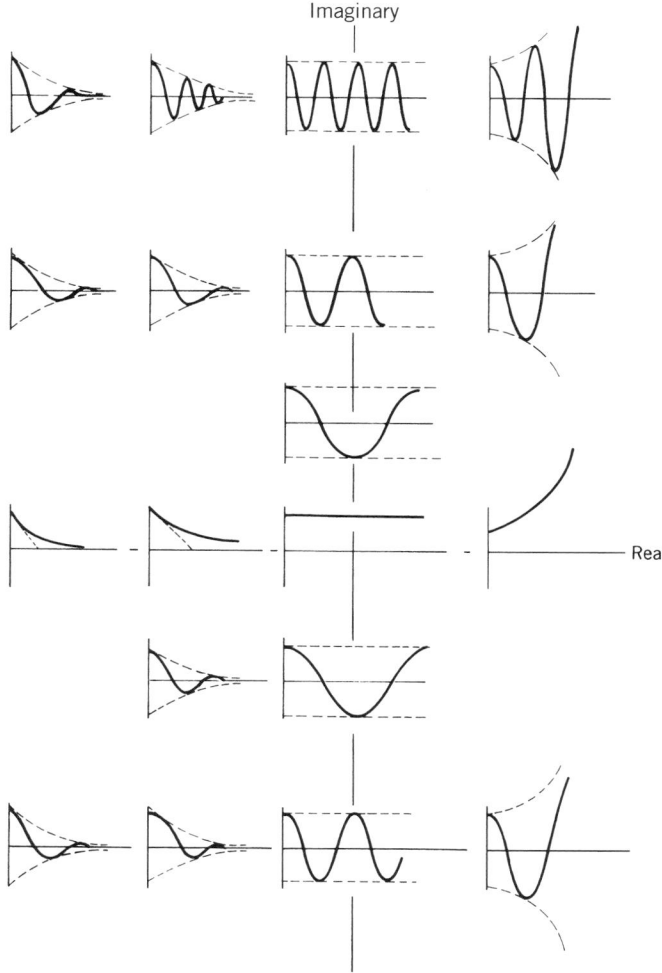

FIGURE 6.12. Correlation of time response with complex eigenvalues.

$$x_1(t) = X_{11}e^{s_1 t} + X_{12}e^{s_2 t} + \cdots + X_{1n}e^{s_n t},$$
$$x_2(t) = X_{21}e^{s_1 t} + X_{22}e^{s_2 t} + \cdots + X_{2n}e^{s_n t},$$
$$\vdots$$
$$x_n(t) = X_{n1}e^{s_1 t} + X_{n2}e^{s_2 t} + \cdots + X_{nn}e^{s_n t}, \tag{6.76}$$

where $X_{11}, X_{21}, X_{31}, \ldots, X_{n1}$ appear in Eq. (6.74) when $s = s_1$; $X_{12}, X_{22}, X_{32}, \ldots, X_{n2}$ appear in Eq. (6.74) when $s = s_2$, and so on. The n^2 values of X in Eq. (6.76) cannot be completely determined by writing Eq. (6.74) for each eigenvalue of s, since in this manner only $n(n-1)$ independent equations will be obtained. The remaining n conditions are represented by the n arbitrary initial conditions

$$x_1(t_0) = x_{10} = X_{11}e^{s_1 t_0} + X_{12}e^{s_2 t_0} + \cdots + X_{1n}e^{s_n t_0},$$

$$x_2(t_0) = x_{20} = X_{21}e^{s_1 t_0} + X_{22}e^{s_2 t_0} + \cdots + X_{2n}e^{s_n t_0},$$

$$\vdots$$

$$x_n(t_0) = x_{n0} = X_{n1}e^{s_1 t_0} + X_{n2}e^{s_2 t_0} + \cdots + X_{nn}e^{s_n t_0}, \qquad (6.77)$$

where $x_{10}, x_{20}, \ldots, x_{n0}$ are the given initial conditions.

Only for very low-order systems is it possible to carry out the operations outlined in Eqs. (6.76) and (6.77) that lead to analytical expressions for the free response. Yet any number of computer simulation programs can numerically solve the unforced equations, given the matrix and the initial conditions. Thus, it is easy to obtain plots of the free response of all the state and output variables if desired.

Furthermore, there are many programs that will calculate the eigenvalues, given only the $[A]$ matrix. Thus, one can understand the qualitative nature of all the components of a system's free response using the ideas behind Figure 6.12 without computing any specific free responses. For example, it is easy to see if the system is unstable by looking for one or more eigenvalues with positive real parts. One can also look for lightly damped oscillations and find their frequencies by using Figure 6.10, and one can find the fastest and slowest response components just by considering the locations of the various eigenvalues in the s-plane. Eigenvalues farthest from the origin of the s-plane in any direction represent the fastest response components, and those closest to the origin represent the slowest response components. Often one can make design decisions about a dynamic system simply by considering the eigenvalue locations.

6.4 FORCED RESPONSE AND FREQUENCY RESPONSE FUNCTIONS

We now turn to the problem of finding solutions to Eqs. (6.1a) or (6.1b) that are compatible with given time functions for $u_1(t), u_2(t), \ldots, u_r(t)$. Again, the principle of superposition will prove useful for the linear system, since it will allow us to find solutions for each input separately and then permit the simple addition of the solutions to find a solution for all the inputs acting simultaneously.

In principle, any time function may be studied as an input, but in practice, sinusoidal forcing is of overwhelming importance. This is partly because periodic forcings may be decomposed into sinusoidal components using the Fourier series and then the effects of the component forcing functions superposed using the principle of superposition, and partly because the testing of real devices is conveniently accomplished in many cases by using sinusoidal inputs at frequencies varied over a range of interest.

For linear systems, it is convenient to represent sinusoidal waves in terms of complex exponentials. A straightforward way to represent sines and cosines may be developed from Eqs. (6.54) and (6.55), but it turns out to be simpler to use the real and imaginary part operators of Eqs. (6.59) and (6.60). Here we will illustrate the use of Eq. (6.59).

Initially, let us return to a first-order system such as the thermal system of Figure 6.1. Since it is hard to imagine the temperature of an oil bath cycling sinusoidally at very high frequencies, let us consider an electrical R–C circuit attached to an oscillatory voltage source as shown in Figure 6.13. The equation for the charge in the capacitor is readily found:

$$\dot{q} = \frac{-1}{RC}q + \frac{E(t)}{R}. \tag{6.78}$$

If we assume that $E(t)$ is sinusoidal with adjustable forcing frequency ω_f,

$$E(t) = E_0 \cos \omega_f t = \text{Re}(E_0 e^{j\omega_f t}), \tag{6.79}$$

then we can search for the *steady-state response*, $q(t)$, that results after any transient response due to the free response and the initial conditions has died away. (Since this system is stable, whatever free response is needed to fit initial conditions will indeed decay in time, leaving only the steady-state or forced response.)

For the linear operations of multiplication by a constant and differentiation or integration with respect to time, it can be shown that it is immaterial whether one first operates on a complex function and then takes the real part of the result or first takes the real part and then performs the operation. Thus, it is possible to suppress the Re operator in Eq. (6.79) until all operations are completed and then take the real part of the final result. To find the forced response of the system, it is only necessary to assume the system variables are all sinusoidal with frequency ω_f. (This assumption is not valid for nonlinear systems.) In the present example, we assume

$$q(t) = \text{Re}(Qe^{j\omega_f t}). \tag{6.80}$$

Substituting Eqs. (6.79) and (6.80) into Eq. (6.78) and suppressing Re, we have

$$j\omega_f Q e^{j\omega_f t} = \frac{-Qe^{j\omega_f t}}{RC} + \frac{E_0}{R}e^{j\omega_f t}, \tag{6.81}$$

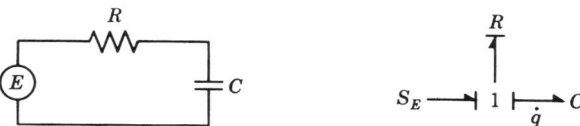

FIGURE 6.13. An electrical circuit analogous to the thermal system of Figure 6.1.

which leads immediately to

$$Q = \frac{E_0/R}{j\omega_f + 1/RC} = \frac{CE_0}{RCj\omega_f + 1}. \tag{6.82}$$

Although we assumed that E_0 was a real sinusoidal amplitude in Eq. (6.79), it turns out that Q in Eq. (6.80) needs to be complex. At very low forcing frequencies, Q is almost purely real,

$$Q \to CE_0 \quad \text{as} \quad \omega_f \to 0; \tag{6.83}$$

but for high frequencies, Q is almost purely imaginary,

$$Q \to \frac{E_0}{Rj\omega_f} = \frac{-jE_0}{R\omega_f} \quad \text{as} \quad \omega_f \to \infty. \tag{6.84}$$

We also note that for high forcing frequencies, the magnitude of Q is proportional to $1/\omega_f$ and thus approaches zero as the frequency approaches infinity.

Let us now plot Q in Eq. (6.82) as a complex number for $0 < \omega_f < \infty$. As an intermediate step, consider the denominator of Q, $j\omega_f + 1/RC$. This is shown in Figure 6.14a. Now Q itself is just E_0/R times the reciprocal of this complex number.

Recall, if you can, that the inverse of a complex number has a magnitude that is the inverse of the magnitude of the number and an angle that is the negative of the angle of the number. Using this fact, one can easily sketch the behavior of $(j\omega_f + 1/RC)^{-1}$ as ω_f varies as in Figure 6.14b. (Perhaps you can convince yourself that the curve in Figure 6.14b is a semicircle.)

Since Q is just E_0/R times the complex number shown in Figure 6.14b, we see that the angle of Q varies from 0 at $\omega_f = 0$ to $-90°$, or $-\pi/2$, as $\omega_f \to \infty$. What this means is that if we think of $E(t)$ in Eq. (6.79) as the real part of a phasor or rotating vector, then $q(t)$ in Eq. (6.80) is represented by a rotating vector that lags behind the vector $E(t)$ by an angle θ (dependent on ω_f) that is somewhere between zero and 90°. In addition, the magnitude of the vector representing q varies from CE_0 to zero as ω_f varies from zero to infinity.

A sketch of the rotating vectors is shown in Figure 6.15 for some particular ω_f. Since Q is shown as having an angle of about $-45°$, when Q is multiplied by $e^{j\omega_f t}$, the result is a vector rotating with speed ω_f but lagging about 45° behind $e^{j\omega_f t}$, since when one multiplies two complex numbers, one must multiply the magnitudes and add the phase angles. The vector representing $E(t)$, however, rotates with $e^{j\omega_f t}$, since E_0 is just a real number (zero angle).

By direct manipulation of Eq. (6.82), or by considering the sketches in Figure 6.14, we can write an expression for the angle of Q,

$$\theta = \angle Q = -\tan^{-1}\frac{\omega_f}{1/RC} = -\tan^{-1} RC\omega_f. \tag{6.85}$$

The angle of Q is often called the *phase angle*, and it is *negative*, or *lagging*, in this case. This notation is related to the rotating vectors of Figure 6.15. The amplitude or

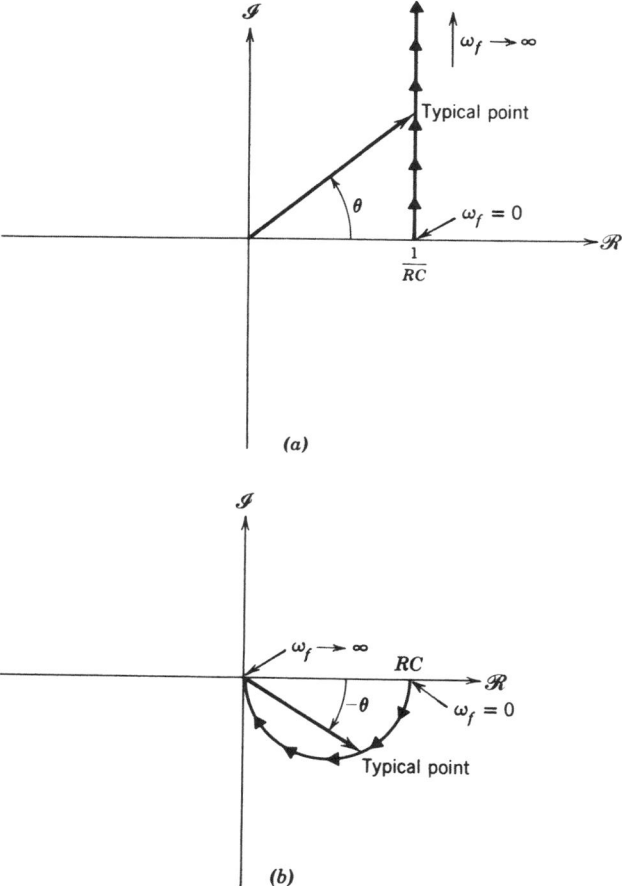

FIGURE 6.14. Plots of (a) $j\omega_f + 1/RC$ and (b) $(j\omega_f + 1/RC)^{-1}$ as ω_f varies.

magnitude of Q is

$$|Q| = \frac{E_0/R}{\left[\omega_f^2 + (1/RC)^2\right]^{1/2}} = \frac{CE_0}{\left[(RC\omega_f)^2 + 1\right]^{1/2}}. \tag{6.86}$$

Note that, except for the real factor E_0/R, both $\angle Q$ and $|Q|$ are essentially plotted for all ω_f in Figure 6.14b.

It is also common to plot $\angle Q$ and $|Q|$ (or $\angle Q/E_0$ and $|Q/E_0|$) separately as a function of ω_f. These plots are called *Bode plots* when $|Q|$ versus ω_f is plotted on log–log scales and when $\angle Q$ is plotted versus $\log \omega_f$. Such plots are shown in Figure 6.16. Bode plots are convenient because they can often be quickly sketched using asymptotic behavior of the functions at low and high frequencies.

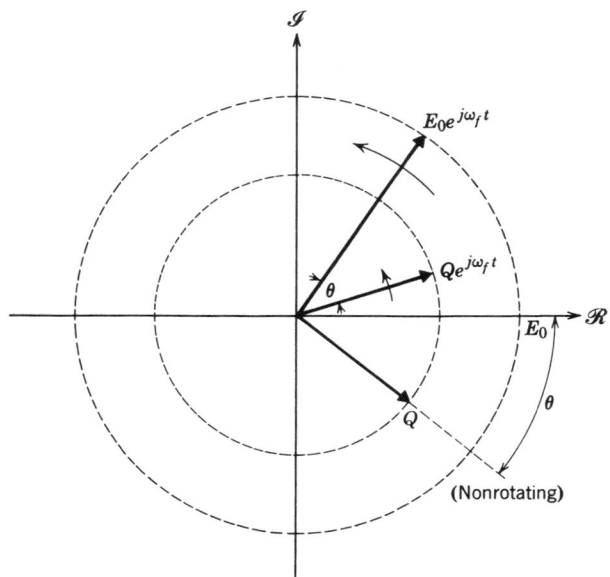

FIGURE 6.15. Rotating vectors representing $E(t)$ and $q(t)$ in Eqs. (6.79) and (6.80).

In our case, we see that Q/CE_0 approaches unity for $\omega_f \to 0$ from Eq. (6.83). This means unit amplitude and zero phase angle. At high frequencies Eq. (6.84) tells us that

$$\frac{Q}{E_0 C} \to -j \frac{1}{(RC)\omega_f} \qquad \text{as} \quad \omega_f \to \infty.$$

This means

$$\frac{\angle Q}{E_0 C} \to -90°, \qquad \left| \frac{Q}{E_0 C} \right| \to \frac{1}{RC}\omega_f^{-1} \tag{6.87}$$

or

$$\log \left| \frac{Q}{E_0 C} \right| \to \log \left(\frac{1}{RC} \right) - \log \omega_f. \tag{6.88}$$

Thus, when $\log |Q/E_0 C|$ is plotted versus $\log \omega_f$, the result will be asymptotic to a straight line with slope -1 at high frequencies.

The low-frequency asymptote and the high-frequency asymptote intersect at

$$RC\omega_f = 1, \qquad \text{or} \quad \tau\omega_f = 1, \tag{6.89}$$

using the definition of the time constant in Eq. (6.20). The frequency $\omega_f = 1/\tau$ is called the *break frequency*, since this is where the low-frequency asymptote breaks into the high-frequency asymptote. At this frequency, the actual value of $|Q/E_0 C|$

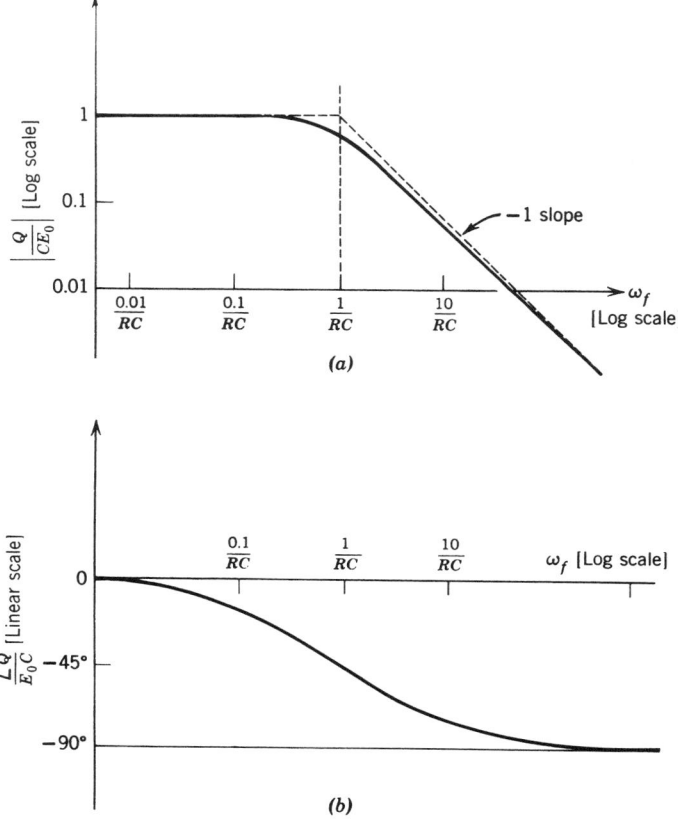

FIGURE 6.16. Bode plots for $Q/CE_0 = 1/(RCj\omega_f + 1)$.

is $1/\sqrt{2}$, as one may see from Eq. (6.86), and $\angle Q/EC$ is $-45°$. For a first-order system, the time constant and its reciprocal, the break frequency, characterize both the speed of the free response and the frequency scale of the sinusoidal response.

We now turn our attention to our second-order example, which has two state variables and two possible input variables. Suppose that the velocity source of Figure 6.3 is an electromagnetic shaker capable of forcing the input velocity $V(t)$ in Eq. (6.23) to be sinusoidal:

$$V(t) = V_0 \cos \omega_f t = \mathrm{Re}(V_0 e^{j\omega_f t}). \tag{6.90}$$

To find a forced solution, we again assume that the state variables will also be sinusoidal with the same frequency as the input quantity:

$$x(t) = \mathrm{Re}(X e^{j\omega_f t}), \tag{6.91}$$

$$p(t) = \mathrm{Re}(P e^{j\omega_f t}). \tag{6.92}$$

Substituting Eqs. (6.90), (6.91), and (6.92) into Eq. (6.23) and neglecting F temporarily, we have

$$j\omega_f X e^{j\omega_f t} = \frac{-1}{m} P e^{j\omega_f t} + V_0 e^{j\omega_f t}, \tag{6.93}$$

$$j\omega_f P e^{j\omega_f t} = kX e^{j\omega_f t} - \frac{b}{m} P e^{j\omega_f t} + bV_0 e^{j\omega_f t}, \tag{6.94}$$

in which the Re operator has been suppressed.

These equations could also be written in a single matrix equation using essentially the same technique that led to Eq. (6.30b):

$$\begin{bmatrix} j\omega_f & 1/m \\ -k & j\omega_f + b/m \end{bmatrix} \begin{bmatrix} X \\ P \end{bmatrix} e^{j\omega_f t} = \begin{bmatrix} 1 \\ b \end{bmatrix} V_0 e^{j\omega_f t}. \tag{6.94a}$$

The problem now is an algebraic problem in which X and P are to be determined. The equations could be satisfied if $e^{j\omega_f t}$ should vanish, but this is not possible [and even $\mathrm{Re}(e^{j\omega_f t})$ vanishes only at particular instants of time], so one attempts to satisfy the equations independent of $e^{j\omega_f t}$:

$$j\omega_f X + \frac{1}{m} P = V_0, \tag{6.95}$$

$$-kX + \left(j\omega_f + \frac{b}{m} \right) P = bV_0. \tag{6.96}$$

In matrix form, the equations become,

$$\begin{bmatrix} j\omega_f & 1/m \\ -k & j\omega_f + b/m \end{bmatrix} \begin{bmatrix} X \\ P \end{bmatrix} = \begin{bmatrix} 1 \\ b \end{bmatrix} \tag{6.96a}$$

in which the coefficient matrix can be recognized as $[j\omega_f I - A]$. That this matrix is the matrix in Eq. (6.30c) with s replaced by $j\omega_f$ should be evident.

Note that there is some similarity between Eqs. (6.95), (6.96), or (6.96a) and Eq. (6.30b) for the free response, but in the present case $j\omega_f$ is given and the right-hand sides of the equations are nonzero. These equations are readily solved as long as the determinant of the coefficients is not zero. For example, using Cramer's rule, we have

$$X = \frac{\begin{vmatrix} V_0 & 1/m \\ bV_0 & j\omega_f + b/m \end{vmatrix}}{\begin{vmatrix} j\omega_f & 1/m \\ -k & j\omega_f + b/m \end{vmatrix}} = \frac{(j\omega_f)V_0}{-\omega_f^2 + (b/m)j\omega_f + k/m}, \tag{6.97}$$

$$P = \frac{\begin{vmatrix} j\omega_f & V_0 \\ -k & bV_0 \end{vmatrix}}{\begin{vmatrix} j\omega_f & 1/m \\ -k & j\omega_f + b/m \end{vmatrix}} = \frac{(bj\omega_f + k)V_0}{-\omega_f^2 + (b/m)j\omega_f + k/m}. \tag{6.98}$$

Note that X and P, the amplitudes of the complex exponential representation of the state-variable forced response, are ratios of two complex numbers that vary with ω_f. The denominator is always the same; it is the same determinant used to generate the characteristic equation (6.31), but with s replaced by $j\omega_f$.

It is not difficult to evaluate X and P for any given ω_f, but it is of even more interest to determine how X and P vary as ω_f varies from zero to infinity. In Figure 6.17a the denominator expression for Eqs. (6.97) and (6.98) is plotted as a complex number as ω_f varies. In Figure 6.17b the reciprocal complex number is plotted. These plots are the second-order analogs of the plots in Figure 6.14 for the first-order system.

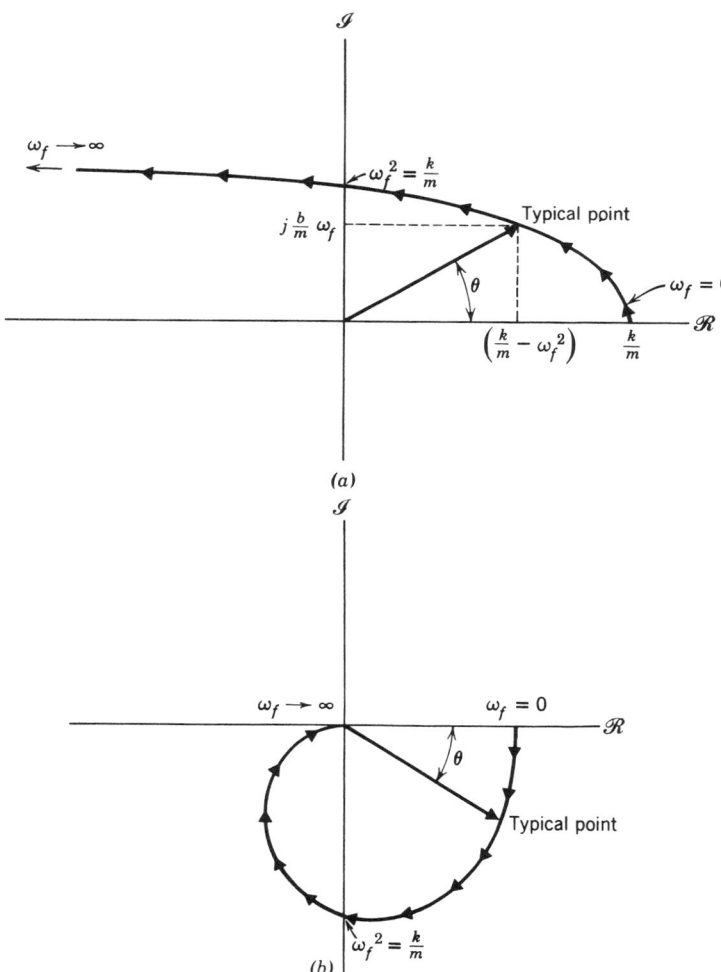

(a)

(b)

FIGURE 6.17. Plots of (a) $k/m - \omega_f^2 + j(b/m)\omega_f$ and (b) $[k/m - \omega_f^2 + j(b/m)\omega_f]^{-1}$ in the complex plane as ω_f varies from $\omega_f = 0$ to $\omega_f \to \infty$.

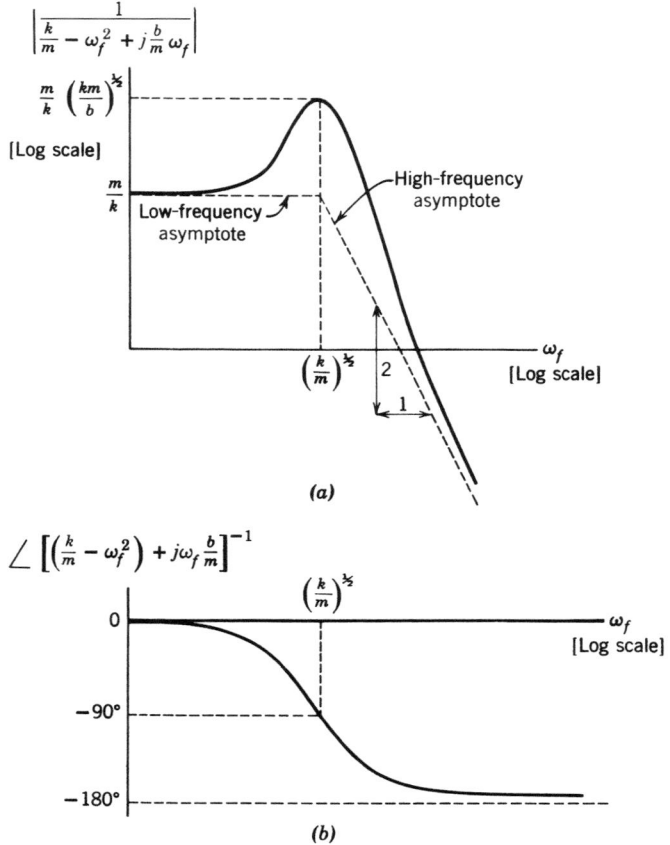

FIGURE 6.18. Bode plots of (a) amplitude and (b) phase of $[(k/m - \omega_f^2) + j(b/m)\omega_f]^{-1}$.

Bode plots for the amplitude and phase of the denominator factor are shown in Figure 6.18. The general shape of the Bode plots may be seen by studying Figure 6.17b, but it is useful to consider some asymptotic cases in order to determine the plots more precisely.

For a start, consider very low frequencies:

$$\lim_{\omega_f \to 0} \frac{1}{k/m - \omega_f^2 + j\omega_f(b/m)} = \frac{1}{k/m} = \frac{m}{k}. \tag{6.99}$$

Therefore, a horizontal asymptote may be drawn on the left side of Figure 6.18a for very low frequencies. (Note that because of the logarithmic scale $\omega_f = 0$ cannot appear on the sketches, but the far left on the diagrams corresponds to very small values of ω_f.)

At the *undamped* natural frequency the expression also simplifies:

$$\frac{1}{\left[(k/m) - \omega_f^2\right] + j\omega_f(b/m)}\Bigg|_{\omega_f=(k/m)^{1/2}}$$

$$= \frac{1}{j(k/m)^{1/2}(b/m)}$$

$$= -j\left(\frac{m}{k}\right)\frac{(mk)^{1/2}}{b} = \frac{m}{k}\frac{(mk)^{1/2}}{b}\angle\frac{-\pi}{2}. \tag{6.100}$$

At the undamped natural frequency the magnitude is as given in Eq. (6.100), and the phase angle is exactly $-90°$, or $-\pi/2$ radians. Finally, for very high frequencies we have

$$\lim_{\omega_f\to\infty}\frac{1}{k/m - \omega_f^2 + j\omega_f(b/m)} = \frac{1}{-\omega_f^2} = \frac{1}{\omega_f^2}\angle -\pi. \tag{6.101}$$

Equation (6.101) indicates that the magnitude of the function varies as ω_f^{-2} and its phase is $-\pi$ radians, or $-180°$, at high frequencies. This indicates that the logarithm of the amplitude is -2 times the logarithm of ω_f—hence the two-to-one negative slope shown on the amplitude plot. The reader should verify that high- and low-frequency asymptotes intersect, as shown in Figure 6.18a, when extended to the undamped natural frequency. The use of asymptotes that plot as straight lines with slope m on a log–log plot, corresponding to terms of the form ω_f^m where m is a positive or negative integer, proves to be of general usefulness.

Now the plots of Figure 6.18 may be used to plot the frequency behavior of X and P from Eqs. (6.97) and (6.98). The numerator functions for X and P (apart from the constant factor V_0) have been plotted in Figure 6.19 using the same concepts as in Figure 6.18. To sketch X and P as a function of ω_f, we must multiply the complex numbers plotted in Figure 6.18 by those in Figure 6.19. This can be done by adding the phase angles and multiplying the amplitudes, or equivalently, *adding the logarithms of the magnitudes* at each ω_f. This process is illustrated in Figures 6.20 and 6.21.

It is also possible to construct asymptotic expressions for X and P directly from their defining relations, Eqs. (6.97) and (6.98), but the method just illustrated indicates how the common denominator expression interacts with different numerators to produce the frequency response functions for all the state variables.

Finally, it remains to interpret the significance of $|X|$, $|P|$, $\angle X$, and $\angle P$ in light of Eqs. (6.91) and (6.92). In Figure 6.22 a rotating phasor representing $V(t) = V_0 e^{j\omega_f t}$ is shown, and the complex number X, constant for any particular ω_f, is also shown. The rotating phasor representing $x(t) = X e^{j\omega_f t}$ has amplitude $|X|$ and an angle $\angle X$ relative to V (shown negative). Thus, in this case, the $x(t)$ phasor lags the $V(t)$ phasor by the phase angle shown in Figure 6.20. Actually, we consider that $V(t)$ is the real part of a rotating phasor (a cosine wave in this case) and $x(t)$ is the real part

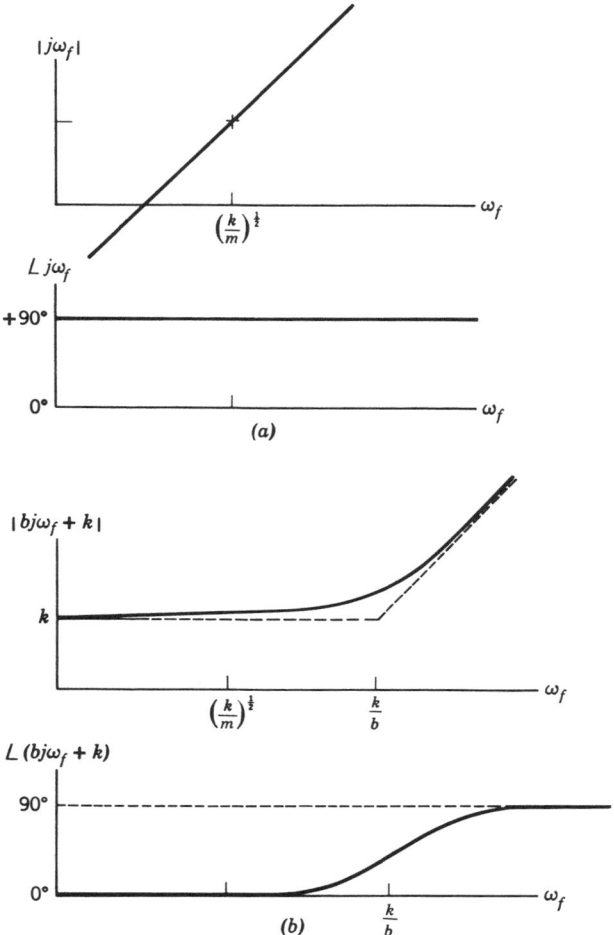

FIGURE 6.19. Bode plots for the numerators of (a) Eq. (6.97) and (b) Eq. (6.98).

of $Xe^{j\omega_f t}$ (a cosine wave with amplitude $|X|$ and a lagging phase angle given by $\angle X$):

$$x(t) = |X| \cos(\omega_f t - \angle X). \qquad (6.102)$$

Of course, it is possible to encounter leading phase angles also.

6.4.1 Normalization of Response Curves

Although, up to now, only the physical parameters m, b, and k have been used in studying the frequency response, it sometimes proves convenient and revealing to nondimensionalize expressions as much as possible. As a first step in this procedure,

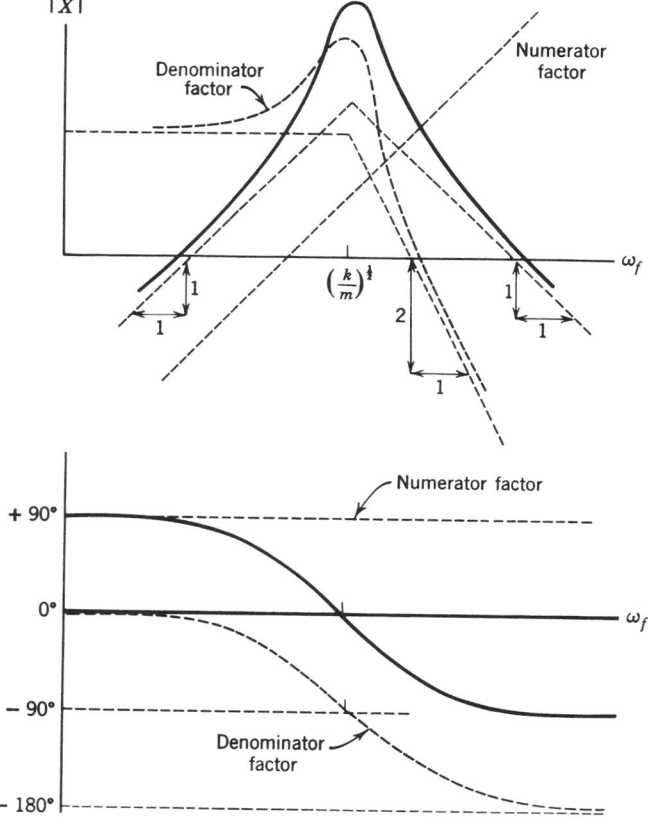

FIGURE 6.20. Bode plot construction for X.

consider rewriting the response quantities X and P in Eqs. (6.97) and (6.98) in terms of damping ratio and undamped natural frequency, Eqs. (6.41) and (6.66):

$$\frac{X}{V_0} = \frac{j\omega_f}{\omega_n^2 - \omega_f^2 + j2\zeta\omega_n\omega_f}, \qquad \frac{P}{V_0} = \frac{m(j2\zeta\omega_n\omega_f + \omega_n^2)}{\omega_n^2 - \omega_f^2 + j2\zeta\omega_n\omega_f}.$$

By defining a frequency ratio

$$\beta \equiv \frac{\omega_f}{\omega_n}, \tag{6.103}$$

further simplification is achieved:

$$\frac{X}{V_0} = \frac{(m/k)^{1/2} j\beta}{(1 - \beta^2) + j2\zeta\beta}, \tag{6.104}$$

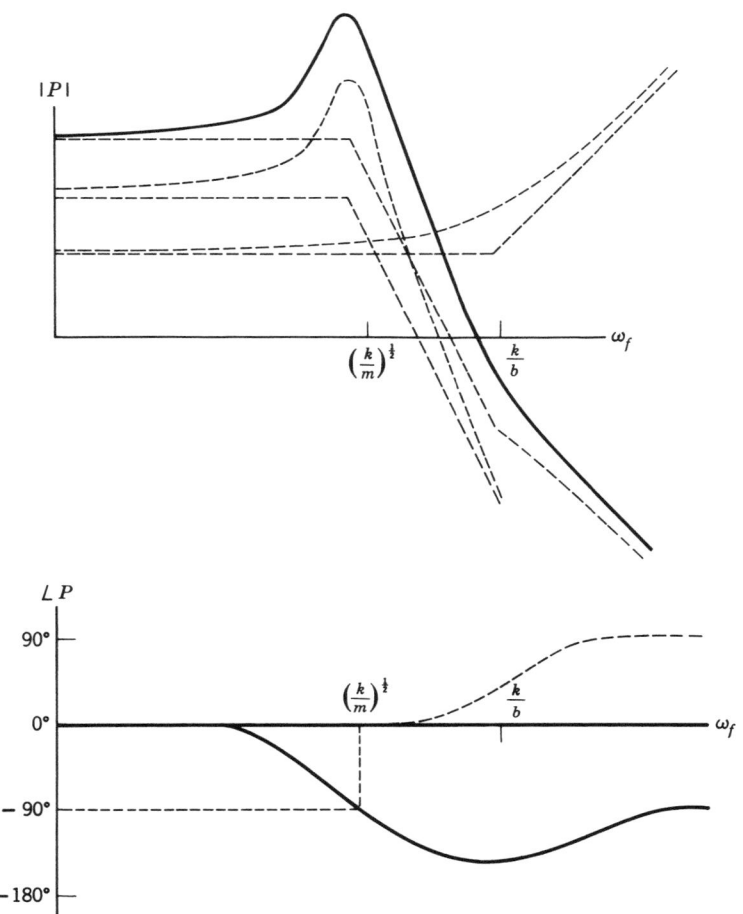

FIGURE 6.21. Bode plot construction for P.

$$\frac{P}{V_0} = \frac{m(j2\zeta\beta + 1)}{(1 - \beta_+^2 j2\zeta\beta)}. \tag{6.105}$$

Since all aspects of the sinusoidal forced response of the system contain the function

$$H(\beta) = \left[(1 - \beta^2) + j2\zeta\beta\right]^{-1}, \tag{6.106}$$

it is useful to show Bode plots for H, as in Figure 6.23. Note that by using β, which essentially involves the definition of nondimensional frequency (or nondimensional time), only ζ remains as a parameter in H. In computing actual frequency response quantities such as X and P, however, some physical parameters are necessary in general, as illustrated in Eqs. (6.104) and (6.105).

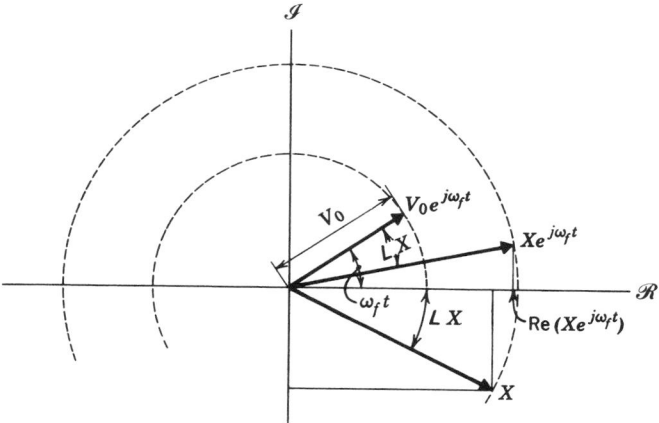

FIGURE 6.22. Representation of $V(t)$ and $x(t)$ in terms of complex exponentials.

6.4.2   The General Case

For any state-space equations of the form (6.1a) the pattern of analysis for computing the sinusoidal response is just the same as has been illustrated in the example.

1. If there are multiple inputs, the responses may be computed individually and later simply summed in order to find the total forced response. Pick one input, say $u_1(t)$, and if it is sinusoidal, represent it in complex exponential form, for example,

$$u_1(t) = \text{Re}(U_1 e^{j\omega_f t}).$$

2. Let each x-variable be represented as a complex exponential, for example,

$$x_1(t) = \text{Re}(X_1 e^{j\omega_f t}).$$

Since only linear operations will be required, one may temporarily suppress the Re operator and simply apply it to the final result.

3. Substituting into Eq. (6.1a), one is faced with the algebraic problem below:

$$(j\omega_f - a_{11})X_1 + (-a_{12})X_2 + \cdots + (a_{1n})X_n = b_{11}U_1,$$
$$(-a_{21})X_1 + (j\omega_f - a_{22})X_2 + \cdots + (-a_{2n})X_n = b_{21}U_1,$$

$$\cdots$$

$$(-a_{n1})X_1 + (-a_{n2})X_2 + \cdots + (j\omega_f - a_{nn})X_n = b_{n1}U_1.$$

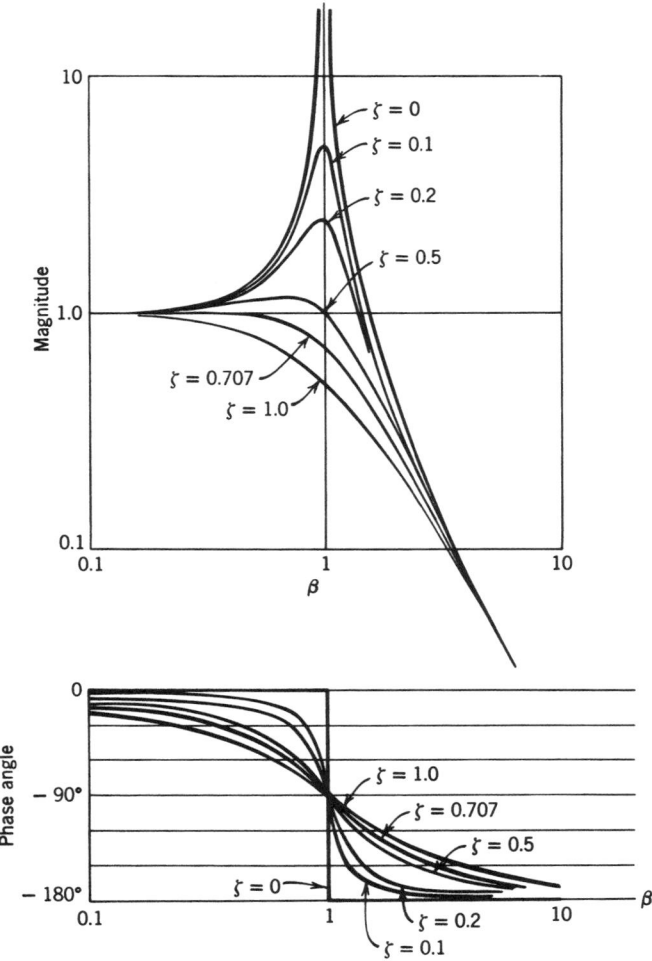

FIGURE 6.23. Bode plots for normalized second-order response function, Eq. (6.106).

In matrix form these equations could be written

$$[j\omega_f I - A][X] = \begin{bmatrix} b_{11} \\ b_{21} \\ \vdots \\ b_{n1} \end{bmatrix} U_1.$$

4. When this equation is solved, using Cramer's rule or any equivalent method, the result is a set of complex numbers for $X_1, X_2, \ldots, X_n$. When these are multiplied by $e^{j\omega_f t}$ and the real part operator is applied, the forced-response computation is complete. Polar plots in the complex plane as in Figures 6.14

and 6.17, or Bode plots such as in Figures 6.16, 6.18, and 6.21, may aid in interpretation of the forced response as a function of the forcing frequency.

6.5   TRANSFER FUNCTIONS

The term *transfer function* is much used in connection with linear systems and usually refers to a generalization of the frequency concepts presented in the previous section. One way to define a transfer function, $H_{y/x}$, between two quantities, x and y, is

$$H_{y/x} \equiv \left. \frac{y}{x} \right|_{x=e^{st}} = H_{y/x}(s), \tag{6.107}$$

where the notation is meant to imply that if the *forced response* of y is computed when $x = e^{st}$, then the ratio y/x will be $H_{y/x}$ and will depend on s. When $s = j\omega_f$, $H_{y/z}(j\omega_f)$ is the function that determines the relative amplitude and phase angle between sinusoidal waveforms representing x and y.

To illustrate the computation of a transfer function, let us compute a transfer function for our second-order example system using f in Eq. (6.24) as the quantity y, and F in Eq. (6.23) as the quantity x. For state equations, transfer function between inputs u and outputs y are commonly computed. For our present purposes we shall consider that V in Eq. (6.23) vanishes, although there is also a transfer function between f and V that could be computed.

If

$$F(t) = e^{st} \tag{6.108}$$

for forced response, it may be assumed that all variables have the same time dependence, but with amplitude factors to be determined:

$$x(t) = Xe^{st}, \tag{6.109}$$

$$p(t) = Pe^{st}. \tag{6.110}$$

Then, from Eq. (6.24),

$$f(t) = kXe^{st} - \frac{b}{m}Pe^{st} = \left(kX - \frac{b}{m}P\right)e^{st}. \tag{6.111}$$

To find X and P, one again substitutes the assumed forms into the state equations:

$$sXe^{st} = -\frac{1}{m}Pe^{st}, \qquad sPe^{st} = kXe^{st} - \frac{b}{m}Pe^{st} + e^{st}.$$

Upon suppressing the factors e^{st}, the following algebraic equations emerge:

$$sX + \frac{P}{m} = 0, \tag{6.112}$$

$$-kX + \left(s + \frac{b}{m}\right) P = 1. \tag{6.113}$$

This problem closely resembles the problem of Eqs. (6.95) and Eq. (6.96) except that s has replaced $j\omega_f$ and the input is F instead of V.

Solving for X and P using Cramer's rule, we have

$$X = \frac{\begin{vmatrix} 0 & 1/m \\ 1 & s+b/m \end{vmatrix}}{\begin{vmatrix} s & 1/m \\ -k & s+b/m \end{vmatrix}} = \frac{-1/m}{s^2 + (b/m)s + k/m}, \tag{6.114}$$

$$P = \frac{\begin{vmatrix} s & 0 \\ -k & 1 \end{vmatrix}}{\begin{vmatrix} s & 1/m \\ -k & s+b/m \end{vmatrix}} = \frac{s}{s^2 + (b/m)s + k/m}. \tag{6.115}$$

Now, using Eq. (6.111), the result is

$$f(t) = \frac{-(k/m) - (b/m)s}{s^2 + (b/m)s + k/m} e^{st}, \tag{6.116}$$

and the desired transfer function, $H_{f/F}(s)$, is

$$H_{f/F}(s) = \frac{f(t)}{F(t)}\bigg|_{F(t)=e^{st}} = \frac{-[(b/m)s + k/m]}{s^2 + (b/m)s + k/m} = \frac{-(2\zeta\omega_n s + \omega_n^2)}{s^2 + 2\zeta\omega_n s + \omega_n^2}. \tag{6.117}$$

Clearly, a transfer function immediately yields a frequency response function if $j\omega_f$ is substituted for s. For example, if

$$F(t) = A\cos\omega_f t = \mathrm{Re}(Ae^{j\omega_f t}), \tag{6.118}$$

then

$$f(t) = \mathrm{Re}\left(H_{f/F}(j\omega_f)Ae^{j\omega_f t}\right). \tag{6.119}$$

6.5.1   Block Diagrams

Block diagrams are commonly used to represent transfer functions, and the block diagram shown in Figure 6.24a indicates that the output signal f is the product of the input signal F and the transfer function $H_{f/F}$.

For signals with the time dependence e^{st}, differentiation is equivalent to multiplication by s, integration to dividing by s. Thus, the transfer function for an integrator is $1/s$. Using this notion, the block diagram of Figure 6.24b represents the state equations (6.23) and (6.24) directly and shows that causally F is an input and f is an output. It is possible to show that Figure 6.24b is equivalent to Figure 6.24a using essentially graphical means. The major tool is the well-known relation shown in Figure 6.25 for combining feedback loops. Although much of classical control

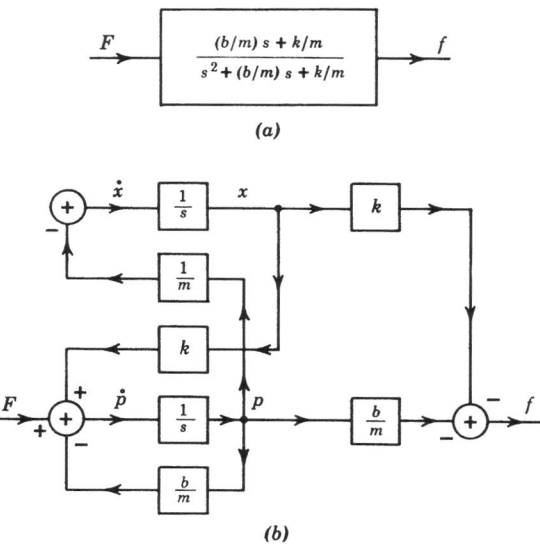

FIGURE 6.24. Block diagrams representing transfer functions between F and f. (*a*) Combined transfer function; (*b*) direct representation of state and output equations.

theory and linear system analysis is concerned with such techniques, it will suffice for present purposes to point out that since bond graphs generate highly organized state equations, only the simple algebra needed to solve equations such as (6.112) and (6.113) is required to generate transfer functions between any input and any output. There are some cases where block diagram reduction techniques provide convenient guides to accomplishing algebraic manipulation, but for large systems computer-automated algebraic methods presented in Reference [2], for example, are far superior to graphical techniques in avoiding human error.

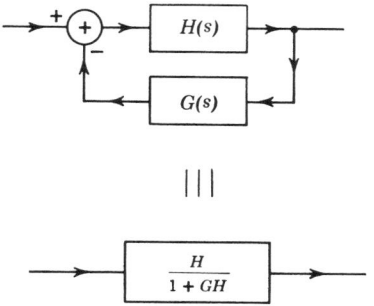

FIGURE 6.25. Block diagram reduction for feedback loops.

6.6 TOTAL RESPONSE

The total response of a system is determined by initial conditions and given input time functions; for linear systems, it is the sum of a free response and a forced response. It is possible to compute the total response using the methods of Sections 6.3 and 6.4, although for most systems of practical interest it is more practical simply to simulate the system using a computer. It is, however, important to understand the nature of the total response to be expected from a simulation.

To understand the role of the initial conditions and the forcing on total response, let us consider the problem of determining the response of the second-order example system to a suddenly applied cosine force. Let

$$F(t) = F_0 \cos \omega_f t = \text{Re}(F_0 e^{j\omega_f t}), \qquad t \geq 0, \tag{6.120}$$

and

$$x(0) = p(0) = 0. \tag{6.121}$$

This example will serve to point out the fact that vanishing initial conditions do not imply vanishing free response when a forcing term is present.

The forced response is easily found, since in computing the transfer function in the previous section, the amplitudes X and P were found when was e^{st}. Equations (6.114) and (6.115) may be modified by multiplying by the amplitude factor F_0 and substituting $s = j\omega_f$ in order to construct the forced solution, x_{forced} and p_{forced}:

$$x_{\text{forced}} = \text{Re}\left(\frac{-F_0/m}{-\omega_f^2 + (b/m)j\omega_f + k/m} e^{j\omega_f t}\right), \tag{6.122}$$

$$p_{\text{forced}} = \text{Re}\left(\frac{-j\omega_f F_0}{-\omega_f^2 + (b/m)j\omega_f + k/m} e^{j\omega_f t}\right). \tag{6.123}$$

Suppose $\omega_f < \omega_n$, so that the phase angle sketched in Figure 6.18 is between $0°$ and $90°$, and $\zeta < 1$, so that there is some amplitude magnification over the amplitude at low frequency. Then

$$x_{\text{forced}} = -A_x \cos(\omega_f t - \varphi), \tag{6.122a}$$

$$p_{\text{forced}} = -A_p \cos(\omega_f t - \varphi). \tag{6.123a}$$

It is easy to see in Eq. (6.122) that x_{forced} is a negative cosine wave, since φ accounts for the phase lag and the numerator contributes a negative sign. In Eq. (6.123) we must first imagine the rotating vector associated with $j\omega_f e^{j\omega_f t}$. This vector is rotated $90°$ *ahead* of the vector $e^{j\omega_f t}$, and its real part is found to describe a negative sine wave. Another way to see this is to note that in a transfer function sense, multiplication by $j\omega_f$ is equivalent to differentiation, and the derivative of $\cos \omega_f t$ is $-\omega_f \sin \omega_f t$. The functions x_{forced} and p_{forced} are sketched in Figure 6.26.

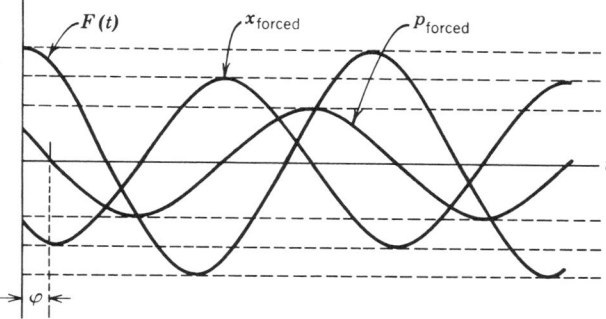

FIGURE 6.26. Sketches of forced-response quantities.

Note that the forced response satisfies (or balances) the forced state equations at all times (even $t < 0$), and its only defect as a complete solution is that at $t = 0$ the forced response does not satisfy the initial conditions of Eq. (6.121). One may think of x_{forced} and p_{forced} as the system response that would have occurred if the cosine force had been applied a long time in the past and if the system had settled down to a steady state. (This interpretation is useful for stable systems that really do settle down.)

Even though x_{forced} and p_{forced} satisfy the forced state equations at all times, the free response satisfies the state equations without F, and thus, if we write

$$x_{total} = x_{free} + x_{forced}, \tag{6.124}$$

$$p_{total} = p_{free} + p_{forced} \tag{6.125}$$

and substitute into the state equations, we find that the equations are still satisfied. Now the amplitudes of the free responses may be adjusted so that the *total* response satisfies the initial conditions. For the given zero initial conditions, Eq. (6.121), this results in

$$x_{free}(0) = -x_{forced}(0), \tag{6.126}$$

$$p_{free}(0) = -p_{forced}(0). \tag{6.127}$$

Equations (6.126) and (6.127) may now be used in Eqs. (6.38) and (6.39) after the forced response at $t = 0$ is extracted from Eqs. (6.122) and (6.123). Then, using Eqs. (6.35) and (6.36), enough conditions have been found to completely determine the free response.

Although the determination of the free response in detail for a particular case is straightforward, even for a simple second-order system the process is rather tedious. The solutions are not hard to sketch, as shown in Figure 6.27. Note that the free responses start at $t = 0$ at values that are just the negative of the values of the forced response and then decay as damped oscillations at the system *damped natural frequency* and with the system damping ratio.

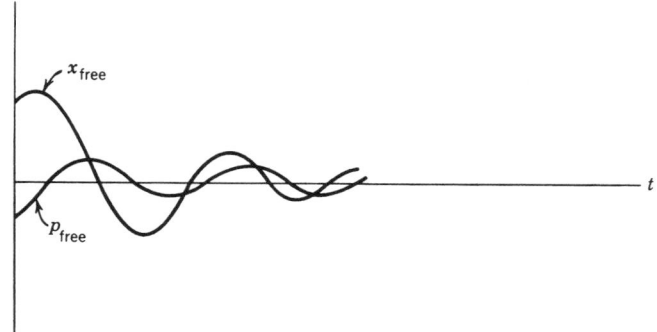

FIGURE 6.27. Sketches of free-response quantities.

When the free and forced responses are summed, the result is as shown in Figure 6.28. Note that at the start of the transient, the solution appears confused because the free response at the system frequency interacts with the forced response at the forcing frequency. After some time, however, the free response becomes very small for this stable system, due to the factor $e^{-\zeta \omega_n t}$, and the forced response dominates. If the system were unstable, of course, the free response would blow up as time increased and the forced response would become unimportant.

The major characteristics worth keeping in mind for the total response of linear systems are as follows:

1. The free-response components of all state variables (and hence, output variables) always have a time behavior dependent on the system eigenvalues; that is, oscillatory response components will be associated with each pair of complex eigenvalues, and a natural frequency and damping ratio may be associated with each pair; exponential response components will be associated with each real eigenvalue, and a time constant may be associated with each such eigenvalue. Changing the initial conditions changes the amplitudes of the free-response components of the total response.

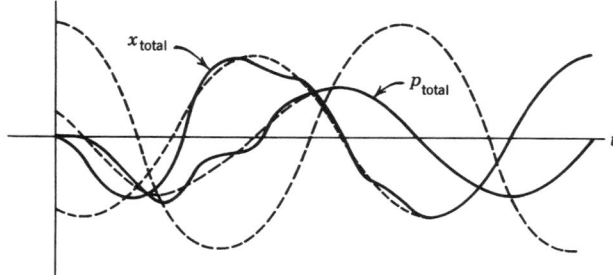

FIGURE 6.28. Sketches of total response.

2. The forced-response components of all state variables all have a time dependence determined by the given inputs. If an input is sinusoidal, all state variables will contain a forced component at the same frequency as the input. The forced components of response are directly proportional to the amplitude of the input and are not influenced by initial conditions.

When simulation is used to find the system response, only the total response will be found, and it is a mental exercise to imagine the splitting of the response into the free and forced parts. If the input is itself transient (in contrast to the case of sinusoidal waves or steps, which go on forever), the concept of splitting the total response into free and forced components though valid, may not be very useful.

It is also important to remember that nonlinear systems may react in much more complicated ways than linear systems. Nonlinear systems excited by sinusoidal inputs can, in contrast to linear systems, respond with periodic solutions containing frequencies other than the exciting frequency, and initial conditions can have a drastic effect on the nature of the response of a nonlinear system to a given forcing function.

6.7   ALTERNATIVE STATE VARIABLES

Although only physically meaningful energy state variables (momenta and displacements) have been used in this book, it is possible and occasionally useful to consider other equivalent state variables. Particularly for linear systems, there are a number of theoretically interesting sets of alternative state variables that can be defined. The state equations for the alternative state variables often exhibit certain properties of the dynamics of the system that are not so obvious in the equations for other state variables. For example, the state equations for the undamped oscillator, given previously,

$$\begin{bmatrix} \dot{x} \\ \dot{p} \end{bmatrix} = \begin{bmatrix} 0 & -1/m \\ k & 0 \end{bmatrix} \begin{bmatrix} x \\ p \end{bmatrix} + \begin{bmatrix} 1 & 0 \\ 0 & 1 \end{bmatrix} \begin{bmatrix} V \\ F \end{bmatrix} \qquad \text{(6.40a) (repeated)}$$

can be transformed to a new set of state equations in the variables x_1 and x_2 by the linear transformation,

$$\begin{bmatrix} x \\ p \end{bmatrix} = \begin{bmatrix} 1 & -1 \\ -j(km)^{1/2} & -j(km)^{1/2} \end{bmatrix} \begin{bmatrix} x_1 \\ x_2 \end{bmatrix} \qquad \text{(6.128)}$$

and the corresponding inverse transformation

$$\begin{bmatrix} x_1 \\ x_2 \end{bmatrix} \begin{bmatrix} 1/2 & j/2(km)^{1/2} \\ 1/2 & -j/2(km)^{1/2} \end{bmatrix} \begin{bmatrix} x \\ p \end{bmatrix}. \qquad \text{(6.129)}$$

The result is a new set of state equations that also could be used to describe the original system model:

$$\begin{bmatrix} \dot{x}_1 \\ \dot{x}_2 \end{bmatrix} = \begin{bmatrix} j(k/m)^{1/2} & 0 \\ 0 & -j(k/m)^{1/2} \end{bmatrix} \begin{bmatrix} x_1 \\ x_2 \end{bmatrix} + \begin{bmatrix} 1/2 & j/(km)^{1/2} \\ 1/2 & -j/(km)^{1/2} \end{bmatrix} \begin{bmatrix} V \\ F \end{bmatrix}. \quad (6.130)$$

In this new set of equations, not only has the new A-matrix become diagonal, but the diagonal elements are exactly the system eigenvalues. These were shown in Eqs. (6.41) and they involve the undamped natural frequency $\omega_n = (k/m)^{1/2}$. This transformation is a particular example of a *similarity transformation* of the original state equations into a *canonical state equation* form using the so-called *modal matrix*. Many texts on automatic control discuss this as well as other transformations. See, for example, Reference [3].

Although Eq. (6.130) is useful in theoretical discussions, the state variables x_1 and x_2 are difficult to interpret physically except through the formalism of the transformation of Eqs. (6.128) and (6.129). Clearly, x_1 and x_2 and are complex, nonphysical variables, whereas the original state variables x and p are real variables and are related to a force and velocity that could be measured. If a state equation such as Eq. (6.130) were to be used to design a control system, it would generally be necessary to use a transformation such as Eq. (6.128) to return to the original state variables to find physical variables to be measured to implement the control scheme.

Another type of transformation is between the n first-order state-space equations that are the basic output of bond graph techniques and nth-order differential equations that are equivalent. One way in which differential equations are sometimes found is by interpreting the variable s in transfer functions as a d/dt operator. If one were given the transfer function in Figure 6.24a or Eq. (6.117), for example, one could write

$$\left(s^2 + \frac{b}{m} s + \frac{k}{m} \right) f = \left(\frac{b}{m} s + \frac{k}{m} \right) F(t), \quad (6.131)$$

and hence the differential equation

$$\ddot{f} + \frac{b}{m} \dot{f} + \frac{k}{m} f = \frac{b}{m} \dot{F}(t) + \frac{k}{m} F(t). \quad (6.132)$$

It is certainly true that Eq. (6.132) would generate the same transfer function that was found from the original state equations. It is also true that f and $\dot{f}$ may be regarded as state variables in some sense. For example, if one calls $\dot{f}$ by a new name, say, g, then Eq. (6.132) may be rewritten as follows:

$$\dot{f} = g,$$

$$\dot{g} = -\frac{b}{m} g - \frac{k}{m} f = \frac{b}{m} \dot{F}(t) + \frac{k}{m} F(t). \quad (6.133)$$

This is almost in the standard form for a state-space equation—only the need to differentiate $F(t)$ seems bothersome. But this too can be avoided by considering the basic input to be $\dot{F}(t) \equiv H(t)$ and defining an extra state variable, X, so that the integral of H will be a state variable:

$$\dot{f} = g,$$

$$\dot{g} = \frac{b}{m}g - \frac{k}{m}f + \frac{b}{m}H(t) + \frac{k}{m}X,$$

$$\dot{X} = H(t). \tag{6.134}$$

Now it is clear that the third-order system of Eq.(6.134) is somewhat artificial, since the block diagram of Figure 6.24*b* or the original state-space equations provided a second-order system that yielded the given transfer function. Although there are techniques to convert Eq. (6.132) from its second-order form into two state equations analogous to the original state equations without introducing superfluous state variables, it is again true that the state variables may be hard to interpret physically. For this reason, it is generally preferable to start with physically motivated state equations and to derive *n*th-order equations if desired, rather than the reverse. Although any *n*th-order equation may be found in the time domain by differentiating and combining the first-order state equations, it is generally simpler to do the algebra in the *s*-domain, as in deriving transfer functions, and then simply to interpret *s* as the derivative operator in order to write the final differential equation.

It should be noted that it is much more difficult to find useful transformations for state variables for the general nonlinear case than for the linear case. This is just an instance of the fact that the concepts discussed in this chapter do not generally apply to nonlinear systems.

REFERENCES

[1] K. Ogata, *System Dynamics*, 3rd ed., Upper Saddle River, NJ: Prentice-Hall, 1998.
[2] J. L. Melsa and S. K. Jones, *Computer Programs for Computational Assistance in the Study of Linear Control Theory*, 2nd ed., New York: McGraw-Hill, 1973.
[3] W. J. Grantham and T. L. Vincent, *Modern Control Systems Analysis and Design*, New York: John Wiley & Sons, Inc., 1993.

PROBLEMS

6-1 Consider Eq. (6.12) with time constant $\tau = RC = 1.5$ s. At $t = 0$ let $T_s = 1500$ °F, and let T_0 be 100 °F and constant. Compute the initial behavior of T_s by using Euler's method, Eq. (6.6). Choose the time step Δt for reasonable accuracy based on τ. Plot T_s versus t for several steps, and show that when the ratio $\Delta t / \tau$ is too large, the integration scheme is very inaccurate. Find an analytic expression for T_s by considering the exponential decay of T_s to T_0.

6-2 Consider the undamped oscillator with natural frequency of 10 rad/s. See Eq. (6.40a). Determine the free response starting from zero initial position and an initial velocity.

Apply Euler's integration method, Eq. (6.6), for a few steps. What would be a reasonable way to pick Δt so that the digital solution would be accurate?

6-3   For the filter circuit shown, write state and output equations with the voltage $E(t)$ as the input and the terminal voltage as the output. Write the characteristic equation from which the eigenvalues could be determined.

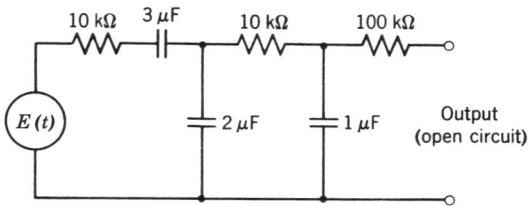

6-4   Consider the bond graph shown below:

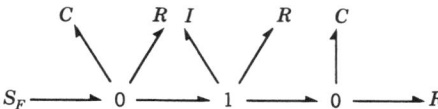

Assuming that all elements are linear, write state equations. Pick two variables that could be output variables, and write output equations for them. Identify the coefficients in Eqs. (6.1a) and (6.2a) with your equation coefficients, as was done in Eq. (6.25). Sketch mechanical, electrical, and hydraulic systems that could be modeled with this bond graph.

6-5   Using equations such as Eq. (6.59) or sketches in the complex plane, demonstrate that

$$\frac{d}{dt}\operatorname{Re}(e^{j\omega t}) = \operatorname{Re}\left(\frac{d}{dt}e^{j\omega t}\right).$$

6-6   If an automobile weighs 3000 lb and has an undamped natural frequency of 1.0 Hz when it oscillates on its main suspension springs, what should the total stiffness of the four shock absorbers be if a damping ratio of 0.707 is to be achieved? Consider the shock absorbers to be linear dampers even though, in practice, shock absorbers are distinctly nonlinear.

6-7   Sketch a frequency response plot for the output voltage divided by the input voltage for the system of Problem 6-3.

6-8   In the system shown, consider $V(t)$ to be a velocity input. Set up state equations for the system, and let the acceleration of the top mass be a system output. (Note that the acceleration is proportional to the total force on the mass.)

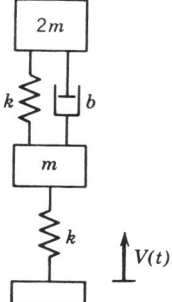

Find the ratio of the acceleration to the input velocity when the velocity is sinusoidal, and sketch the frequency response plot.

6-9 If a first-order system is excited by a cosine wave starting from a zero initial condition at $t = 0$, sketch the forced response, free response, and total response as was done in Figures 6.26–6.28.

6-10 By labeling signals and writing equations implied by the blocks, demonstrate the validity of the block diagram identity of Figure 6.25.

6-11 Using Figure 6.25, reduce Figure 6.24b to Figure 6.24a.

6-12 The rotor of the motor shown is connected through a flexible shaft to the pinion, which drives the rack and load mass M. Assuming the pinion may be modeled as having no rotary inertia, construct a bond graph for the free motion of the system. Find the eigenvalues of the system and interpret them.

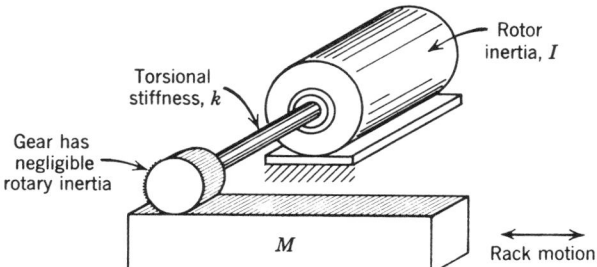

6-13 Figures a and b below depict two linear filters that are supposed to remove unwanted high-frequency noise from a voltage signal $E(t)$. In Figure a the output voltage measured when no current is allowed to flow is e_a; in Figure b the output is e_b. All resistors have the same resistance R, and all capacitors have the capacitance C.

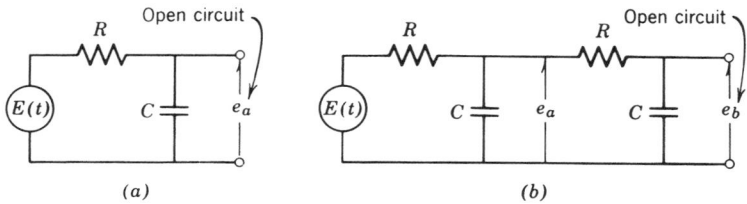

(a) (b)

(a) For Figure a derive state equations, and relate the output, e_a, to the state variables. For Figure b find state equations, and relate both e_a and e_b to the state variables.

(b) For Figure a find the transfer function $H_{e_a/E}(s)$ relating e_a and E. For Figure b find transfer functions for $H_{e_a/E}$ and $H_{e_b/E}(s)$ e_a and e_b.

(c) Should the two transfer functions for e_a in Figures a and b not be equal? Should the transfer function for e_b not be the square of the transfer function for e_a? If so, why?

6-14 Consider the bond graphs and state equations below.

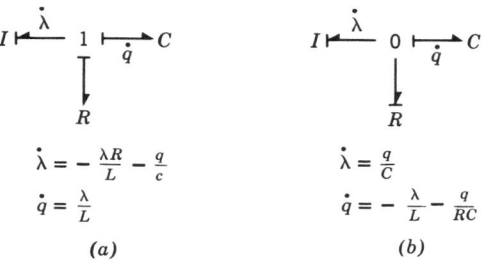

$$\dot{\lambda} = -\frac{\lambda R}{L} - \frac{q}{c}$$
$$\dot{q} = \frac{\lambda}{L}$$

$$\dot{\lambda} = \frac{q}{C}$$
$$\dot{q} = -\frac{\lambda}{L} - \frac{q}{RC}$$

(a) (b)

(a) Draw the corresponding circuits.

(b) Describe qualitatively the effect in both cases when $R \rightarrow 0$ and $R \rightarrow \infty$.

(c) Find relations for the system eigenvalues in both cases, using the following procedure:

   i. Let $\dot{x} = Ax$ be the state equation.

   ii. If $x = Xe^{st}$, then $sX = AX$, or $[sI - a][X] = [0]$ if $[I]$ is the unit matrix.

   iii. The eigenvalues are values of s for which $\det[sI - A] = 0$, since only when the determinant vanishes can X be nontrivial.

Note that A is given, so all that is required is to form $sI - A$ and set its determinant equal to zero. The results should square with (b).

6-15 The sketches below show a schematic diagram of a rotating eccentric vibrator and a "mechanical network" representation of the device. The assumption is that the motor produces a simple harmonic relative horizontal velocity of

$$V \cos \omega t = \tau \omega \cos \omega t = \mathrm{Re}(V e^{j\omega t})$$

between the rotating mass m and the main mass M. Because of the assumed constraints, we do not discuss vertical forces or vertical motion.

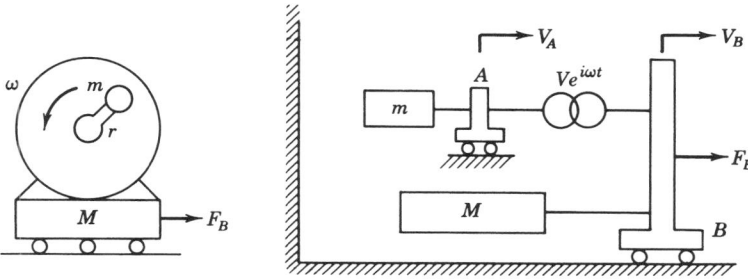

(a) Construct a bond graph for the system, and fully augment it. Assume F_B is a known force applied to the system.

(b) Write a state equation for the system, and write output equations for V_A and V_B.

(c) Verify, using the results of (b), that the following terms sometimes used to describe the exciter are valid when $V e^{j\omega t}$ is the complex representation of $V \cos \omega t$ for the velocity source:

   i. The *internal mobility* looking back at B is the ratio of velocity to force when the velocity generator is stopped and locked in position:

$$\frac{V_B}{F_B} = \frac{1}{j\omega(M + m)}.$$

   ii. The *blocked force* output at B is F_B when some external agent prevents motion of B:

$$F_{B(\mathrm{blocked})} = j\omega m V.$$

   iii. The *free-velocity* output at B is the velocity at B when there is no force F_B:

$$V_{B(\mathrm{free})} = \frac{m}{M + m} V.$$

   iv. A general network theorem states that the internal mobility is the quotient of the free velocity and the blocked force. See H. H. Skilling, *Electrical Engineering Circuits*, New York: John Wiley & Sons, Inc., 1957, p. 339. Verify the theorem for this example.

6-16 Reinterpret the bond graph in Figure 6.3b as if it referred to an electric circuit with parameters R, C, and L. After showing the circuit, redraw the Bode plots of Figure 6.18 showing the significant points on the plot in terms of the electrical rather than the mechanical parameters.

6-17 A transfer function between an output Y and an input X is

$$\frac{Y(x)}{X(s)} = \frac{G(s^2 + \omega_0^2)}{s^2 + 2\zeta\omega_n s + \omega_n^2} \qquad \omega_n < \omega_0.$$

(a) Let $s = j\omega$ and write down the complex frequency response function.

(b) If $x(t)$ is a harmonic input,

$$x = x\cos(\omega t);$$

then, for a linear system, we know

$$Y = Y\cos(\omega t + \varphi).$$

Derive expressions for Y and φ.

(c) Sketch Y/X versus ω for low, medium, and high ζ, and imagine a physical system that would behave like this.

6-18 The system below is supposed to isolate the mass, m, from ground motion, $v_i(t)$. The mass sits on a fluid-filled displacer that is attached to an air com-

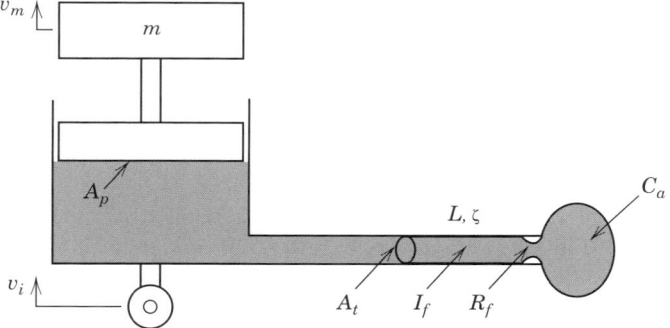

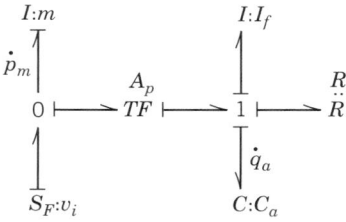

pliance using a long, fluid-filled tube. The tube has area A_t, length L, and fluid density ζ. The nominal air volume is V. Construct a bond graph for this device and show that yours is the same as the one given.

(a) Derive the state equations. There is some derivative causality, so use procedures from Chapter 5 to deal with this problem.

(b) Derive the transfer function relating output, v_m, to input, v_i.

(c) Compare this transfer function with the one from Problem 6-17. What do you think of this device as a vibration isolator?

(d) From your knowledge of how the fluid inertia, I_f, relates to physical system parameters, can you suggest design changes that will improve the high-frequency isolation?

6-19 For the two devices below, construct bond graph models, derive state equations, and derive transfer functions relating $v_m(s)$ to $v_i(s)$. Turn these transfer functions into frequency response functions and sketch $|v_m/v_i|$ versus frequency. Comment on how these devices compare with respect to vibration isolation.

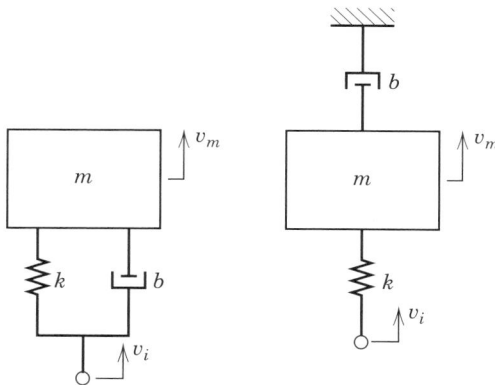

6-20 In control systems, an error is generated between a desired output and an actual output, and this error is passed through a controller filter, the output of which is an effort of flow that moves the system in a direction that reduces the error. A typical control filter is the PID controller which produces a component of activator output proportional (P) to the error, proportional to the integral (I) of the error, and proportional to the derivative (D) of the error. In the s-domain, this is given by

$$G_c(s) = K_P + K_D s + \frac{K_I}{s}.$$

It is proposed to use the PID controller in a position control system. The physical system is modeled as shown below, where we have control of the force, F_c.

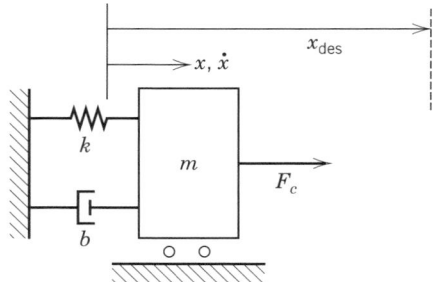

(a) Construct a bond graph model of this system and derive the transfer function relating mass position, x, to force, F_c. Call this transfer function $G_p(s)$.

(b) A block diagram of the entire control system is shown. Derive the closed-loop transfer function relating the output, x, to the desired position, x_{des}.

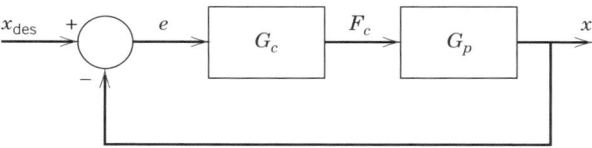

(c) Compare the system eigenvalues without control (open loop) to those when the controller is operating (closed loop). Show how the controller gains can change the eigenvalues.

(d) Sketch the response of x to a step change in x_{des} and comment on the ability of the PID controller to produce a reasonable response.

6-21 The figure shows a relatively simple electric circuit and a numbered bond graph. The circuit has two energy storage elements, an inductor and a capacitor, so unless there is differential causality involved, this should be a second-order system. Assume that all elements are linear, with coefficients as shown in general or physical form in the table to the right of the bond graph.

(a) Apply causality to the bond graph and write the state equations in matrix form using either version of the coefficients. The results should be of the form of Eq. (6.1b) and the matrices should resemble those in Eq. (6.25).

(b) Find the characteristic equation as was done in Eqs. (6.31a) and (6.31b) by setting the determinant of the matrix $[sI - A]$ to zero.

(c) By matching your characteristic equation coefficients to the coefficients in the version shown in Eq. (6.65), determine the undamped natural frequency, ω_n, and the damping ratio, ζ, in terms of the physical parameters L, C, R_a, and R_b.

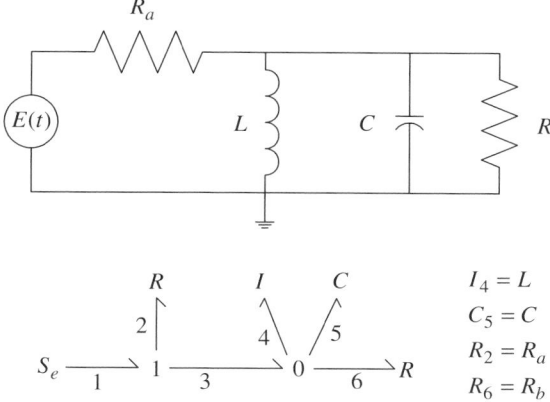

6-22 The figure below shows a simple mechanical system consisting of a mass, two dashpots, and a velocity source with friction between the mass and the stationary base. The dashpots and the ground friction are represented as if they are linear friction elements with coefficients B_2, B_3, and B_5. The input forcing function is the velocity $V_6(t)$, which can be represented in general terms as the source flow $f_6(t)$.

(a) First, convince yourself that the bond graph does represent the system. Then apply causality and write the state equation using general effort and flow variables and generalized coefficients such as, I_1, R_2, and so on. Based on the equation, determine the time constant in terms of the generalized coefficients as well as the corresponding physical coefficients, M, B_2, and so forth.

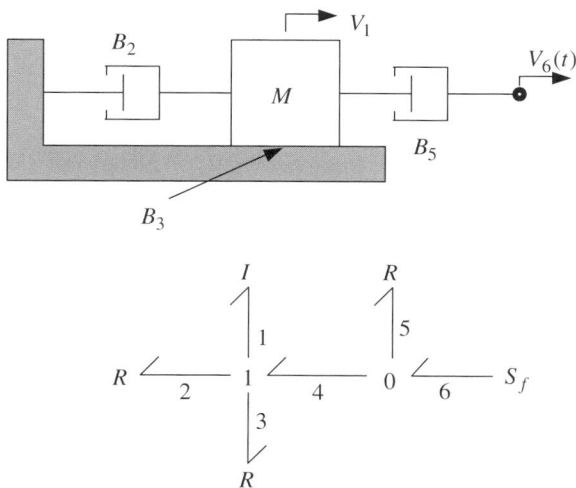

(b)   Now suppose that the applied velocity, $V_6(t)$, is replaced by an applied force, $F_6(t)$. (The applied force is represented in general terms as the effort $e_6(t)$. It is too bad that force and flow both start with F. In part a, $f_6 = V_6$, but here e_6 is the force F_6.)

Again apply causality and write the state equation. What is the time constant now under the new type of forcing?

7

MULTIPORT FIELDS AND JUNCTION STRUCTURES

In the first part of this book, dynamic models for a variety of physical systems were constructed using the 1-port elements —R, —C, —I, S_e—, S_f—; the 2-port elements —TF— and —GY—; and the 3-port 0- and 1-junctions. In this chapter, *fields*, which are multiport generalizations of —R, —C, and —I elements, are introduced, and the concept of a *junction structure*, which is an assemblage of power-conserving elements,

$$-0-, \quad -1-, \quad -TF-, \quad \text{and} \quad -GY-,$$

is presented. Using fields and junction structures, one may conveniently study systems containing complex multiport components using bond graphs. In fact, bond graphs with fields and junction structures prove to be a most effective way to handle the modeling of complex multiport systems, combining both structural detail and clarity with ease of visualization. Succeeding chapters demonstrate the application of the elements introduced here to the modeling of systems containing multiport devices.

7.1 ENERGY-STORING FIELDS

In Chapter 3 it was shown that the elements —C and —I can store energy and can give back stored energy without loss. This is true no matter what the constitutive laws for these elements happen to be. Any single-valued functional relationship between effort and displacement defines an energy-conservative capacitor, and any single-valued relationship between flow and momentum defines an energy-conservative inertia element. The multiport generalizations of —C and —I, which we shall call

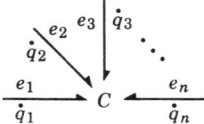

FIGURE 7.1. The symbol of an n-port C-field.

C-fields and I-fields, respectively, will also be conservative. As we shall see, the conservation of energy in the multiport cases imposes a constraint on the constitutive laws for C- and I-fields.

7.1.1 C-Fields

The symbol for a C-field is simply the letter C, with as many bonds as the C-field has ports. An n-port C-field is shown in Figure 7.1. Note that the flow at the ith port has been labeled $\dot{q}_i$ and that an inward sign convention is assumed. The energy $\mathbf{E}$ stored in the C-field can then be expressed as

$$\mathbf{E} = \int_{t_0}^{t} \sum_{i=1}^{n} (e_i f_i) \, dt = \int_{t_0}^{t} \sum_{i=1}^{n} e_i \dot{q}_i \, dt$$

$$= \int_{q_0}^{q} \sum_{i=1}^{n} e_i(\mathbf{q}) \, dq_i = \int_{q_0}^{q} \mathbf{e}(\mathbf{q}) \, d\mathbf{q} = \mathbf{E}(\mathbf{q}), \qquad (7.1)$$

in which the identity $f_i \, dt = dq_i$ has been used, and column vectors of efforts and flows have been defined thus:

$$\mathbf{q} \equiv \begin{bmatrix} q_1 \\ q_2 \\ q_3 \\ \vdots \\ q_n \end{bmatrix}, \qquad \mathbf{e} \equiv \begin{bmatrix} e_1 \\ e_2 \\ e_3 \\ \vdots \\ e_n \end{bmatrix}, \qquad (7.2)$$

and $\mathbf{e} \, d\mathbf{q}$ represents the scalar or inner product, which could also be represented as $\mathbf{e}^t \, d\mathbf{q}$ using the transpose of the column matrix $\mathbf{e}$.

At time t_0, $\mathbf{q} = \mathbf{q}_0$, and at time t, $\mathbf{e} = \mathbf{e}(\mathbf{q})$ represents the constitutive laws of the C-field evaluated at $\mathbf{q}(t)$, the instantaneous value of the vector of displacements.

Before going on to study the properties of general C-fields as defined by Eq. (7.1), it may be worthwhile to exhibit some relatively simple C-fields as they arise in practice. Some components of systems are described naturally by C-fields. The beam of Figure 7.2, for example, might be part of a vibratory system with masses or other elements attached at the locations indicated by the F_1, $\dot{x}_1$ and F_2, $\dot{x}_2$ ports. If one neglects the mass of the beam (which is permissible if the system frequencies are sufficiently low) and if one neglects at least temporarily the power losses associated

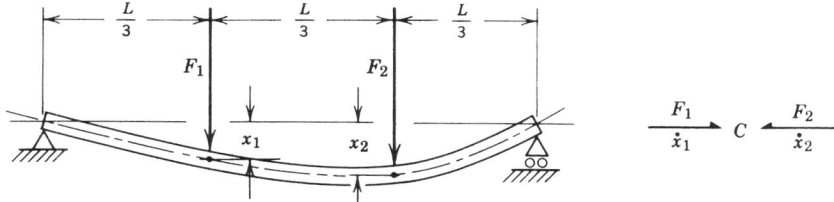

FIGURE 7.2. An elastic beam deformed by the action of two forces and represented as a 2-port C-field. Uniform beam: elastic modulus E; area moment of inertia I.

with inelastic material behavior and the support pivots, then the beam is a purely elastic structure and can be represented as a 2-port C-field. The beam then may be represented by constitutive relations among F_1, F_2, x_1, and x_2.

Virtually all engineering undergraduates could find the constitutive laws for this C-field at some point in their careers, although many have forgotten how. A convenient way to represent the beam is through the use of *superposition* to add up the effects on x_1 and x_2 of the forces F_1 and F_2. This procedure (See Reference [1], for example) works only because the beam is represented for a *linear* range of strains and deflections. The result is conveniently expressed in a matrix form:

$$\begin{bmatrix} x_1 \\ x_2 \end{bmatrix} = \frac{L^3}{243EI} \begin{bmatrix} 4 & \frac{7}{2} \\ \frac{7}{2} & 4 \end{bmatrix} \begin{bmatrix} F_1 \\ F_2 \end{bmatrix}. \tag{7.3}$$

This form of the constitutive laws, which gives displacements as a function of efforts, is the *compliance form*. As long as the matrix in Eq. (7.3) is invertible, one could solve the equation for F_1 and F_2 in terms of x_1 and x_2. This would be the *stiffness form* for the constitutive laws.

When an element is described from the beginning as a set of effort–displacement relations at n ports, as in the beam example, we shall call the model of the element an *explicit field*. Generally, it is much more convenient to treat explicit-field elements as fields, although in some cases it would be possible to find an equivalent system of 1-port energy-storing elements bonded together with junction structure elements that would have the same port constitutive laws as the field. On the other hand, when systems are assembled, it often happens that a group of 1-port energy-storing elements, say —Cs, are interconnected with

$$-\overset{\mid}{0}-, \quad -\overset{\mid}{1}-, \quad \text{and} \quad -TF-$$

elements. Such a subsystem often can be usefully treated as a field. Such a field will be called an *implicit field*, since the field constitutive laws at the external ports must be deduced from the constitutive laws of the elements constituting the field. An example is shown in Figure 7.3. Except for the sign convention on one internal bond, the electrical and mechanical systems shown in Figure 7.3 have the same bond graph. If one imposes integral causality on the 1-port C-elements as in Figure 7.3a,

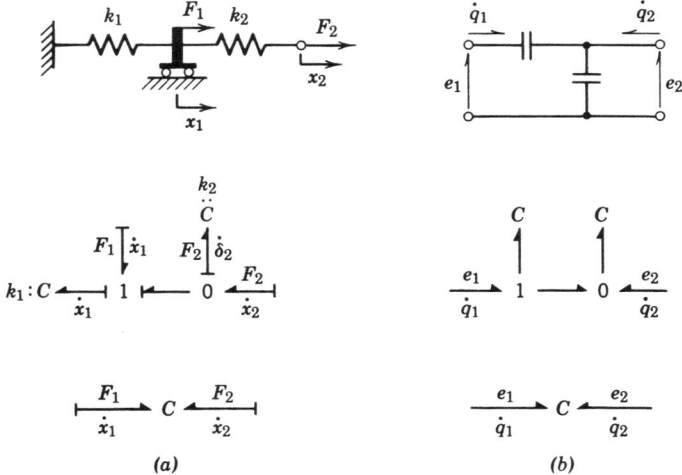

(a)   **(b)**

FIGURE 7.3. A C-field composed of 1-port C-elements and junction elements. (*a*) A mechanical example; (*b*) an electrical example.

then it is clear that the state variables are x_1 and δ_2, namely, the deflections of the two springs. Thus it is not immediately clear that the C-field representation in terms of x_1 and x_2 is possible. Using the mechanical example, let us work out the C-field constitutive laws.

The causal strokes indicate that we can find F_1 and F_2 as output variables in terms of the state variables x_1 and δ_2:

$$F_1 = k_1 x_1 - k_2 \delta_2, \qquad F_2 = 0 x_1 + k_2 \delta_2, \tag{7.4}$$

where, for the sake of preserving simplicity in the example, the springs have been assumed to be linear. The state equations are

$$\dot{x}_1 = \dot{x}_1(t) \quad \text{(given)}, \tag{7.5}$$

$$\dot{\delta}_2 = \dot{x}_2(t) - \dot{x}_1(t), \tag{7.6}$$

in which $\dot{x}_1$ and $\dot{x}_2$ are determined by some system external to the C-field and thus function as input variables.

Using Eq. (7.6), we see that δ_2 may be expressed in terms of the C-field variables, x_1 and x_2, by means of a time integration:

$$\delta_2 = x_2 - x_1 + \text{const.} \tag{7.7}$$

Referring to Figure 7.3*a*, it can be seen that x_1 and x_2 can be defined such that when $x_1 = x_2 = 0$, $F_1 = F_2 = \delta_2 = 0$. Thus the integration constant in Eq. (7.7) can be set to zero when x_1 and x_2 represent deviations from the positions at which the springs are unstretched. Using Eq. (7.7), we can eliminate δ_2 from Eq. (7.4). Then

$$\begin{bmatrix} F_1 \\ F_2 \end{bmatrix} = \begin{bmatrix} k_1 + k_2 & -k_2 \\ -k_2 & k_2 \end{bmatrix} \begin{bmatrix} x_1 \\ x_2 \end{bmatrix}, \tag{7.8}$$

which represents the constitutive laws of a 2-port C-field in stiffness matrix form. [The reader may benefit from carrying out the steps required to express the constitutive laws of the electrical C-field of Figure 7.3b in a form analogous to Eq. (7.8).]

An inspection of the constitutive laws of Eqs. (7.3) and (7.8) shows that for these linear C-fields, the *compliance matrix* [Eq. (7.3)] and the *stiffness matrix* [Eq. (7.8)] are symmetric. In general, compliance matrices and stiffness matrices, which are inverses of each other, are always symmetric. This can be proven in a form valid for even nonlinear C-fields by studying the stored-energy function (7.1).

A change in any component of $\mathbf{q}$, say Δq_i, will produce a change in $\mathbf{E}$, say $\Delta \mathbf{E}$. By a direct inspection of Eq. (7.1), one can deduce that the coefficient relating $\Delta \mathbf{E}$ to Δq_i is, in fact, e_i. Thus, the partial derivative of $\mathbf{E}$ with respect to q_i is just e_i, or

$$\frac{\partial \mathbf{E}}{\partial q_i} = e_i(\mathbf{q}), \qquad i = 1, 2, \ldots, n. \tag{7.9}$$

Because $\mathbf{E} = \mathbf{E}(\mathbf{q})$ is supposed to represent the stored-energy function, which we assume to be a single-valued scalar function of the vector $\mathbf{q}$, if $\mathbf{E}$ is smooth enough to have second derivatives, then

$$\frac{\partial e_i}{\partial q_j} = \frac{\partial^2 E}{\partial q_j \partial q_i} = \frac{\partial e_j}{\partial q_i}, \qquad i, j = 1, 2, 3, \ldots, n, \tag{7.10}$$

which shows how $e_i(\mathbf{q})$ and $e_j(\mathbf{q})$ are constrained by the existence of the stored-energy function, $\mathbf{E}$. In other words, not every set of constitutive laws expressing efforts in terms of displacements could come from an energy-storing, or energy-conservative, C-field. Physically, it is clear that elastic structures, capacitor networks, and the like cannot supply more energy than the amount previously stored, so Eq. (7.10) will apply to such devices.

In the linear case, Eq. (7.10) implies that stiffness matrices are necessarily symmetrical for any conservative C-field. If we use k_{ij} for stiffness matrix coefficients and $\mathbf{k}$ for the matrix itself, then the C-field constitutive laws for an n-port can be written thus:

$$e_i = \sum_{j=1}^{n} k_{ij} q_j,$$

or

$$\mathbf{e} = \mathbf{kq},$$

and Eq. (7.8) is a particular example. The stored energy is

$$\mathbf{E(q)} = \frac{1}{2} \sum_{i=1}^{n} \sum_{j=1}^{n} k_{ij}q_i q_j = \frac{1}{2}\mathbf{q^t kq}. \qquad (7.11)$$

Application of Eq. (7.10) to Eq. (7.11) yields

$$k_{ij} = k_{ji}, \quad \text{or} \quad \mathbf{k^t = k}, \qquad (7.12)$$

and since compliance matrices such as that in Eq. (7.3) are inverses of stiffness matrices and the inverse of a symmetric matrix is also symmetric, we conclude that compliance matrices must also be symmetric. The equations (7.12), which are often called Maxwell's reciprocal relations, are most easily derived from the more general nonlinear relations (7.10), but graphical derivations based on special integration paths are often used for the linear case. See Reference [2], for example.

The reciprocal relations for C-fields and other energy-storing fields are more useful and informative in some cases than it might at first appear. In later chapters dealing with a variety of applications, these relations will prove to be very important. For now, we show only a few examples that are less obvious than our first ones. The system of Figure 7.4 shows how the end of a cantilever beam can interact with a rotational and a translational port. Using beam superposition tables or any other method to find the beam deflections, the compliance matrix may be found and inverted to find the stiffness matrix:

$$\begin{bmatrix} F \\ \tau \end{bmatrix} = EI \left[\begin{array}{c|c} 12/L^3 & -6/L^2 \\ \hline -6/L^2 & 4/L \end{array} \right] \begin{bmatrix} x \\ \theta \end{bmatrix}. \qquad (7.13)$$

The symmetry of this matrix serves as a useful check on the correctness of the computations leading up to it. Furthermore, we can make statements such as the following. The number of newtons (the force) required to keep x at zero when $\theta = 0.1$ rad is equal to the number of newton-meters (the torque) required to keep θ at zero when $x = 0.1$ m. Or, if $F = 1$ N and $\tau = 0$, then θ in radians will be numerically equal to x in meters when $F = 0$ and $\tau = 1$ N-m. (The latter statement is based on the symmetry of the compliance matrix.)

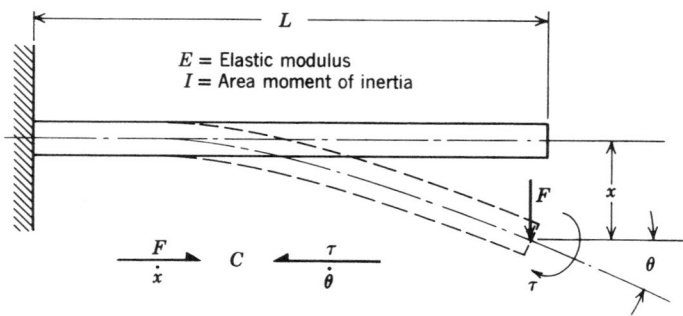

FIGURE 7.4. Cantilever beam represented as a C-field.

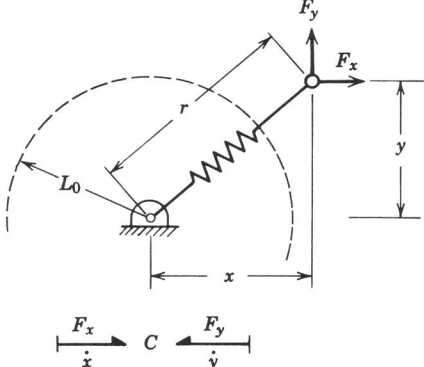

FIGURE 7.5. A nonlinear C-field representation of a linear spring, one end of which can move in a plane.

Nonlinear C-fields are common in certain application areas. Here we show how a mechanical element that behaves as a linear 1-port in some restricted cases can be modeled as a nonlinear 2-port C-field when two-dimensional motion is permitted. The element, shown schematically in Figure 7.5, is a simple linear spring attached to ground at one end and deflecting under the action of forces F_x and F_y at the other end. It is possible to express F_x and F_y in terms of x and y and thus to consider this device as a C-field. The constitutive relations are

$$F_x = kx - k(x^2 + y^2)^{-1/2} L_0 x,$$
$$F_y = ky - k(x^2 + y^2)^{-1/2} L_0 y, \qquad (7.14)$$

where k is the spring constant and L_0 is the free length of the spring.

A check on (7.14) is given by Eq. (7.10):

$$\frac{\partial F_x}{\partial y} = -k(x^2 + y^2)^{-3/2} L_0 x \, 2y, \qquad (7.15)$$

$$\frac{\partial F_y}{\partial x} = -k(x^2 + y^2)^{-3/2} L_0 y \, 2x. \qquad (7.16)$$

Because the expressions on the right-hand sides in Eqs. (7.15) and (7.16) are identical, the constitutive laws (7.14) do represent a nonlinear conservative C-field.

Actually, it is easiest to derive Eqs. (7.14) by using energy methods. The stored energy **E** is easily expressed in terms of the radius r:

$$\mathbf{E} = \frac{1}{2} k(r - l_0)^2 = \frac{1}{2} k \left(r^2 - 2rl_0 + l_0^2 \right),$$

and, using $r^2 = x^2 + y^2$, we have

$$\mathbf{E} = \frac{1}{2} k \left[x^2 + y^2 - 2l_0 \left(x^2 + y^2 \right)^{1/2} + l_0^2 \right]. \tag{7.17}$$

When Eq. (7.17) is used according to Eq. (7.9) to derive Eq. (7.14), the reciprocal relations, Eqs. (7.15) and 7.16, must follow immediately.

7.1.2   Causal Considerations for C-Fields

As in the case of 1-port C-elements, we may distinguish between integral and derivative causality for multiport C-fields, but multiport fields also admit mixed integral–derivative causalities. The completely integral causality form is

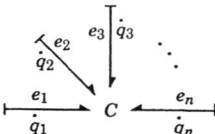

and the constitutive laws may be stated thus:

$$e_i = \Phi_{C_i}^{-1}(q_1, q_2, \ldots, q_n), \qquad i = 1, 2, \ldots, n. \tag{7.18}$$

The completely derivative causality form is

and the constitutive laws may be restated in the form

$$q_i = \Phi_{C_i}(e_1, e_2, \ldots, e_n), \qquad i = 1, 2, \ldots, n. \tag{7.19}$$

In Eqs. (7.18) and (7.19), the generally nonlinear functions $\Phi_{C_i}^{-1}$ and Φ_{C_i} are named in analogy with their 1-port counterparts from Chapter 3. In mechanical systems the compliance form corresponds to derivative causality and the stiffness form to integral causality. When the ports are numbered so that the first j ports have integral causality and the remainder have derivative causality, then a mixed causal form appears as

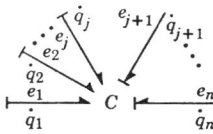

and the constitutive laws are

$$e_i = \Phi_i(q_1, q_2, \ldots, q_j, e_{j+1}, \ldots, e_n), \qquad i = 1, 2, \ldots, j; \qquad (7.20)$$

$$q_k = \Phi_k(q_1, q_2, \ldots, q_j, e_{j+1}, \ldots, e_n), \qquad k = j+1, \ldots, n. \qquad (7.21)$$

The question of which causal forms are possible for an energy-storing field is an interesting one. For an explicit field that is specified by a set of constitutive laws in a particular causal form, the question is an algebraic one in which each causal form involves solving the given relationships again in terms of different input and output variables. For nonlinear fields, it can be very hard to decide whether this will be possible, and even for linear fields the problems are not trivial. Of course, to convert from completely integral to completely derivative causality or the reverse, one needs only to invert a matrix in the linear case. Since there are straightforward tests of matrices to determine whether or not the inverse exists, it is not hard to decide, for example, whether given a field in derivative causality the integral causality form is permissible. It is harder, however, to establish a universal rule for testing all the mixed causal forms that an n-port field can exhibit.

For implicit fields, on the other hand, the rules of causality can often be used on the elements of the field to deduce which causal forms of the field are possible. As a very simple example, consider the implicit field formed by a 1-port C-element and a 0-junction, as shown in Figure 7.6. This system is readily treated as a 2-port C-field in integral causality (see Figure 7.6a). Reading the implicit-field causality, we may find e_1 and e_3, assuming that the 1-port C is linear:

$$e_1 = e_3 = \frac{q_3}{C_3}, \qquad (7.22)$$

$$e_2 = e_3 = \frac{q_3}{C_3} \qquad (7.23)$$

Now we wish to convert from q_3 as a state variable to the field state variables, q_1 and q_2. Reading the bond graph, we have

$$\dot{q}_3 = \dot{q}_1 + \dot{q}_2, \qquad (7.24)$$

or

$$q_3 = q_1 + q_2 + \text{const.} \qquad (7.25)$$

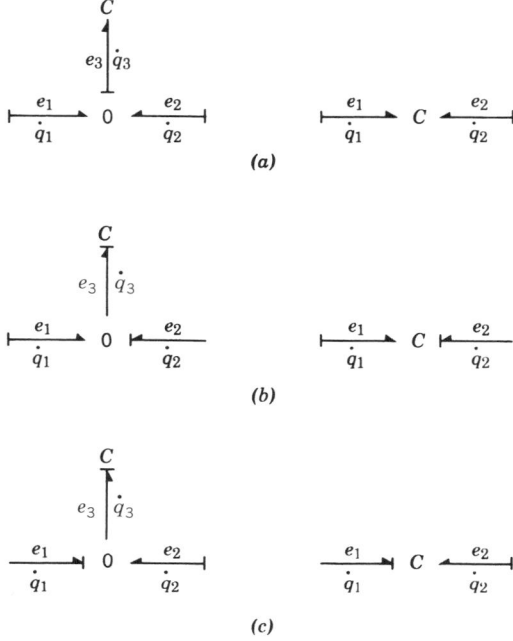

FIGURE 7.6. A simple implicit field. (*a*) Integral causality; (*b*) and (*c*) mixed causality. (Derivative causality for this field is not possible.)

If for simplicity we define q_1, q_2, and q_3 so that the integration constant vanishes in Eq. (7.25), then the field equations follow by substitution in Eqs. (7.22) and (7.23):

$$
\begin{bmatrix} e_1 \\ e_2 \end{bmatrix} = \begin{bmatrix} 1/C_3 & 1/C_3 \\ 1/C_3 & 1/C_3 \end{bmatrix} \begin{bmatrix} q_1 \\ q_2 \end{bmatrix}. \tag{7.26}
$$

There are two ways to see that the completely derivative causal form is not possible: (1) the rules of causality when applied to the implicit field of Figure 7.6 do not allow the imposition of both e_1 and e_2 as inputs to the 0-junction and (2) the determinant of the matrix in Eq. (7.26) clearly vanishes, indicating that Eq. (7.26) cannot be solved for q_1, q_2 in terms of e_1 and e_2.

On the other hand, two mixed causal forms are possible, as shown in Figures 7.6*b* and *c*. For the form of Figure 7.6*b*, the relations are derived by reading the implicit-field bond graph:

$$
e_1 = e_2, \tag{7.27}
$$

$$
\dot{q}_2 = -\dot{q}_1 + \dot{q}_3 = -\dot{q}_1 + \frac{d}{dt}(C_3 e_3) = -\dot{q}_1 + \frac{d}{dt}(C_3 e_2). \tag{7.28}
$$

Equation (7.28) must be integrated in time, and if q_1 and q_2 are properly defined, the integration constant can be made to vanish:

$$q_2 = -q_1 + C_3 e_2. \tag{7.29}$$

In matrix form, the mixed causal form is

$$\begin{bmatrix} e_1 \\ q_2 \end{bmatrix} = \left[\begin{array}{c|c} 0 & 1 \\ \hline -1 & C_3 \end{array}\right] \begin{bmatrix} q_1 \\ e_2 \end{bmatrix}. \tag{7.30}$$

Note that the integral form, Eq. (7.26), is symmetric, as it must be according to the energy argument, but mixed forms such as Eq. (7.30) will generally have anti-symmetric terms.

Although almost any nontrivial system may be considered to contain an energy-storing field (as we have just seen, even a single —C and a 0-junction can be treated as a 2-port field), it is often not worthwhile to consider the manipulation of a bond graph into a form in which an implicit field is obvious. An exception to this general rule occurs when, in the course of assigning causality, it is found that one or more elements must be assigned derivative causality. Often this occurrence signals the presence of an implicit field that can usefully be converted into explicit form, thus simplifying further equation formulation. Of course, the methods of equation formulation presented in Chapter 5 for the cases involving derivative causality permit equation formulation without identifying fields. Although there is no less labor involved in converting an implicit field with derivative causality into explicit form, as we do here, this process may increase one's insight into the causes and nature of derivative causality. Also, the explicit-field representation can be used repeatedly, thus avoiding the necessity of handling derivative causality in each system of which the field is a part.

As an example, consider the electrical subsystem shown in Figure 7.7a. The bond graph of Figure 7.7b does not appear to have any unusual characteristics, but a little experimentation with causal assignments will show that the three —C elements cannot all have integral causality. What this means is that the three displacements—in this case, three electrical charge variables—cannot be independent. In Figure 7.7c, the C-field is singled out for study. In this form, the charges q_2 and q_6, for example, can play the roles of state variables, and the charge q_4 will be statically related to q_2 and q_6. Let us now find the explicit-field representation shown in Figure 7.7d, in which the charge variables q_1 and q_7 will play the roles of state variables.

Reading the graph of Figure 7.7c, we find that the port output variables, e_1 and e_7, can be expressed in terms of the state variables, q_2 and q_6 (for simplicity, we assume all the —C elements are linear, although our operations could be carried out for nonlinear —C elements as well):

$$e_1 = e_2 = \frac{q_2}{C_2}, \qquad e_7 = e_6 = \frac{q_6}{C_6}. \tag{7.31}$$

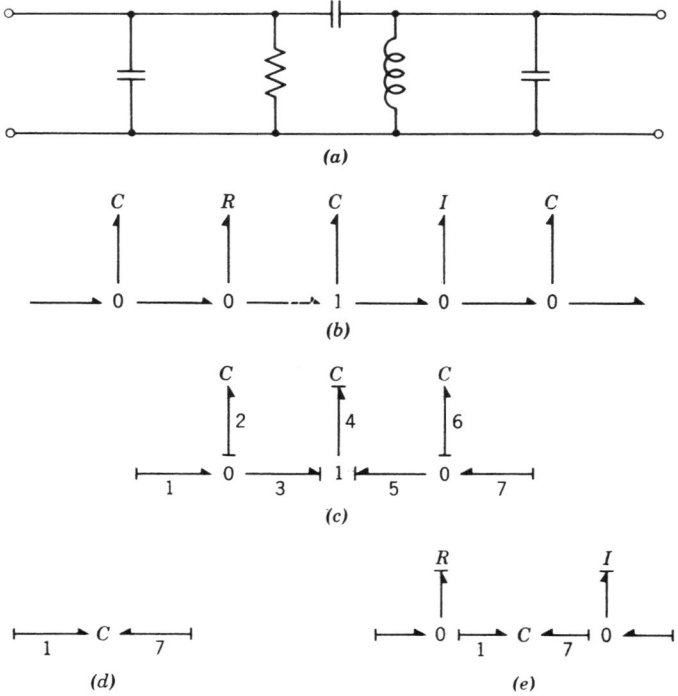

FIGURE 7.7. System containing an implicit C-field. (*a*) Section of an electrical network; (*b*) bond graph; (*c*) implicit field identified; (*d*) explicit-field representation; (*e*) bond graph of (*a*) using explicit-field representation.

The state equations may be used to relate q_2 and q_6 and q_7:

$$\dot{q}_2 = f_1 - f_3 = f_1 - f_4 = f_1 - \frac{d}{dt}q_4$$

$$= f_1 - \frac{d}{dt}C_4 e_4 = f_1 - \frac{d}{dt}C_4(e_3 + e_5)$$

$$= f_1 - \frac{d}{dt}C_4(e_2 + e_6) = f_1 - \frac{d}{dt}C_4\left(\frac{q_2}{C_2} + \frac{q_6}{C_6}\right). \tag{7.32}$$

Similarly,

$$\dot{q}_6 = f_7 - \frac{d}{dt}C_4\left(\frac{q_2}{C_2} + \frac{q_6}{C_6}\right). \tag{7.33}$$

These equations may be rearranged in matrix form:

$$
\begin{bmatrix} (C_2 + C_4)/C_2 & C_4/C_6 \\ C_4/C_2 & (C_6 + C_4)/C_6 \end{bmatrix} \begin{bmatrix} \dot{q}_2 \\ \dot{q}_6 \end{bmatrix} = \begin{bmatrix} f_1 \\ f_7 \end{bmatrix} \tag{7.34}
$$

Basically, we wish to integrate Eq. (7.34) in order to relate q_2 and q_6 to q_1 and q_7, but, because of the differential causality, a matrix inversion will be required. The result is

$$
\begin{bmatrix} \dot{q}_2 \\ \dot{q}_6 \end{bmatrix} = \frac{1}{C_2 C_4 + C_2 C_6 + C_4 C_6} \begin{bmatrix} C_2 C_6 + C_2 C_4 & -C_2 C_4 \\ -C_4 C_6 & C_2 C_6 + C_4 C_6 \end{bmatrix} \begin{bmatrix} \dot{q}_1 \\ \dot{q}_7 \end{bmatrix}, \tag{7.35}
$$

where $\dot{q}_1 = f_1, \dot{q}_7 = f_7$.

Since Eq. (7.35) is a relation between derivatives of charge variables, it can be integrated in time to yield the desired relations between the charges. Again, one must consider the charges at some initial time in order to evaluate integration constants. If the system is assembled out of initially uncharged capacitors, then q_2, q_6, q_1, and q_7 can all vanish initially and the integration constants also vanish. Then Eq. (7.35) can be integrated by simply removing the dots over the qs. When this is done, a substitution into Eq. (7.31) yields the explicit-field equations:

$$
\begin{bmatrix} e_1 \\ e_7 \end{bmatrix} = \frac{1}{C_2 C_4 + C_2 C_6 + C_4 C_6} \begin{bmatrix} C_6 + C_4 & -C_4 \\ -C_4 & C_2 + C_4 \end{bmatrix} \begin{bmatrix} q_1 \\ q_7 \end{bmatrix}. \tag{7.36}
$$

This relation is the constitutive law for the 2-port C-field of Figure 7.7d in integral causality form. Note that we have arranged an inward sign convention for the C-field. This means that Eq. (7.36) should have a symmetric matrix, and it does. Now, the explicit-field representation, which does not have derivative causality problems, can be used in the original system, as shown in Figure 7.7e. The same field might appear in many systems, and the explicit representation would eliminate the need to perform algebraic manipulations associated with derivative causality in each system.

Finally, it should be mentioned that it is sometimes convenient to reduce implicit fields to an explicit form, regardless of whether or not derivative causality is involved. Figure 7.8 shows two examples in which 1-ports, 2-ports, and 3-ports are connected and form implicit fields. In Figure 7.8a, the presence of the field is signaled by the derivative causality required on one of the $—C$ elements. In Figure 7.8b, no derivative causality is required, but by defining an explicit field, one can reduce the number of state variables required.

Reading the bond graph of Figure 7.8a, one finds

$$
F_4 = F_2 = \frac{a}{b} F_1 = \frac{a}{b} k_1 X_1,
$$

$$
\dot{X}_1 = \frac{a}{b} V_2 = \frac{a}{b} (V_4 - V_3) = \frac{a}{b} \left(V_4 - \frac{d}{dt} \frac{F_3}{k_3} \right) = \frac{a}{b} \left(V_4 - \frac{d}{dt} \frac{F_2}{k_3} \right)
$$

$$
= \frac{a}{b} \left[V_4 - \frac{d}{dt} \left(\frac{a}{b} \frac{F_1}{k_3} \right) \right] = \frac{a}{b} \left[V_4 - \frac{d}{dt} \left(\frac{a}{b} \frac{k_1 X_1}{k_3} \right) \right], \tag{7.37}
$$

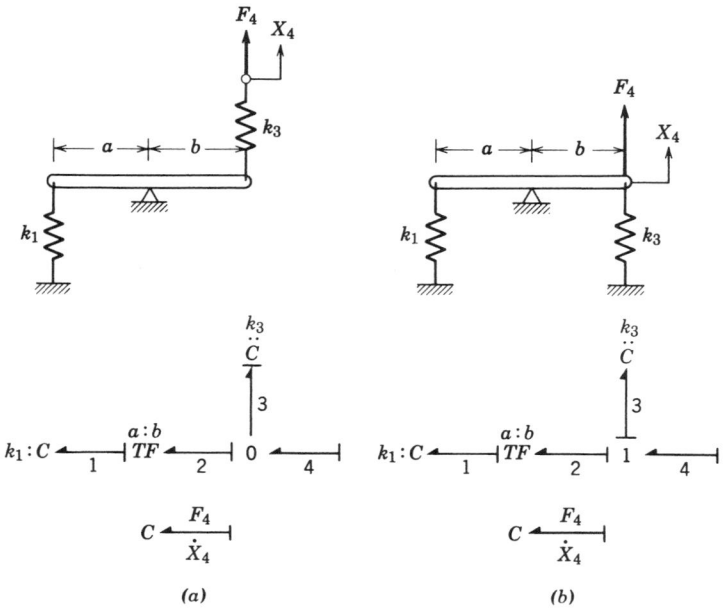

FIGURE 7.8. Two implicit fields. (a) A 1-port field with internal derivative causality; (b) 1-port field without internal derivative causality.

or

$$\dot{X}_1 = \frac{b^2 k_3}{a^2 k_1 + b^2 k_3} \frac{a}{b} V_4,$$

or

$$X_1 = \frac{b^2 k_3}{a^2 k_1 + b^2 k_3} \frac{a}{b} X_4, \tag{7.38}$$

assuming that X_1 and X_4 represent deflections away from equilibrium. Substituting Eq. (7.38) into Eq. (7.37), the characteristics of an equivalent 1-port compliance field are found:

$$F_4 = \frac{a^2 k_1 k_3}{a^2 k_1 + b^2 k_3} X_4. \tag{7.39}$$

The system of Figure 7.8b is easily described, since no derivative causality is involved:

$$F_4 = F_2 + F_3 = \frac{a}{b} F_1 + F_3 = \frac{a}{b} k_1 X_1 + k_3 X_3. \tag{7.40}$$

Two state equations are involved:

$$\dot{X}_1 = \frac{a}{b}V_2 = \frac{a}{b}V_4, \tag{7.41}$$

$$\dot{X}_3 = V_4, \tag{7.42}$$

but both equations can be integrated to yield

$$X_1 = \frac{a}{b}X_4, \qquad X_3 = X_4, \tag{7.43}$$

again assuming that integration constants vanish, that is, that $X_1 = X_3 = X_4 = 0$ represents the condition when all springs are unstretched. Substituting Eqs. (7.43) into Eq. (7.40), the equivalent 1-port C-field constitutive law is found:

$$F_4 = \left(\frac{a^2}{b^2}k_1 + k_3\right)X_4. \tag{7.44}$$

Both C-field representations (7.39) and (7.44) are practically useful, the first because the algebra associated with different causality is solved once and for all, and the second because an essentially trivial extra state equation has been integrated once and for all.

7.1.3   *I*-Fields

Inertial fields are strictly analogous to the capacitive fields just discussed. Instead of constitutive laws relating efforts to displacements, inertial elements have constitutive laws relating flows to momenta. All the results for C-fields will hold for I-fields if flows are substituted for efforts and momenta are substituted for displacements.

For example, the energy stored in the n-port I-field shown in Figure 7.9 is just

$$\begin{aligned}
\mathbf{E} &= \int_{t_0}^{t} \sum_{i=1}^{n} f_i e_i \, dt \\
&= \int_{t_0}^{t} \sum_{i=1}^{n} f_i \dot{p}_i \, dt \\
&= \int_{\mathbf{p}_0}^{\mathbf{p}} \sum_{i=1}^{n} f_i(p) \, dp_i \\
&= \int_{\mathbf{p}_0}^{\mathbf{p}} \mathbf{f}(\mathbf{p}) \, d\mathbf{p} = \mathbf{E}(\mathbf{p}),
\end{aligned} \tag{7.45}$$

in which, as in Eq. (7.1), column vectors for the flows and momenta have been defined:

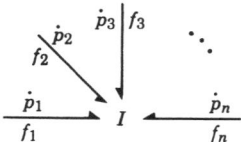

FIGURE 7.9. The symbol for an n-port I-field.

$$\mathbf{p} \equiv \begin{bmatrix} p_1 \\ p_2 \\ \vdots \\ p_n \end{bmatrix}, \qquad \mathbf{f} \equiv \begin{bmatrix} f_1 \\ f_2 \\ \vdots \\ f_n \end{bmatrix}. \qquad (7.46)$$

The analogs of Eq. (7.9) are

$$\frac{\partial \mathbf{E}}{\partial p_i} = f_i(\mathbf{p}), \qquad i = 1, 2, \ldots, n, \qquad (7.47)$$

and the reciprocity relations are

$$\frac{\partial f_i}{\partial p_j} = \frac{\partial^2 \mathbf{E}}{\partial p_j \partial p_i} = \frac{\partial f_j}{\partial p_i}. \qquad (7.48)$$

Equation (7.48), which is valid for nonlinear fields, carries the implication for the linear cases that mass matrices, inductance matrices, and the like must be symmetric because energy must be conserved.

In common with C-fields, I-fields occur in both implicit and explicit forms. Often the form which an I-field takes depends on the system modeler's point of view, and it is important to realize that manipulations of fields may result in practically useful, simplified means of representing system components. In mechanics, for example, the concept of a rigid body implies that elemental masses are constrained to move in ways such that the distances between them do not vary. This means that all rigid bodies are I-fields, and in Chapter 9, the general means of describing mechanical systems containing rigid bodies will be discussed in some detail. For now, a single example may suffice to show how an analyst could construct either an implicit- or an explicit-field representation of a rigid body.

The long, thin, rigid bar of Figure 7.10a has area A, mass density ρ, and length L. Thus its total mass m is

$$m = \rho A L, \qquad (7.49)$$

and its centroidal moment of inertia J about an axis perpendicular to the long dimension is

$$J = \frac{mL^2}{12}. \qquad (7.50)$$

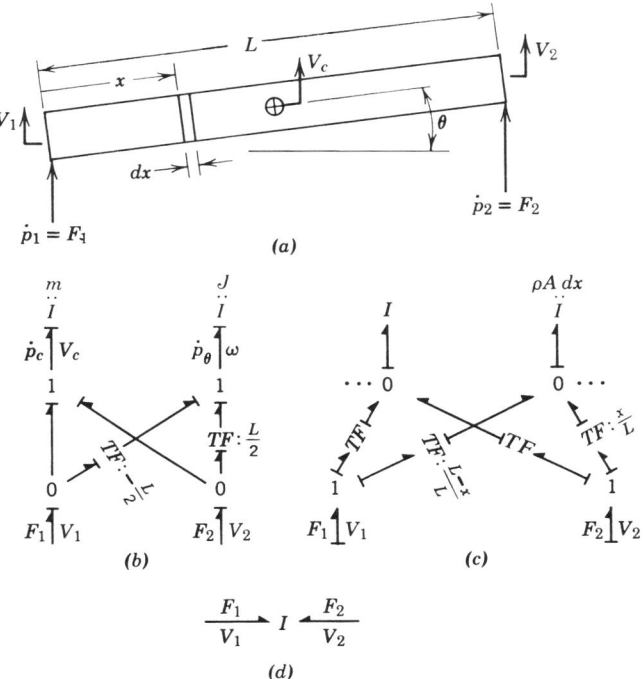

FIGURE 7.10. *I*-field representation of a rigid body in plane motion. (*a*) Schematic diagram; (*b*) implicit field using total mass and centroidal moment of inertia; (*c*) implicit field using differential elements; (*d*) explicit field.

If we consider plane motion of the bar and allow only vertical motion of the center of mass and a small angular rotation relative to a horizontal axis, and if we consider two ports at the end of the bar with forces F_1 and F_2 and velocities V_1 and V_2, then this rigid body can be described by a linear 2-port *I*-field.

One way to find the constitutive laws for the *I*-field is to describe the motion first in terms of the velocity of the center of mass, V_c, and the angular velocity, $\omega = \dot{\theta}$. The net force on the bar is then the rate of change of the linear momentum, and the momentum is related to V_c by the mass ($V_c = p_c/m$). Similarly, the net torque about the center of mass is the rate of change of the angular momentum, and the angular momentum is related to $\omega = \dot{\theta}$ by the moment of inertia ($\omega = p_\theta/J$). In Figure 7.10b, these constitutive laws are represented by the two 1-port inertial elements. The remainder of the graph serves to relate V_1 and V_2 to V_c and ω, and also to relate the net force and net torque to F_1 and F_2. (The part of the graph involving 0- and 1-junctions and transformers is a special kind of junction structure that appears frequently in mechanics and that will be discussed further in Chapter 9.)

From the bond graph (or the schematic diagram) we may now write the output equations:

$$V_1 = V_c - \frac{L}{2}\omega = \frac{p_c}{m} - \frac{L}{2}\frac{p_\theta}{J}, \tag{7.51}$$

$$V_2 = V_c + \frac{L}{2}\omega = \frac{p_c}{m} + \frac{L}{2}\frac{p_\theta}{J}, \tag{7.52}$$

where p_c and p_θ designate the linear and angular momentum variables, respectively. The state equations are

$$\dot{p}_c = F_1 + F_2, \tag{7.53}$$

$$\dot{p}_\theta = -\frac{L}{2}F_1 + \frac{L}{2}F_2. \tag{7.54}$$

In order to have an explicit field at the external ports, these equations must be integrated in time. If p_c and p_θ (and hence, V_c and ω) vanish at the initial time in the integration, then the possible integration constants vanish, and the results are

$$p_c = p_1 + p_2 \tag{7.55}$$

$$p_\theta = -\frac{L}{2}p_1 + \frac{L}{2}p_2. \tag{7.56}$$

Substituting Eqs. (7.55) and (7.56) into Eqs. (7.51) and (7.52), the result is an explicit I-field constitutive law:

$$\begin{bmatrix} V_1 \\ V_2 \end{bmatrix} = \begin{bmatrix} (1/m) + (L^2/4J) & (1/m) - (L^2/4J) \\ (1/m) - (L^2/4J) & (1/m) + (L^2/4J) \end{bmatrix} \begin{bmatrix} p_1 \\ p_2 \end{bmatrix}. \tag{7.57}$$

A different, more fundamental, but less convenient approach to this problem will generate an explicit I-field representation directly. We consider the bar to be a rigid massless rod on which are attached an infinite number of masses of value $\rho A\,dx$ at the generic position x. The velocity V_x of such a mass is

$$V_x = \frac{L - x}{L}V_1 + \frac{x}{L}V_2. \tag{7.58}$$

The bond graph of Figure 7.10c shows how these elemental masses are related to the external ports. Since the forces F_1 and F_2 are the sums of forces generated by the elemental masses, the derivative causality shown is convenient. Reading the bond graph, we have F_1 and F_2 expressed as the sum (integral) of the components due to the elemental masses:

$$F_1 = \int_0^L \frac{L - x}{L}\frac{d}{dt}[(\rho A\,dx)V_x] = \int_0^L \rho A\frac{L - x}{L}\left(\frac{L - x}{L}\dot{V}_1 + \frac{x}{L}\dot{V}_2\right)dx, \tag{7.59}$$

$$F_2 = \int_0^L \frac{x}{L}\frac{d}{dt}[(\rho A\,dx)V_x] = \int_0^L \rho A\frac{x}{L}\left(\frac{L - x}{L}\dot{V}_1 + \frac{x}{L}\dot{V}_2\right)dx, \tag{7.60}$$

where Eq. (7.58) has been used. The result of the integration in x is

$$\begin{bmatrix} F_1 \\ F_2 \end{bmatrix} = \begin{bmatrix} \dot{p}_1 \\ \dot{p}_2 \end{bmatrix} = \frac{\rho A L}{6} \begin{bmatrix} 2 & 1 \\ 1 & 2 \end{bmatrix} \begin{bmatrix} \dot{V}_1 \\ \dot{V}_2 \end{bmatrix}, \qquad (7.61)$$

or, if we agree to define velocities such that p_1 and p_2 vanish when V_1 and V_2 vanish, then

$$\begin{bmatrix} p_1 \\ p_2 \end{bmatrix} = \frac{\rho A L}{6} \begin{bmatrix} 2 & 1 \\ 1 & 2 \end{bmatrix} \begin{bmatrix} V_1 \\ V_2 \end{bmatrix}. \qquad (7.62)$$

This is the constitutive law for the explicit field of Figure 7.10d in derivative causality (or mass matrix) form. It is left as an exercise for the reader to show that, using Eqs. (7.49) and (7.50), the constitutive laws of Eqs. (7.57) and (7.62) are the same except for the different causality.

Electrical circuits containing mutually interacting coils can be conveniently represented using explicit I-fields. In order to write the constitutive laws for such fields, one requires not only a sign convention for currents and voltages, but also a convention dealing with the relative orientation of the coils. One such convention is shown in the circuit diagrams of Figures 7.11a and b as dots placed near the ends of the

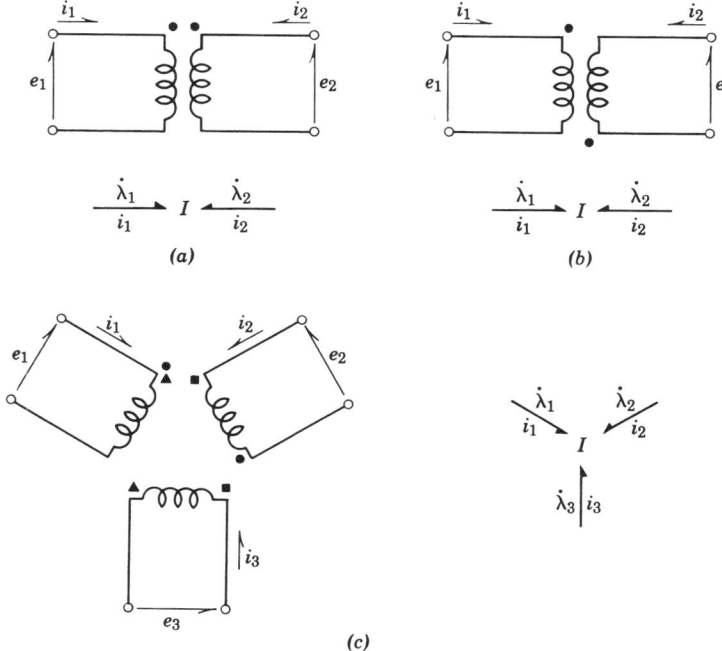

FIGURE 7.11. Mutual inductance in electrical systems. (a) and (b) 2-port I-fields with different coil orientation; (c) 3-port I-field.

coils. The idea is that if the currents, i_1 and i_2, are defined so that when positive, they both enter or both leave dotted ends of their coils, then the mutual-inductance effects will be positive. If, on the other hand, the dots are placed as in Figure 7.11b, so that one of the currents enters through a dotted end and the other leaves through a dotted end, then the mutual-inductance effects are negative. The convention is most readily illustrated for the linear case, but it applies equally well for the nonlinear case. For example, if one denotes self-inductance coefficients by the letter L and mutual-inductance coefficients by the letter M, as is the usual convention, then the system of Figure 7.11a has the constitutive law

$$\begin{bmatrix} \lambda_1 \\ \lambda_2 \end{bmatrix} = \begin{bmatrix} L_1 & M_{12} \\ M_{12} & L_2 \end{bmatrix} \begin{bmatrix} i_1 \\ i_2 \end{bmatrix}. \tag{7.63}$$

Note that this field representation is in derivative causality form and is symmetric, as it must be. The field of Figure 7.11b has the constitutive law

$$\begin{bmatrix} \lambda_1 \\ \lambda_2 \end{bmatrix} = \begin{bmatrix} L_1 & -M_{12} \\ -M_{12} & L_2 \end{bmatrix} \begin{bmatrix} i_1 \\ i_2 \end{bmatrix}. \tag{7.64}$$

For three or more interacting coils, it is rather awkward to show the orientations. In Figure 7.11c, the dots have become circles, squares, and triangles to indicate the signs of the three mutual-inductance coefficients. The field laws are readily written:

$$\begin{bmatrix} \lambda_1 \\ \lambda_2 \\ \lambda_3 \end{bmatrix} = \begin{bmatrix} +L_1 & -M_{12} & -M_{13} \\ -M_{12} & +L_2 & +M_{23} \\ -M_{13} & +M_{23} & +L_3 \end{bmatrix} \begin{bmatrix} i_1 \\ i_2 \\ i_3 \end{bmatrix}. \tag{7.65}$$

In the bond graph I-field representation, the convention indicated by the circles, squares, and triangles is evident in the sign pattern of the mutual-inductance coefficients. Note that with the inward sign convention shown, the self-inductance terms must be positive for physical coils. Also, independently of the coil orientation convention, the inductance matrix and its inverse will be symmetric if the I-field is conservative.

When energy-storing fields are used in systems, it sometimes happens that explicit representations such as Eq. (7.65) must be algebraically manipulated. For example, if the coils of Figure 7.11c are interconnected, as shown in Figure 7.12a, then the rules of causality indicate that integral causality cannot be applied to all three I-field ports. If there were no mutual inductance, the system would be as shown in Figure 7.12b. This system has one element in differential causality, but the formulation methods of Chapter 5 would handle the system without trouble. In fact, this system is just the dual of the C-field of Figure 7.7. On the other hand, when mutual-inductance effects are considered, the system appears as shown in Figure 7.12c. If one assigns integral causality on bonds 1 and 3, then bond 2 must have derivative causality.

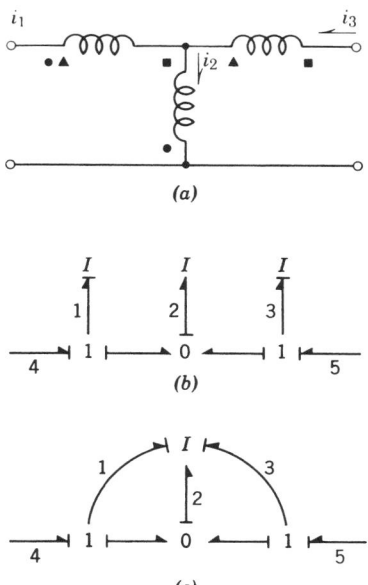

FIGURE 7.12. A network of interacting coils. (*a*) The coils of Figure 7.11 interconnected; (*b*) system bond graph if mutual inductance is absent; (*c*) system bond graph with *I*-field.

It is not particularly difficult to switch the constitutive law of Eq. (7.65) into the mixed causal form of Figure 7.12*c*. Rewriting parts of Eq. (7.65), we have

$$
\begin{bmatrix} \lambda_1 \\ \lambda_3 \end{bmatrix} = \begin{bmatrix} L_1 & -M_{13} \\ -M_{13} & L_3 \end{bmatrix} \begin{bmatrix} i_1 \\ i_3 \end{bmatrix} + \begin{bmatrix} -M_{12} \\ M_{23} \end{bmatrix} [i_2]. \tag{7.66}
$$

Inverting this relation to yield i_1 and i_3 in terms of λ_1, λ_3, and i_2 gives

$$
\begin{bmatrix} i_1 \\ i_3 \end{bmatrix} = \frac{1}{L_1 L_3 - M_{13}^2} \begin{bmatrix} L_3 & M_{13} & L_3 M_{12} - M_{13} M_{23} \\ M_{13} & L_1 & M_{13} M_{12} - L_1 M_{23} \end{bmatrix} \begin{bmatrix} \lambda_1 \\ \lambda_3 \\ i_2 \end{bmatrix}. \tag{7.67}
$$

Substituting Eq. (7.67) into the second equation of (7.65), we can also obtain the third equations for λ_2 in terms of λ_1, λ_3, and i_2:

$$
\lambda_2 = \lambda_2(\lambda_1, \lambda_3, i_2). \tag{7.68}
$$

When mutual-inductance effects are present, the writing of state equations is not quite as straightforward as when there is no mutual inductance. Basically, this is because of the role of i_2 in Eq. (7.67). Starting with the equation for λ_1,

$$\lambda_1 = e_4 - e_2 = e_4 - \frac{d}{dt}\lambda_2(\lambda_1, \lambda_3, i_2). \qquad (7.69)$$

Now,

$$i_2 = i_1 + i_3, \qquad (7.70)$$

and we can use Eq. (7.67) to find i_1 and i_3 (and hence i_2) in terms of λ_1, λ_3, and i_2. But now we are in an algebraic loop reminiscent of those discussed Chapter 5 when the imposition of integral causality did not suffice to determine the causality of all bonds. Using Eqs. (7.70) and (7.67), we can express i_2 in terms of itself and then solve this algebraic equation for i_2 in terms of λ_1, λ_3 in order to compute the state equation (7.69). Failure to do this will result in an endless looping through the equations in an attempt to eliminate i_2.

Although one could write state equations for the system in the manner outlined above, there is a simpler way to handle the system. In Figure 7.13, it is shown that the imposition of complete differential causality [which corresponds to Eq. (7.65) anyway] allows one to compute the explicit 2-port field without solving any algebraic equations. Reading the graph, and using Eq. (7.65), we have

$$e_4 = e_1 + e_2 = \frac{d}{dt}(\lambda_1 + \lambda_2)$$

$$= \frac{d}{dt}[(L_1 - M_{12})i_1 + (L_2 - M_{12})i_2 + (M_{23} - M_{13})i_3]$$

$$= \frac{d}{dt}[(L_1 - M_{12})i_4 + (L_2 - M_{12})(i_4 + i_5) + (M_{23} - M_{13})i_5],$$

or

$$\lambda_4 = (L_1 + L_2 - 2M_{12})i_4 + (L_2 - M_{12} - M_{13} + M_{23})i_5 \qquad (7.71)$$

upon integration. Similarly,

$$\lambda_5 = (L_2 - M_{12} - M_{13} + M_{23})i_4 + (L_2 + L_3 + 2M_{23})i_5. \qquad (7.72)$$

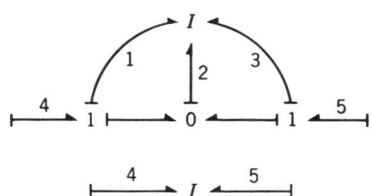

FIGURE 7.13. Reduction of system of Figure 7.12 to explicit 2-port I-field form.

These equations represent the constitutive laws for the 2-port I-field in derivative causality form. The all-integral-causality form follows by inverting the equations. Thus, λ_4 and λ_5 can readily be used as the two independent state variables. Clearly, the 2-port I-field is much more convenient to use in equation formulation than the 3-port field and its associated junction structure. Such studies of fields are, of course, as useful for C-field problems as they are for I-fields.

7.1.4   Mixed Energy-Storing Fields

There are occasions when an energy-storing device cannot be described as a C-field or an I-field but rather acts as a C-field at some ports and an I-field at others. Within a single-energy domain, the need for such an element is far from evident, but when transducers are studied in Chapter 8, it will be seen that many transducers are essentially energy conservative but are not pure I-fields or C-fields. Even in the case of transducers, it has been a common practice to arrange the analogies between variables in the two energy domains linked by the transducer in such a way that the transducer is a pure field. There is no real need to do this, however, and in bond graphs we prefer to retain our identification of effort, flow, displacement, and momentum quantities for all energy domains. We need, then, to discuss what will be called IC-fields. Not only is this policy just as convenient as the policy of switching analogies to suit the problem at hand; it is required in principle for some systems. There is no way, for example, to pick an effort–flow identification for an electromechanical system containing both a movable-plate capacitor and a solenoid transducer such that both transducers would be described as pure I- or C-fields.

Figure 7.14 shows the general n-port IC-field. The ports are numbered so that the first j-ports ($1 \le j \le n$) are inertial in character and the ports from $j + 1$ to n are capacitive in character. In integral causality form as shown, the first j state variables are momenta and the remainder are displacements. The stored energy, $\mathbf{E}$, is a function of this mixed set of state variables:

$$\mathbf{E} = \int^t \sum_{i=1}^n e_i f_i \, dt = \int^\mathbf{p} \sum_{i=1}^j f_i \, dp_i + \int^\mathbf{q} \sum_{k=j+1}^n e_k \, dq_k, \qquad (7.73)$$

where $\mathbf{p}$ represents the vector of momenta and $\mathbf{q}$ represents the vector of displacements. From Eq. (7.73), we see that

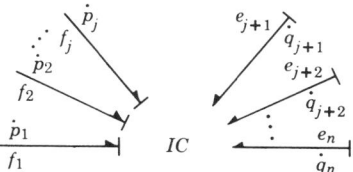

FIGURE 7.14. The general mixed energy-storage field.

$$f_i = \frac{\partial \mathbf{E}}{\partial p_i}, \qquad i = 1, 2, \ldots, j, \tag{7.74}$$

$$e_k = \frac{\partial \mathbf{E}}{\partial q_k}, \qquad k = j + 1, j + 2, \ldots, n. \tag{7.75}$$

The reciprocity conditions for the constitutive laws can be easily derived by computing second partial derivatives of $\mathbf{E}$. For example, the second partial derivatives of $\mathbf{E}$ with respect to a momentum from the first set of ports and a displacement from the second set of ports yields

$$\frac{\partial f_i}{\partial q_k} = \frac{\partial^2 E}{\partial q_k \partial p_i} = \frac{\partial e_k}{\partial p_i}, \tag{7.76}$$

where $1 \leq i \leq j, \ j + 1 \leq k \leq n$. These results should be compared with the results for pure C-fields and I-fields [Eqs. (7.1), (7.10), (7.45), and (7.46)]. Physical examples involving IC-fields will be found in Chapter 8.

7.2 RESISTIVE FIELDS

An R-field is an n-port, the constitutive laws of which relate the n-port efforts and the n-port flows by means of static (or algebraic) functions. This definition includes power-conservative elements such as 0- and 1-junctions and elements containing sources as R-fields, but in practice most R-fields studied dissipate power. Both explicit and implicit R-fields exist. Explicit multiport R-fields arise frequently in modeling nonlinear devices. Implicit fields arising from the interconnection of 1-port resistances, transformers, gyrators, and junction 3-ports may be conveniently represented in R-field form as relations between external port efforts and flows.

The causality for R-fields usually is determined by the source and energy-storing elements in the system. Two fundamental causal patterns are shown in Figure 7.15. The constitutive laws for the *resistance* causality shown in Figure 7.15a may be written

$$e_i = \Phi_{R_i}(f_1, f_2, \ldots, f_n), \qquad i = 1, 2, \ldots, n. \tag{7.77}$$

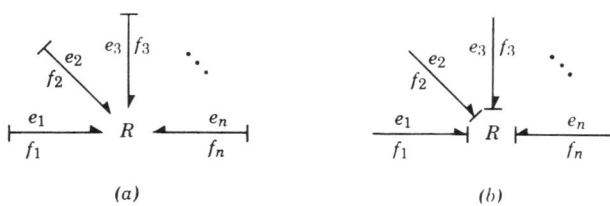

(a)   (b)

FIGURE 7.15. The general R-field. (a) Resistance form; (b) conductance form.

The *conductance* causality shown in Figure 7.15*b* has constitutive laws that may be represented by the following:

$$f_i = \Phi_{R_i}^{-1}(e_1, e_2, \ldots, e_n), \qquad i = 1, 2, \ldots, n. \tag{7.78}$$

In addition to these two fundamental causal forms, a large number of other forms are possible in which some of the bonds are causally oriented as in Figure 7.15*a* and the remainder are oriented as in Figure 7.15*b*. Although *R*-fields are basically neutral with respect to causality, we shall encounter some constitutive laws for *R*-fields that cannot be unique for certain causalities (this happens, of course, only in the nonlinear case).

Although there is no stored energy function for *R*-fields as there is for energy-storing fields, and hence there is no simple way to show that *R*-field constitutive laws are constrained, still, for special classes of *R*-fields, some useful properties of the constitutive laws may be found. We will illustrate some of these properties by example.

First, let us consider linear *R*-fields which contain no sources and no gyrators. A typical example is shown in Figure 7.16. This *R*-field will accept resistance causality on the three external ports, as shown in Figure 7.16*b*. Reading the bond graph, the constitutive laws are readily derived:

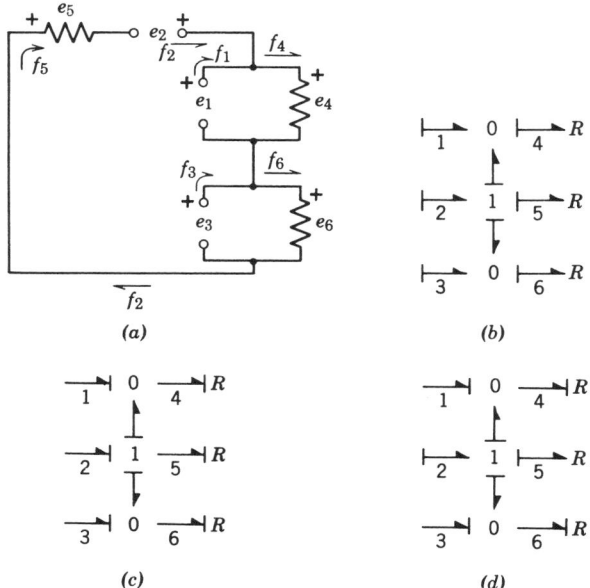

FIGURE 7.16. An implicit *R*-field. (*a*) Circuit graph; (*b*) bond graph showing resistance causality; (*c*) bond graph showing conductance causality; (*d*) bond graph showing mixed causality.

$$
\begin{bmatrix} e_1 \\ e_2 \\ e_3 \end{bmatrix} = \begin{bmatrix} R_4 & R_4 & 0 \\ R_4 & R_4 + R_5 + R_6 & R_6 \\ 0 & R_6 & R_6 \end{bmatrix} \begin{bmatrix} f_1 \\ f_2 \\ f_3 \end{bmatrix}. \tag{7.79}
$$

As the bond graph of Figure 7.16c demonstrates, this R-field will also accept conductance causality. So, by inverting Eq. (7.79) or, more simply, by reading the bond graph of Figure 7.16c, the constitutive laws in conductance form may be found:

$$
\begin{bmatrix} f_1 \\ f_2 \\ f_3 \end{bmatrix} = \begin{bmatrix} 1/R_4 + 1/R_5 & -1/R_5 & 1/R_5 \\ -1/R_5 & 1/R_5 & -1/R_5 \\ 1/R_5 & -1/R_5 & 1/R_5 + 1/R_6 \end{bmatrix} \begin{bmatrix} e_1 \\ e_2 \\ e_3 \end{bmatrix}. \tag{7.80}
$$

Notice that both Eqs. (7.79) and (7.80) are symmetric. These forms may be called "Onsager" forms in analogy to the well-known Onsager reciprocal relations of irreversible thermodynamics [3]. Onsager proposed his reciprocity conditions for variables called "affinities" and "fluxes" that are analogous to the efforts and flows of bond graphs. In general, implicit R-fields composed of linear 1-port resistances, 0- and 1-junctions, and transformers will obey Onsager reciprocity; that is, when expressed in resistance or conductance form, the matrix in the constitutive laws will be symmetric. On the other hand, for mixed causality, the matrix will have antisymmetrical terms, and if gyrators are present, the Onsager reciprocity relations do not hold. Thus the reciprocity of energy-storing fields, which may be called Maxwell reciprocity, is more general than Onsager reciprocity. A perfectly reasonable explicit R-field characterization expressed in resistance or conductance form can be unsymmetrical. One may think of the unsymmetrical part of the matrices as arising from gyrational effects that introduce antisymmetrical terms, although in an explicit field one cannot identify a gyrator unless one can find some implicit field that has the same port constitutive laws as the explicit field.

Using our example, let us first show that an R-field in mixed causality has antisymmetric terms. A possible mixed causal pattern is shown in Figure 7.16d. Reading the bond graph, the constitutive laws are as follows:

$$
\begin{bmatrix} f_1 \\ e_2 \\ f_3 \end{bmatrix} = \begin{bmatrix} 1/R_4 & -1 & 0 \\ 1 & R_5 & 1 \\ 0 & -1 & 1/R_6 \end{bmatrix} \begin{bmatrix} e_1 \\ f_2 \\ e_3 \end{bmatrix}, \tag{7.81}
$$

in which the antisymmetric terms are evident. Such a form, which results when some of the input and output variables are interchanged for an R-field that obeys the Onsager reciprocal relations in resistance or conductance causality, is sometimes called a Casimir form [4].

It may be tempting to conclude that if R-fields are described with either resistance or conductance causality, they will exhibit Onsager reciprocity, and if they are described with a mixed causality, they will show a Casimir form. But it is easy to show that if gyrators are allowed in an implicit R-field, then this conclusion is false.

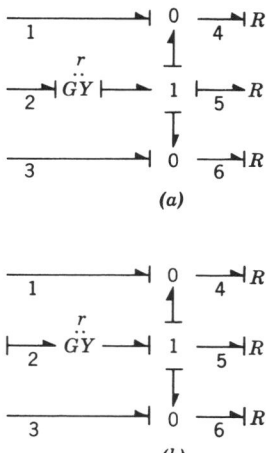

FIGURE 7.17. Modified version of the R-field of Figure 7.15. (a) Conductance causality; (b) mixed causality.

Consider, for example, the R-field of Figure 7.17, which is formed from the field of Figure 7.16 by adding a single gyrator. In the conductance causality shown in Figure 7.17a, the field exhibits a Casimir form rather than an Onsager form:

$$\begin{bmatrix} f_1 \\ f_2 \\ f_3 \end{bmatrix} = \begin{bmatrix} 1/R_4 & -1/r & 0 \\ 1/r & R_5/r^2 & 1/r \\ 0 & -1/r & 1/R_6 \end{bmatrix} \begin{bmatrix} e_1 \\ e_2 \\ e_3 \end{bmatrix}, \tag{7.82}$$

in which r is the gyrator parameter. Similarly, for the mixed causality of Figure 7.17b, the R-field is symmetric:

$$\begin{bmatrix} f_1 \\ e_2 \\ f_3 \end{bmatrix} = \begin{bmatrix} 1/R_4 + 1/R_5 & -r/R_5 & 1/R_5 \\ -r/R_5 & r^2/R_5 & -r/R_5 \\ 1/R_5 & -r/R_5 & 1/R_5 + 1/R_6 \end{bmatrix} \begin{bmatrix} e_1 \\ f_2 \\ e_3 \end{bmatrix}. \tag{7.83}$$

A general R-field is not representable in an Onsager or a Casimir form. A simple example is shown in Figure 7.18. Its constitutive laws in resistance form are

$$\begin{bmatrix} e_1 \\ e_2 \end{bmatrix} = \begin{bmatrix} R_3 + R_4 & r + R_4 \\ -r + R_4 & R_4 + R_5 \end{bmatrix} \begin{bmatrix} f_1 \\ f_2 \end{bmatrix}, \tag{7.84}$$

in which a symmetrical part due to the elements which obey Onsager reciprocity and an antisymmetrical part due to the gyrator can be recognized. If Eq. (7.84) were

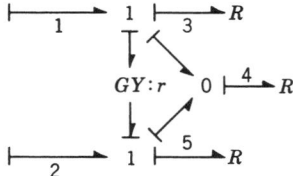

FIGURE 7.18. An R-field containing a gyrator.

presented with numerical values for the parameters, it would have a rather undistinguished appearance, being neither an Onsager nor a Casimir form.

Note that if positive resistance parameters are used, then all of the example fields will dissipate energy for any possible port conditions. This is true because the 1-port resistors can only dissipate power, and the junction elements and the 2-ports, —TF— and —GY—, conserve power. A way to check for power dissipation given an explicit linear field is to check whether the matrix of coefficients is positive definite or not. In the resistance form, for example, if **e** is the column vector of efforts, **f** the column vector of flows, and **R** the resistance matrix, then the power **P** is

$$\mathbf{P} = \mathbf{f}^t\mathbf{e},$$

and the constitutive law for the field is

$$\mathbf{e} = \mathbf{R}\mathbf{f},$$

so that

$$\mathbf{P} = \mathbf{f}^t\mathbf{R}\mathbf{f}, \tag{7.85}$$

where t stands for the transpose of a vector or matrix.

Equation (7.85) shows that power will be dissipated for any **f** if **R** is positive definite. If one can only say that no power should be created, then **R** must be positive semidefinite; that is, there may be some finite flows for which zero power is dissipated. Using the rules for checking a matrix for positive definiteness, the reader may demonstrate that all the examples of R-fields given above are at least positive semidefinite. In this regard, it is useful to remember that the antisymmetric part of any matrix contributes nothing to the definiteness of the matrix. This correlates with the idea that the antisymmetric terms due to the presence of a gyrator represent no power generation or dissipation in a field such as that associated with Eq. (7.84).

7.3   MODULATED 2-PORT ELEMENTS

The 2-port elements —TF— and —GY— are linear, power-conserving elements, the usefulness of which has been demonstrated in previous chapters. Here we discuss the modulated transformer, —MTF—, and modulated gyrator, —MGY—, which are nonlinear, power-conserving generalizations of —TF— and —GY—. Basically, the parameters of the elements —TF— and —GY— are allowed to be a function of some parameter, say ξ, in the modulated 2-ports. The usual symbols and constitutive laws or these elements are shown below:

$$
\begin{array}{cc}
m(\xi) & r(\xi) \\
\downarrow & \downarrow \\
\xrightarrow[1]{} MTF \xrightarrow[2]{}, & \xrightarrow[1]{} MGY \xrightarrow[2]{}, \\
m(\xi)e_1 = e_2, & e_1 = r(\xi)f_2, \\
f_1 = m(\xi)f_2, & r(\xi)f_1 = e_2.
\end{array}
\tag{7.86}
$$

Notice that for both elements the power $e_1 f_1$ is always equal to the power $e_2 f_2$ no matter what the value of $m(\xi)$ or $r(\xi)$ may be. Also, the parameters m and r are shown as changing by means of a signal or active bond rather than by means of a power bond. Thus, it is characteristic of these modulated-parameter elements that the parameters change value without a directly associated power flow.

The modulated elements will prove useful for modeling certain classes of systems in later chapters. In Figure 7.19, two examples are shown that illustrate typical uses of the modulated elements. The modulated transformer is particularly useful in describing mechanical systems moving though large angles. The rigid, pivoted bar of Figure 7.19a may be described thus:

$$\tau = (l \cos \theta) F, \tag{7.87}$$

$$(l \cos \theta)\omega = V, \tag{7.88}$$

where the variables are defined in the figure.

These constitutive relations may be derived by applying the laws of mechanics to the linkage. Equation (7.87) is a torque equilibrium statement, and Eq. (7.88) is an angular velocity–velocity relation which could be derived by time-differentiating the equation

$$l \sin \theta = x. \tag{7.89}$$

The bond graph of Figure 7.19b represents Eqs. (7.87) and (7.88), where the transformer modulus is $l \cos \theta$ and θ plays the role of ξ in the general equations (7.86). It is a peculiarity of the *MTF*s of mechanics that the modulus varies with the displacements at the ports. For this reason, these *MTF*s are sometimes called *displacement-modulated transformers*.

Figure 7.19c shows an electromechanical system in which a field current i_f is responsible for establishing a magnetic field $B(i_f)$ in a gap. If then a current-carrying conductor of length l moves with velocity V, as shown, it is a consequence of Faraday's law that a voltage e will be induced along the conductor according to the following relation:

$$e = B(i_f)lV. \tag{7.90}$$

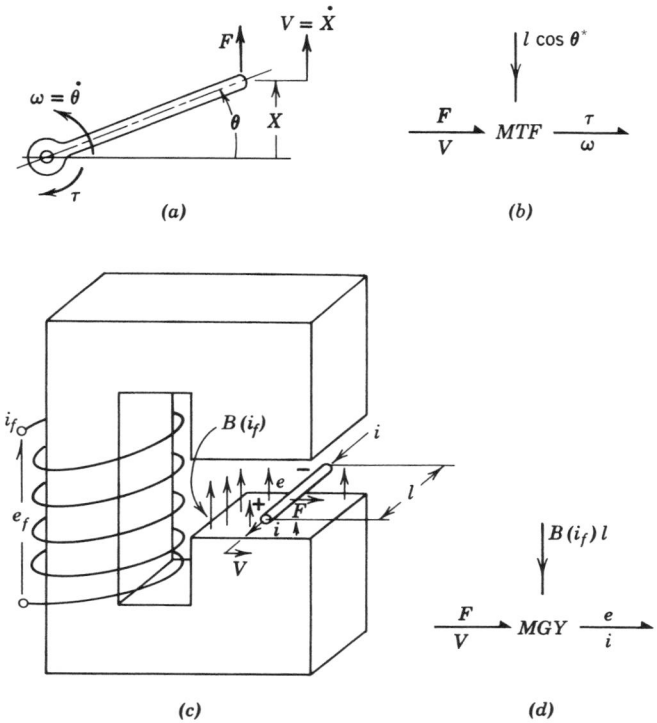

FIGURE 7.19. Examples of modulated 2-port elements. (*a*) Mechanical linkage; (*b*) modulated transformer for (*a*); (*c*) electromechanical system; (*d*) modulated gyrator for (*c*).[*Modulus is defined unambiguously by Eqs. (7.87) and (7.88)].

The force F required to move the conductor is found from the Lorentz force law to be

$$B(i_f)li = F. \tag{7.91}$$

The modulated gyrator of Figure 7.19*d* represents these laws, with $B(i_f)l$ representing the modulus r and i_f the parameter ξ in the general form of Eq. (7.86). Note that $ei = FV$ (as long as the electrical and the mechanical power are measured in the same units), and i_f can be changed with no power associated with the *MGY*. On the other hand, i_f is also associated with a self-inductance effect and coil resistance, so that there is power flow and energy storage associated with i_f. The point is that no energy is associated with changes in Bl from the point of view of the *MGY*. The modulated gyrator will be used in modeling voice coils, electrical motors, and similar devices in Chapter 8.

Since the sign conventions and causal restrictions for the modulated elements are the same as for the —*TF*— and —*GY*— elements studied earlier, there is little new to be said here. A few words of caution are in order, however. Since the modulated elements incorporate a pure signal interaction, it is quite possible to construct bond graphs that do not have physical interpretations by imagining that the modulus of a transformer or gyrator can be a function of any variable at all. Just as it is easy to make incorrect block diagrams, signal flow graphs, or other signal descriptions which violate power and energy constraints that exist in real physical systems, so too is it rather easy to assume that an element can be modeled using a modulated 2-port when, in fact, the element may really be a true 3-port or when the physical system may not allow the modulus to be a function of the signal assumed.

To show just one example of the trap into which one may sometimes fall, consider the rack-and-pinion system shown in Figure 7.20. Clearly, if the pinion is small ($r \ll l$), then the rack is nearly a lever with a lever ratio of $(l/2 - r\theta)/(l/2 + r\theta)$ if the rack is centered when $\theta = 0$. Furthermore, if we rotate the pinion to some position and then hold θ fixed, the rack will function almost as a —TF— with the transformer modulus set by the value of θ. Thus, it seems reasonable to represent the system as in Figure 7.20b. However, this representation is false. The reason is that, whenever F_1 and F_2 are not zero, θ cannot be changed with no power. In fact, the torque τ is proportional to F_1 or F_2. Thus, the device is a true 3-port as shown in

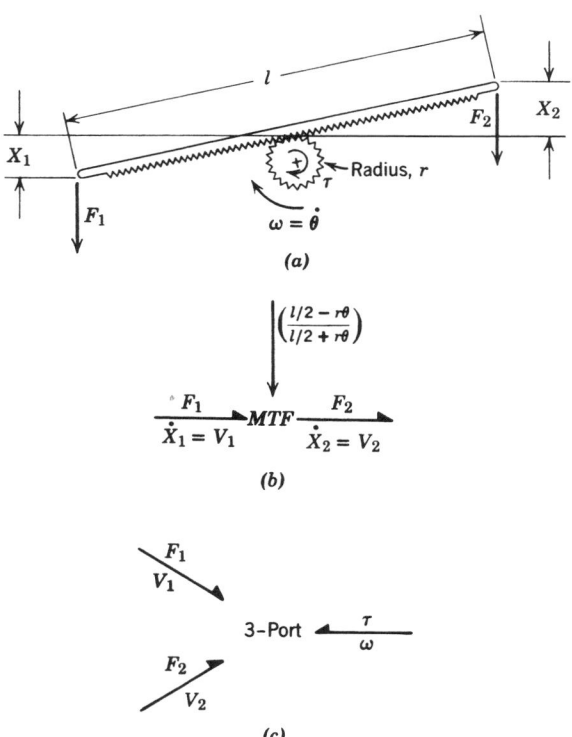

FIGURE 7.20. Rack-and-pinion system. (*a*) Schematic diagram; (*b*) incorrect bond graph; (*c*) correct bond graph.

Figure 7.20c. It can be modeled using *MTF* elements and 0- and 1-junctions as a multiport modulated transformer, as will be shown in Chapter 9, but the representation of Figure 7.20b is fundamentally incorrect.

7.4   JUNCTION STRUCTURES

Junction structures, which are assemblages of 0- and 1-junctions, transformers, and gyrators, are the energy switchyards which enforce the constraints among parts of dynamic systems. No power is dissipated or generated in a junction structure, so the *net* power into a junction structure at its ports always vanishes. Since junction structures do provide relations between efforts and flows at their ports, they are special types of *R*-fields that never dissipate power. One might expect, for linear junction structures, to find the Onsager and Casimir forms that were found for general *R*-fields. These forms are indeed found, but junction structures typically cannot accept all possible causal assignments at their ports, so some forms for the constitutive laws simply cannot exist.

Some example junction structures are shown in Figure 7.21. Note that all these structures are shown with an inward sign convention, so that the sum of the port

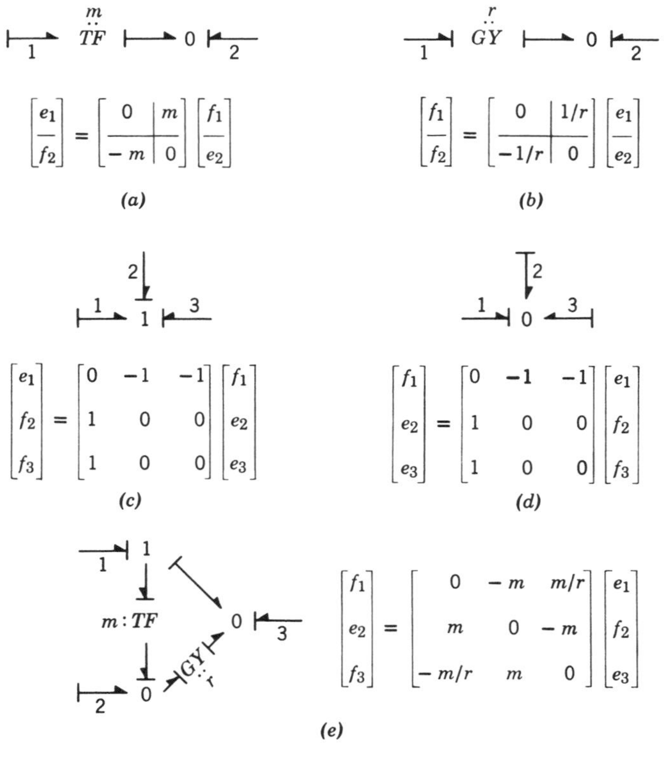

$$\begin{bmatrix} e_1 \\ \hline f_2 \end{bmatrix} = \begin{bmatrix} 0 & m \\ \hline -m & 0 \end{bmatrix} \begin{bmatrix} f_1 \\ \hline e_2 \end{bmatrix}$$

(a)

$$\begin{bmatrix} f_1 \\ \hline f_2 \end{bmatrix} = \begin{bmatrix} 0 & 1/r \\ \hline -1/r & 0 \end{bmatrix} \begin{bmatrix} e_1 \\ \hline e_2 \end{bmatrix}$$

(b)

$$\begin{bmatrix} e_1 \\ f_2 \\ f_3 \end{bmatrix} = \begin{bmatrix} 0 & -1 & -1 \\ 1 & 0 & 0 \\ 1 & 0 & 0 \end{bmatrix} \begin{bmatrix} f_1 \\ e_2 \\ e_3 \end{bmatrix}$$

(c)

$$\begin{bmatrix} f_1 \\ e_2 \\ e_3 \end{bmatrix} = \begin{bmatrix} 0 & -1 & -1 \\ 1 & 0 & 0 \\ 1 & 0 & 0 \end{bmatrix} \begin{bmatrix} e_1 \\ f_2 \\ f_3 \end{bmatrix}$$

(d)

$$\begin{bmatrix} f_1 \\ e_2 \\ f_3 \end{bmatrix} = \begin{bmatrix} 0 & -m & m/r \\ m & 0 & -m \\ -m/r & m & 0 \end{bmatrix} \begin{bmatrix} e_1 \\ f_2 \\ e_3 \end{bmatrix}$$

(e)

FIGURE 7.21. Constitutive laws for some simple junction structures.

powers must vanish. This means that, in each case, the matrix relating inputs to outputs must be antisymmetric; that is, zeros must appear on the main diagonal, and the *ij*th component must be the negative of the *ji*th component.

For systems without gyrators and in mixed resistance–conductance causality, such as Figures 7.21*a*, *c*, and *d*, the relations are Casimir forms (but with zeros on the main diagonal). The system of Figure 7.20*b* is in conductance form but, because it contains a gyrator, does not obey Onsager reciprocity and is instead antireciprocal. Figure 7.21*e* is antisymmetric partly because of mixed causality (which contributes two antisymmetric terms) and partly because of the gyrator (which contributes the other two).

Since an *R*-field that contains no gyrators must obey Onsager reciprocity in resistance or conductance form, and any junction structure must have antisymmetric constitutive laws, we conclude that junction structures without gyrators cannot accept resistance or conductance causality on all ports.

7.5   MULTIPORT TRANSFORMERS

Another interesting viewpoint about junction structures involves the transformations of variables that they provide. Consider, for example, the junction structure of Figure 7.22. It transforms the variables e_3 and e_4 into the variables e_1 and e_2:

$$\begin{bmatrix} e_1 \\ \hline e_2 \end{bmatrix} = \begin{bmatrix} -1 & -1 \\ \hline -m_1 & -m_2 \end{bmatrix} \begin{bmatrix} e_3 \\ \hline e_4 \end{bmatrix}. \tag{7.92}$$

But this transformation of efforts is accompanied by a transformation of flows:

$$\begin{bmatrix} f_3 \\ \hline f_4 \end{bmatrix} = \begin{bmatrix} 1 & m_1 \\ \hline 1 & m_2 \end{bmatrix} \begin{bmatrix} f_1 \\ \hline f_2 \end{bmatrix}, \tag{7.93}$$

and, since the junction structure is accomplishing the transformations, the transformations must conserve power. Rearranging the equations, the antisymmetric form

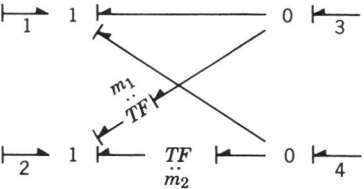

FIGURE 7.22. A junction structure that is a 2×2 bilateral transformation of variables.

characteristic of power conservation appears:

$$
\begin{bmatrix} e_1 \\ \overline{e_2} \\ \overline{f_3} \\ f_4 \end{bmatrix} = \left[\begin{array}{cc|cc} 0 & 0 & -1 & -1 \\ 0 & 0 & -m_1 & -m_2 \\ \hline 1 & m_1 & 0 & 0 \\ 1 & m_2 & 0 & 0 \end{array} \right] \begin{bmatrix} f_1 \\ \overline{f_2} \\ \overline{e_3} \\ e_4 \end{bmatrix}. \tag{7.94}
$$

Such a structure is usefully considered to be a multiport generalization of a 2-port transformer. It is appropriate to change from an "all inward" sign convention which has been used in studying general junction structures to a "through" sign convention for the external ports, as shown in Figure 7.23. With this type of sign convention, the signs in Eq. (7.92) change and the transformation equations can be written in a form analogous to the form used for 2-port transformers:

$$
\begin{bmatrix} e_1 \\ e_2 \end{bmatrix} = \left[\begin{array}{c|c} 1 & 1 \\ \hline m_1 & m_2 \end{array} \right] \begin{bmatrix} e_3 \\ e_4 \end{bmatrix}, \tag{7.92a}
$$

$$
\left[\begin{array}{c|c} 1 & m_1 \\ \hline 1 & m_2 \end{array} \right] \begin{bmatrix} f_1 \\ f_2 \end{bmatrix} = \begin{bmatrix} f_3 \\ f_4 \end{bmatrix} \tag{7.93a}
$$

In this example, it may be seen that the matrix which relates the efforts is simply transposed to form the matrix which transforms the flows. In general, a multiport transformer is characterized by a matrix (and its transpose), just as a 2-port transformer is characterized by a modulus that appears in both the effort and the flow relations. (One can even think of the modulus of a 2-port TF as a 1×1 matrix which is equal to its transpose.)

Although this example involves a 2×2 matrix since there are two ports on either side of the transformer, there can be n ports on one side of the transformer and m ports on the other side. Then the matrix will be $n \times m$ and its transpose will be $m \times n$.

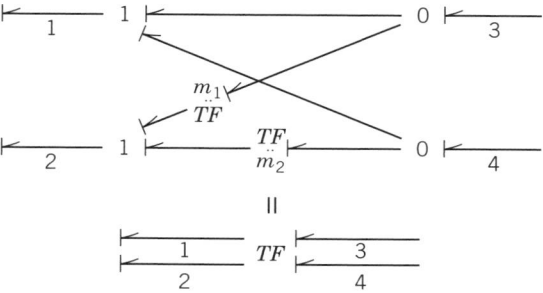

FIGURE 7.23. Structure of Figure 7.22 with modified sign half-arrows at ports 1 and 2.

It is easy to prove that transformations involving a matrix and its transpose imply that power flowing into one side of a multiport transformer must equal the power flowing out of the other side [7]. Let $[e_1]$, $[f_1]$ be column vectors of efforts and flows on one side of a multiport transformer and $[e_2]$, $[f_2]$ be vectors of efforts and flows on the other side. Then let $[M]$ be a transformation matrix of appropriate dimensions so that the equations analogous to Eqs. (7.92a) and (7.93a) are

$$[e_1] = [M][e_2], \tag{7.95}$$

$$[M'][f_1] = [f_2]. \tag{7.96}$$

By transposing Eq. (7.96) and postmultiplying both sides by $[e_2]$, the result is

$$[f_1]^t[M][e_2] = [f_2]^t[e_2], \tag{7.97}$$

or, upon using Eq. (7.95),

$$[f_1]^t[e_1] = [f_2]^t[e_2], \tag{7.98}$$

which simply states that the sum of the port powers on the number 1 side equals the sum of the port powers on the number 2 side. With the new sign convention this means the net power flowing in on one side equals the power flowing out of the other side.

It is interesting that in the above proof, not only can there be any number of ports on the two sides of the transformer, but there is no requirement that the elements of the matrix be constant. Thus the considerations apply not only to multiport transformers with constant matrices which occur in many types of systems but also for multiport modulated transformers in which the elements of the matrix change with time. An important application of multiport modulated transformers is in mechanics, where geometric nonlinearities often arise. This topic will be explored in depth in Chapter 9, but here we merely point out that when geometric constraints or transformations among displacements are converted to velocity relations, they always can be represented by multiport modulated transformers or transformer junction structures. Furthermore, when the velocity relationships have been determined and incorporated in a matrix of functions, the transpose of the matrix will automatically relate the forces.

Consider, for example, the transformation between rectangular coordinates and polar coordinates, as shown in Figure 7.24. (This will be involved in an introductory example in Chapter 9.) The relationships between r and θ and x and y are rather complicated nonlinear functions which can be expressed in several forms. For example, to convert from r and θ to x and y we can write

$$x = r\cos\theta, \qquad y = r\sin\theta. \tag{7.99}$$

The inverse relationship is

$$r = (x^2 + y^2)^{1/2}, \qquad \theta = \tan^{-1}(y/x). \tag{7.100}$$

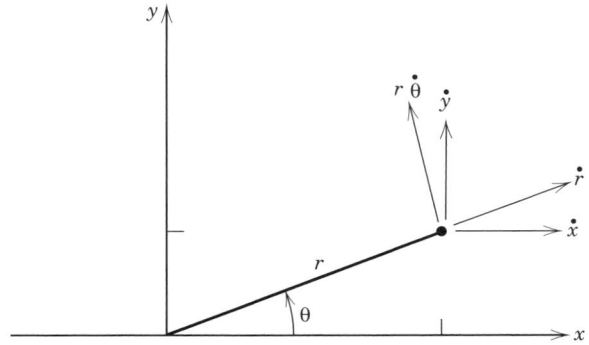

FIGURE 7.24. Rectangular and polar coordinates.

These displacement relationships can be arbitrarily complicated, but the corresponding velocity relationships have a particular structure. Starting with Eq. (7.99) for example, we find

$$\frac{dx}{dt} = \frac{\partial x}{\partial r}\frac{dr}{dt} + \frac{\partial x}{\partial \theta}\frac{d\theta}{dt},$$

and

$$\frac{dy}{dt} = \frac{\partial y}{\partial r}\frac{dr}{dt} + \frac{\partial y}{\partial \theta}\frac{d\theta}{dt}. \tag{7.101}$$

Working this out, we find a relation among flow variables,

$$\begin{bmatrix} \dot{x} \\ \dot{y} \end{bmatrix} = \begin{bmatrix} \cos\theta & -r\sin\theta \\ \hline \sin\theta & r\cos\theta \end{bmatrix} \begin{bmatrix} \dot{r} \\ \dot{\theta} \end{bmatrix}. \tag{7.102}$$

If this transformation is represented by a modulated multiport transformer, then we know that the transformation of the corresponding efforts (forces or torques) must involve the transpose of the matrix in Eq. (7.102):

$$\begin{bmatrix} \cos\theta & \sin\theta \\ \hline -r\sin\theta & r\cos\theta \end{bmatrix} \begin{bmatrix} F_x \\ F_y \end{bmatrix} = \begin{bmatrix} F_r \\ \tau \end{bmatrix}. \tag{7.103}$$

The validity of Eq. (7.103) is easily checked by imagining forces F_x and F_y acting on the point in Figure 7.24 and then computing the radial force and the torque exerted by these forces. Note that instead of a representation as a multiport transformer, as in Figure 7.25, one could create a junction structure version involving four 2-port MTFs in the manner of Figure 7.23. Because of the implied causality, the 0- and 1-junctions in Figure 7.23 would interchange places.

It is not always necessary to derive velocity relationships by differentiating displacement relations. In this example, one could find $\dot{r}$ and $\dot{\theta}$ by direct considera-

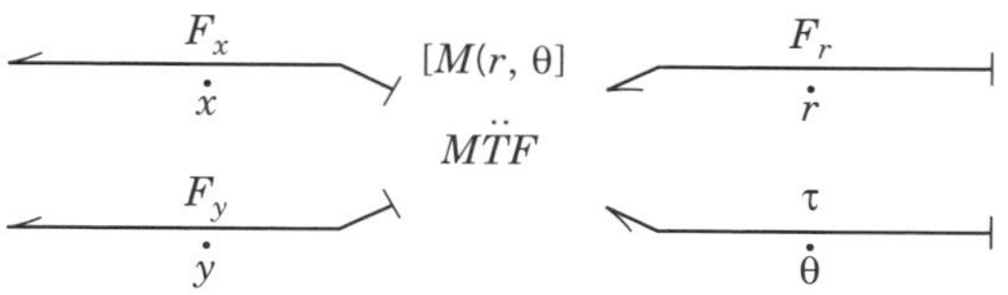

$$\xleftarrow{\quad F_x \quad} \quad [M(r, \theta)] \quad \xleftarrow{\quad F_r \quad}$$

FIGURE 7.25. An *MTF* representation of the polar-to-rectangular coordinate transformation.

tion of the velocity components sketched in Figure 7.24. The result is

$$
\begin{bmatrix} \dot{r} \\ \dot{\theta} \end{bmatrix} =
\begin{bmatrix} \cos\theta & \sin\theta \\ -\sin\theta/r & \cos\theta/r \end{bmatrix}
\begin{bmatrix} \dot{x} \\ \dot{y} \end{bmatrix}
\tag{7.104}
$$

which implies

$$
\begin{bmatrix} \cos\theta & -\sin\theta/r \\ \sin\theta & \cos\theta/r \end{bmatrix}
\begin{bmatrix} F_r \\ \tau \end{bmatrix} =
\begin{bmatrix} F_x \\ F_y \end{bmatrix}.
\tag{7.105}
$$

This representation reverses the causality shown in Figure 7.25 and involves the inverse of the matrices used in Eqs. (7.102) and (7.103). Of course, if this were not a transformer with equal numbers of ports on both sides, the matrix would not be square, it would not be possible to reverse causality, and there would be no inverse matrix.

A final example will make some points regarding geometric constraints in mechanical systems. The rod in Figure 7.26 is rigid and acted upon by three forces. The force F_3 is perpendicular to the rod and F_1 and F_2 are horizontal and vertical, respectively. The corresponding velocities are V_1, V_2, and V_3. Geometrically, there is only a single degree of freedom, so if one velocity is known, the remaining two can be determined.

The *MTF* shows that V_1 and V_2 are determined by V_3 and that F_3 is determined by F_1 and F_2. By consideration of the kinematics of the system, one can derive the

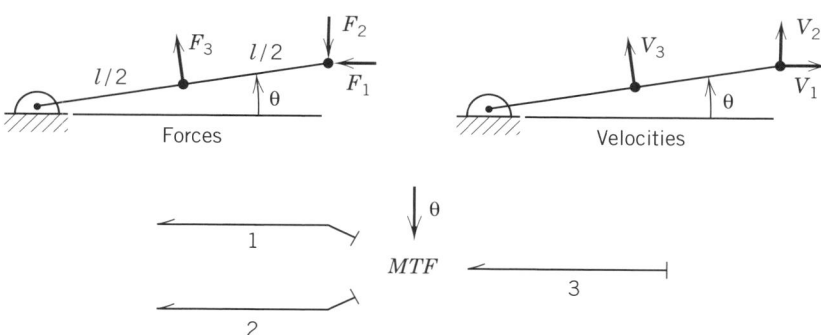

FIGURE 7.26. A rigid rod with three applied forces.

following velocity relations directly:

$$\begin{bmatrix} V_1 \\ V_2 \end{bmatrix} = \begin{bmatrix} -2\sin\theta \\ 2\cos\theta \end{bmatrix} [V_3].$$ (7.106)

The transpose of the matrix in Eq. (7.106) relates the forces,

$$[-2\sin\theta | 2\cos\theta] \begin{bmatrix} F_1 \\ F_2 \end{bmatrix} = [F_3].$$ (7.107)

The fact that the angle θ appears in the transformer matrices while $\dot{\theta}$ is not one of the flows involved points up an interesting aspect of mechanical systems involving modulated transformers. In principle, the coefficients in the transformer matrices are functions of displacements associated with flows at the transformer ports. This is obvious if displacement relations such as Eq. (7.99) are differentiated as shown in Eq. (7.101) to produce equations such as (7.102) and (7.103). In the present case, θ has been used even though $\dot{\theta}$ was not a flow of interest. In this case, θ must be considered to be a state variable just like the ps and qs associated with the Is and Cs in the system. This means that the *MTF* forces us to write an equation for $\dot{\theta}$ even though there may be no C-element for which θ would be a state variable. Since V_3 determined all the velocities of the rod, it also determines $\dot{\theta}$ so we must only write a separate state equation,

$$\dot{\theta} = \frac{V_3}{l/2},$$ (7.108)

to complete the system formulation.

Often, C-elements attached to modulated transformers automatically make the integrals of some port flows into displacement state variables. These displacements are then available for use in computing the variable coefficients in the transformation matrices. In this example, however, $\dot{\theta}$ does not appear on any bond so the extra state equation (7.108) is necessary.

As will become clear in Chapter 9, there are often many alternative ways to describe the geometrical constraints in complex mechanical systems. Some choices may lead to straightforward formulations while others may be nearly intractable. These difficulties are inherent in mechanics and arise whether bond graph methods or other techniques are used.

REFERENCES

[1] S. H. Crandall and N. C. Dahl, eds., *An Introduction to the Mechanics of Solids*, New York: McGraw-Hill, 1959, p. 378.

[2] S. H. Crandall, D. C. Karnopp, E. F. Kurtz, and D. C. Pridmore-Brown, *Dynamics of Mechanical and Electro-mechanical Systems*, New York: McGraw-Hill, 1968, pp. 220, 294, 296.

[3] I. Prigogine, *Introduction to the Thermodynamics of Irreversible Processes*, Springfield, IL: C. C. Thomas, 1955.

[4] J. Meixner, "Thermodynamics of Electric Networks and the Onsager–Casimir Reciprocal Relations," *J. Math. Phys.*, **4**, 154 (1963).

[5] G. Kron, *Tensor Analysis of Networks*, New York: Wiley, 1939.

[6] R. C. Rosenberg, "State Space Formulation for Bond Graph Models of Multiport Systems," *Trans. ASME, J. Dyn. Syst. Meas. Control*, **93**, Ser. G, No. 1, 35–40 (Mar. 1971).

[7] D. C. Karnopp, "Power-Conserving Transformations: Physical Interpretations and Applications Using Bond Graphs," *J. Franklin Inst.*, **288**, No. 3, 175–201 (Sept. 1969).

PROBLEMS

7-1 Three linear springs are attached to massless carts, as shown. Make a bond graph for this system, and manipulate the relations for the implicit C-field into the relation for the explicit 1-port C-field shown. Assume that when $X = 0$, all the springs are relaxed.

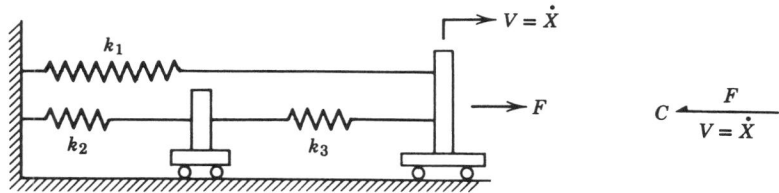

7-2 Two linear springs are pin-joined together as shown. Consider *small* motions, x and y, and small spring extensions, e_1, e_2. Show that the implicit field shown represents the system, and find the transformer moduli. Convert to an explicit-field form at the x and y ports.

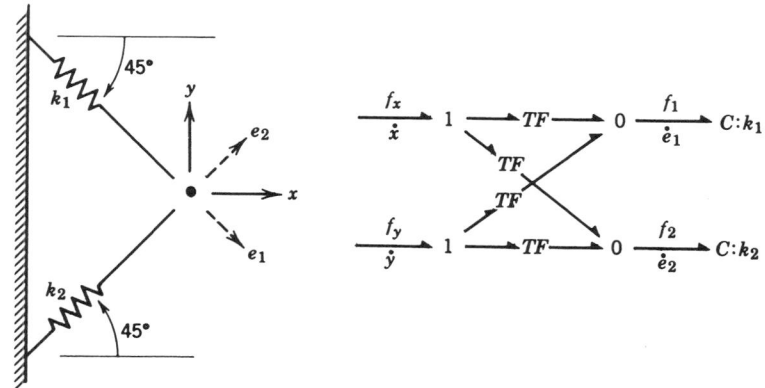

7-3   Three concentrated mass points are mounted on a massless rigid bar. Consider plane vertical motion such that the bar moves through only small angles. By expressing the motion of each mass in terms of $V_1 = \dot{X}_1$ and $V_2 = \dot{X}_2$, show that this system may be represented as a 2-port, implicit I-field. Find the constitutive relations for the I-field as seen at the F_1, F_2 ports.

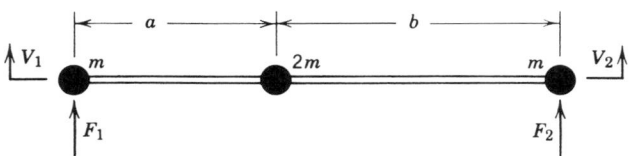

7-4   Three equal pipes carrying an incompressible fluid join at a tee junction. If each pipe has fluid inertia $\rho l / A$, show an implicit I-field representation. Demonstrate that all three 1-port inertial elements cannot simultaneously have integral causality. Find a 3-port, explicit I-field representation.

7-5   Write the equations of motion for the system shown, assuming the constitutive laws for the transformer are given by Eq. (7.63).

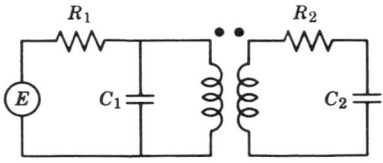

7-6   For the R-field shown, show causalities that would result in (a) Onsager forms, and (b) Casimir forms.

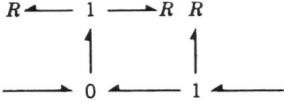

7-7   Construct an implicit R-field that is expressible in neither an Onsager nor a Casimir form, and write out its constitutive laws as was done in Eq. (7.84).

7-8   Prove that if a resistance matrix were antisymmetric (with zeros on the main diagonal), then the net power dissipated would always be zero, assuming an inward sign convention on all external bonds.

7-9 Two bond graph representations are shown in a and b below:

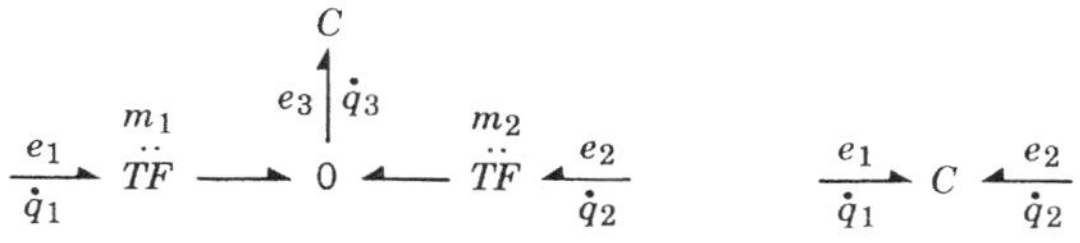

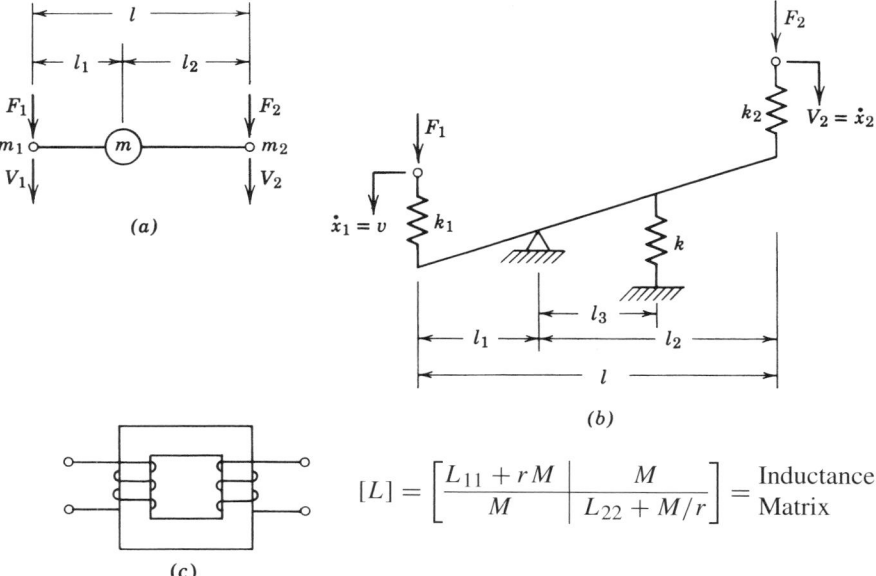

(a) (b)

In a assume $q_3 = m_1 q_1 + m_2 q_2$ and $e_3 = C_3^{-1}(q_3)$, a nonlinear capacitor. If the representation of b is to be valid, and the 2-port C-field is energy conservative, then

$$\frac{\partial e_1}{\partial q_2} = \frac{\partial e_2}{\partial q_1}.$$

Prove that this expression is valid for the system of a. Extend the proof to the case in which the TFs in a become MTFs and q_3 is a general function of q_1 and q_2, $q_3 = q_3(q_1, q_2)$.

7-10 Study of some 2-port fields:

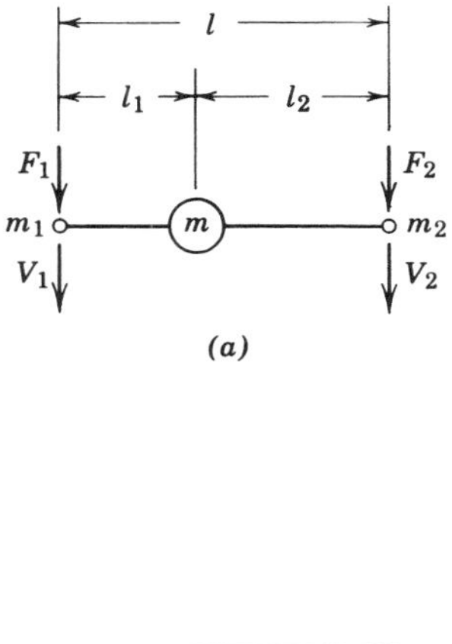

(a)

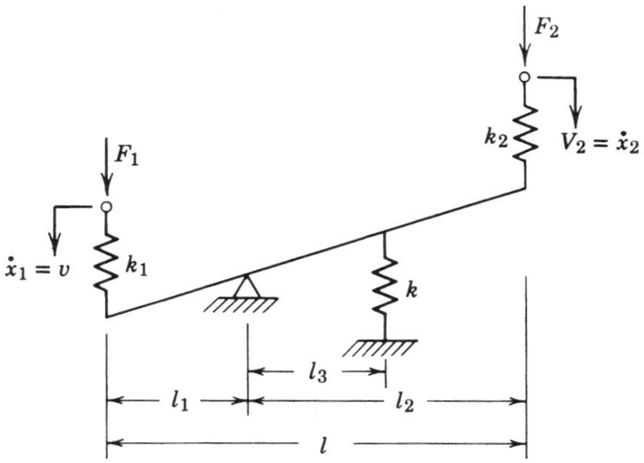

(b)

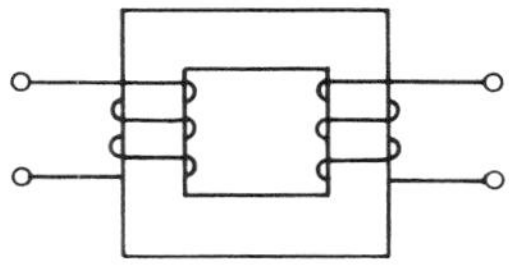

(c)

$$[L] = \left[\begin{array}{c|c} L_{11} + rM & M \\ \hline M & L_{22} + M/r \end{array}\right] = \text{Inductance Matrix}$$

(i) Draw (a), (b), and (c) as 2-port fields, and for (a) and (b) show the fields in terms of 0, 1, TF, I, C elements.

(ii) Show that the mass matrix for (a) is

$$\left[\begin{array}{c|c} m_1 + (ml_2^2/l^2) & m(l_1l_2/l^2) \\ \hline m(l_1l_2/l^2) & m_2 + (ml_1^2/l^2) \end{array}\right].$$

Show that the compliance matrix for (b) is

$$\left[\begin{array}{c|c} 1/k_1 + (1/k)(l_2^2/l_3^2) & -(1/k)(l_1l_2/l_3^2) \\ \hline -(1/k)(l_1l_2/l_3^2) & 1/k_2 + (1/k)(l_2^2/l_3^2) \end{array}\right].$$

Compare these results with the inductance matrix for case (c).

(iii) Show that

for (a):
if $m_1 \ll m, m_2 \ll m$,
then
$(l_1/l_2)p_1 = p_2$,
$v_1 = (l_1/l_2)v_2 + (l^2/ml_2^2)p_2$.

for (c):
if $L_{11} \ll M, L_{22} \ll M$,
then
$\lambda_1 = r\lambda_2$,
$ri_1 = i_2 + \lambda_1/M$.

for (b):
if $k_1 \gg k, k_2 \gg k$,
then
$(l_2/l_1)x_1 = x_2$,
$F_1 = (l_2/l_1)F_2 + (kl_3^2/l_1^2)x_1$.

(iv) Show that for (a) if $m \to \infty$, for (b) if $k \to 0$, and for (c); if $M \to \infty$, all systems become —TF—.

(v) Show that saturation of the iron in (c) is analogous to the limiting of motion by stops in (b), and thus d-c voltages will not pass in (c) and d-c velocities will not pass in (b).

7-11 Consider the bond graphs below:
System 1:

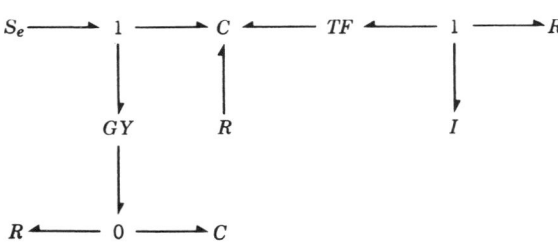

System 2:

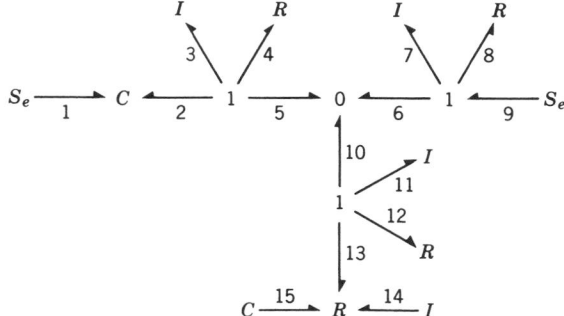

(a) If the graphs are not in the standard form for sign conventions, insert extra 0- or 1-junctions to bring them into the standard form.

(b) Add causality to the system, and list the state vectors.

7-12 Consider a linear, multiport C-field with stiffness matrix K. The constitutive relations are then

$$e = Kq, \tag{i}$$

where e and q are n-vectors and K is an $n \times n$ matrix. The stored energy $\mathbf{E}$ is then

$$\mathbf{E} = \frac{1}{2}q^t Kq. \tag{ii}$$

The text argues that if energy is conserved, then K in Eq. (i) must be symmetrical, that is, $K = K^t$.

(a) Convince yourself that any square matrix may be decomposed into a symmetrical part and an unsymmetrical part. For example,

$$K = K_s^t + K_a; \qquad K_s = \frac{K + K^t}{2}, \qquad K_a = \frac{K - K^t}{2};$$

$$K_s = K_s^t, \qquad K_a = -K_a^t.$$

(b) Using a 2×2 example with

$$K = \begin{bmatrix} k_{11} & k_{12} \\ k_{21} & k_{22} \end{bmatrix}, \qquad k_{12} \neq k_{21},$$

show that $\mathbf{E}(q)$ depends on the symmetrical part of K, that is , that

$$\frac{1}{2}q^t Kq = \frac{1}{2}q^t K_s q,$$

and that, hence, $e = K_s q$ if the constitutive laws are derived by differentiation of the stored-energy function.

7-13 The linear R-field is shown in a mixed-causal form, that is, the input and output port variables are related by a matrix K thus:

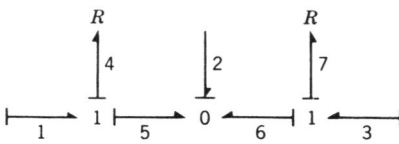

$$\begin{bmatrix} e_1 \\ f_2 \\ e_3 \end{bmatrix} = K \begin{bmatrix} f_1 \\ e_2 \\ f_3 \end{bmatrix}$$

(a) What can you predict about K before writing out the R-field equations?

(b) Suppose the R-field were forced to accept resistance or conductance causality. What characteristics would the resistance and conductance matrices have?

(c) Using causality, prove that the resistance matrix does not exist.

(d) Write out the matrix K and the conductance matrix to verify your predictions in (a) and (b). Using the equations for mixed-causal form, show that the equations cannot be manipulated into the resistance causality form.

7-14 Find a mechanical system having the implicit C-field shown in Figure 7.7c.

7-15 Consider a lightweight horizontal rod on which three masses are fastened, one each at both ends and one at the midpoint. Assume that the end masses move vertically in such a way that the angle of the rod with the horizontal remains small. Show that this system has a bond graph similar to the bond graph shown in Figure 7.12b for interconnected self-inductances.

8

TRANSDUCERS, AMPLIFIERS, AND INSTRUMENTS

This chapter deals primarily with models of devices that link two subsystems in two distinct energy domains. In some cases, the efficiency of power transduction is important. Motors, generators, pumps, and transmissions, for example, usually are designed so that they can transduce energy without much loss, at least when operating normally. Instruments and amplifiers, on the other hand, are designed to operate at low power efficiency. An ideal instrument would extract information from a system without power absorption and could communicate at finite power to another system. An amplifier accepts an input signal at near-zero power level and influences another system in response to the input signal at finite power.

The high-efficiency transducers are usually passive; that is, they contain no sources of power. The instruments and amplifiers are often active in the sense that they need a power supply in order to satisfy the first law of thermodynamics. Most practical transducers are themselves rather complex systems that can be modeled in detail by bond graphs. On the other hand, designers of large systems containing transducers as components cannot afford the luxury of modeling transducers in detail and must use good approximate models. In this chapter, we will show the main features of several types of transducers and demonstrate a philosophy of modeling in which nonideal effects may be progressively added to an idealized basic model. We concentrate first on passive transducers and then briefly study instruments and amplifiers.

Passive transducers which contain loss elements can, on the average, only transmit less power than they receive. Some transducers do, however, have the capability of storing energy so that they can temporarily deliver excess power. We call the two types of passive transducers *power transducers* and *energy-storing transducers*.

8.1   POWER TRANSDUCERS

Ideal power transducers were introduced in Chapter 4 as transformers and gyrators. Devices such as hydraulic rams, positive-displacement pumps and motors, permanent-magnet d-c motors and generators, and the like behave roughly as power-conserving elements. Real devices, of course, do exhibit power losses and also contain energy-storing mechanisms and associated inertial and capacitance effects. Although clever design of transducers can result in rather small values for these parasitic effects, they still impose an ultimate limit on the performance of any real device. When a very accurate model of a real transducer is desired, it will almost invariably be necessary to replace linear elements of the model with more accurate nonlinear ones. The process of starting with a highly idealized model, adding parasitic elements, and replacing linear elements with nonlinear ones to achieve increasingly accurate transducer models will be illustrated for some typical power transducers.

In Figure 8.1, two hydraulic ram configurations are shown schematically, and a series of bond graph models are also shown. The model of Figure 8.1b for the basic ram has been used in Chapter 4. This transformer model instantaneously and without loss of power transduces the hydraulic power, PQ, into mechanical power, FV. In some cases, this model may be quite adequate for a system analysis. On the other hand, in real pistons the mass effects and frictional losses associated with piston rings, packing, or tight fits to prevent leakage can be appreciable. The bond graph of Figure 8.1c essentially provides for a loss in the force F due to the force required to accelerate the piston and the frictional loss.

The mass effect of the piston is quite straightforward, and in many cases, when the piston rod is stiff and connects directly with a load mass, the piston mass and the load mass may simply be lumped together. But the friction force is more complex. A friction force may be small, and therefore negligible for some purposes, but the next simplest representation, the linear friction force, is virtually never an accurate representation of mechanical friction. In Figure 8.2, several possible friction force laws are shown. The linear law is often used in order to study systems using the analytically convenient linear methods, but usually the friction coefficient is left as a variable so that the model can be "tuned" to reproduce experimental results. Real friction usually includes a component of dry friction, represented by $F_f = F_0 \operatorname{sgn} V$, where sgn V is $+1$ if $V > 0$ and -1 if $V < 0$. This law is not as convenient to work with as the linear law, particularly near $V = 0$. A more complicated phenomenon is sometimes called "stiction." It is easily observed that it takes more force to start a block lying on a table moving than to keep it moving at low speeds. The general friction law shown in Figure 8.2 attempts to model this phenomenon. When this type of friction is present, there is a tendency for the system to chatter. This is fine if the system is a violin string and bow but is distinctly unpleasant if the system is chalk on a blackboard, a machine tool and workpiece, or a computer simulation program with numerical stability problems. The experienced system modeler treads a fine line between realistic but intractable friction models and useful but possibly oversimplified models.

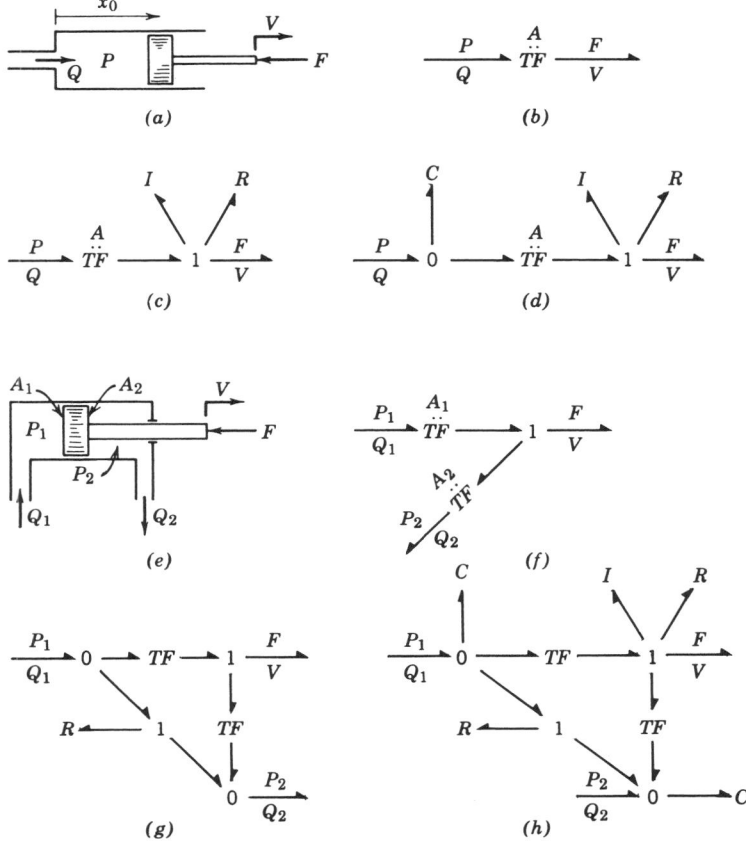

FIGURE 8.1. Hydraulic rams. (*a*) Basic ram; (*b*), (*c*), and (*d*) models of basic ram; (*e*) hydraulic cylinder; (*f*), (*g*), and (*h*) models of hydraulic cylinder.

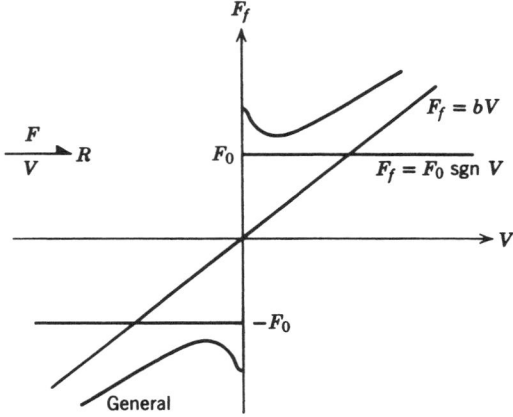

FIGURE 8.2. Several mechanical friction laws.

Another phenomenon that is important in high-performance hydraulic systems involves the compliance of the working fluid. Since hydraulic fluids are not usually very compliant, small density changes accompany large pressure changes, and a linear constitutive relation usually suffices. One may define the bulk modulus β as the coefficient relating the pressure P and the change in volume, ΔV, of a mass of fluid that occupies volume V_0 when $P = 0$:

$$P = \beta \frac{\Delta V}{V_0}. \tag{8.1}$$

Actually, β, which is expressed in pressure units, varies somewhat with mean pressure, temperature, amount of air in the working fluid, and so on, so a more useful form of Eq. (8.1) is

$$P = P_0 + \beta \frac{V}{V_0} = P_0 + \frac{\beta}{V_0} \int^t Q \, dt, \tag{8.2}$$

where P_0 is an equilibrium pressure, V_0 is a nominal volume, and $Q = \dot{V}$ is the volume flow rate into the nominal volume. Then β is more obviously the slope of a nonlinear constitutive law relating pressure and the compressed volume of the fluid. The bond graph of Figure 8.1d shows a —C element that models this compliance effect. Clearly, there is an oscillatory phenomenon that occurs as energy is exchanged between the mass of the piston and the compliance of the hydraulic fluid. Note that for the linear theory of Eq. (8.2), one must pick a nominal volume, $V_0 = Ax_0$, where x_0 is an average position of the piston (see Figure 8.1a). Thus, the hydraulic fluid stiffness is high when x_0 is small and gets lower as x_0 is increased.

Figure 8.1e shows a 3-port transducer in which, because of the area of the piston rod, the areas that transduce the two pressures, P_1 and P_2, into force components are not equal. Also, the flows, Q_1 and Q_2, are not equal. The bond graph of Figure 8.1f represents this system in its most highly idealized form. In Figure 8.1g the hydraulic leakage resistance across the piston has been included, and in Figure 8.1h mechanical inertia and friction and hydraulic compliance effects have also been modeled. As one may see from this example, even the simplest transducer models can become rather complex when many nonideal effects are accounted for.

Another useful example of power transducers is the d-c motor or generator. The basic ideal transducer is shown in Figure 8.3a. The power in the armature circuit, $e_a i_a$, is transduced to shaft power, $\tau \omega$, when the device operates as a motor and the power flow reverses during generator action. The field port establishes a magnetic field, which provides the coupling between electric variables and mechanical variables for the individual conductors on the rotor. For a permanent-magnet motor, this field is constant, but for the separately excited motor shown in Figure 8.3a, the field is a function of the field current i_f. The commutator of the motor essentially maintains the field due to the armature current in a direction perpendicular to the field generated by i_f. For this reason, although i_f influences the transduction, there is virtually no back effect on e_f due to e_a or i_a. Thus, the field port influence is shown acting on an activated bond.

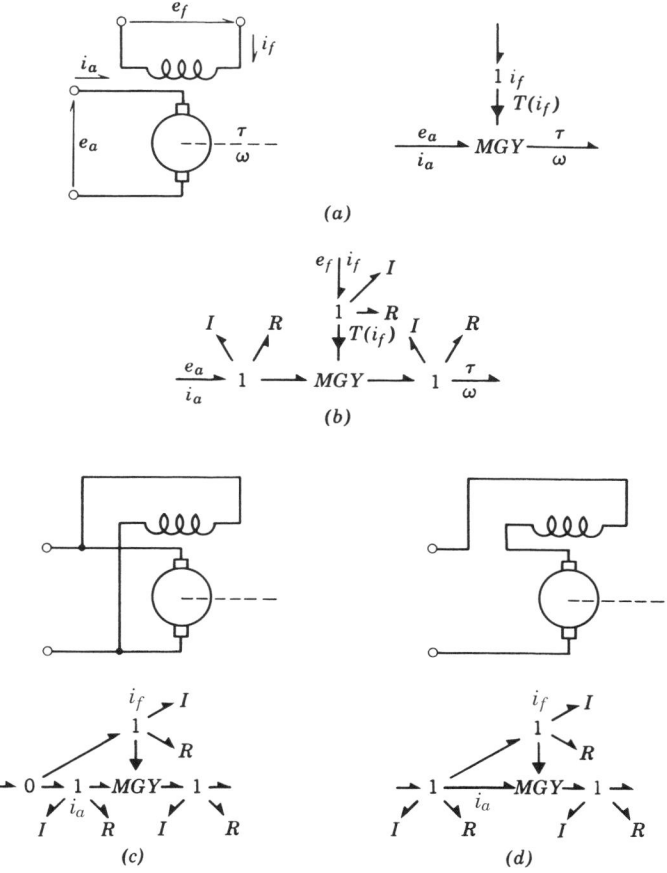

FIGURE 8.3. Direct-current electric motor–generator models. (*a*) Basic ideal transducer; (*b*) self-inductances, resistances, inertia, friction, and windage effects added; (*c*) shunt motor; (*d*) series motor.

The equations corresponding to Figure 8.3*a* are as follows:

$$e_a = T(i_f)\omega, \qquad T(i_f)i_a = \tau, \qquad e_f = 0. \tag{8.3}$$

Note that the transduction coefficient T, which is a gyrator parameter, can assume two different values if electrical and mechanical powers are measured in different units. Since that is the case, it may be worthwhile to rewrite Eq. (8.3) as

$$e_a = T_{em}\omega, \qquad T_{me}i_a = \tau, \tag{8.3a}$$

with T_{em} expressed in volts per radian per second and T_{me} in foot-pounds per ampere, for example. Power conservation is then

$$e_a i_a = \frac{T_{em}}{T_{me}} \tau \omega, \tag{8.4}$$

with T_{em}/T_{me} relating volt-amperes to foot-pounds per second. One must not assume, however, that because T_{me} and T_{em} differ numerically, the device is not power conserving. In the SI system, one volt-ampere is identical to one newton-meter per second, so that T_{em} is equal to T_{me}.

The parameter T involves the strength of the field as well as the number and effective lengths of the armature conductors that interact with the field. (See the discussion of the modulated gyrator in Chapter 7 for a discussion of this type of interaction with simplified geometry.) In some cases, it is reasonable to assume that the field is proportional to i_f. Then

$$T(i_f) \cong A i_f, \tag{8.5}$$

where A is a constant. Thus, if i_f is a constant, torque is proportional to i_a, and if i_a is a constant, torque is proportional to i_f. In such special cases, the device behaves linearly, even though it is basically nonlinear in a multiplicative sense. In real devices, saturation of the magnetic material limits the field, so that the relation between T and i_f does not remain linear for large currents. Also, magnetic hysteresis effects, when present, mean that the field is not a single-valued function of i_f but rather depends on the previous history of magnetization. (The permanent-magnet motor is an extreme example of this—even though there is no i_f, a field exists, and hence $T = $ const.)

In most cases, the ideal transducer must be supplemented with loss and energy storage elements to model practically important effects. In Figure 8.3b a common motor model is shown. The self-inductance and resistance of the field and armature coils are included, as well as the moment of inertia of the rotor and a mechanical resistor that models bearing losses and windage from the rotor and any built-in cooling fans. The electrical I- and R-elements are typically assumed to be linear unless a very accurate model is required. The mechanical resistor is rarely linear in reality, but sometimes an effective linear resistor will give good results if the mechanical friction is small enough that its detailed nature is not critical in determining system performance.

Two common 2-port motors can be constructed from the separately excited motor model. These are the shunt-wound and series-wound motors of Figure 8.3c and d. Note that, in Figure 8.3d, i_a and i_f are identical, so that one could simplify the bond graph by combining the field and armature inductances and resistances.

Before leaving the subject of power transducers, let us consider two examples in which the ideal transducer is fairly complicated even when loss and parasitic-energy-storage effects are neglected. In Figure 8.4a an alternating-current (a-c) generator employing a permanent magnet is shown in highly schematic form. As the square coil rotates in the uniform B-field, there is an interaction relating the electric power variables, e and i, and the mechanical variables, τ and ω.

One of the constitutive laws of the device may be found by using the basic definition of the flux linkage λ. The amount of flux linked by the current path is B times

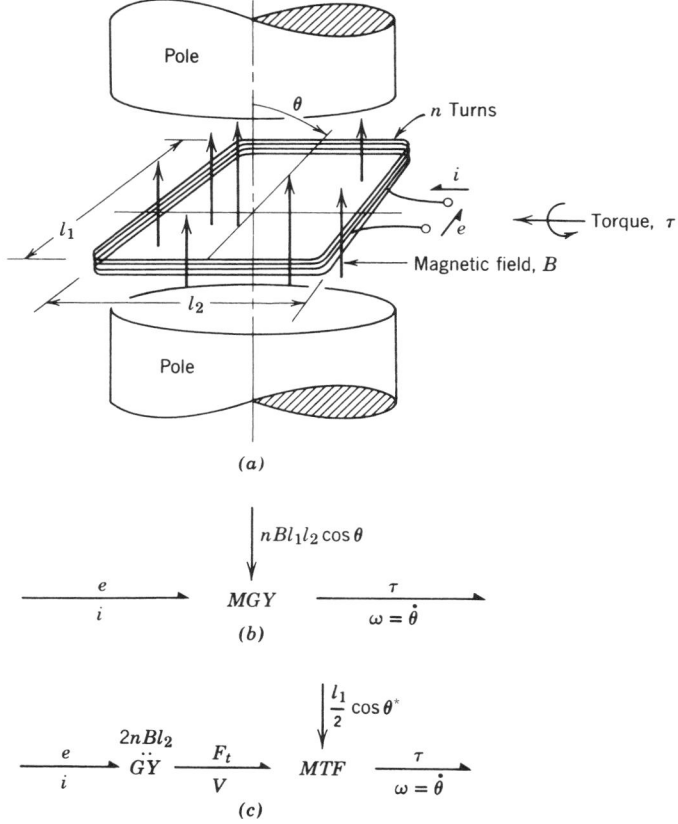

FIGURE 8.4. Elementary a-c generator. (*a*) Schematic diagram; (*b*) bond graph using *MGY*; (*c*) bond graph using *GY* and *MTF*. [*Use of modulus is defined by Eqs. (8.13) and (8.14).]

the projected area of the coil times the number of turns in the coil, n:

$$\lambda = Bl_2l_1(\sin\theta)n. \tag{8.6}$$

Using the fact that $\dot{\lambda} = e$, one may differentiate Eq. (8.6) to find a relation between e and $\omega = \dot{\theta}$:

$$e = (nBl_1l_2\cos\theta)\omega. \tag{8.7}$$

Then, noting that power must be conserved according to $ei = \tau\omega$, the remaining constitutive law is

$$(nBl_1l_2\cos\theta)i = \tau, \tag{8.8}$$

where the variables are expressed in metric units so that ei and $\tau\omega$ are both in the same units of power.

The constitutive equations of the device, Eqs. (8.7) and (8.8), are embodied in the bond graph of Figure 8.4b. Note that the modulating function involves θ, the integral of the local flow ω.

Another way to derive the laws for this device involves computing the force F on the lengths of wire that cut the flux lines. Note that the lengths of wire associated with l_2 cut flux lines, while the lengths associated with l_1 do not. If F is the force perpendicular to the direction of B, and V is the velocity in this direction, then for a typical conductor of length l_2, the basic equations as discussed in Chapter 7 are

$$e_1 = Bl_2 V, \tag{8.9}$$

$$Bl_2 i_1 = F, \tag{8.10}$$

where e_1 and i_1 are the voltage and current associated with a single length of conductor. There are, however, $2n$ such lengths of conductor, so that the terminal voltage is $2ne_1$:

$$e = 2nBl_2 V. \tag{8.11}$$

Also, each length of conductor has the same current, and each produces a force that ultimately will add to produce the torque τ. Calling the total torque producing force F_t, we have

$$2nBl_2 i = F_t. \tag{8.12}$$

Now, the cutting velocity V is

$$V = \left(\frac{l_1}{2} \cos \theta \right) \omega, \tag{8.13}$$

and the relation between F_t and τ is

$$\left(\frac{l_1}{2} \cos \theta \right) F_t = \tau. \tag{8.14}$$

Thus, the entire device can be represented by a gyrator with modulus $2nBl_2$ and a modulated transformer with modulus $(l_1 \cos \theta)/2$, as shown in Figure 8.4c. Clearly, the two representations of Figures 8.4b and c, that is, those of Eqs. (8.7) and (8.8) and Eqs. (8.11), (8.12), (8.13), and (8.14), are equivalent. Loss and energy-storing elements could be added to this power-conserving model.

There are many analogies between rotary electromechanical devices and hydromechanical devices. Just as a d-c motor with multiple windings and a commutator functions as a gyrator, a pump with several pistons and a porting arrangement functions essentially as a transformer. The field port of a d-c motor allows modulation of the gyrator parameter, and a stroke control on a pump, if it exists, allows

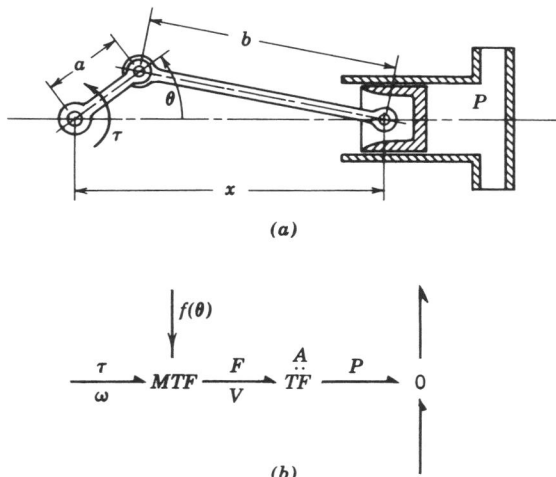

FIGURE 8.5. Crank and piston transducer. (*a*) Schematic diagram; (*b*) bond graph.

modulation of the transformer ratio. The a-c machines are often physically simpler but functionally more complex. The a-c generator of Figure 8.4 is similar to the single piston and crank arrangement of Figure 8.5, which could form part of a pump if suitable valving were added. Although the relation between the piston force and velocity and the pressure and flow of the hydraulic fluid is readily represented by a transformer, the relation between the torque τ, the angular speed ω, and the other power variables is more complex.

One way to find a relation between ω and the piston speed V is to start with a relation between x and θ (see Figure 8.5a). Working out the geometry, one finds

$$x = a \cos \theta + (b^2 - a^2 \sin^2 \theta)^{1/2}. \tag{8.15}$$

If $\dot{x} = V$ and $\dot{\theta} = \omega$, then the result of differentiating Eq. (8.15) is

$$V = [-a \sin \theta - (b^2 - a^2 \sin^2 \theta)^{-1/2} a^2 \sin \theta \cos \theta] \omega. \tag{8.16}$$

Since $FV = \tau \omega$ if F is the force on the piston, the remaining constitutive relation must be

$$[-a \sin \theta - (b^2 - a^2 \sin^2 \theta)^{-1/2} a^2 \sin \theta \cos \theta] F = \tau. \tag{8.17}$$

Recognizing the complicated function of θ as a transformer modulus, and calling it $f(\theta)$ for convenience, one may represent the device by the bond graph of Figure 8.5b. Note the similarity between this bond graph and that of Figure 8.4c. It should be evident from these examples that considerations of power conservation can greatly aid in modeling power transducers.

8.2 ENERGY-STORING TRANSDUCERS

The transducers of the previous section were modeled, in ideal form, by junction-structure elements. They were energy conservative, but beyond that, power in one domain was instantaneously transduced into another domain. In this section, we study transducers which are also ideally energy conservative but in which energy storage plays an indispensable role. Thus, for these energy-storing transducers, energy from one domain may be stored and released in another domain at a later time.

The transducer models in this section are based on C-fields, I-fields, and mixed IC-fields, which were discussed in Chapter 7. Here, we merely discuss some example systems and use the results of Chapter 7 without much discussion. As in the previous section, models of real transducers can be assembled from ideal models supplemented with loss and dynamic elements to account for effects that are present in real devices but not accounted for in the ideal transducer.

A typical energy-storing transducer of practical interest is the condenser microphone or electrostatic loudspeaker. The sketch of Figure 8.6a shows roughly how these devices may be constructed. A capacitor is formed by mounting a movable plate near a rigidly mounted plate and providing electrical connections as one would in a conventional parallel-plate capacitor. In practice, the movable plate might be a thin diaphragm or a tightly stretched membrane on which a thin layer of conducting

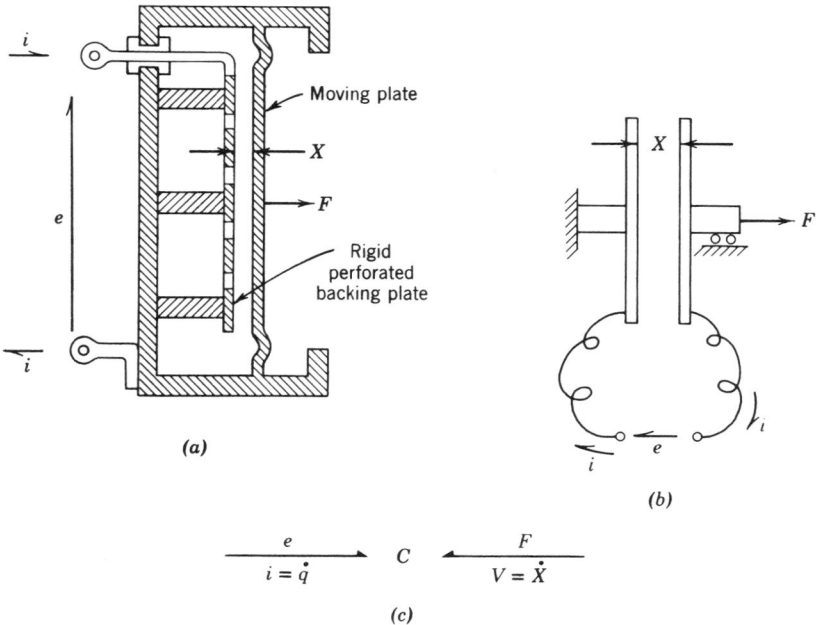

FIGURE 8.6. The movable-plate capacitor. (*a*) Sketch of microphone or speaker; (*b*) schematic diagram; (*c*) bond graph representation.

material is fixed. Such distributed-parameter "plates" could actually move in complicated ways, but for simplicity we will consider a system with only one mechanical coordinate, X, as shown in Figure 8.6b. (One may think of X as the displacement of the first normal-mode shape of a diaphragm for those cases in which the contribution of higher modes may be neglected.) The force F, which may be due to acoustical pressure in practical cases, is defined so that the mechanical power is $F\dot{X} = FV$, where V is the (generalized) velocity of the moving plate. The electrical power is, of course, ei.

When X is fixed, the device is an ordinary electrical capacitor, and when the charge q is fixed, then F depends on X as in a mechanical spring. In general, we may hypothesize that both e and F depend on q and X:

$$e = e(q, X), \tag{8.18}$$

$$F = F(q, X). \tag{8.19}$$

But the constitutive relations above are certainly not arbitrary, since the device can at best conserve energy if we temporarily ignore loss effects. The stored energy, $\mathbf{E}$, is readily computed:

$$\mathbf{E}(t) = \mathbf{E}_0 + \int_0^t (ei + FV)\, dt$$

$$= \mathbf{E}_0 + \int_{0,0}^{q,X} e(q, X)\, dq + F(q, X)\, dX = \mathbf{E}(q, X), \tag{8.20}$$

where $\mathbf{E}_0$, which will generally be assumed to vanish, represents an initial energy at $t = 0$ or when $q = X = 0$.

A close inspection of Eq. (8.20) shows that the constitutive laws of Eqs. (8.18) and (8.19) can be recovered from $\mathbf{E}(q, X)$,

$$e = \frac{\partial \mathbf{E}}{\partial q}, \qquad F = \frac{\partial \mathbf{E}}{\partial X}, \tag{8.21}$$

and the required relation between the two constitutive laws is

$$\frac{\partial e}{\partial X} = \frac{\partial^2 \mathbf{E}}{\partial X\, \partial q} = \frac{\partial F}{\partial q}. \tag{8.22}$$

This result is the integrability or Maxwell reciprocity conditions which the laws of the device must satisfy in order that energy be conserved. These considerations mirror those discussed in Chapter 7 for general C-fields, so the bond graph representation is simply that shown in Figure 8.6c. The constitutive laws for the C-field are Eqs. (8.18) and (8.19), which must obey Eq. (8.22).

A useful approximation to the constitutive laws may be found by assuming that the device is *electrically linear*, that is, that a capacitance C may be defined for every

X and that

$$e = \frac{q}{C(X)} \tag{8.23}$$

is the form of Eq. (8.18). Noting that from physical reasoning $F = 0$ when $q = 0$, we may evaluate **E** using Eq. (8.20) by letting $q = 0$, taking X to any particular value, and then charging the capacitor with $X = $ const (or $dX \equiv 0$). Then the integral in Eq. (8.20) is just

$$\mathbf{E}(q, X) = \int_0^q \frac{q}{C(X)} dq = \frac{q^2}{2C(X)}. \tag{8.24}$$

The force law corresponding to Eq. (8.23) is then

$$F = \frac{\partial \mathbf{E}}{\partial X} = \frac{q^2}{2} \frac{d\,[C(X)]^{-1}}{dX}, \tag{8.25}$$

so that F may be determined from measurements of the variation of C with X. [For an ideal parallel-plate capacitance as sketched in Figure 8.6b, the law would be $C(X) = \varepsilon A / X$, where ε is the dielectric constant of the medium between the plates and A is the area of the plates.] The use of **E** to find the force law is far more convenient than a direct calculation or an experimental measurement.

Note that even when one assumes that this device is electrically linear, the force law is decidedly nonlinear. In use, the device is normally subjected to both a high polarizing voltage and a fluctuating signal voltage. The plates tend to move together and to short the electrical circuit. This is prevented by the mechanical spring of the diaphragm, which supplies a force in the direction of F (Figure 8.6b). In addition, all real diaphragms have mass and exhibit some energy loss when in motion. Such effects are readily modeled by adding elements to the C-field of Figure 8.6c. This device is discussed in more detail in Reference [1], Chapter 6.

The electrical solenoid shown in Figure 8.7 can serve as a prototype of a mixed IC-field transducer. The device consists simply of a coil of wire in which a soft-iron

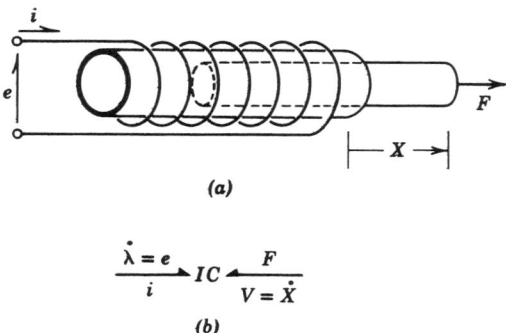

FIGURE 8.7. The solenoid. (a) Sketch of device; (b) bond graph representation.

slug can freely slide. When current flows in the coil, the slug is pulled into the coil. In analyzing this device, it is clear that from the electrical port, the coil will certainly exhibit the characteristic self-inductance effects of any coil, although the position of the coil, X, will presumably affect the electrical behavior. From the point of view of the mechanical port, it seems clear that the force on the slug, F, will depend on X, that is, the device will possess some of the properties of a mechanical spring, although the electrical variables will also affect F. If we assume that the current in the coil depends on the flux linkage λ as well as X, and that F also depends on λ and X,

$$i = i(\lambda, X), \tag{8.26}$$

$$F = F(\lambda, X), \tag{8.27}$$

then the stored energy $\mathbf{E}$ is

$$\mathbf{E} = \mathbf{E}_0 + \int_0^t (ie + FV)\,dt = \mathbf{E}_0 + \int_{0,0}^{\lambda,X} i\,d\lambda + F\,dX, \tag{8.28}$$

where $\dot{\lambda} = e$ and $\dot{X} = V$ and we will assume that $\mathbf{E}_0 = 0$.

By direct examination of Eq. (8.28), one may see that

$$i = \frac{\partial \mathbf{E}}{\partial \lambda}, \qquad F = \frac{\partial \mathbf{E}}{\partial X}, \tag{8.29}$$

and

$$\frac{\partial i}{\partial X} = \frac{\partial^2 \mathbf{E}}{\partial X\,\partial \lambda} = \frac{\partial F}{\partial \lambda}, \tag{8.30}$$

which is the integrability condition or Maxwell reciprocal condition constraining Eqs. (8.26) and (8.27). The bond graph of Figure 8.7b represents this ideal transducer. For mnemonic purposes the electrical port is shown impinging on the I and the mechanical port on the C.

Further insight into the device may be gained by assuming that the device is electrically linear. We assume that an inductance, $L(X)$, exists that relates i and λ for any position of the slug. Thus Eq. (8.26) becomes

$$i = \frac{\lambda}{L(X)}. \tag{8.31}$$

Also, on physical grounds, when i and λ vanish, F must also vanish, so that in evaluating $\mathbf{E}$ in Eq. (8.28), we may establish the slug at some particular position, X, with $\lambda = 0$, without doing any work on the device and then hold X fixed while λ is brought to its final value. During the change in λ, $dX = 0$, so only electrical energy is stored. The energy is then

$$\mathbf{E}(\lambda, X) = \int_0^\lambda \frac{\lambda}{L(X)}\,d\lambda = \frac{\lambda^2}{2L(X)}. \tag{8.32}$$

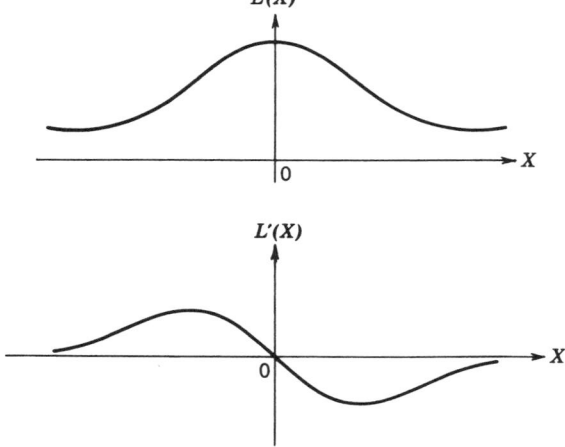

FIGURE 8.8. Inductance functions for the solenoid.

The constitutive law corresponding to Eq. (8.27) can now be found using Eq. (8.29):

$$F = \frac{\lambda^2}{2} \frac{d\,[L(X)]^{-1}}{dX} = -\frac{\lambda^2}{2} \frac{L'}{L^2}, \tag{8.33}$$

where L' is dL/dX. The general forms for $L(X)$ *and* $L'(X)$ are sketched in Figure 8.8. From these sketches and Eq. (8.33), one can see that the slug will experience a force tending to center it, but even when the simple law (8.31) is assumed, the corresponding force law is rather complex and inherently nonlinear in both the electrical and the mechanical variables.

The *IC*-field concept is particularly useful for electromagnetic devices, which often involve interacting magnetic fields associated with moving parts. Alternating-current motors and generators, for example, typically involve coils that rotate with respect to each other. The electrical ports of these devices are inertial, and the rotary mechanical port is capacitive when the device is described as an *IC*-field. Examples of such devices appear in the problems and in Reference [1], Chapter 6.

8.3   AMPLIFIERS AND INSTRUMENTS

The central idea behind the words "amplifier" and "instrument" is a low-power or a one-way interaction without back effect. The description of ideal amplifiers and instruments is functional rather than physical, and thus the physical, bilateral power interactions of bond graphs must be degenerated into signal interactions by the use of activated bonds in order to represent these devices. One assumes frequently that an ideal amplifier supplies an output power variable such as a voltage or current at finite power in response to an input signal at essentially zero power. Similarly, an instrument is supposed to extract information about some variable without affecting the system in which the variable appears and to transmit the information, often at finite power levels.

Clearly, ideal instruments and amplifiers violate even more physical laws than the ideal transducers discussed previously. For example, the usual functional descriptions violate even the first law of thermodynamics, since finite output power is somehow produced from zero input power. In reality, of course, most amplifiers have a readily identifiable power supply that is built into the functional relationships. The bond graphs of Figure 8.9 show the power supply and show that power is not created from a vacuum.

On the other hand, the active bond that indicates a signal flow with no associated power can only be approximate. At the microscopic level, the Heisenberg uncertainty principle is essentially a statement that signal interactions without back effect are impossible, but at a macroscopic level we know that amplifiers with extremely high power gain can be built and that instruments that have virtually no effect on the observed system are available in many cases. But no absolute statements about the appropriateness of an active bond representation can be made.

Probably every engineering student has had the experience of thinking of some real instrument as an ideal signal transducer and amplifier only to find that, in some cases, the attachment of the instrument to some system greatly distorted the behavior of the system to be measured. An oscilloscope, for example, has a high but finite input impedance. Therefore, it cannot be expected to measure voltages in a system having impedances of the same order of magnitude as its own without a large effect on those voltages due to the current flowing to the instrument. Similarly, the final stage of a hydraulic amplifier of the size, say, of a ship steering engine requires sizable power levels at its input. The input can be considered to be an active bond only under restricted circumstances, for example, if the next to last stage of the amplifier

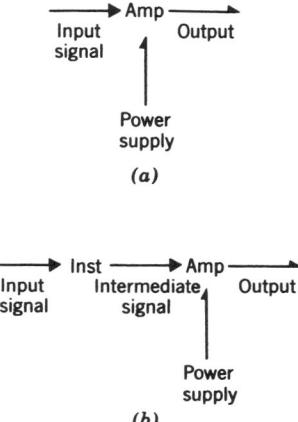

FIGURE 8.9. Amplifiers and instruments. (*a*) Basic bond graph representation of an amplifier; (*b*) instrumentation system with signal level transducer and associated amplifier.

is sized so that the load of the input to the last stage does not cause a significant effect on the response of the next to last stage.

In low-power applications, power efficiency is generally not very important, and amplifiers are designed to provide a drastic decoupling of input and output back effects. In high-power applications, on the other hand, amplifiers must usually be treated more physically, since it is not feasible to build such high-power-gain components that the input to an amplifier stage can reasonably be considered an active signal. Thus, it is a modeling decision whether or not to represent an amplifier or instrument with an active bond, and such a representation must be justified for each system of which the real device is to be a part.

With these caveats in mind, let us consider some useful bond graph representations of ideal elements, incorporating active bonds. The simplest type of amplifier model consists of a signal-controlled effort or flow source. In Figure 8.10a, the ubiquitous electrical voltage amplifier is shown in bond graph form. The output voltage e_0 is assumed to be a static or dynamic function of the input voltage e_i. In the static, linear case, there is a voltage gain G and

$$e_0 = Ge_i(t). \tag{8.34}$$

In the dynamic case, the output voltage may be related to the input voltage by a differential equation, or in the common linear case by means of a transfer function. In the latter case, one may find internal state variables for the amplifier that yield the desired transfer function. Generally, these state variables are nonphysical and merely serve to provide state equations equivalent to the specified frequency-domain representation of the amplifier. Such state equations may be used in time-domain analysis of the complete system.

The input voltage e_i in Figure 8.10a is shown as coming from a 0-junction as a signal on an active bond. In writing the current sum relation for all the bonds incident on the 0-junction, one assumes that there is no current associated with the active bond. Thus, the ideal amplifier does not affect the 0-junction from which it obtains its input voltage. At the output, a finite current i_0 may exist, but e_0 is not affected by i_0. Thus, the power gain is infinite in this ideal case.

In well-designed systems, amplifiers may indeed function as controlled sources, but in some cases, the controlled-source assumption is made even when a more detailed and physical model would be preferable. In Figure 8.10b, for example, a vi-

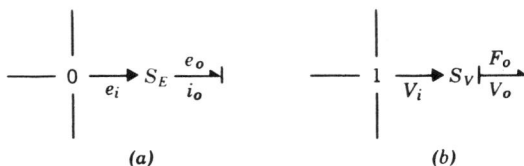

(a)   *(b)*

FIGURE 8.10. Amplifier models using controlled sources. (*a*) Voltage-controlled voltage source; (*b*) vibration shaker shown as a velocity source.

bration shaker is modeled as a controlled velocity source in which a desired velocity time history is generated as V_i and a complex servomechanism system is supposed to enforce $V_0(t)$ at a port of a test system in response to $V_i(t)$. In many cases, the reaction force does affect the servo, so that V_0 does not faithfully track V_i. In such a case, the simple model of Figure 8.10b is clearly inadequate, but the simplicity of controlled-source models often tempts system engineers to use them, particularly when the dynamics of the real devices have not been well explored or documented.

Many physical devices function essentially as amplifiers or transducers but are inadequately modeled as simple controlled sources. For example, a gasoline engine clearly amplifies the power of a human or automatic controller. The torque–speed curves of an engine, as sketched in Figure 8.11a, are drastically modified by the position of the throttle valve, which has been indicated by the angle θ. Also, the power required to move the throttle valve is often extremely small, not only in comparison with the engine power, but also in comparison with spring forces, inertial forces, and pivot-friction forces in the throttle linkage. Thus, it may be reasonable to consider that the torque associated with movement of the throttle linkage is negligible, but we cannot merely assume that θ controls a source, since the torque τ is always a function of the angular velocity of the output shaft, ω. Since torque is related to speed, one may describe the engine as a resistor, albeit an unusual one in which power is normally supplied rather than dissipated. (The power supply in the gasoline has been absorbed into the torque–speed curves.) Thus, the amplifier representation of the engine can be shown as a controlled resistance, as in Figure 8.11b.

The active bond indicating that there is no torque on the throttle linkage corresponding to $\dot\theta$ is equivalent to a signal, so one may use block diagram notation to show that $\dot\theta$ is integrated into the signal θ that controls the resistor representing the engine. In general, when amplifiers and instruments are part of a system, the use of block diagrams for showing dynamic relations for active-bond signals in a bond graph can be very useful.

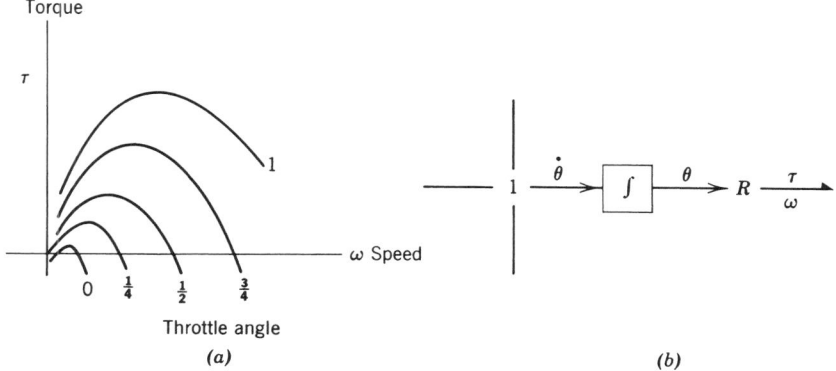

FIGURE 8.11. Static model of gasoline engine as amplifier. (a) Torque-speed curves as a function of throttle linkage angle θ; (b) amplifier model of engine using controlled resistance.

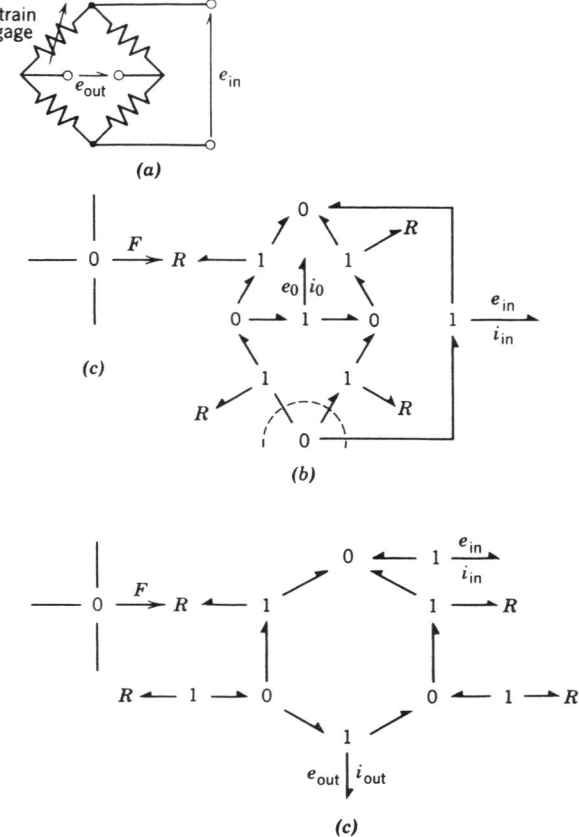

FIGURE 8.12. Strain-gage instrument. (*a*) Circuit diagram; (*b*) bond graph; (*c*) simplified bond graph after choice of ground voltage.

As another example of the use of a controlled resistance, consider the strain-gage instrument shown in Figure 8.12. The idea behind a resistance strain gage is that the electrical resistance of the gage is a function of the mechanical strain of the member to which the gage is attached. In many applications, the member functions as an elastic element that relates strain to stress and, ultimately, to force in the structure. Thus, when the strain gage is put into the bridge circuit of Figure 8.12*a*, and the bridge is supplied with the voltage e_i, then the voltage e_0 reacts to the force F in the structure. A bond graph representation of the basic instrument can be constructed out of a force-controlled resistance and the bond graph version of a bridge circuit, as shown in Figures 8.12*b* and *c*. Of course, one could continue to simplify this instrumentation system. For example, if e_i were supplied by a constant source, then one could reduce the system to a force-modulated resistance at the output port with the voltage supply built into the resistance relation. Furthermore, if an amplifier were connected to the output port, it could be arranged to supply an output voltage as a function of the force. Thus, the entire system might simplify to a force-controlled voltage source.

Although virtually any transducer might be arranged to act as an instrument under certain conditions and, in conjunction with a power supply, could function as an amplifier, variable-resistance elements are particularly important. In this category fall most electronic devices such as transistors, vacuum tubes, and the like, and the important valve-controlled hydromechanical devices. Perhaps a final example will help indicate how such devices may be modeled in bond graph terms.

Valves are devices in which the position of a mechanical part influences the hydraulic resistance. This change in resistance can be converted to a change in pressure drop across the valve, and thus one may deduce the position of the movable member; that is, the valve can be made into a position-indicating instrument. Also, by moving the valve, large amounts of fluid power may be controlled, and the valve may be made into an amplifier. It is this latter application that we now consider.

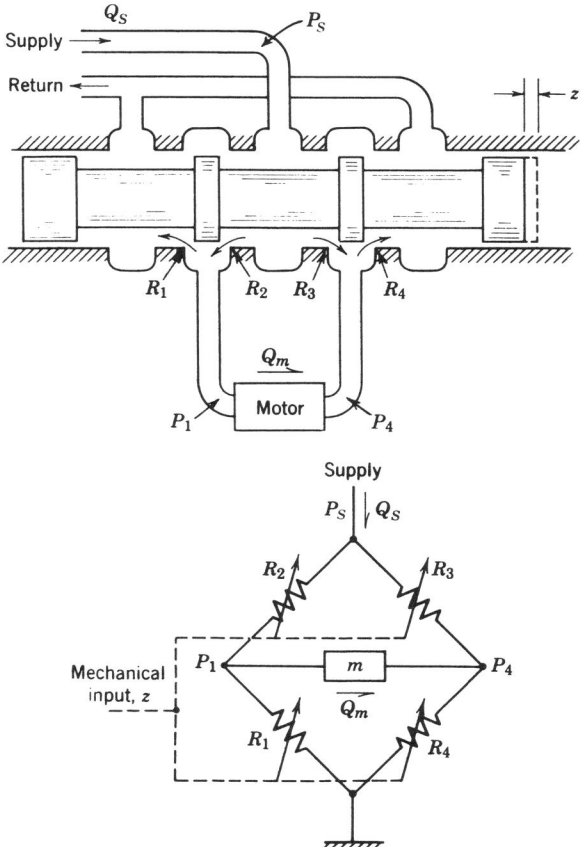

FIGURE 8.13. The general four-way valve and its equivalent circuit.

The so-called four-way valve shown in Figure 8.13 is common in hydraulic power systems. Actually, four resistances in the valve are modulated by the valve spool position z simultaneously. When the valve is connected to a pressure supply and a load (typically a hydraulic ram or a positive-displacement rotary hydraulic motor), a large flow of hydraulic power, $P_m Q_m$, is controlled by a small amount of power associated with the movement of the spool, $\dot{z}$. As shown in Figure 8.13, the hydraulic circuit is a bridge.

Such systems involving hydraulic valves are complicated by the intrinsic nonlinearity of hydraulic resistors. These systems have been extensively studied, however, and rather simple models may often suffice. From Chapter 7 of Reference [2], for example, we find the curves of Figure 8.14, which give the relation between motor flow Q_m and motor pressure P_m ($P_m = P_1 - P_4$ in Figure 8.13) for various values of the valve spool position z. Clearly, this amplifier is represented as a displacement-modulated resistance, and the curves of Figure 8.14 are the constitutive laws of the device. (Various characteristic curves for different valve geometries are given in Reference [2].) The bond graphs of Figure 8.15 show how the amplifier may be represented. In Figure 8.15a, the supply pressure is merely absorbed into the constitutive laws of the resistor. In Figure 8.15b, the valve is shown in 2-port resistance form. In this case, the effect on the amplifier of operating with various supply pressures can be studied. In either model a complex physical system is represented in a simple functional form that is valid only under restricted conditions. Naturally, more accurate physically based models of the device may be made using standard bond graph methods, but such models will be more complex than the functional model, and they

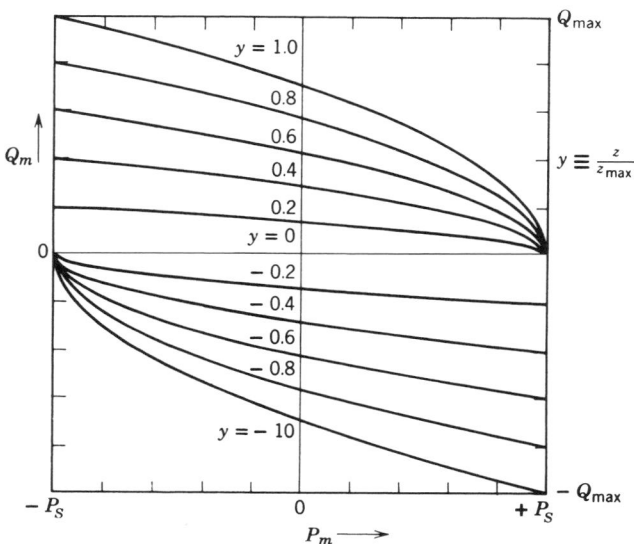

FIGURE 8.14. Pressure flow characteristics for a four-way valve.

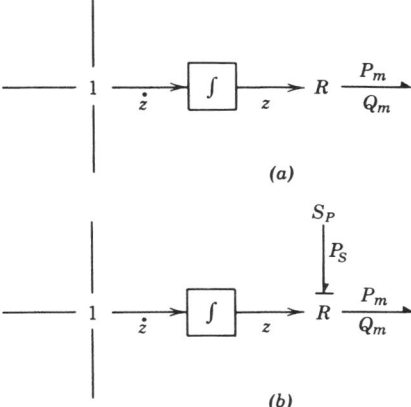

FIGURE 8.15. Bond graph models of the four-way hydraulic valve. (*a*) Displacement-modulated resistance with P_s built into constitutive law; (*b*) representation showing pressure supply.

may not add much to the usefulness of the overall system model. For this reason, the functional model may at least serve the purposes of a first system analysis. Later, a more refined model may prove desirable.

8.4 BOND GRAPHS AND BLOCK DIAGRAMS FOR CONTROLLED SYSTEMS

Feedback control systems are widely used to modify the dynamic behavior of engineering systems. They require *sensors* to measure some aspects of the system response, *signal processors* to realize the control law, and *actuators* to affect the system. The sensors and actuators are usually very similar to instruments and amplifiers and thus are typically modeled using active bonds. The signal processor, which is usually a special-purpose analog or digital computer, is normally described by its action on certain signals without regard to any physical power required.

Naturally, a sensor or an actuator is really a physical system and could be described by a bond graph—as could, in principle, a computer. For a well-designed system this is neither necessary nor desirable, and it is common practice to use signal flow graphs, block diagrams, or computer programs to describe the information processing which goes on between the sensor output and the actuator input in a controlled system. When the sensors or actuators have dynamics which cannot be neglected, they can be incorporated either as part of the physical system or, more artificially, as part of the dynamic laws of the control system itself.

For controlled systems, then, it is advantageous to use composite representations in which the physical system is represented using bond graphs and the controller is represented as a signal processor. We illustrate the choices of representation through an example. The signal processing will be represented through the use of block diagrams of various degrees of specificity. The combination of bond graphs and block diagrams is particularly useful when a continuous system simulation program which can accept bond graph models directly is available. (See Chapter 13.) In such a program the bond graph equations are generated automatically and the block diagram or control algorithm only needs to be added. Alternatively one could represent even the physical system in block diagram or equivalent form, since an augmented bond graph is nothing more than a very compact representation of the information in a block diagram or equation set. For simple systems, the resulting combined block diagram may yield insight, but for more complex systems, the result may be too complicated to be of much use.

Figure 8.16 shows a schematic diagram of an electronically controlled automotive suspension featuring a fast-acting load leveler and a semi-active damper. See References [3–5] for details on the systems. For simplicity, the system shown is only a "quarter car" model with the mass M representing the body and m the wheel. The velocity V_0 represents vertical velocity inputs from roadway unevenness, and k is a linearized tire spring constant. A sensor is supposed to generate an approximation to V, the absolute vertical velocity of the car body, and another measures X, the wheel-to-body deflection.

The controller drives one actuator which provides the relative velocity V_c of the main suspension spring attachment point as a conventional load leveler does. (The sketch shows the mechanical equivalent of systems which might actually be realized using pneumatic or hydro-pneumatic means.) The other controller output is F_c, a

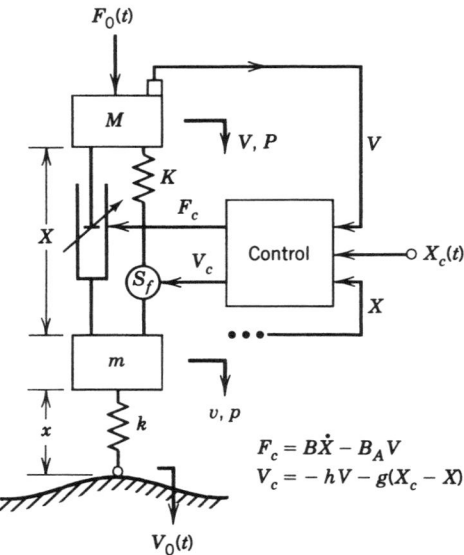

FIGURE 8.16. Quarter car model with electronically controlled suspension system incorporating a semi-active damper and a fast load leveler.

command force which is to be realized by a semi-active damper. The damper is a variable resistor—for example, a hydraulic shock absorber with electromechanical valves. Obviously such a device cannot provide an arbitrary F_c, since its force times its relative velocity $\dot{X}$ must always represent power dissipation. Thus, in the model, a nonlinear function Φ will be used to relate the actual damper force to F_c and $\dot{X}$. The details of Φ depend on the philosophy of the particular semi-active damper law and the physical construction of the device.

In this simple case, the schematic diagram is explicit enough that from it alone equations for the physical system could be written easily; but often this is not the case. A bond graph model combined with a schematic bock diagram is shown in Figure 8.17. In this representation, it is clear that the physical system model is completely defined. Active bonds show that it is assumed that V and X can be measured by the sensors with no significant dynamics. The $1/s$ block merely indicates that the $\dot{X}$-signal on the active bond is integrated to obtain X. (The symbol s represents the Laplace transform generalized frequency variable, so $1/s$ is the transfer function of an integrator.) This is an instance in which an active bond is treated exactly as a block diagram signal. The controller has a command input $X_c(t)$ which is related to the desired ride height and could come from another supervisory controller. (The ride height might vary with travel speed for aerodynamic reasons or because of driver manual inputs.)

The controller output V_c drives an ideal velocity source. Actuator dynamics could be included if desired by replacing the ideal source with a physical model of the

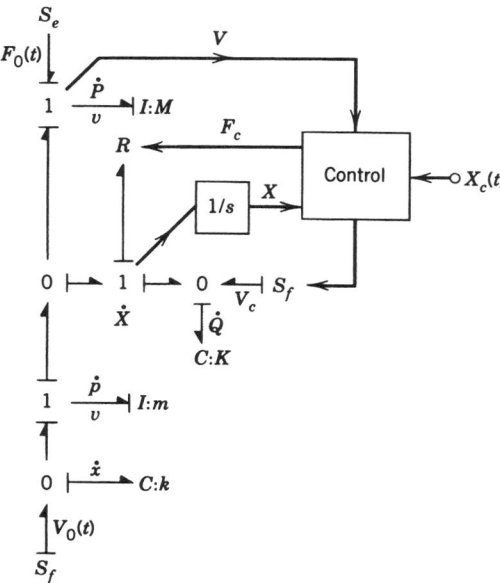

FIGURE 8.17. Bond graph for system of Figure 8.16 using active bonds for controller input and output signals.

actuator—perhaps a voltage source acting on a model of an electric motor and gear-box. Alternatively, some lag or delay might be incorporated in the controller block diagram or algorithm to model the actuator dynamics more accurately. The signal F_c modulates the R-element representing the controlled shock absorber. Again, engineering judgment is required to decide whether or not actuator dynamics in the semi-active damper should be included.

To simulate the system of Figure 8.17, one would only need to represent the bond graph without active bonds to a bond graph processor and then add to the program the relationships relating F_c and V_c to V, X, and X_c from a control law such as the one shown in Figure 8.16. In fact, for that rather simple control law, an explicit bond graph block diagram can be shown. See Figure 8.18. In this combined diagram, virtually all relationships are clear except for the semi-active damper. Here only the command force to the damper is shown, but what the actual force will be must still be specified.

Since the bond graph of Figure 8.18 is equivalent to a block diagram, it is possible to convert to a complete block diagram representation. As shown in Figure 8.19, this results in a very explicit but complex representation because all of the internal efforts and flows are shown. Despite some theoretical advantages to such a block diagram, few control engineers would be enthusiastic over it.

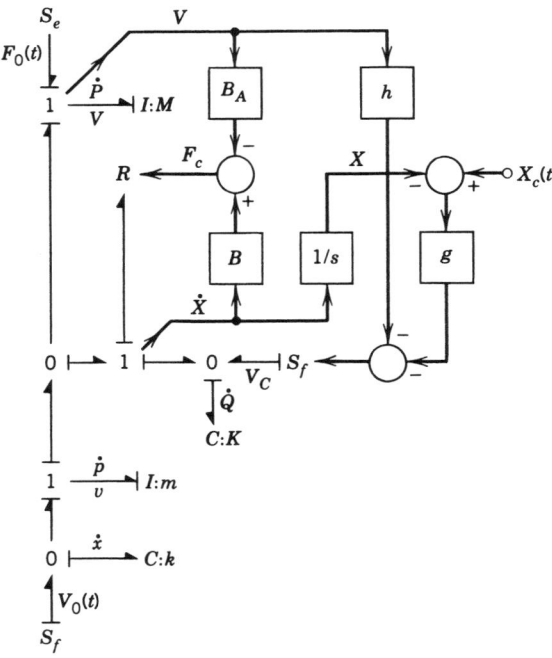

FIGURE 8.18. Bond graph incorporating block diagram for controller.

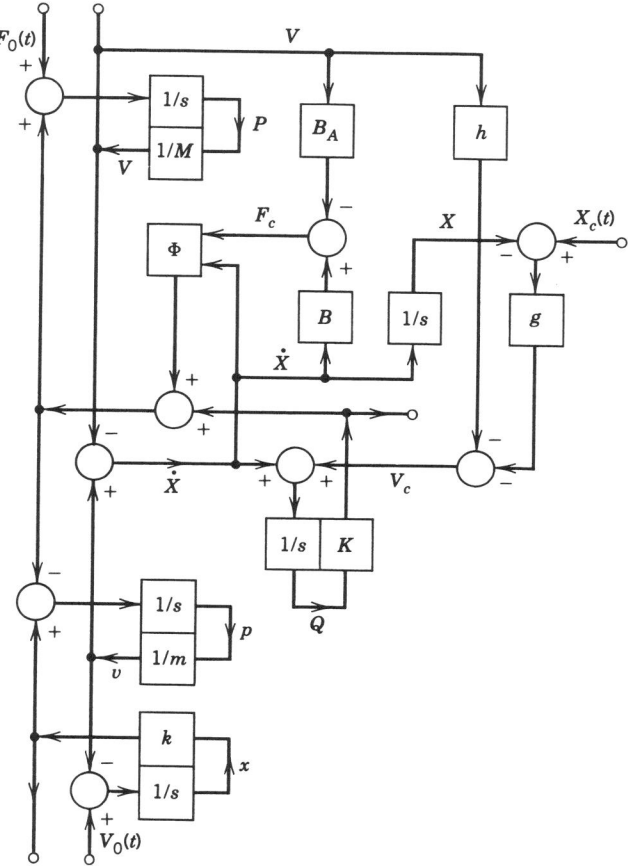

FIGURE 8.19. Block diagram derived directly from bond graph and combined with controller diagram.

By combining some linear operations and rearranging the diagram of Figure 8.19, a diagram more to the liking of control engineers can be developed. It is shown in Figure 8.20. Here the command X_c, response X, and disturbances F_0 and V_0 are shown together with feedback loops from V and X in a conventional layout. Certainly such a diagram helps one to understand the control aspects of the system. However, the distinction between finite-power and zero-power signals has been lost completely.

The lesson is that every system representation from schematic diagrams to mixed bond graph signal diagrams to pure signal graphs or computational diagrams has its place. Success in system dynamics and control is likely to come to those who can use all types of representations.

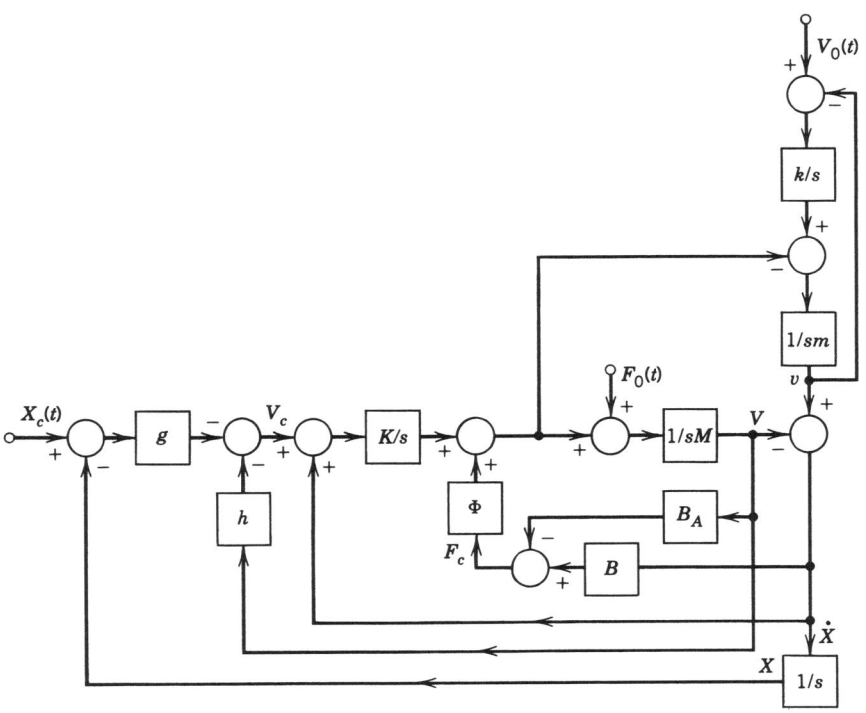

FIGURE 8.20. Block diagram of Figure 8.19 simplified and rearranged into typical feedback-controlled form.

REFERENCES

[1] S. H. Crandall, D. C. Karnopp, E. F. Kurtz, and D. C. Pridmore-Brown, *Dynamics of Mechanical and Electromechanical Systems*, Melbourne, FL: Krieger, 1968.

[2] J. F. Blackburn, G. Reethof, and J. L. Shearer, *Fluid Power Control*, Cambridge, MA: MIT Press, 1960.

[3] D. Karnopp, "Active Suspensions Based on Fast Load Levelers," *Vehicle Syst. Dynam.*, **16**, No. 5–6, 335–380 (1987).

[4] D. Karnopp, "Active Damping in Road Vehicle Suspension Systems," *Vehicle Syst. Dynam.*, **12**, No. 6, 291–311 (1983).

[5] D. Karnopp, M. J. Crosby, and R. A. Harwood, "Vibration Control Using Semi-active Force Generators," *Trans. ASME J. Eng. Ind.*, **96**, Ser. B, No. 2, 619–626 (1974).

PROBLEMS

8-1 The diagram shows a positioning system using a separately excited d-c motor. Physical system variables and parameters are identified:

θ_0 = output position angle;

e_{in} = input voltage;

e_a = output voltage of linear amplifier;

i_a = motor armature current;

i_f = motor field current, assumed constant;

K_a = gain of linear amplifier, assumed to have no significant time constants;

R_a = resistance of armature winding;

L_a = inductance of armature winding;

J = inertial load;

β = viscous–damping constant;

K_T = torque constant of motor;

K_v = back–emf constant of motor (emf = electromotive force).

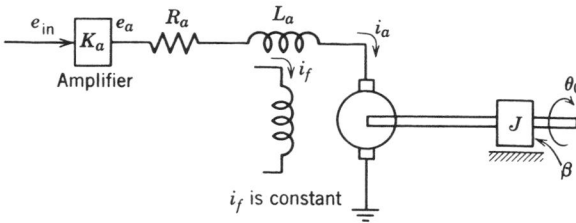

The differential equations that govern the dynamics of the system are

$$J\ddot{\theta}_0 + \beta\dot{\theta}_0 = K_T i_a, \qquad L_a \dot{i}_a + R_a i_a = V_a - K_v \dot{\theta}_0.$$

(a) Construct a bond graph for the system.

(b) Write state-space equations, and verify that your equations are equivalent to those listed above.

(c) Compare the two methods for analyzing this system. For example, is the system third or second order? Are K_T and K_v related in any way?

8-2   Consider the seismometer sketched below:

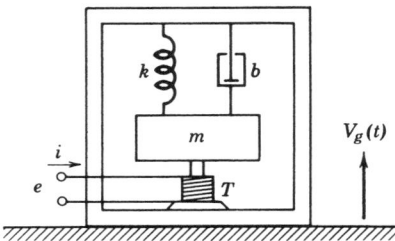

The input is ground motion, $V_g(t)$, and an electrical transducer using a permanent magnet moving in a coil reacts to the relative motion between the case and the seismic mass m.

(a)   Construct a bond graph for the device, leaving the electrical port as a free bond and neglecting coil resistance and inductance.

(b)   Assume that the device is connected to a voltage amplifier, so that $i \cong 0$. Find the transfer function between $V_g(t)$ and e.

(c)   It is sometimes preferable to use current rather than voltage as a signal for reasons connected with noise pickup. Suppose that a current amplifier is connected to the seismometer terminals, so that $e \cong 0$. Show that when the coil resistance is neglected, one can still find a transfer function between V_g and i, even though the system state equations degenerate.

(d)   Reconsider the case of (c) when the coil resistance R_c is not neglected.

8-3   Consider the solenoid shown in Figure 8.7, but include the mass m of the moving element and a Coulomb friction force F_f, which the constitutive law

$$F_f = F_0 \operatorname{sgn} V = \frac{F_0 V}{|V|},$$

where F_0 is the magnitude of the friction force. Show a bond graph for this transducer, and write state equations for it assuming electrical linearity and a voltage source input.

8-4   The float shown is cylindrical and is immersed in water in a cylindrical container. The cross-sectional area of the float is not negligible with respect to the tank area. The pressure P at the tank inlet is γh, where γ is the weight density of the water and h is the height of the water. The volume of water, V, is the time integral of the flow rate Q. If we assume that

$$P = P(V, x), \qquad F = F(V, x),$$

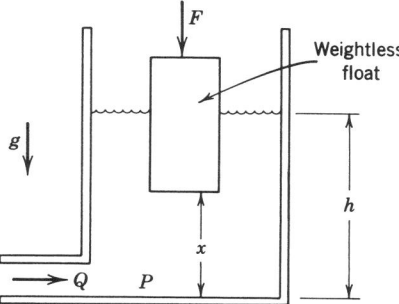

what type of bond graph element is this device? Are the functions (P and F) related in any way? Can you sketch them?

8-5 An electrostatic loudspeaker system is shown in which a high charging voltage E and a signal voltage $e(t)$ are applied through a current-limiting resistor R. The moving plate or membrane may be assumed to have effective mass m and a mechanical spring constant k. (The mechanical spring represents the combined effect of membrane tension and the air spring formed by the sealed cabinet.) The effect of the acoustic loading is shown by the bond graph fragment, in which A is the effective area of the speaker and P represents the acoustic overpressure. The I and R elements may be adjusted at any single frequency to match the impedance of the air:

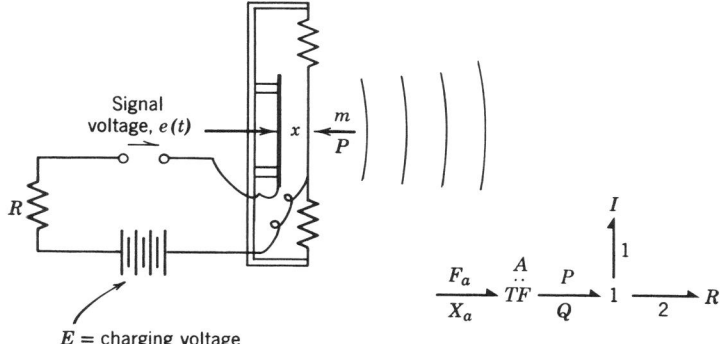

Assuming the basic transducer is an electrically linear device with capacitance $C(x) = \varepsilon A/x$, where ε is the dielectric constant for the air in the cabinet, construct a bond graph for the system. Let the mechanical spring be relaxed at $x = x_0$ when $E = 0$. Write the state equations for the system, letting the acoustic inertia and resistance be called I_1 and R_2, respectively.

8-6   The basic transduction mechanism for electrical alternators and motors may be understood by studying the ideal energy-storing transducer shown below, consisting of a fixed and a moving coil. Assume the two currents and the torque τ are related to two flux-linkage variables λ_1, λ_2 and the angular position θ:

$$i_1 = i_1(\lambda_1, \lambda_2, \theta), \qquad i_2 = i_2(\lambda_1, \lambda_2, \theta), \qquad \tau = \tau(\lambda_1, \lambda_2, \theta).$$

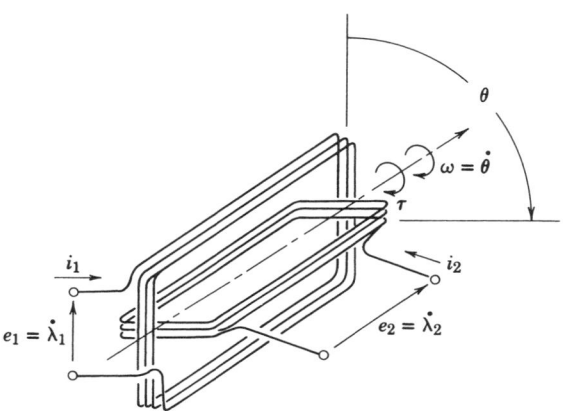

For the *electrically linear* case, it is conventional to specify self- and mutual-inductance parameters; for example,

$$\begin{bmatrix} L_1 & L_0 \cos\theta \\ L_0 \cos\theta & L_2 \end{bmatrix} \begin{bmatrix} i_1 \\ i_2 \end{bmatrix} = \begin{bmatrix} \lambda_1 \\ \lambda_2 \end{bmatrix},$$

where L_0, L_1, L_2 are constants and $L_1 L_2 \geq L_0^2$.

(a)   What kind of bond graph element describes this device?

(b)   Derive the torque relation from the inductance matrix above by computing the stored energy at fixed θ when λ_1 and λ_2 are brought from zero to final values and then differentiating the energy function with respect to θ (see Reference [1], pp. 322–323).

8-7   A model of a doorbell chime is shown that features a solenoid, return spring, frictional force, and a striker and chime. The schematic diagram attempts to depict the physical effects to be modeled for the case in which the button switch is closed. Make a bond graph for the system, including all physical effects. Let the striker–chime interaction be modeled by a nonlinear spring with zero force until the striker contacts the chime. Write the equations of

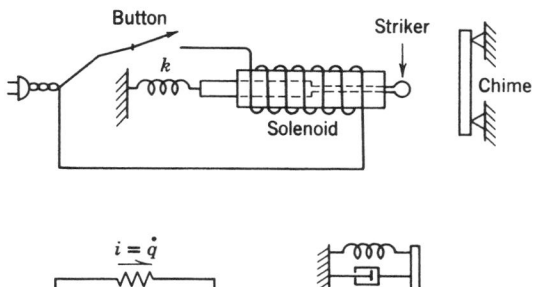

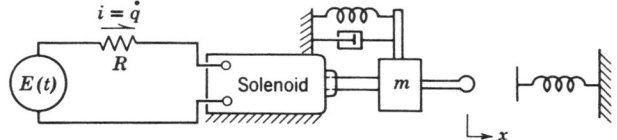

motion for this device, assuming electrical linearity for the solenoid. Leave your results in functional form, that is, let $L(x)$ be the inductance and $F(x)$ be the striker spring force. Sketch the shape of $L(x)$ and $F(x)$.

8-8 A speed control system uses a d-c motor with the following parameters:

$$L_f = \text{field inductance;}$$

$$R_f = \text{field resistance;}$$

$$L_a = \text{armature inductance;}$$

$$R_a = \text{armature resistance;}$$

$$J = \text{total moment of inertia;}$$

$$B = \text{rotary dashpot coefficient;}$$

$$T(i_f) = \text{transduction coefficient (see Figure 8.3).}$$

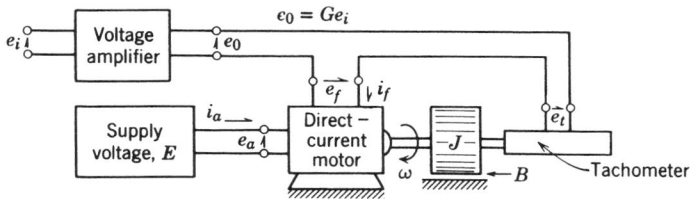

The control voltage, e_i, drives an amplifier that functions as a voltage-controlled voltage source:

$$\xrightarrow{e_i} S_e \xrightarrow{e_0}, \qquad e_0 = Ge_i(t).$$

The tachometer is a permanent-magnet device that functions as a instrument, so that its mechanical bond may be activated:

$$1_\omega \longrightarrow -\ddot{G}Y \overset{K_t}{\underset{i_t}{\frac{e_t}{\quad}}}, \qquad e_t = K_T \omega$$

Write a bond graph for the system, write state equations, and find the transfer function between e_i and ω when $i_a = $ const.

8-9 An a-c bicycle generator is shown in which a small wheel of the diameter d bears upon the bicycle wheel of diameter D. Assume the generator functions exactly like the one shown in Figure 8.4, although in reality it is more common for a magnet to rotate inside a fixed coil. Set up a bond graph representation that models the coil self-inductance and resistance and the bulb resistance and that would suffice to predict how the bulb voltage would vary with forward speed,

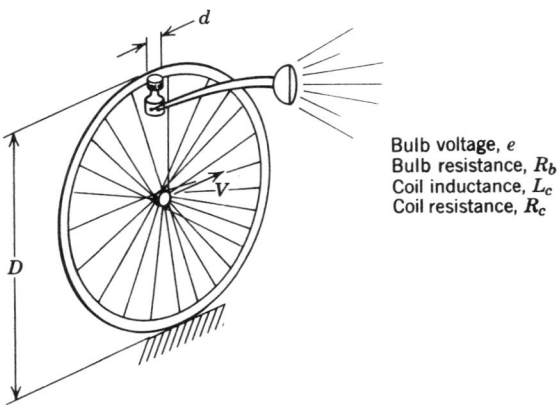

Bulb voltage, e
Bulb resistance, R_b
Coil inductance, L_c
Coil resistance, R_c

Explain the self-regulation feature of the L–R circuit that allows the bulb voltage to rise less than proportionally to V due to frequency response effects.

8-10 It is desired to study the response of the generating system shown to changes in throttle setting, $\theta(t)$, and load resistance, $R(t)$. Use the bond graph model of Figure 8.11 to represent the engine, and let the load resistance be controlled by an active bond, $-R\leftarrow$.

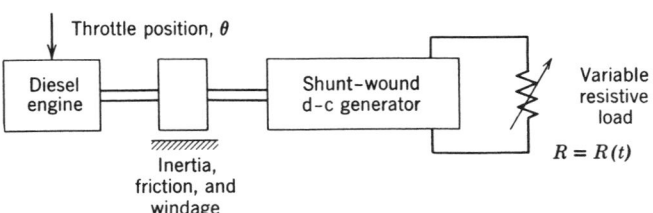

Show a bond graph for the system, and write the equations of motion in functional form. See Figure 8.3.

8-11   A simple pump is sketched in part a below. The check valves are represented by nonlinear resistors with characteristics such as that shown in part b.

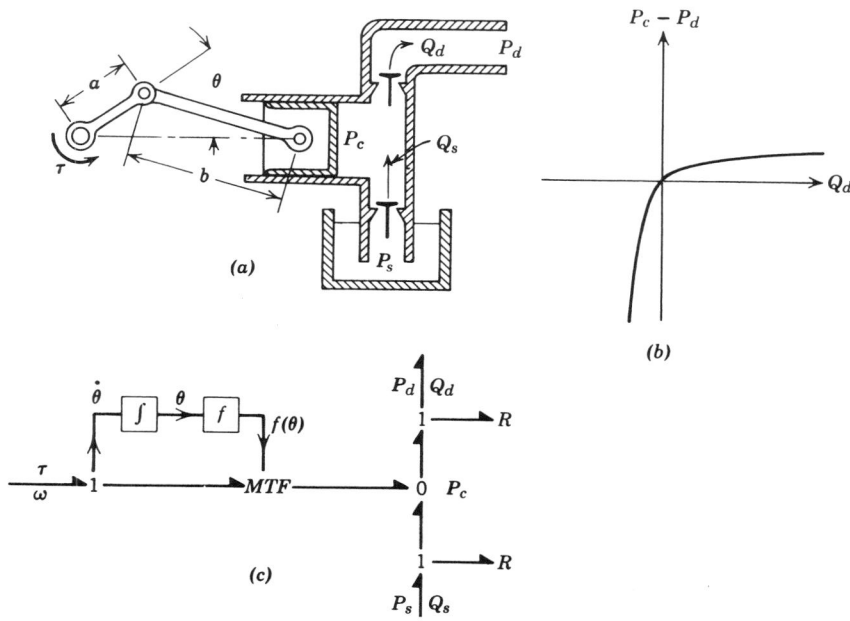

(a)   Verify the bond graph shown in part c, and compute the modulation function $f(\theta)$.

(b)   Suppose the speed, $\omega = \dot{\theta}$, is essentially constant. Sketch how the discharge flow and the suction flow vary as functions of time.

(c)   You should be able to see that this pump acts like an electrical half-wave rectifier circuit. Invent a pump that acts like a full-wave rectifier circuit, and show a bond graph for your pump.

8-12   Consider the high-performance speed control system shown below. The idea is to drive a hydraulic pump with a shunt-wound d-c motor and to control the speed ω_2 by stroking the bypass valve in a hydrostatic transmission. The valve stroke is $x(t)$, which may be changed manually or, ultimately, automatically by a servocontrol system. The system should have fast response, since if the valve is slammed shut, the hydraulic motor pump will suddenly be directly coupled so that energy stored in the rotating inertia J_1 will be available to accelerate J_2 very rapidly.

(a)   Draw a bond graph for the system including all the effects shown in the sketch. Augment the bond graph, and list the state variables required.

(b)   Either write out the state equations or construct a block diagram corresponding to the bond graph. Identify the input variables to your dynamic equations.

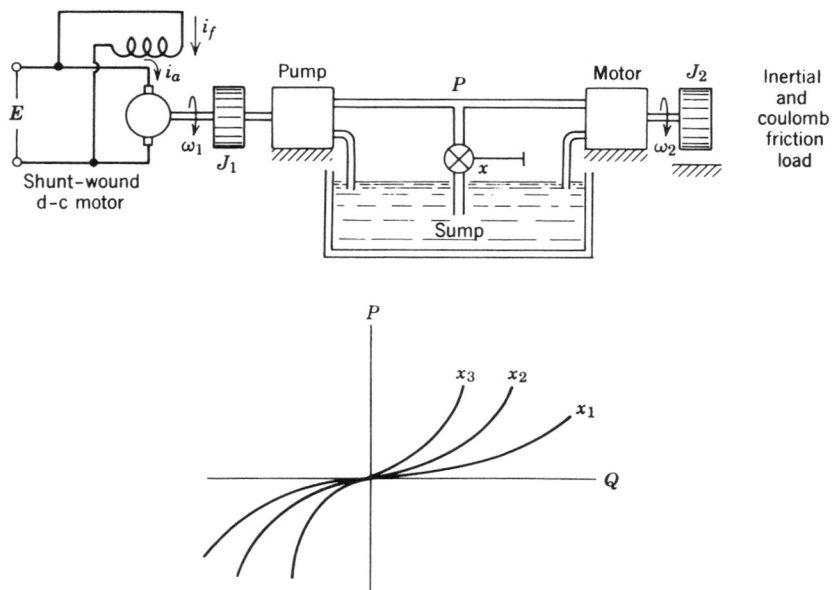

The basic motor characteristic is

$$T(i_f)i_a = \tau, \qquad e_a = T(i_f)\omega_1.$$

The pressure–flow law for a valve depends on the stroke x, that is,

$$P = A(x)Q|Q| = A(x)Q^2 \operatorname{sgn} Q.$$

The model should include armature and field inductance and resistance. The constitutive laws for a pump and hydraulic motor are:

(1)   for a pump,

$$\tau_p = \alpha_p P_p, \qquad \alpha_p \omega_p = Q_p;$$

(2)   for a motor,

$$\tau_m = \alpha_m P_m, \qquad \alpha_m \omega_m = Q_m.$$

Neglect leakage in the pump and motor and compressibility in the oil lines.

8-13 In the diagram, a conventional hydropneumatic suspension system such as those found on certain automobiles has been made active by the addition of a valve. A simple model of the valve is indicated in which the flow of oil, Q, is simply proportional to the relative displacement of the suspension, x. We assume a very high-pressure source of oil (not shown) so that the valve can act as a source of flow modulated by x,

$$\xrightarrow[x]{} S_Q \mapsto,$$

independent of the pressure in the suspension. With this system, the "static deflection" of the suspension is always zero for any gain g, but as g is increased, it is not so clear that the system will remain stable.

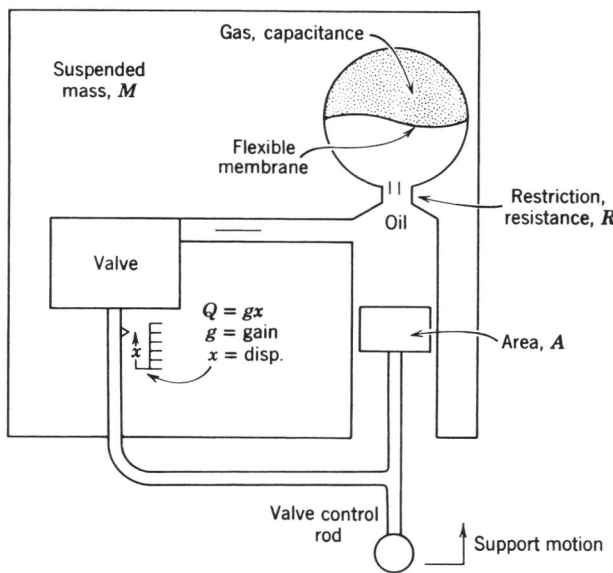

(a) Find a bond graph for the system.

(b) Decide on appropriate system inputs, and write a set of state-space equations for the system.

(c) Assuming linearized characterizations for all elements of the system, set up an expression that will yield system eigenvalues for any particular numerical values of parameters.

8-14 A servomechanism has been constructed by connecting a spool valve and hydraulic ram with feedback linkage:

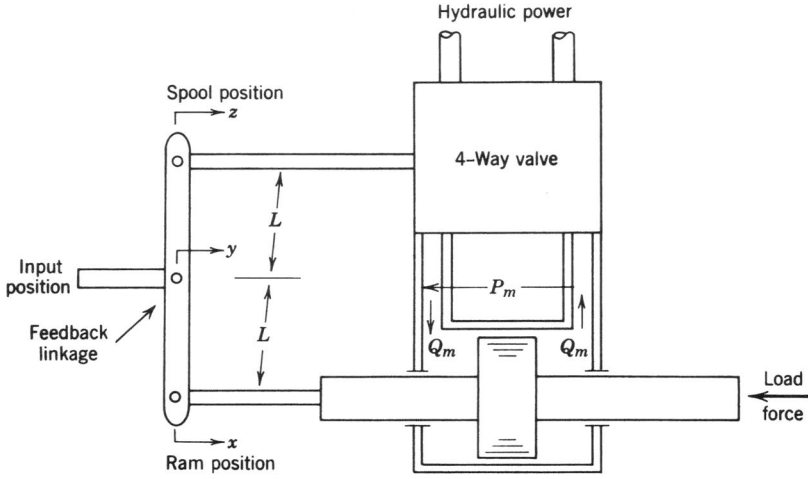

Using the text discussion and Figures 8.13, 8.14, and 8.15, construct a bond graph for this device. Assume the following:

(a)   The force required to move the valve is $F_r = F_0 \, \text{sgn} \, \dot{z}$, where $F_0 = $ const.

(b)   The load force is inertial and resistive, and oil compressibility is included.

(c)   The linkage is light and frictionless and moves only through very small angles.

(d)   The working area of the ram is A, and the supply pressure is P_s. Using your bond graph, answer the following questions:

   (i)   What is the force required to move the input, and what is the maximum load force possible?

   (ii)   With zero load force, what is the maximum velocity of the ram?

   (iii) Write the equations of motion, leaving the valve constitutive laws in a general functional form, for example, $Q_m = Q_m(z, P_m)$.

8-15 Add a control system block diagram to the bond graph for the motor shown in Problem 8-1 for two cases:

(a)   A speed control system such that $e_{in} = g(\dot{\theta}_c - \dot{\theta}_0)$, in which g is a gain and $\dot{\theta}_c$ is a command angular speed.

(b)   A position control in which $e_{in} = -g\dot{\theta}_0 + h(\theta_c - \theta_0)$, g and h are gains, and θ_c is an angular position command signal.

   In each case, note the order of the controlled system. Write the state equations for (b).

8-16 A mass–spring–damper system has an equation of motion

$$m\ddot{x} + b\dot{x} + kx = f,$$

where f is an applied force. A standard way to make a block diagram for such a system results in the following:

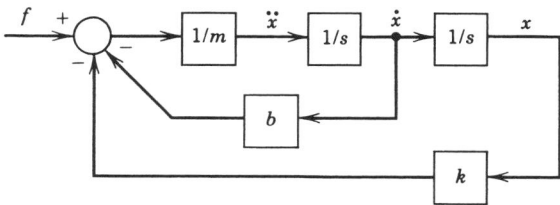

Make a bond graph for this system and convert it directly into a block diagram in the fashion of Figure 8.19. Then rearrange the block diagram and show that it is equivalent to the one shown above.

8-17 Consider an electrically linear parallel-plate capacitor as described by Eqs. (8.23) and (8.25). Rewrite the force law in the form

$$F = \frac{-q^2 C'}{2C^2}, \qquad C' = \frac{dC(X)}{dX},$$

in analogy to the solenoid force law in Eq. (8.33).

(a) Use Eq. (8.23) to find the force for the case when the voltage is held constant at the value e_0 by eliminating q in favor of e_0.

(b) Using the approximate expression $C(X) = \varepsilon A / X$ mentioned below Eq. (8.25), show that when q is constant, the force does not vary with X under this assumption, but if the voltage is held constant, the force does vary with X.

8-18 A butterfly condenser has a capacitance which varies with the angular position of a rotor, θ, rather than with a linear position X as in Problem 8-17. This requires a torque, τ, in analogy with the force F for a parallel-plate capacitor. Assuming the capacitance varies with θ according to

$$C(\theta) = C_0 + C_1 \cos 2\theta$$

with

$$C_0 = 15 \times 10^{-12} \ F, \qquad C_1 = 10 \times 10^{-12} \ F,$$

and assuming the condenser is attached to a 1000-V source, compute the magnitude of the torque resulting at an angle when it is a large as possible.

8-19 An inertial actuator is a device in which, by appropriate acceleration of an internal mass, a prescribed reaction force can be generated. By attaching the actuator to a structure, the structure motion can be controlled.

The device shown is voice coil driven, where only the winding resistance is important on the electrical side. The mass m is attached to the base with spring k and damper b.

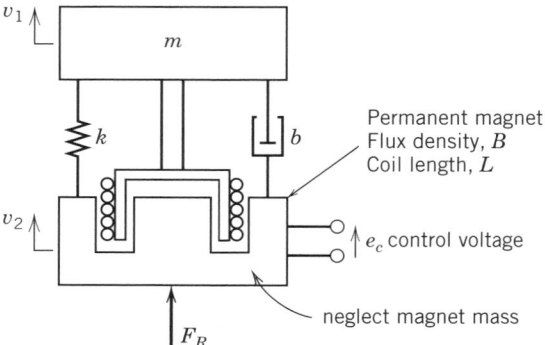

Here, F_R is the output force. Assume the base has prescribed motion, $v_2(t)$. Construct a bond graph model and derive state equations. Derive the output equation for F_R.

8-20 From the state equations of Problem 8-19, derive the transfer functions relating F_r to v_2 and F_r to e_c. Derive

$$\frac{F_r}{v_2} = G_{Fv}(s) \quad \text{and} \quad \frac{F_r}{e_c} = G_{Fc}(s).$$

8-21 Since Problem 8-19 is treated as a linear system, we can express the output force in terms of the transfer functions from Problem 8-20 as

$$F_R(s) = G_{Fv}v_2 + G_{Fc}e_c.$$

Sometimes we desire the reaction force to mimic the effect of a damper attached between the base and inertial ground. In other words, we desire

$$F_R = b_c v_2,$$

where b_c is a controller gain. Derive the ideal control filter that will yield the desired reaction force from the inherent dynamics of the device.

8-22 Shown below is the device from Problem 8-19 attached to a structure consisting of a mass, m_s, spring, k_s, and damper, b_s. A force, F_d, acts upon the structure mass. Construct a bond graph model for this total system and derive state equations.

Derive the open-loop ($e_c = 0$) transfer function relating output, v_2, to input, F_d. Using the control transfer function from Problem 8-21, derive the closed-loop transfer function between v_2 and F_d. Does the result have a component similar to a damper to ground?

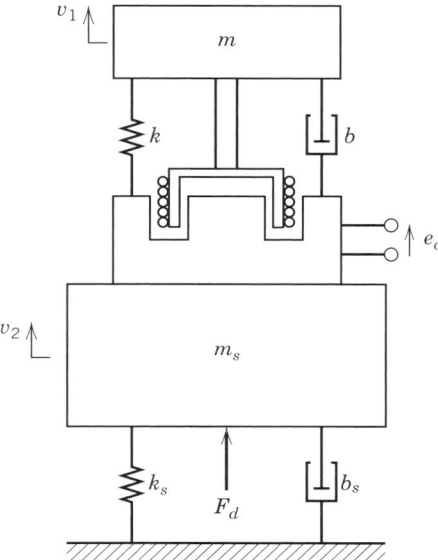

8-23   Repeat Problems 8-19 through 8-22, but assume that the voice coil is current driven, i_c, rather than voltage driven.

9

MECHANICAL SYSTEMS WITH NONLINEAR GEOMETRY

By now the reader should be somewhat familiar, and perhaps even comfortable, with the use of bond graphs for representing all kinds of interacting dynamic systems. We know that bond graphs straightforwardly indicate, through causality assignment, the state variables and the ease with which component models can be assembled into overall system models. We know that algebraic problems may exist, and methods of dealing with algebraic problems were presented in an earlier chapter. In a later chapter, computational procedures are presented, and, as the reader may have surmised, causality also indicates the ease with which a computable model can be constructed. In fact, some algebraic problems may be so complicated, involving so many variables in noninvertable, nonlinear functions, that a computable model is virtually impossible to formulate under the modeling assumptions active during the original model construction.

This chapter deals with multidimensional rigid-body mechanics and the inclusion of such complex mechanics in overall system models. Vehicle systems are a primary example of the engineering use of nonlinear mechanics, although there are many others. It turns out that it is possible to cast rigid-body mechanics into a bond graph format such that the resulting bond graph fragment can be used in the construction of overall models. We can literally add suspension components, tires, engines and engine mounts, brakes, and so forth to a rigid-body frame model in a manner that is the computational equivalent to bolting the hardware together. What follows is not simple. However, with some persistence the reader will be rewarded with the ability to construct reasonably low-order system models which include three-dimensional (3-D) rigid-body motion and which are straightforwardly formulated into state equations, ready for computation.

9.1 MULTIDIMENSIONAL DYNAMICS

Figure 9.1 shows a general rigid body both translating and rotating in space. Inertial axes X, Y, Z are shown, and axes x, y, z are attached to the body, at its center of mass, and aligned with the principal axes of the body. With respect to these body-fixed coordinates, the rotational inertial properties remain invariant and the products of inertia are all zero. While these body-fixed coordinates are not the best from which to view the body motion, they are practical coordinates for the computation of body motion.

At the instant shown, the body has absolute velocity $\mathbf{v}$ and absolute angular velocity $\boldsymbol{\omega}$. These vectors have been cast into three mutually perpendicular components: v_x, v_y, v_z and ω_x, ω_y, ω_z. Newton stated that the net force $\mathbf{F}$ acting on the body changes its momentum:

$$\mathbf{F} = \frac{d}{dt}\mathbf{p}, \tag{9.1}$$

where

$$\mathbf{p} = m\mathbf{v}. \tag{9.2}$$

If $\mathbf{v}$ is expressed with respect to a rotating frame, then

$$\mathbf{F} = \left.\frac{\partial \mathbf{p}}{\partial t}\right|_{\text{rel}} + \boldsymbol{\omega} \times \mathbf{p}, \tag{9.3}$$

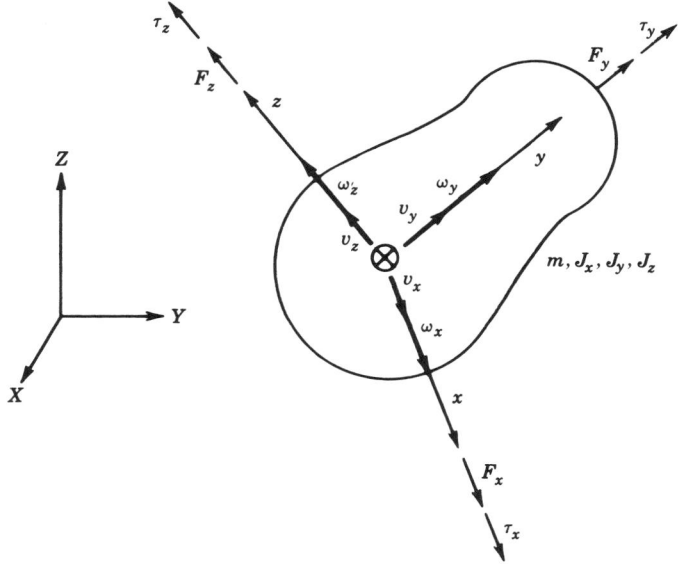

FIGURE 9.1. Body in general 3-D motion.

where

$$\frac{\partial \mathbf{p}}{\partial t}\bigg|_{rel}$$

indicates the rate of change of momentum relative to the moving frame.

An angular momentum law analogous to Eq. (9.1) can be derived if the net torque acting on the body τ and the angular momentum $\mathbf{h}$ are evaluated with respect to either a fixed point, if one exists for the body, or the center of mass,

$$\tau = \frac{d}{dt}\mathbf{h}. \tag{9.4}$$

If the x–y–z axis system is assumed to be aligned with the principal axes of the body, then the angular momentum is related to the angular velocity by

$$\mathbf{h} = \mathbf{J}\omega, \tag{9.5}$$

in which $\mathbf{J}$ is a diagonal matrix of the principal moments of inertia, J_x, J_y, J_z. For $\mathbf{h}$ expressed with respect to the rotating frame,

$$\tau = \frac{\partial \mathbf{h}}{\partial t}\bigg|_{rel} + \omega \times \mathbf{h}. \tag{9.6}$$

Using the right-hand rule, it is straightforward to write down the component equations for (9.3) and (9.6). They are

$$F_x = m\dot{v}_x + m\omega_y v_z - m\omega_z v_y, \tag{9.7}$$

$$F_y = m\dot{v}_y + m\omega_z v_x - m\omega_x v_z, \tag{9.8}$$

$$F_z = m\dot{v}_z + m\omega_x v_y - m\omega_y v_x, \tag{9.9}$$

and

$$\tau_x = J_x\dot{\omega}_x + \omega_y J_z\omega_z - \omega_z J_y\omega_y, \tag{9.10}$$

$$\tau_y = J_y\dot{\omega}_y + \omega_z J_x\omega_x - \omega_x J_z\omega_z, \tag{9.11}$$

$$\tau_z = J_z\dot{\omega}_z + \omega_x J_y\omega_y - \omega_y J_x\omega_x. \tag{9.12}$$

These nonlinear differential equations are known as Euler's equations. They have no general solution and can only be solved analytically for some special cases. If they were solved, then we would know v_x, v_y, v_z, ω_x, ω_y, ω_z, with respect to a frame that will be aiming in a different direction at every instant, which makes interpreting the body motion a bit difficult. Furthermore, the force and torque components (coming from attached systems) must be aligned with the body-fixed coordinates in order to use these equations. This will never occur naturally, so the utility of these equations is questionable at this stage of development.

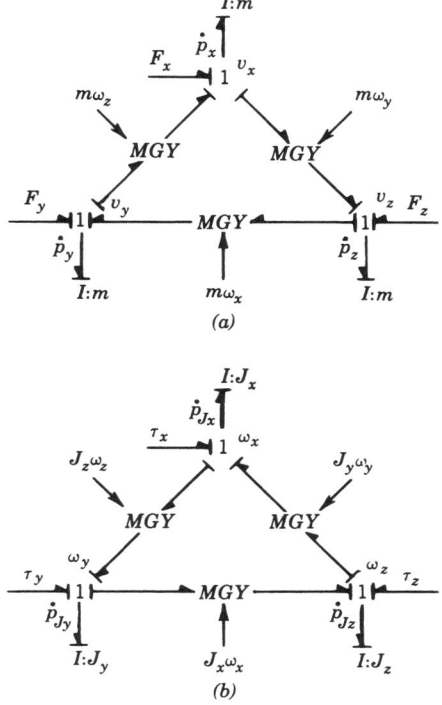

FIGURE 9.2. Bond graphs for 3-D rigid-body motion.

If we treat the cross-product terms as forces in Eqs. (9.7), (9.8), and (9.9) and as torques in Eqs. (9.10), (9.11), and (9.12), then an elegant bond graph representation results for translation and rotation, as shown in Figure 9.2. The reader should add the forces and torques at the respective 1-junctions and be convinced that Euler's equations are, in fact, represented by these bond graph fragments. These gyrator ring constructions appropriately introduce the momenta p_x, p_y, p_z and angular momenta p_{J_x}, p_{J_y}, p_{J_z} as state variables. The modulated gyrator (—MGY—) elements represent the cross-product terms in Eqs. (9.7)–(9.12), and the gyrator moduli are time varying and depend on the angular momentum state variables.

As long as external forces F_x, F_y, F_z and torques τ_x, τ_y, τ_z, are causal inputs to the bond graph fragments, integral causality exists for all I-elements. If we agree to always use these fragments with forces and torques as causal inputs and with velocities and angular velocities as causal outputs, then we can write the state equations, once and for all, for these submodels, ready to be appropriately connected to any external systems in the process of constructing an overall model. We make this agreement, and write the state equations as

$$\dot{p}_x = F_x + m\omega_z \frac{p_y}{m} - m\omega_y \frac{p_z}{m}, \tag{9.13}$$

$$\dot{p}_y = F_y + m\omega_x \frac{p_z}{m} - m\omega_z \frac{p_x}{m}, \tag{9.14}$$

$$\dot{p}_z = F_z + m\omega_y \frac{p_x}{m} - m\omega_x \frac{p_y}{m}, \tag{9.15}$$

$$\dot{p}_{J_x} = \tau_x + J_y\omega_y \frac{p_{J_z}}{J_z} - J_z\omega_z \frac{p_{J_y}}{J_y}, \tag{9.16}$$

$$\dot{p}_{J_y} = \tau_y + J_z\omega_z \frac{p_{J_x}}{J_x} - J_x\omega_x \frac{p_{J_z}}{J_z}, \tag{9.17}$$

$$\dot{p}_{J_z} = \tau_z + J_x\omega_x \frac{p_{J_y}}{J_y} - J_y\omega_y \frac{p_{J_x}}{J_x}, \tag{9.18}$$

where

$$\omega_x = p_{J_x}/J_x, \tag{9.19}$$

$$\omega_y = p_{J_y}/J_y, \tag{9.20}$$

$$\omega_z = p_{J_z}/J_z. \tag{9.21}$$

A shorthand notation for the bond graphs of Figure 9.2 is shown in Figure 9.3.

Coordinate Transformations Since it is unlikely that external forces and torques will be conveniently lined up with the continuously changing principal directions, and since the body motion is difficult to interpret with respect to the body-fixed coordinates, it is necessary to transfer from body-fixed coordinates to other more convenient frames through a series of coordinate transformations. There are many possible coordinate transformations which can take us from body-fixed coordinates to inertial coordinates, the most familiar probably being the transformation involving Euler angles. The transformation used depends largely upon the system being analyzed: Spinning bodies of revolution, typical of gyroscopic motion, are most conveniently analyzed using Euler angles, while multibody systems, such as robotic arms, call for some other transformation.

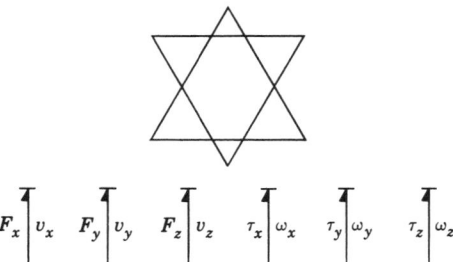

FIGURE 9.3. Shorthand representation of 3-D rigid-body mechanics.

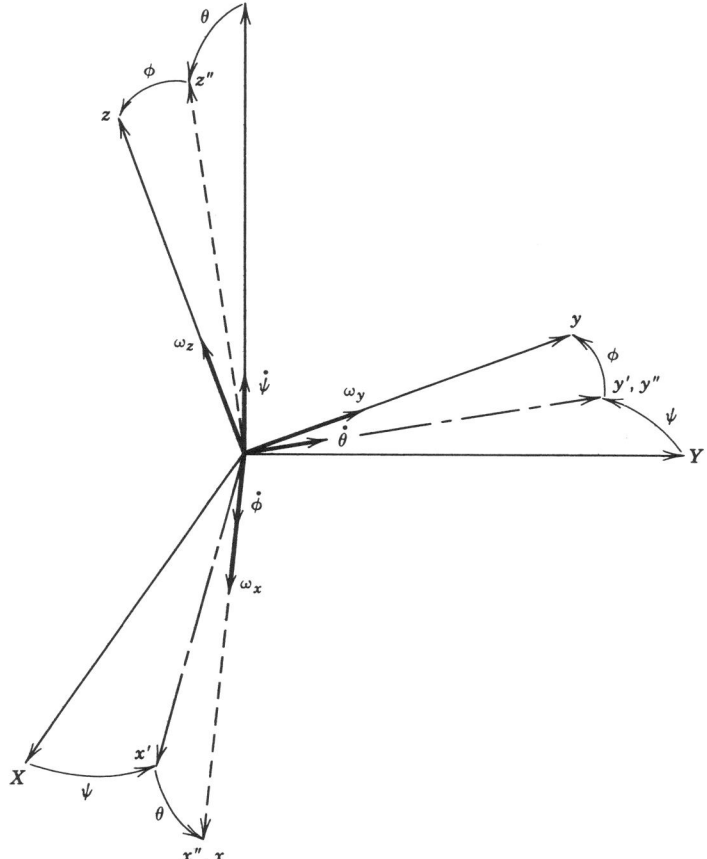

FIGURE 9.4. Cardan angle coordinate transformation.

The concept of a coordinate transformation will be developed here using ground vehicles as an example, and we shall introduce the Cardan angles, which correspond to the familiar yaw, pitch, and roll angles of an automobile. Figure 9.4 shows an inertial frame (X, Y, Z), a body-fixed orientation (x, y, z), and two intermediate frames $(x', y', z'$ and $x'', y'', z'')$. The body-fixed frame is arrived at by first rotating about Z through angle ψ (yaw), yielding the frame x', y', z'. We next rotate about the y'-axis through the angle θ (pitch), yielding the x'', y'', z'' axes. Finally, we rotate about the x''-axis through the angle ϕ (roll), yielding the instantaneous body-fixed frame, x, y, z.

Let us now assume that body-fixed angular velocities ω_x, ω_y, ω_z are known (outputs, by agreement, from the gyrator-ring bond graph fragments), and write down the angular velocity components in the intermediate frames and finally in the inertial frame:

$$\omega_{x''} = \omega_x, \tag{9.22}$$

$$\omega_{y''} = \omega_y \cos\phi - \omega_z \sin\phi, \tag{9.23}$$

$$\omega_{z''} = \omega_y \sin\phi + \omega_z \cos\phi; \tag{9.24}$$

$$\omega_{x'} = \omega_{x''} \cos\theta + \omega_{z''} \sin\theta, \tag{9.25}$$

$$\omega_{y'} = \omega_{y''}, \tag{9.26}$$

$$\omega_{z'} = -\omega_{x''} \sin\theta + \omega_{z''} \cos\theta; \tag{9.27}$$

$$\omega_X = \omega_{x'} \cos\psi - \omega_{y'} \sin\psi, \tag{9.28}$$

$$\omega_Y = \omega_{x'} \sin\psi + \omega_{y'} \cos\psi, \tag{9.29}$$

$$\omega_Z = \omega_{z'}. \tag{9.30}$$

We can write these relationships in matrix form as

$$\begin{bmatrix} \omega_{x''} \\ \omega_{y''} \\ \omega_{z''} \end{bmatrix} = \begin{bmatrix} 1 & 0 & 0 \\ 0 & \cos\phi & -\sin\phi \\ 0 & \sin\phi & \cos\phi \end{bmatrix} \begin{bmatrix} \omega_x \\ \omega_y \\ \omega_z \end{bmatrix}, \tag{9.31}$$

$$\begin{bmatrix} \omega_{x'} \\ \omega_{y'} \\ \omega_{z'} \end{bmatrix} = \begin{bmatrix} \cos\theta & 0 & \sin\theta \\ 0 & 1 & 0 \\ -\sin\theta & 0 & \cos\theta \end{bmatrix} \begin{bmatrix} \omega_{x''} \\ \omega_{y''} \\ \omega_{z''} \end{bmatrix}, \tag{9.32}$$

$$\begin{bmatrix} \omega_X \\ \omega_Y \\ \omega_Z \end{bmatrix} = \begin{bmatrix} \cos\psi & -\sin\psi & 0 \\ \sin\psi & \cos\psi & 0 \\ 0 & 0 & 1 \end{bmatrix} \begin{bmatrix} \omega_{x'} \\ \omega_{y'} \\ \omega_{z'} \end{bmatrix}. \tag{9.33}$$

Thus, if we know ω_x, ω_y, ω_z, then we can determine the angular velocity components in all frames, including the inertial frame. If we denote the transformation matrices as

$$\boldsymbol{\Phi} = \begin{bmatrix} 1 & 0 & 0 \\ 0 & \cos\phi & -\sin\phi \\ 0 & \sin\phi & \cos\phi \end{bmatrix}, \tag{9.34}$$

$$\boldsymbol{\Theta} = \begin{bmatrix} \cos\theta & 0 & \sin\theta \\ 0 & 1 & 0 \\ -\sin\theta & 0 & \cos\theta \end{bmatrix}, \tag{9.35}$$

$$\boldsymbol{\Psi} = \begin{bmatrix} \cos\psi & -\sin\psi & 0 \\ \sin\psi & \cos\psi & 0 \\ 0 & 0 & 1 \end{bmatrix}, \tag{9.36}$$

then

$$\begin{bmatrix} \omega_X \\ \omega_Y \\ \omega_Z \end{bmatrix} = \boldsymbol{\Psi}\boldsymbol{\Theta}\boldsymbol{\Phi} \begin{bmatrix} \omega_x \\ \omega_y \\ \omega_z \end{bmatrix}. \tag{9.37}$$

It should be realized that the exact same relationship exists for transforming velocity components in the body-fixed directions to velocity components in the inertial directions. Thus

$$
\begin{bmatrix} v_X \\ v_Y \\ v_Z \end{bmatrix} = \boldsymbol{\Psi\Theta\Phi} \begin{bmatrix} v_x \\ v_y \\ v_z \end{bmatrix}.
\tag{9.38}
$$

This transformation from body-fixed components to inertial components is a power-conserving transformation and is shown as a bond graph in Figure 9.5. The modulated transformers, —*MTF*—, indicate the three transformations $\boldsymbol{\Phi}$, $\boldsymbol{\Theta}$, $\boldsymbol{\Psi}$, which take the body-fixed velocity and angular velocity components through the intermediate frames and produce the inertial velocity and angular velocity components.

We now make use of the power-conserving nature of this transformation of coordinates. Since no energy is stored or dissipated by the transformation, power must be instantaneously conserved. This was shown in Chapter 7 to result in the following relationship among the forces and torques:

$$
\begin{bmatrix} F_x \\ F_y \\ F_z \end{bmatrix} = (\boldsymbol{\Psi\Theta\Phi})^t \begin{bmatrix} F_X \\ F_Y \\ F_Z \end{bmatrix},
\tag{9.39}
$$

$$
\begin{bmatrix} \tau_x \\ \tau_y \\ \tau_z \end{bmatrix} = (\boldsymbol{\Psi\Theta\Phi})^t \begin{bmatrix} \tau_X \\ \tau_Y \\ \tau_Z \end{bmatrix},
\tag{9.40}
$$

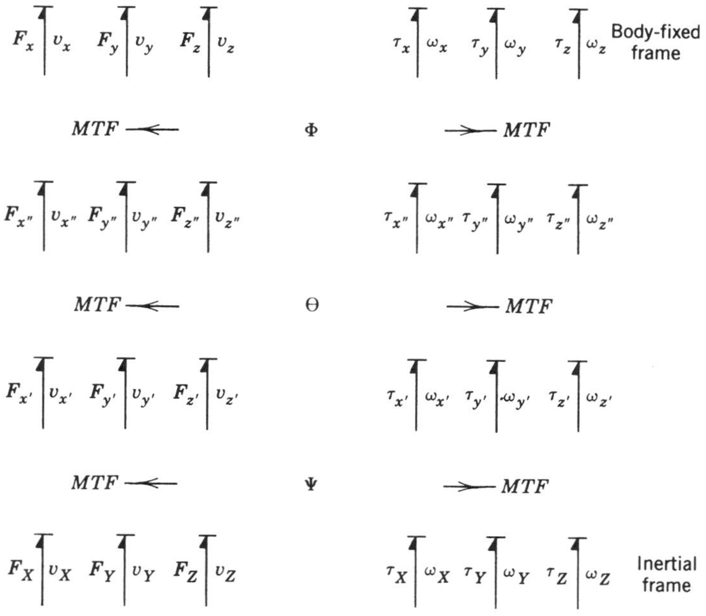

FIGURE 9.5. Bond graph of coordinate transformation.

or

$$\begin{bmatrix} F_x \\ F_y \\ F_z \end{bmatrix} = \Phi^t \Theta^t \Psi^t \begin{bmatrix} F_X \\ F_Y \\ F_Z \end{bmatrix}, \tag{9.41}$$

$$\begin{bmatrix} \tau_x \\ \tau_y \\ \tau_z \end{bmatrix} = \Phi^t \Theta^t \Psi^t \begin{bmatrix} \tau_X \\ \tau_Y \\ \tau_Z \end{bmatrix}. \tag{9.42}$$

Thus, the transformation indicated in Figure 9.5 not only transforms the body-fixed velocities and angular velocities to inertial components; it also transforms inertial forces and torques into the body-fixed directions. This bilateral transformation is indicated by the causality in Figure 9.5.

Let us agree to always use the coordinate transformation of Figure 9.5 with the causality shown. With this agreement we can now connect the bond graph of Figure 9.5 to the rigid-body dynamics of Figure 9.3, and we now have the capability of formulating a complete state representation of 3-D rigid-body dynamics. This model can accept forces and torques in inertial coordinate directions, properly align these efforts into the instantaneous principal directions, and then, through integration of first-order equations, deliver velocities and angular velocities in principal directions and transform these to inertial directions. This is a very nice package.

We should note that the angles ϕ, θ, and ψ are needed to perform the coordinate transformations. It is relatively simple to relate $\dot{\phi}$, $\dot{\theta}$, and $\dot{\psi}$ to the body-fixed components ω_x, ω_y, and ω_z. From Figure 9.4,

$$\omega_x = \dot{\phi} - \dot{\psi} \sin \theta, \tag{9.43}$$

$$\omega_y = \dot{\theta} \cos \phi + \dot{\psi} \cos \theta \sin \phi, \tag{9.44}$$

$$\omega_z = -\dot{\theta} \sin \phi + \dot{\psi} \cos \theta \cos \phi. \tag{9.45}$$

Equations (9.44) and (9.45) can be solved for $\dot{\theta}$ and $\dot{\psi}$:

$$\dot{\theta} = \cos \phi \, \omega_y - \sin \phi \, \omega_z, \tag{9.46}$$

$$\dot{\psi} = \frac{\sin \phi}{\cos \theta} \omega_y + \frac{\cos \phi}{\cos \theta} \omega_z. \tag{9.47}$$

Then, from (9.43),

$$\dot{\phi} = \omega_x + \sin \phi \frac{\sin \theta}{\cos \theta} \omega_y + \cos \phi \frac{\sin \theta}{\cos \theta} \omega_z. \tag{9.48}$$

Equations (9.46), (9.47), and (9.48) are three additional state equations which must be integrated along with the others so that ϕ, θ, and ψ are continuously available as modulating variables.

A very nice shorthand notation for the rigid-body mechanics and coordinate transformation is shown in Figure 9.6. For computational purposes, the building blocks

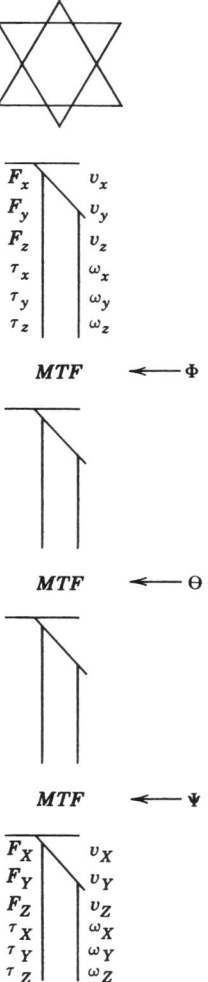

FIGURE 9.6. Shorthand bond graph for the complete 3-D mechanics plus coordinate transformation.

of Figure 9.6 could not be better organized, nor more ready for interaction with external dynamic systems. However, remember our agreement. The causality shown in Figures 9.3 and 9.6 must be maintained.

9.2   KINEMATIC NONLINEARITIES IN MECHANICAL DYNAMICS

This section is addressed to the following problem: Given a schematic representation of a mechanical system plus accompanying verbal and mathematical descriptions of the parts, find a convenient way to model the system so as to predict its dynamic behavior.

The importance of the problem hardly needs emphasizing. It has been worked on in essentially modern form at least since the time of Newton and has attracted the attention of many notable scientists, including Hamilton and Lagrange. The development since the time of Newton has relied, not upon pictorial or graphical representation, but upon analytically oriented notation in the form of operators and equations of various types. Since we now know how to formulate mechanics problems in a variety of ways, it is fair to say that the basic problem has been solved [1, 2].

The principal purpose of this section is to present and develop the use of certain bond graph forms as standard models in mechanics, thereby bringing the study of this very important class of generally nonlinear problems into the multiport-systems pattern. We shall study mechanical systems involving the large-scale motion of particles and rigid bodies. A variety of techniques are used, including the selection by the modeler of key variables for formulation, the determination of certain required transformations by simple analytic means, and the combination of all the parts into a unified representation by a bond graph.

The development begins with the simplest type of system, involving straightforward coupling between inertial and compliance elements.

9.2.1   The Basic Modeling Procedure

Let us start by considering the result we would like to achieve. For a given problem in mechanics involving particles, rigid bodies, and elastic elements, we wish to obtain a set of first-order differential equations in terms of variables that yield physical insight into system behavior. We anticipate that the equations will be coupled and, if possible, explicit (i.e., one derivative in each equation), having the following form:

$$C\text{-field:} \qquad \mathbf{f}_C = \phi_C(\mathbf{q}_C), \qquad (9.49)$$

$$I\text{-field:} \qquad \mathbf{v}_I = \phi_I(\mathbf{p}_I), \qquad (9.50)$$

$$\text{Junction structure:} \qquad \dot{\mathbf{q}}_C = [\mathbf{T}_{CI}(\mathbf{q}_C)]\,\mathbf{v}_I \qquad (9.51)$$

$$\dot{\mathbf{p}}_I = \left[-\mathbf{T}_{CI}^t(\mathbf{q}_C)\right]\mathbf{f}_C, \qquad (9.52)$$

where $\mathbf{f}_C$ is the set of forces defining the potential co-energy, $\mathbf{q}_C$ is the set of displacements defining the potential energy, $\mathbf{v}_I$ is the set of (inertial) velocities defining the kinetic co-energy; $\mathbf{p}_I$ is the set of momenta defining the kinetic energy, and $\dot{\mathbf{q}}_C$ and $\dot{\mathbf{p}}_I$ are the time derivatives of $\mathbf{q}_C$ and $\mathbf{p}_I$, respectively.

Inspection of the equations indicates that Eq. (9.49) arises directly from the compliance elements' constitutive laws. The set of relations may be linear or nonlinear and coupled or decoupled according to the nature of the C-field. Equation (9.50) is obtainable directly from the inertia elements in the system. The vector $\mathbf{v}_I$ is defined with respect to an inertial frame and represents translational velocities of the centers of mass of rigid bodies and angular velocities with respect to nonrotating coordinate systems.

Typically, the most difficult aspect of mechanics is the generally nonlinear coupling of C-fields and I-fields introduced by the geometry of large-scale motions. Such coupling is represented in Eqs. (9.51) and (9.52) by the transformation array $\mathbf{T}_{CI}(\mathbf{q}_C)$ and its transpose, and in bond graphs by multiport displacement-modulated transformers as discussed in Section 7.5. For now, let us assume that each element of $\mathbf{T}_{CI}$ is a scalar function of all the elements of the vector $\mathbf{q}_C$, in principle.

Before we discuss an example, let us observe that Eqs. (9.49)–(9.52) may be combined to eliminate $\mathbf{f}_C$ and $\mathbf{v}_I$, giving

$$\dot{\mathbf{q}}_C = [\mathbf{T}_{CI}(\mathbf{q}_C)]\,\mathbf{\Phi}_I(\mathbf{p}_I), \tag{9.53}$$

$$\dot{\mathbf{p}}_I = \left[-\mathbf{T}_{CI}^t(\mathbf{q}_C)\right]\mathbf{\Phi}_C(q_C). \tag{9.54}$$

These are the equations of a nonlinear conservative system expressed in terms of state variables $\mathbf{q}_C$ and $\mathbf{p}_I$. They can be modified easily to include nonconservative force effects.

Bond graph models representing the type of system discussed thus far are shown in Figure 9.7. The compliances are represented by the coupled C-field in part a and the "M" C-elements in part b. The inertias are represented by the coupled I-field in part a and the "P" I-elements in part b. The transformation coupling is represented by the M-port $\times$ P-port MTF element, and the 1-junctions are introduced for the sake of clarity and to allow a "through" sign convention for the MTF. Notice that the moduli of the MTF depend on the vector $\mathbf{q}_C$.

As an illustration of the type of formulation in which we are interested, consider the nonlinear oscillator of Figure 9.8a. The mass is free to slide on the rod and is

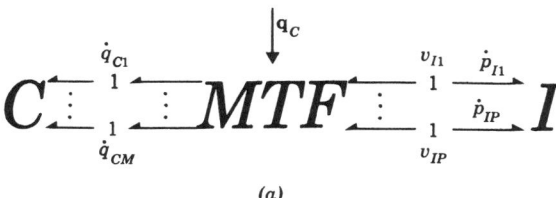

(a)

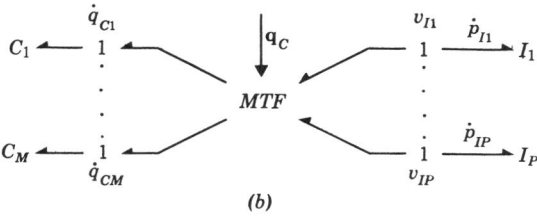

(b)

FIGURE 9.7. Symbolic bond graph model of basic nonlinear conservative system. (a) Coupled C- and I-fields; (b) 1-port C- and I-fields.

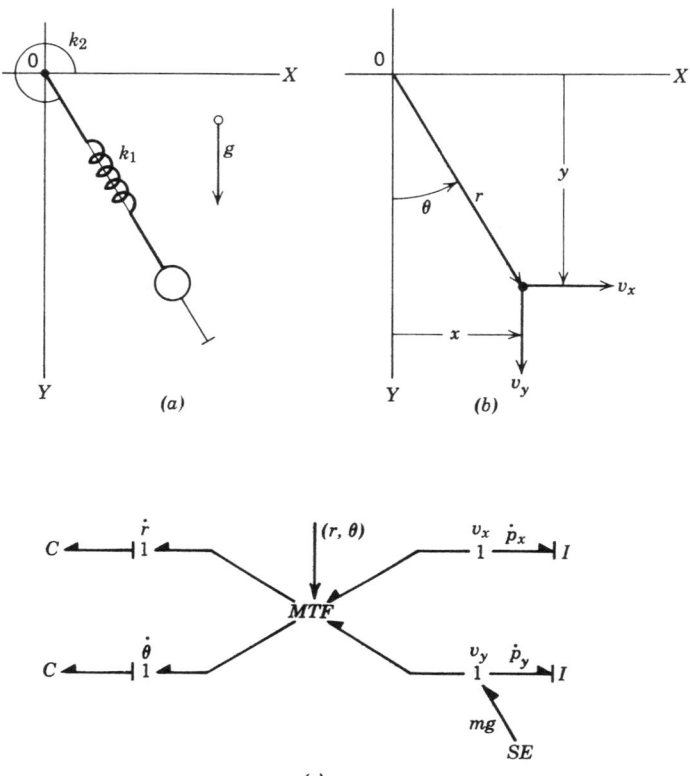

FIGURE 9.8. A nonlinear mechanical oscillator. (*a*) Schematic diagram; (*b*) key geometric quantities defined; (*c*) bond graph model.

constrained by a spring. A second spring constrains the rod as it rotates in the X–Y plane. Key geometric variables are identified in part *b* as r, θ, x, y, v_x, and v_y. Constitutive relations for the linear springs and inertias, in the form of Eqs. (9.49) and (9.50), are given by Eqs. (9.55) and (9.56), respectively:

$$F = k_1(r - R), \tag{9.55a}$$

$$\tau = k_2\theta, \tag{9.55b}$$

$$v_x = m^{-1}p_x, \tag{9.56a}$$

$$v_y = m^{-1}p_y, \tag{9.56b}$$

where R is the free length of the rod spring, F is the rod spring force, τ is the torque of the torsional spring on the rod about an axis through the origin normal to the X–Y plane, and k_1 and k_2 are spring constants. Both the C-field and the I-field are linear and decoupled.

If we identify r and θ as elements of the vector $\mathbf{q}_C$, and v_x and v_y as elements of the vector $\mathbf{v}_I$, the array $\mathbf{T}_{CI}(\mathbf{q}_C)$ may be found by studying the velocity components as shown in Section 7.5,

$$\dot{r} = (\sin\theta)v_x + (\cos\theta)v_y, \tag{9.57a}$$

$$\dot{\theta} = \left(\frac{\cos\theta}{r}\right)v_x + \left(-\frac{\sin\theta}{r}\right)v_y, \tag{9.57b}$$

and

$$\mathbf{T}_{CI}(\mathbf{q}_C) = \begin{bmatrix} \sin\theta & \cos\theta \\ \dfrac{\cos\theta}{r} & -\dfrac{\sin\theta}{r} \end{bmatrix}. \tag{9.58}$$

It remains for us to calculate $\dot{\mathbf{p}}_I$ in terms of the spring forces. By appropriate resolution of the force F and torque effect τ in the directions of X and Y (i.e., p_x and p_y), we obtain

$$\dot{p}_x = (-\sin\theta)F + \left(-\frac{\cos\theta}{r}\right)\tau, \tag{9.59a}$$

$$\dot{p}_y = (-\cos\theta)F + \left(\frac{\sin\theta}{r}\right)\tau + mg. \tag{9.59b}$$

Careful inspection of Eqs. (9.59) shows that indeed the transformation array $\mathbf{T}_{CI}(\mathbf{q}_C)$ is embedded as the negative transpose, namely,

$$-\mathbf{T}_{CI}^{t}(\mathbf{q}_C) = \begin{bmatrix} -\sin\theta & -\dfrac{\cos\theta}{r} \\ -\cos\theta & \dfrac{\sin\theta}{r} \end{bmatrix}. \tag{9.60}$$

The influence of gravity is merely added in the appropriate way (i.e., to directly influence p_y) and is shown in Figure 9.8c as an effort source, *SE*.

At this point, the thoughtful reader might well ask, "If we have obtained $\mathbf{T}_{CI}(\mathbf{q}_C)$ as in Eq. (9.58) once, is it really necessary to obtain it by separate development again, as in Eq. (9.60)?" The answer is no, and a systematic procedure that takes note of this result is the next topic. First we should combine Eqs. (9.57) and (9.59), eliminating F, τ, v_x, and v_y by Eqs. (9.55) and (9.56). The system state equations are

$$\dot{r} = (\sin\theta)m^{-1}p_x + (\cos\theta)m^{-1}p_y, \tag{9.61a}$$

$$\dot{\theta} = \left(\frac{\cos\theta}{r}\right)m^{-1}p_x + \left(-\frac{\sin\theta}{r}\right)m^{-1}p_y, \tag{9.61b}$$

$$\dot{p}_x = (-\sin\theta)k_1(r - R) + \left(-\frac{\cos\theta}{r}\right)k_2\theta, \tag{9.61c}$$

$$\dot{p}_y = (-\cos\theta)k_1(r - R) + \left(\frac{\sin\theta}{r}\right)k_2\theta + mg. \tag{9.61d}$$

A bond graph representation of the complete system is given in Figure 9.8c. The fields are shown explicitly, and the junction-structure transformation is implied by the two-by-two *MTF*. The minus signs in Eqs. (9.59) and (9.60) arise because of the sign conventions at the 1-junctions at the right side of Figure 9.8c.

Definitions of Key Geometric Quantities As the first step in specifying a procedure for constructing bond graph models in mechanics, we must identify key variables. The approach of this chapter is geometric, meaning that we shall use displacement and velocity quantities to organize the system, with forces and momenta constrained by *MTF*s.

Two key vectors have already been described. They are $\mathbf{q}_C$, the vector of displacements that define the potential energy, and $\mathbf{v}_I$, the vector of velocities that define the kinetic co-energy. The vector $\mathbf{q}_C$ is directly associated with the C-field, and the vector $\mathbf{v}_I$ with the I-field.

If it were always possible to relate $\mathbf{v}_I$ to $\dot{\mathbf{q}}_C$ in terms of $\mathbf{q}_C$ as nicely as was done in the previous example, then mechanics problems would not be the *bête noire* they typically are. It is often the case that there are more velocities and displacements in the problem than the coupling constraints permit to be independent. To treat this situation, it will be useful to identify another vector, $\mathbf{q}_k$, the kinematic displacement vector or generalized coordinate vector in the language of Lagrangian mechanics. Basically, the elements of $\mathbf{q}_k$ are necessary and sufficient to fix the configuration of the system at any instant. Therefore the vector $\mathbf{q}_C$ can be found in terms of $\mathbf{q}_k$, and the necessary velocity relations for $\dot{\mathbf{q}}_C$ can be suitably evaluated, as we show next. Furthermore, if the vector $\dot{\mathbf{q}}_k$ represents a necessary and sufficient set of velocities to fix all the motions at any instant, then the system is holonomic.* We shall consider only holonomic systems here, but a slight extension to the development would permit treatment of nonholonomic systems. If $\mathbf{q}_k$ describes all motions, then the vector $\mathbf{v}_I$ must be expressible in terms of $\dot{\mathbf{q}}_k$, given $\mathbf{q}_k$.

Therefore, the modeling process begins with the identification of three key geometric vectors: $\mathbf{q}_k$, the vector of kinematic displacements, or generalized coordinates; $\mathbf{q}_C$, the vector of C-field displacements; and $\mathbf{v}_I$, the vector of I-field velocities.

In Figure 9.8b the kinematic displacement vector could have been chosen to contain (r, θ) or (x, y), or even other pairs, such as (θ, x). Notice that if

$$\mathbf{q}_k = \begin{bmatrix} r \\ \theta \end{bmatrix},$$

then

$$\mathbf{q}_k = \mathbf{q}_C = \begin{bmatrix} r \\ \theta \end{bmatrix},$$

that is, $\mathbf{q}_k$ would be identical to $\mathbf{q}_C$. That would be convenient for many reasons. On

---

*See, for example, Reference [4].

the other hand, if

$$\mathbf{q}_k = \begin{bmatrix} x \\ y \end{bmatrix},$$

then

$$\dot{\mathbf{q}}_k = \begin{bmatrix} \dot{x} \\ \dot{y} \end{bmatrix} = \begin{bmatrix} v_x \\ v_y \end{bmatrix} = \mathbf{v}_I,$$

that is, $\dot{\mathbf{q}}_k$ would be identical to $\mathbf{v}_I$. That would be convenient for many reasons, too. To a significant extent, the wise choice of $\mathbf{q}_k$ to balance the "needs" of $\mathbf{q}_C$ and $\mathbf{v}_I$ can ease the formulation of equations for a complicated system. In this regard, experience is the best teacher. It is an important aspect of the method being presented that a range of choices and some of their implications are made available in clear fashion.

As an example of a problem involving some significant choices, consider the spring–pendulum system shown in Figure 9.9a. The pivot point of the pendulum is constrained to slide on the Y-axis. The pendulum swings in the X–Y plane. A number of geometric quantities are shown in Figure 9.9b.

The configuration of the system can be specified at any instant by the position of the pivot (y_k) and the angle of the pendulum (α). Other choices are possible, such as y_m and α. However, let us use

$$\mathbf{q}_k = \begin{bmatrix} y_k \\ \alpha \end{bmatrix}. \tag{9.62}$$

The potential energy of the system resides in the spring, and y_s is an obvious way to specify it. Hence,

$$\mathbf{q}_C = [y_s]. \tag{9.63}$$

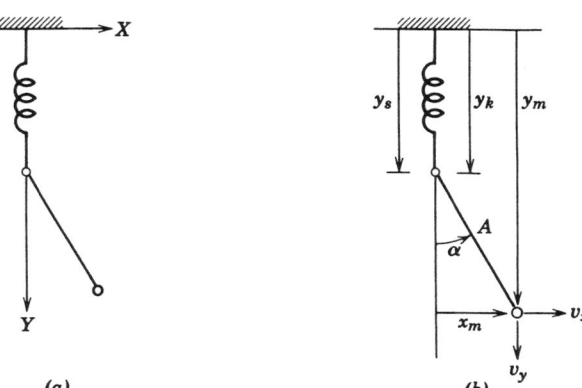

FIGURE 9.9. A mass–spring oscillator example. (a) Schematic diagram; (b) key geometric quantities.

Finally, the kinetic co-energy is associated with the particle (the rod being assumed massless), so

$$\mathbf{v}_I = \begin{bmatrix} v_x \\ v_y \end{bmatrix}. \tag{9.64}$$

A bond graph identifying the variables and the C- and I-fields is shown in Figure 9.10a. It will become standard practice to write a column of 1-junctions for $\dot{\mathbf{q}}_C$, a column for $\dot{\mathbf{q}}_k$, and a column for $\mathbf{v}_I$. The next steps are to develop transformation relations among the vectors and to represent them in the bond graph.

Calculating Velocity Transformations One way to calculate velocity transformations is first to write displacement relations and then to differentiate them [5]. The relations between $\mathbf{q}_k$ and $\mathbf{q}_C$ may be written as

$$\mathbf{q}_C = \phi_{Ck}(\mathbf{q}_k), \tag{9.65}$$

where ϕ_{Ck} is a set of relations defining each element of $\mathbf{q}_C$ in terms of the elements of $\mathbf{q}_k$. By differentiating each relation in Eq. (9.65) with respect to time, we get

$$\dot{\mathbf{q}}_C = \frac{\partial \phi_{Ck}(\mathbf{q}_k)}{\partial \mathbf{q}_k} \dot{\mathbf{q}}_k = \mathbf{T}_{Ck}(\mathbf{q}_k)\dot{\mathbf{q}}_k. \tag{9.66}$$

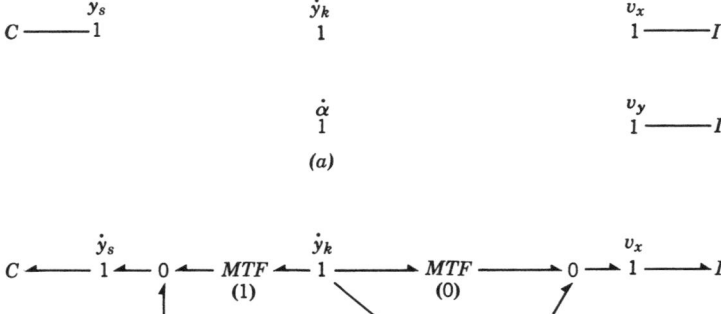

FIGURE 9.10. Explicit bond graph model for the mass–spring oscillator. (a) Key geometric variable sets; (b) insertion of the transformations T_{Ck} and T_{Ik}.

The *i*th relation is

$$\dot{q}_{Ci} = \sum_{j=1}^{N} \frac{\partial \phi_{Cki}}{\partial q_{kj}} \dot{q}_{kj}, \tag{9.67}$$

where there are N displacements in $\mathbf{q}_k$.

For example, in the pendulum problem of Figure 9.9, $\mathbf{q}_C$ is related to $\mathbf{q}_k$ as follows, where the vectors are given by Eqs. (9.62) and (9.63):

$$y_s = 1 y_k + 0\alpha; \tag{9.68}$$

then

$$\dot{y}_s = 1 \dot{y}_k + 0\dot{\alpha} \tag{9.69}$$

and

$$\mathbf{T}_{Ck}(q_k) = [\,1 \quad 0\,]. \tag{9.70}$$

In this case, $\mathbf{T}_{Ck}$ does not depend on $\mathbf{q}_k$ explicitly.

As the next step, we express the inertia-velocity vector $\mathbf{v}_I$ in terms of $\mathbf{q}_k$ and $\dot{\mathbf{q}}_k$, thereby obtaining $\mathbf{T}_{Ik}(\mathbf{q}_k)$. Generally it is easiest to do this by first creating an inertia-displacement vector $\mathbf{q}_I$, chosen so that

$$\dot{\mathbf{q}}_I = \mathbf{v}_I. \tag{9.71}$$

Then, we may write

$$q_I = \phi_{Ik}(\mathbf{q}_k), \tag{9.72}$$

from which we can calculate $\mathbf{v}_I$ as

$$\mathbf{v}_I = \dot{\mathbf{q}}_I = \left(\frac{\partial \phi_{Ik}}{\partial \mathbf{q}_k} \right) \dot{\mathbf{q}}_k = [\mathbf{T}_{Ik}(\mathbf{q}_k)]\,\dot{\mathbf{q}}_k. \tag{9.73}$$

The *i*th relation is

$$v_{Ii} = \sum_{j=1}^{N} \left(\frac{\partial \phi_{Iki}}{\partial q_{kj}} \right) \dot{q}_{kj}, \tag{9.74}$$

where there are N elements in $\mathbf{q}_k$.

In the pendulum example, we may write

$$\mathbf{q}_I = \begin{bmatrix} x \\ y \end{bmatrix} \quad \text{and} \quad \dot{\mathbf{q}}_I = \mathbf{v}_I = \begin{bmatrix} \dot{x} \\ \dot{y} \end{bmatrix} = \begin{bmatrix} v_x \\ v_y \end{bmatrix},$$

so $\mathbf{q}_I$ is related to $\mathbf{q}_k$ by

$$x = A \sin \alpha, \tag{9.75a}$$

$$y = y_k + A \cos \alpha. \tag{9.75b}$$

From Eqs. (9.75) we obtain T_{Ik} by differentiating, namely,

$$v_x = \dot{x} = (0)\dot{y}_k + (A\cos\alpha)\dot{\alpha}, \tag{9.76a}$$

$$v_y = \dot{y} = (1)\dot{y}_k + (-A\sin\alpha)\dot{\alpha}, \tag{9.76b}$$

and

$$\mathbf{T}_{Ik}(\mathbf{q}_k) = \begin{bmatrix} 0 & A\cos\alpha \\ 1 & -A\sin\alpha \end{bmatrix}. \tag{9.77}$$

In this case, the transformation depends on one element of $\mathbf{q}_k$, that is, α.

Establishing the Junction Structure   Since the basic pattern of both the transformations $\mathbf{T}_{Ck}$ and $\mathbf{T}_{Ik}$ is that the elements of $\dot{\mathbf{q}}_k$ are multiplied by constants or functions of $\mathbf{q}_k$ and added together, we may represent these transformations by 0-junctions and 2-port *MTF*s. The *MTF*s are used to multiply by the appropriate function, and the 0-junctions are used to add up velocity terms, as was discussed in Section 7.5.

Referring to the partial bond graph of Figure 9.10a, $\mathbf{T}_{Ck}$ and $\mathbf{T}_{Ik}$ of Eqs. (9.70) and (9.77) are introduced in the graph of Figure 9.10b between the appropriate sets of 1-junctions. We now have a basic bond graph model of the spring–pendulum system of Figure 9.9a.

Now consider that we wish to add two more dynamic effects—dissipation in the pivot joint and a gravity force. The dissipation torque arises directly from the relative motion $\dot{\alpha}$ and is shown by the R-element appended to that junction in Figure 9.11. The gravity force acts on v_y and is appended directly to the v_y-junction as a constant source of force of magnitude mg. We observe that the fields C and R can be linear or nonlinear. The junction structure is often nonlinear due to the geometric coupling represented by *MTF* moduli. In practice, the difficulty of generating state equations depends somewhat upon the nature of the nonlinearities [6].

The bond graph model of Figure 9.11 has been reproduced in Figure 9.12a with *MTF*s having zero moduli eliminated. In part b, additional simplifications have been made, and causality has been added. In this case, we have been lucky and there are

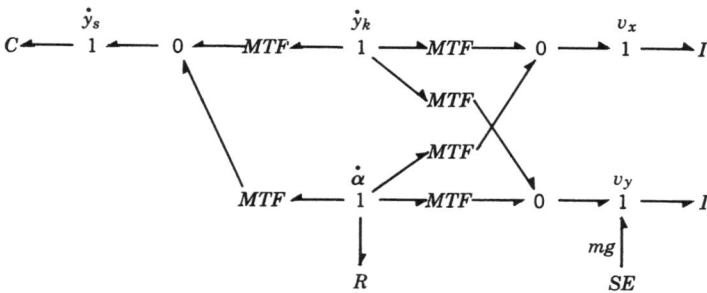

FIGURE 9.11. The mass–spring oscillator of Figure 9.4 with dissipation and gravity effects added.

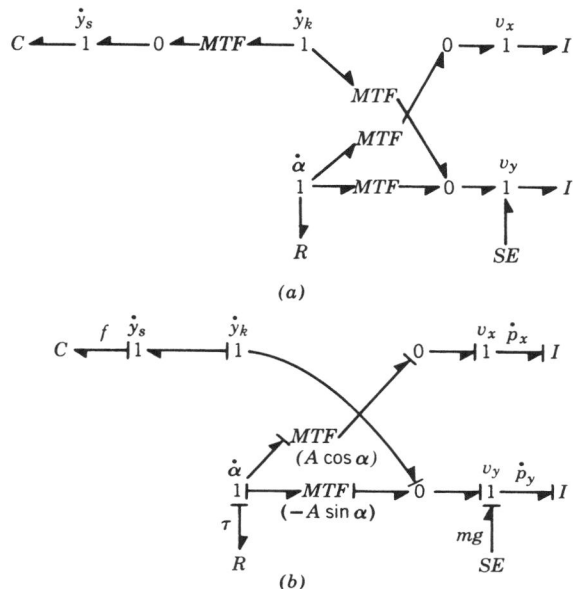

FIGURE 9.12. Simplification of the oscillator bond graph. (*a*) *MTF*s with zero modulus removed; (*b*) further simplification of the graph.

no algebraic loops or cases of derivative causality. The one peculiarity is that α is not a state variable in the C-field and yet is necessary because of the MTFs. The causal strokes show that $\dot{\alpha}$ can be found from p_x by inverting Eq. (9.76a). The C-, I-, and R-field relations are

$$f = k(y_s - Y) \qquad \text{(spring)}, \tag{9.78a}$$

$$v_x = m^{-1} p_x \qquad \text{(mass)}, \tag{9.78b}$$

$$v_y = m^{-1} p_y \qquad \text{(mass)}, \tag{9.78c}$$

$$\tau = R\dot{\alpha} \qquad \text{(dissipation)}, \tag{9.78d}$$

respectively, where Y is the free length of the spring in Eq. (9.78a), m is the mass in Eqs. (9.78b,c), and R of Eq. (9.78d) is the viscous dissipation parameter in the joint.

The connection or junction-structure relations are, in initial form,

$$\dot{p}_x = (-\tan \alpha) f - \frac{R}{A \cos \alpha} \dot{\alpha}, \tag{9.79a}$$

$$\dot{p}_y = -f + mg, \tag{9.79b}$$

$$\dot{y}_k = (\tan \alpha) v_x + v_y, \tag{9.79c}$$

$$\dot{\alpha} = \frac{1}{A \cos a} v_x. \tag{9.79d}$$

If we use Eqs. (9.78) to eliminate f, τ, v_x, and v_y from Eqs. (9.79), we get a momentum–displacement set of state equations.

$$\dot{p}_x = (-\tan\alpha)k(y_k - Y) - \frac{Rm^{-1}}{(A\cos\alpha)^2}p_x, \tag{9.80a}$$

$$\dot{p}_y = -k(y_k - Y) + mg, \tag{9.80b}$$

$$\dot{y}_k = (\tan\alpha)m^{-1}p_x + m^{-1}p_y, \tag{9.80c}$$

$$\dot{\alpha} = \frac{1}{A\cos\alpha}m^{-1}p_x. \tag{9.80d}$$

One could eliminate p_x in terms of v_x and v_y, to get a velocity–displacement set of equations,

$$\dot{v}_x = (-\tan\alpha)\frac{k}{m}(y_k - Y) - \frac{R}{m(A\cos\alpha)^2}v_x, \tag{9.81a}$$

$$\dot{v}_y = -\frac{k}{m}(y_k - Y) + g, \tag{9.81b}$$

$$\dot{y}_k = (\tan\alpha)v_x + v_y, \tag{9.81c}$$

$$\dot{\alpha} = \left(\frac{1}{A\cos\alpha}\right)v_x. \tag{9.81d}$$

Summary of the Procedure   The procedure for obtaining a bond graph model may be summarized in the following series of steps.

1. Identify the key vectors, $\mathbf{q}_k$, $\mathbf{q}_C$, and $\mathbf{v}_I$ (or $\mathbf{q}_I$). Write the 1-juctions corresponding to $\dot{\mathbf{q}}_k$, $\dot{\mathbf{q}}_C$, and $\mathbf{v}_I$.
2. Obtain the displacement transformation relating $\mathbf{q}_C$ to $\mathbf{q}_k$. Differentiate with respect to time to obtain a velocity transformation relating $\dot{\mathbf{q}}_C$ to $\dot{\mathbf{q}}_k$ in terms of $\mathbf{q}_k$. Then

$$\dot{\mathbf{q}}_C = \mathbf{T}_{Ck}(\mathbf{q}_k)\dot{\mathbf{q}}_k.$$

Write the results into the bond graph using *MTF* and 0-junction elements.
3. Obtain the velocity transformation relating $\mathbf{v}_I$ to $\dot{\mathbf{q}}_k$. Do this directly, or by relating $\mathbf{q}_I$ to $\mathbf{q}_k$ and differentiating with respect to time (being sure that $\dot{\mathbf{q}}_I = \mathbf{v}_I$). Then

$$\mathbf{v}_I = \mathbf{T}_{Ik}(\mathbf{q}_k)\dot{\mathbf{q}}_k.$$

Write the results into the bond graph using *MTF* and 0-junction elements.
4. Append the C-field and I-field elements to the $\dot{\mathbf{q}}_C$ and $\mathbf{v}_I$ junctions as indicated.

5. Append dissipation effects, force sources, and geometric constraints not yet included, as appropriate. If necessary construct additional velocities, using the transformational method.

6. Simplify the bond graph by removing *MTF*s with zero modulus, making *MTF*s with unit modulus into direct bonds, and combining 2-port junctions, when power directions permit.

The procedure outlined above is often successful at deriving a bond graph that will yield state equations fairly easily, but it also can lead to a bond graph with algebraic loops or derivative causality. For example, a little experimentation with the general bond graph for the example shown in Figure 9.11 will reveal that when integral causality is applied using the sequential causal assignment procedure, arbitrary causality must be assigned to some *MTF* bonds. This indicates an algebraic loop problem. This did not happen in Figure 9.12 because two zero-modulus *MTF*s could be eliminated.

A particularly vexing problem occurs whenever two rigid bodies are constrained in a nonflexible way. The constraint might be enforced by a pin joint between the bodies or the bodies might be forced to slide on one another in some way, for example. In any case, the degrees of freedom of the two bodies are reduced by the constraints, and this always leads to derivative causality which can be particularly difficult because geometric nonlinearities are typically involved.

In three dimensions, a rigid body has six degrees of freedom represented by the six inertial elements in Figure 9.2. When two bodies are connected rigidly, the three linear and three angular velocities for the two bodies cannot all be independent, meaning that not all the *I*-elements can have integral causality. In the simpler case of plane motion, rigid bodies have only three degrees of freedom, but again whenever two bodies are linked, derivative causality will arise. Thus, it is important to realize that it is not enough to be able to assemble a correct bond graph for a mechanical system, but it is also important to have a scheme for drawing useful results from any bond graph model.

The field of multibody systems is concerned with exactly the problem of describing interconnected rigid bodies, and a variety of formulation and analysis techniques have been developed to treat these systems. Here we will just illustrate a variety of techniques which are particularly applicable to bond graph models of systems which contain constrained rigid bodies as subsystems. For more details on some of these techniques, see References [12–17].

9.2.2 Multibody Systems

The description and analysis of multibody systems have been the subject of research for a long time. When a number of rigid bodies are constrained by different types of joints and yet move through large angles in three dimensions, the kinematic relationships can become very complex. Specialists in multibody systems have developed a number of techniques for handling this type of mechanical dynamic system, and it would be presumptuous to suggest that bond graphs would always be a better choice as a descriptive mechanism than the alternative established techniques. Although bond graphs certainly can be used to study purely mechanical multibody systems (see Reference [17], for example), it is probably more logical to use a bond graph approach when the system to be studied contains other components which are particularly well treated using bond graphs.

As an elementary example, consider the system of Figure 9.13, which consists of an electric motor which moves a cart through a pulley and belt arrangement on which an upside-down pendulum is mounted. The object is to design a control system which varies the voltage on the motor in such a way that the pendulum is stabilized in the upright position at any desired cart position x.

As shown in the figure, the bond graph representation of the drive system is straightforward, but the rigid constraint between the cart and the bar will certainly cause derivative causality problems which are complicated if the angle ϕ_1 is not restricted to be small. This system, neglecting any details of the drive, was one example used in Reference [12] to argue that the so-called descriptor form for multibody equations is particularly good at making the multibody system structure evident.

We will follow the development of Reference [14] to show that one bond graph approach to multibody systems is closely related to the multibody system approach leading to equations in the descriptor form. The first step is to consider bond graph elements representing the rigid-body dynamics without any consideration of the constraints among the bodies. The coordinates x for the cart and x_1, y_1, and ϕ_1 for the bar are defined with respect to an inertial frame. Thus, the four I-elements at the right

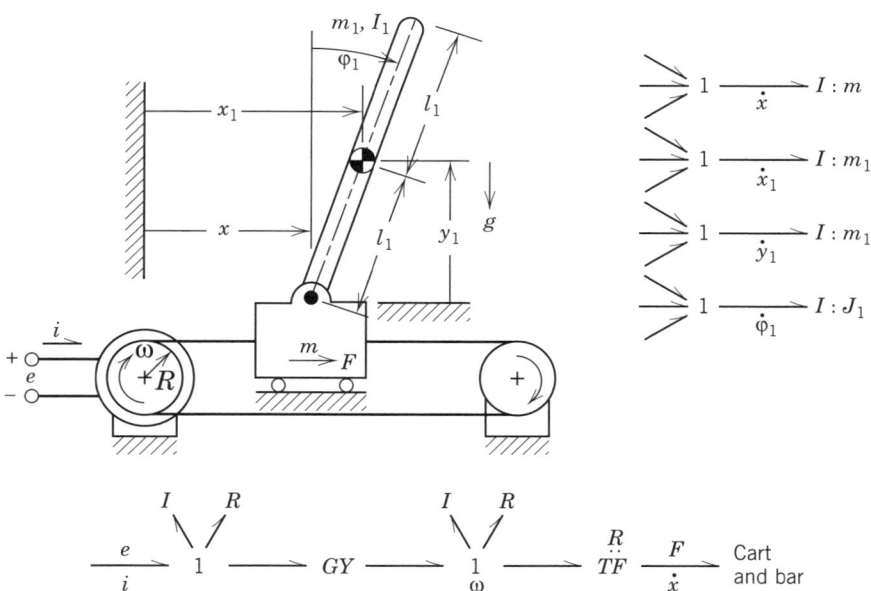

FIGURE 9.13. Control system experiment, including an upside-down pendulum mounted on a massive cart.

of Figure 9.13 properly represent the laws which state that the sums of all applied forces equal the rate of change of the linear momenta and that the velocity components of the centers of mass are the momenta divided by the masses. Also, the sum of the moments about the center of mass of the bar is equal to the rate of change of the angular momentum. For this plane motion case, the angular velocity divided by the centroidal moment of inertia gives the angular velocity of the bar. (As noted previously, the bond graph representing three-dimensional motion of unconstrained rigid bodies is considerably more complicated.)

In order to constrain the four I-elements in Figure 9.13 and thus to create a bond graph for the bar and cart subsystem to be attached to the drive train bond graph, we first define gaps δ_1 and δ_2 in Figure 9.14. These horizontal and vertical gaps between the bar pivot and the cart pivot should vanish. Furthermore, forces across the gaps λ_1 and λ_2 will be necessary in order to assure that δ_1 and δ_2 do indeed vanish. The use of λ to represent forces associated with constraints is intended to remind one that in the multibody formalism of Reference [12], these gap forces are Lagrange multipliers.

The gaps are easily related to the position coordinates of the rigid bodies in Figure 9.14:

$$\delta_1 = x_1 - x - \ell_1 \sin \phi_1, \qquad \delta_2 = y_1 - \ell_1 \cos \phi_1. \qquad (9.82)$$

By differentiating the gap expressions, one can relate the gap velocities to the body velocities:

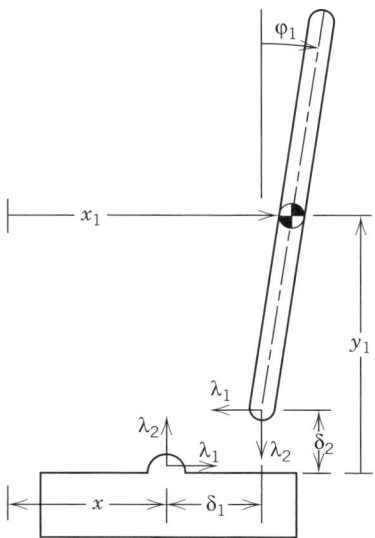

FIGURE 9.14. The bar and cart of Figure 9.13 showing gaps, δ_1 and δ_2, which should vanish, and the associated forces, λ_1 and λ_2.

$$
\begin{bmatrix} \dot\delta_1 \\ \dot\delta_2 \end{bmatrix} = \left[\begin{array}{cc|c|c} -1 & 1 & 0 & -\ell_1\cos\phi_1 \\ \hline 0 & 0 & 1 & \ell_1\sin\phi_1 \end{array} \right] \begin{bmatrix} \dot x \\ \dot x_1 \\ \dot y_1 \\ \dot\phi_1 \end{bmatrix}.
\tag{9.83}
$$

This relationship can be represented in bond graph terms by a multiport displacement-modulated transformer, as was discussed in Chapter 7. Figure 9.15 shows the transformer attached to the four I-elements representing the two unconstrained bodies.

The transpose of the matrix in Eq. (9.83) then automatically relates the gap forces λ_1 and λ_2 to effective forces at the four I-elements:

$$
\begin{bmatrix} F_x \\ F_{x_1} \\ F_{y_1} \\ \tau_1 \end{bmatrix} = \left[\begin{array}{c|c} -1 & 0 \\ \hline 1 & 0 \\ \hline 0 & 1 \\ \hline -\ell_1\cos\phi_1 & \ell_1\sin\phi_1 \end{array} \right] \begin{bmatrix} \lambda_1 \\ \lambda_2 \end{bmatrix}.
\tag{9.84}
$$

In Figure 9.15, the drive system force F and the gravity force $m_1 g$ have also been applied. Following the causal marks, some dynamic equations can be written:

$$
\begin{aligned}
\dot p_x &= F - F_x = F + \lambda_1, & \dot x &= p_x/m; \\
\dot p_{x_1} &= -F_{x_1} = -\lambda_1, & \dot x_1 &= p_{x_1}/m_1; \\
\dot p_{y_1} &= -m_1 g - F_{y_1} = -m_1 g - \lambda_2, & \dot y_1 &= p_{y_1}/m_1; \\
\dot p_{\phi_1} &= -\tau_1 = (\ell_1\cos\phi_1)\lambda_1 - (\ell_1\sin\phi_1)\lambda_2, & \dot\phi_1 &= p_{\phi_1}/J_1.
\end{aligned}
\tag{9.85}
$$

The multibody descriptor form for the mechanical subsystem is just Eq. (9.85) slightly rewritten:

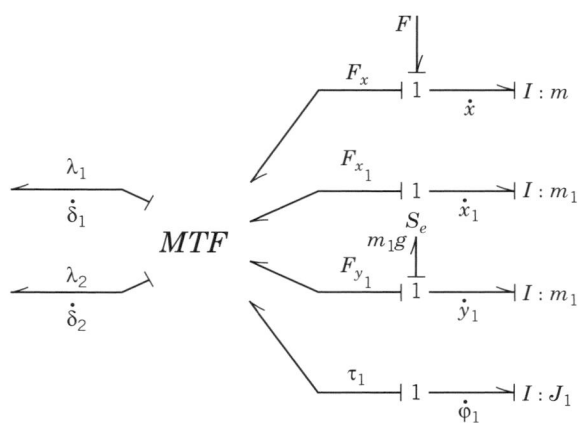

FIGURE 9.15. Bond graph relating gap velocities to body velocities.

$$m\ddot{x} = F + \lambda_1,$$
$$m_1\ddot{x}_1 = -\lambda_1,$$
$$m_1\ddot{y}_1 = -m_1 g - \lambda_2,$$
$$J_1\ddot{\phi}_1 = (\ell_1 \cos\phi_1)\lambda_1 - (\ell_1 \sin\phi_1)\lambda_2.$$

(9.86)

Equations (9.85) or (9.86) are straightforward except that the gap forces λ_1 and λ_2 are unknown. In the conventional multibody formulation, the extra information is supplied by requiring δ_1 and δ_2 to vanish in Eq. (9.82).

Although this formulation is mathematically complete, it represents a differential-algebraic set of equations, and such systems are not particularly easy to solve numerically. The bond graph of Figure 9.15 suggests an alternative approach. See References [16, 17].

Suppose that instead of requiring that δ_1 and δ_2 vanish exactly, we modeled the pivot interaction between the two bodies as if relatively stiff springs existed which would generate the λ forces whenever the gaps were not zero. In one sense, this may seem to be more realistic, since no truly rigid bodies exist. However, if realistic bearing stiffnesses were to be used, one would in all probability create a model with very high vibrational natural frequencies, which would necessitate short time steps in any numerical simulation. It is better philosophically to consider springs and dampers inserted across gaps as artificial devices to enforce constraints approximately and to do this with explicit differential equations instead of differential-algebraic equations. The idea is to experiment with spring stiffness to find the lowest stiffnesses which yield sufficiently small values of δ_1 and δ_2 in actual system operation. This will yield the longest possible time steps in simulation and the shortest simulation times.

The complete system bond graph for the system of Figure 9.13 is shown in Figure 9.16. In essence, the bond graph of the drive system shown in Figure 9.13 has been combined with the multibody bond graph of Figure 9.15 with the addition of R- and C-elements to generate the constraint forces λ_1 and λ_2 as a function of δ_1, δ_2, $\dot{\delta}_1$, and $\dot{\delta}_2$. Increasing the stiffness of the C-elements will have the effect of producing larger forces when the gaps are nonzero and thus reducing the gaps. Since the introduction of springs implies that vibrational motion is possible, it is useful also to introduce artificial R-elements to provide damping. Again, some experimentation with damping parameter values will be necessary to achieve a reasonable result.

The observant reader may have noticed that where the drive train bond graph of Figure 9.13 is attached to the multibody bond graph of Figure 9.15, an extra C-element has been inserted. The reason is that if the two bond graph fragments are simply joined, the motor rotary inertia and the cart inertia are coupled by a transformer relating rotary to linear motion variables. This is another instance of rigid coupling between inertia elements that always induces derivative causality. In this case, it would be simple to eliminate this inconvenience by defining a single equivalent rotary or linear inertia element or to define an I-field using the methods of Chapter 7. Instead, this rigid constraint has been replaced with an approximate one by defining an equivalent C-element. This element could be considered to represent

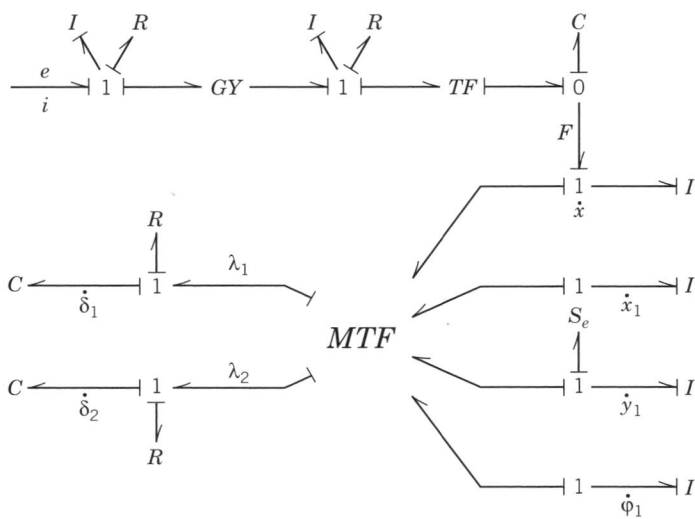

FIGURE 9.16. Bond graph for the system of Figure 9.13.

the belt drive flexibility or simply a device for making the entire model capable of having integral causality.

Finally, one may note that it would be easy to write down the state equations for all the I- and C-elements in Figure 9.16. One must remember, however, that in this multibody technique, the displacement-modulated transformer uses the displacements of inertia elements to express the constraint or gap equations. In bond graphs, displacement variables for inertia elements are not normally necessary as state variables so some extra state equations will be necessary.

The transformer matrices for the example shown in Eqs. (9.83) and (9.84) actually need one displacement, ϕ_1, so strictly speaking only the very last equation in Eq. (9.85) would need to be included. For this type of system, it would probably be useful to include state equations for all displacements x, x_1, y_1, and ϕ_1 in any case, as indicated in Eq. (9.85). These equations do increase the order of the system, but their inclusion has only a minor effect on the simulation time for a computer study of the system since three of the four displacements are not coupled to the remainder of the dynamic system.

To illustrate how this method of enforcing constraints among interconnected bodies can be extended, we briefly consider a second bar attached to the system already studied, as shown in Figure 9.17. By defining gaps δ_3 and δ_4 and corresponding forces λ_3 and λ_4 pertaining to the second set of constraints relating the position of the top of bar 1 to the bottom of bar 2, we can derive two more relations to supplement Eq. (9.82) which still applies:

$$\delta_3 = x_2 - \ell_2 \sin \phi_2 - x_1 - \ell_1 \sin \phi_1,$$
$$\delta_4 = y_2 - \ell_2 \cos \phi_2 - y_1 - \ell_1 \cos \phi_1. \qquad (9.87)$$

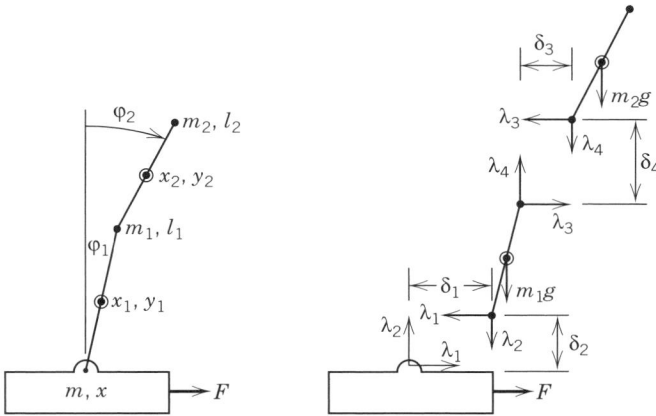

FIGURE 9.17. Double upside-down pendulum system.

A time differentiation of Eq. (9.87) results in equations relating $\dot\delta_3$ and $\dot\delta_4$ to $\dot x_1$, $\dot y_1$, $\dot\phi_1$, $\dot x_2$, $\dot y_2$, and $\dot\phi_2$, analogous to Eqs. (9.83). The transpose of the matrix relates six force components to λ_3 and λ_4, and both sets of relations are embodied in a 2×6 port modulated transformer, as shown in Figure 9.18. In this case, both bars have three degrees of freedom, so six generalized velocities are involved when the two bodies are constrained.

Again, there is the possibility of relating λ_3 and λ_4 to forces acting on the I-elements and requiring δ_3 and δ_4 to vanish in Eq. (9.87), which completes the multibody formulation in descriptor form. Figure 9.18 shows the alternative of using spring and damping elements to enforce the constraint in an approximate manner.

The idea of using artificial C- and R-elements to enforce constraints and thus to avoid derivative causality or differential-algebraic equations may appear to be a "brute-force" approach. This may be true, but first, a brute-force approach which is effective should not be discounted and, second, it has been argued that this approach is in many cases superior to the alternatives [17].

9.2.3 Lagrangian or Hamiltonian *IC*-Field Representations

In Section 9.2.1, a procedure for constructing a bond graph for mechanical systems which could be described by generalized coordinates was presented. An example was worked out and illustrated in Figures 9.9–9.11. In this case, using normal bond graph causality, it was found that explicit differential equations could be written fairly easily despite geometric nonlinearities in the kinematic relationships involved. In many cases, however, the resulting bond graph, while formally correct, is intractable due to algebraic loops and derivative causality. Lagrange's equations are particularly useful for such situations, so in this section, it will be shown how a mechanical subsystem can be represented by a special type of *IC*-field, the equations of which are

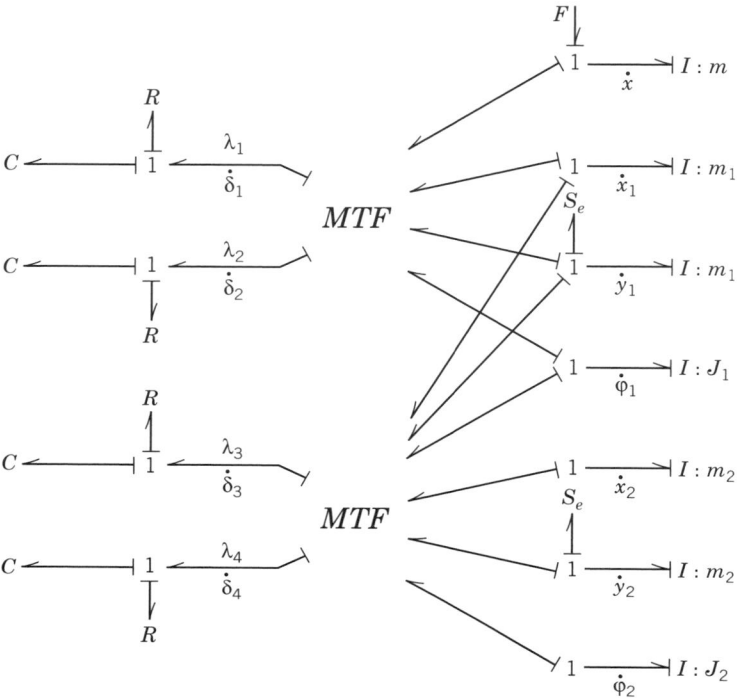

FIGURE 9.18. Bond graph for the double upside-down pendulum system.

Lagrange's equations in a Hamiltonian form. This often is the simplest way to derive equations for a nonlinear mechanical subsystem.

Consider a holonomic system which can be described by a vector of generalized coordinates, $\mathbf{q}_k$, as described in Section 9.2.1. The kinetic energy T of all the particles and rigid bodies in the system can then be written as a function of the variables in $\mathbf{q}_k$ and their rates of change, $\dot{\mathbf{q}}_k$. Then the standard form for Lagrange's equation is

$$\frac{d}{dt}\frac{\partial T}{\partial \dot{q}_i} - \frac{\partial T}{\partial q_i} = E_i, \qquad (9.88)$$

where q_i is the ith generalized coordinate in $\mathbf{q}_k$ (the subscript k has been dropped for simplicity since will be dealing only with the "kinematic" generalized coordinate displacements in this section); E_i includes all generalized forces for the ith coordinate, including those which could be derived from a potential energy function.

Equation (9.88) leads to coupled second-order equations, but a simple change leads to twice as many first-order equations. This Hamiltonian form not only is more useful for computation but also is more compatible with the bond graph formalism.

First, define generalized momenta p_i corresponding to the displacements q_i by the expressions

$$p_i \equiv \frac{\partial T}{\partial \dot{q}_i}. \tag{9.89}$$

Then each equation in Eq. (9.88) may be rewritten in momentum form:

$$\dot{p}_i = \frac{\partial T}{\partial q_i} + E_i$$

or

$$\dot{p}_i = e'_i + E_i, \qquad e'_i \equiv \frac{\partial T}{\partial q_i}. \tag{9.90}$$

State equations for the q_i come by inverting Eq. (9.89) to solve for the $\dot{q}_i$ as functions of the p_i and q_i. As shown in Reference [15], Eq. (9.89) always has a special form. If $\mathbf{p}_k$ is the vector of generalized momenta, Eq. (9.89) can be represented thus:

$$\mathbf{p}_k = \mathbf{M}(\mathbf{q}_k, t)\dot{\mathbf{q}}_k + \mathbf{a}(q_k, t), \tag{9.91}$$

where $\mathbf{M}$ is a symmetric matrix and the vector $\mathbf{a}$ only occurs if the system includes time-varying velocity sources. Then the state equations for the q_i are formally

$$\dot{\mathbf{q}}_k = \mathbf{M}^{-1}(\mathbf{q}_k, t)\,[\mathbf{p}_k - \mathbf{a}(\mathbf{q}_k, t)]. \tag{9.92}$$

In the worst case, the inversion of the "mass" matrix $\mathbf{M}$ may have to be done repeatedly as a simulation proceeds, but in some cases it can be done only once—if $\mathbf{M}$ is a constant—or analytically if $\mathbf{M}$ is not completely coupled.

Note that Eqs. (9.90) and (9.92) are just generalizations of the type of equations we normally obtain for systems including I- and C-elements when bond graphs are used.

Equations (9.90) and (9.92) are elegantly summed up in the bond graph of Figure 9.19, which could be used to represent a complex mechanical subsystem connected to other elements readily described in bond graph terms. Of course, the e'_i force terms could even include forces derivable from potential energy, V, if e'_i in Eq. (9.90) were extended thus:

$$e'_i = \frac{\partial T}{\partial q_i} - \frac{\partial V}{\partial q_i}. \tag{9.93}$$

This would leave only nonpotential forces in the E_i. Often it is easy to put forces from elastic elements or gravity in the bond graph itself, and this may show system structure better than representing everything in the IC equations.

Although the equations for the IC-field may appear complex, one must remember that when many inertial elements are highly constrained, only a few degrees of freedom may be necessary to describe a subsystem which would otherwise have many

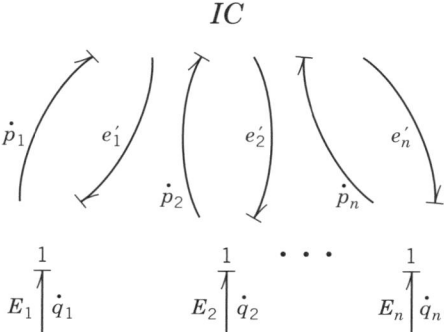

FIGURE 9.19. An *IC*-field representation of the Hamiltonian form of Lagrange's equations.

cases of derivative causality. Furthermore, many cases of derivative causality are only local in nature so that a few *IC*-fields in combination with a number of elements in the bond graph with no causal problems may result in a useful model of a large and complex system.

An Example System The system sketched in Figure 9.20 is elementary but yet shows a number of the difficulties associated with inertial elements in mechanics. It is not hard to create a bond graph for this system by considering the *x*, *y*, and *z* coordinates of the particle and the corresponding inertial space velocities $\dot{x}$, $\dot{y}$, and $\dot{z}$ as the flows on three simple *I*-elements. The torsion spring and a rotary inertia both have $\dot{\alpha}$ as their flow variable, and friction at the pendulum pivot has $\dot{\theta}$ as its flow variable.

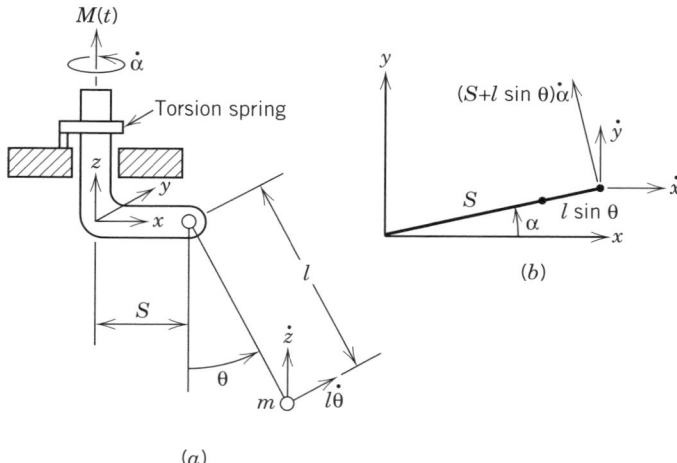

FIGURE 9.20. Example system. (*a*) Three dimensional view; (*b*) view from above.

An obvious way to enforce the geometric constraints in the problem is to consider α and θ as generalized coordinates which then determine x, y, and z:

$$x = (S + \ell \sin \theta) \cos \alpha,$$
$$y = (S + \ell \sin \theta) \sin \alpha,$$
$$z = -\ell \cos \theta. \tag{9.94}$$

Notice that time is not explicitly involved so that the **a** in Eqs. (9.91) and (9.92) will be missing.

The velocities $\dot{x}$, $\dot{y}$, and $\dot{z}$ are found by differentiating Eq. (9.94):

$$\begin{bmatrix} \dot{x} \\ \dot{y} \\ \dot{z} \end{bmatrix} = \begin{bmatrix} -(S + \ell \sin \theta) \sin \alpha & \ell \cos \theta \cos \alpha \\ (S + \ell \sin \theta) \cos \alpha & \ell \cos \theta \sin \alpha \\ 0 & \ell \sin \theta \end{bmatrix} \begin{bmatrix} \dot{\alpha} \\ \dot{\theta} \end{bmatrix}. \tag{9.95}$$

Using these relationships, a bond graph can be assembled as shown in Figure 9.21. Either 2-port *MTF*s with 0- and 1-junctions can be used or a multiport *MTF* as shown with the 3×2 matrix in Eq. (9.95) as its transformation matrix. This bond graph is not easy to use, however, because if two *I*-elements are put into integral causality, the remaining *I*-elements will be forced into derivative causality. In the causality shown in Figure 9.21, $\dot{\alpha}$ and $\dot{x}$ are determined by *I*-elements in integral causality. By a manipulation of Eq. (9.95), $\dot{y}$ and $\dot{z}$ could be related to $\dot{\alpha}$ and $\dot{x}$.

To develop an *IC*-field representation using α and θ as generalized coordinates, we first write the kinetic energy either with the help of Eq. (9.95) or, more simply, directly from the sketch of Figure 9.20:

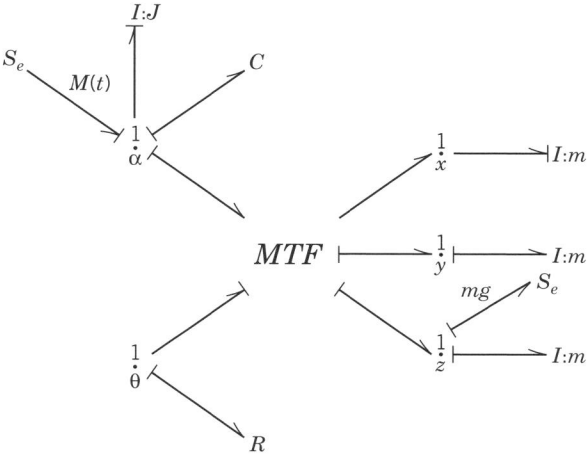

FIGURE 9.21. Bond graph for the system of Figure 9.20.

$$T = \frac{1}{2}m\left\{[(S + \ell \sin\theta)\dot{\alpha}]^2 + (\ell\dot{\theta})^2\right\} + \frac{1}{2}J\dot{\alpha}^2, \tag{9.96}$$

where J is the moment of inertia about the z-axis.

The generalized momenta are

$$p_\alpha = \frac{\partial T}{\partial \dot{\alpha}} = m(S + \ell \sin\theta)^2\dot{\alpha} + J\dot{\alpha}, \tag{9.97}$$

$$p_\theta = \frac{\partial t}{\partial \dot{\theta}} = m\ell^2\dot{\theta}. \tag{9.98}$$

In this case, the inverted forms corresponding to Eq. (9.92) are simply

$$\dot{\alpha} = \frac{p_\alpha}{m(S + \ell \sin\theta)^2 + J}, \tag{9.99}$$

$$\dot{\theta} = \frac{p_\theta}{m\ell^2}. \tag{9.100}$$

The remaining equations of motion are

$$\dot{p}_\alpha = \frac{\partial T}{\partial \alpha} + E_\alpha = E_\alpha, \tag{9.101}$$

$$\dot{p}_\theta = \frac{\partial T}{\partial \theta} + E_\theta = m\dot{\alpha}^2(S + \ell \sin\theta)\ell \cos\theta + E_\theta. \tag{9.102}$$

Note that Eqs. (9.99)–(9.102) are essentially explicit first-order equations suitable for machine solution. One could use Eq. (9.99) to eliminate $\dot{\alpha}$ in Eq. (9.102) in favor of the state variables p_α and θ, but if the equations are integrated in the order given, this step is not really necessary in a computational sense since the numerical value of $\dot{\alpha}$ will be known before it is needed in Eq. (9.102).

Figure 9.22 shows a bond graph for the system in which an IC-field represents the mechanical subsystem and the S_e-, C-, and R-elements supply the generalized

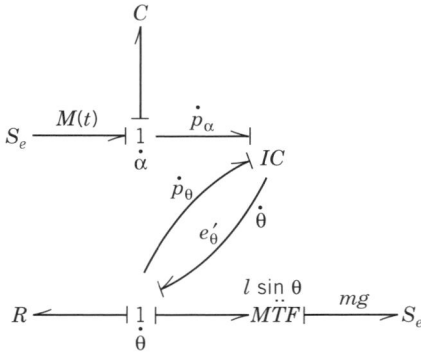

FIGURE 9.22. An IC-field representation for the example system of Figure 9.20.

forces (torques) E_α and E_θ. Note that the gravity force acts through an *MTF* which in Figure 9.21 is hidden in the multiport *MTF*. It is associated with the last line of Eq. (9.95). Also, since

$$e'_\alpha \equiv \frac{\partial T}{\partial \alpha} \equiv 0, \qquad (9.103)$$

the bond conjugate to the $\dot{p}_\alpha$ bond is not needed, and, except for the torsion spring, α is not needed as a state variable.

Conclusions Many problems with derivative causality can be solved by defining *I*- or *C*-fields for subsystems, as shown in Chapter 7, but mechanical inertia elements pose a particularly difficult set of problems when nonlinear geometric constraints are involved. The extension of the *I*-field idea to an *IC*-field, in the spirit of Lagrange's or Hamilton's equations, eliminates the derivative causality at the possible expense of an inertia matrix inversion which at worst may have to be made numerically and repeatedly during a computer simulation.

9.3   APPLICATION TO VEHICLE DYNAMICS

The bond graph fragment of Figure 9.6 is the fundamental building block for all rigid-body mechanics. Different transformations may be substituted, depending upon the rigid-body system being modeled, but the fundamental structure remains unchanged. In this section the use of this building block for constructing very sophisticated, yet still low-order models of ground vehicles is discussed.

Figure 9.23 shows a vehicle with a body-fixed frame attached at its center of mass and aligned with the principal directions. The vehicle body is considered rigid, with mass m_b and principal moments of inertia J_x, J_y, and J_z. It is a four-wheeled vehicle with pneumatic tires, front-wheel steering, front-wheel drive, and four-wheel independent suspension. It is a more or less conventional automobile. Through use of bond graphs, we can break the modeling effort into modeling of component pieces. We can then assemble the pieces, having been tested individually, into a system model.

Tire and Suspension Pneumatic tires are extremely complicated in the way forces are generated at the contact patch. We will not attempt to develop this material thoroughly here and will keep the complexity within the context of demonstrating the procedure for building system models using 3-D rigid-body mechanics. The interested reader should see Reference [8] for a complete treatment of pneumatic tires.

Figure 9.24 shows a tire acted upon by forces at the ground contact patch and with an attached suspension unit. The tire is of mass m_t, which is representative of the unsprung mass associated with the wheel assembly, and the tire is compliant, indicated by the stiffness, k_t. It should be noted that the vertical compliance of a tire is nonlinear; however, for discussions here it is convenient to refer to a stiffness parameter. The realistic behavior can, of course, be included in the final computational model.

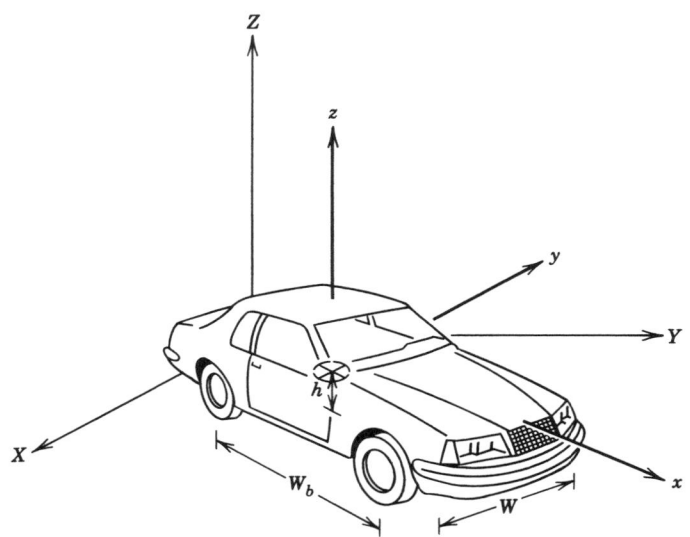

FIGURE 9.23. Ground vehicle with body-fixed frame.

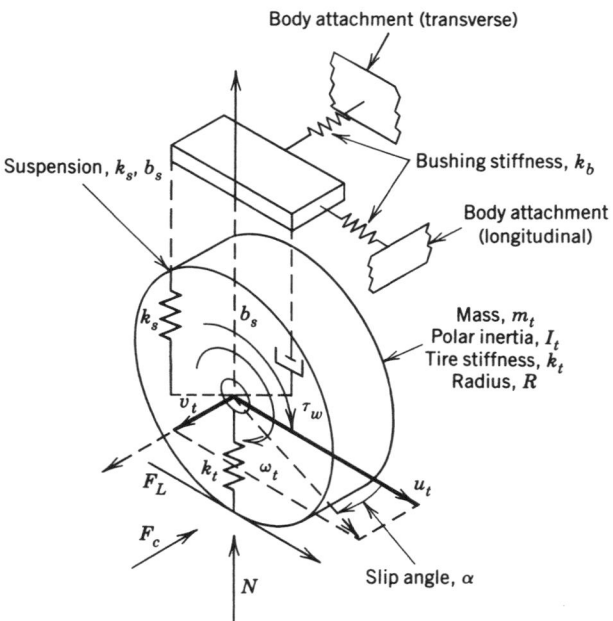

FIGURE 9.24. Schematic of tire and suspension.

At the instant shown, the forward velocity of the tire is u_t, its transverse or sideways velocity is v_t, and its angular velocity is ω_t. It is acted upon by a longitudinal force F_L, a cornering force F_c, and a normal force N through the contact patch. The wheel torque τ_w is also shown in the figure, and it represents the torque from the drive train or torque due to braking.

The longitudinal force F_L is dependent upon the longitudinal *slip* of the tire. The slip s is expressed as a percentage and is calculated from

$$s = \frac{|R\omega_t - u_t|}{R\omega_t} \times 100 \tag{9.104}$$

during acceleration and

$$s = \frac{|u_t - R\omega_t|}{u_t} \times 100 \tag{9.105}$$

during braking. The slip determines the friction coefficient μ, through a functional dependence shown qualitatively in Figure 9.25a, and the force is given by

$$F_L = \mu N. \tag{9.106}$$

It is interesting that the tires generate a maximum force at some finite slip, and then the force decreases as the slip increases. When slip equals 100%, the wheel is spin-

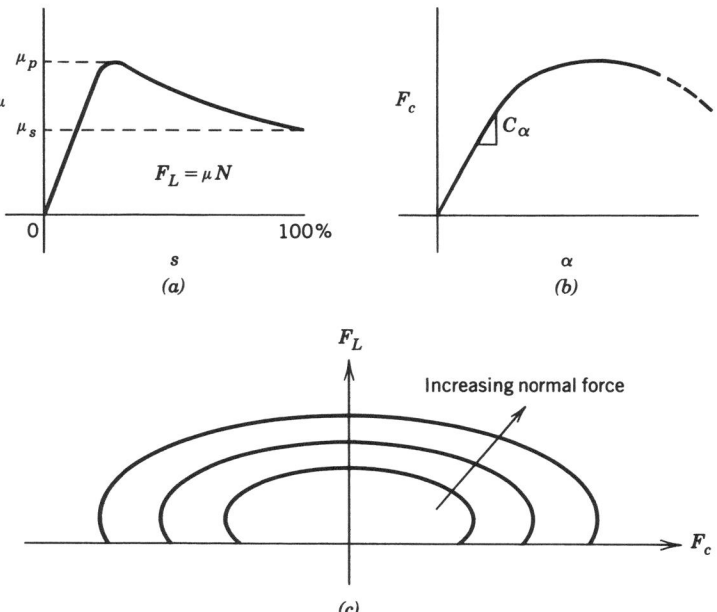

FIGURE 9.25. Qualitative force generation in pneumatic tires. (*a*) Longitudinal force; (*b*) cornering force.

ning with no forward velocity (as on ice), or else the wheel is locked, but the vehicle is skidding with some forward velocity. The curve of Figure 9.25a is strongly dependent on road surface conditions and also the cornering force F_c.

When a tire is steered and the vehicle turns, the absolute velocity of the wheel center is not directed along the plane of the wheel. The tire deforms, and a side force F_c is developed. The slip angle α of the tire is defined to be the angle between the absolute velocity of the wheel and the plane of the wheel. This angle is shown in Figure 9.24 and can be calculated from

$$\alpha = \tan^{-1} \frac{v_t}{u_t}. \tag{9.107}$$

The cornering force is functionally related to α, as shown in Figure 9.25b. The linear part of the curve is characterized by the *cornering stiffness* C_α of the tire. The curve of Figure 9.25b is dependent upon road surface conditions and upon the longitudinal force. The tradeoff between F_L and F_c is shown qualitatively in Figure 9.25c. For our discussion it is only important to realize that if α and s are both known, then F_L and F_c can be computed, regardless of the complexity of their interrelationship.

Figure 9.26 shows a bond graph of the wheel–tire system. The longitudinal dynamics in part a, the cornering dynamics in part b, and the vertical dynamics in part c. The wheel is, in part, another rigid body that could be represented in a manner similar to the vehicle body. Figure 9.26 shows the wheel dynamics decoupled. For the wheel it is reasonable to neglect cross coupling, since the angular velocity components about the vertical and horizontal longitudinal axes are very small. We are also neglecting gyroscopic effects due to the high angular velocity ω_t of the tire.

In Figure 9.26a the torque from the drive train or brakes is an input to the bond graph fragment, and a force on the body is an output. The modulated R-element represents the tire slip behavior and can be as complicated as necessary for the intended use of the model. In Figure 9.26b the cornering dynamics are represented. They too possess a modulated R-element for the cornering force generation. This fragment also outputs a force to the body. The vertical dynamics of Figure 9.26c allow for roadway unevenness under the tire contact patch, and this fragment also contains the suspension elements. A force on the body is an output from this fragment.

Notice in Figure 9.26 that integral causality exists for all models. Thus state equations are straightforwardly derived. These equations will accept drive train torque τ as an input, and they will have forces as outputs. These forces then become inputs into the body model. Notice also that each wheel will have an identical but individual representation.

Drive Train Model The final subsystem that requires discussion is the drive train, consisting of the engine, transmission, and differential (front only for this discussion). The drive train is not an essential element for simulating all vehicle systems. If we choose, we can simply specify the torque τ in the bond graph of Figure 9.26a. For completeness, we present the drive train model here.

Figure 9.27 shows, schematically, the drive train of a front-wheel drive vehicle. The engine power passes through a transmission and then is split by the differen-

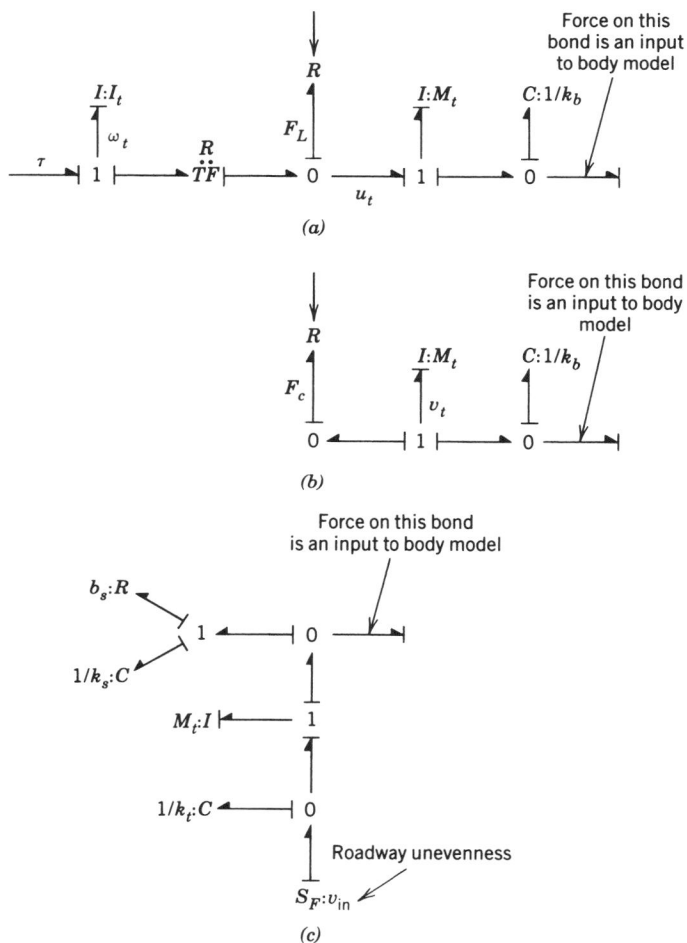

FIGURE 9.26. Bond graph of wheel–tire system. (*a*) Longitudinal dynamics; (*b*) transverse dynamics; (*c*) vertical dynamics.

tial and transmitted to the right and left front wheels. The differential applies equal torque, left and right, but permits the right and left wheels to have different angular velocities. A bond graph fragment for the drive train is also shown in Figure 9.27. In part *b*, no drive line compliance is included. The engine is modeled as a modulated torque source, and torques τ_L and τ_R are outputs from the model. In Figure 9.27*c*, drive line compliance is included. Torques are still outputs from the model. But this time the engine must be modeled as a modulated source of angular velocity. The output torques are, of course, the inputs to the wheel–tire dynamics of Figure 9.26.

Conclusions Three-dimensional rigid-body mechanics have been developed using bond graphs. A very elegant representation of Euler's equations resulted, but a pre-

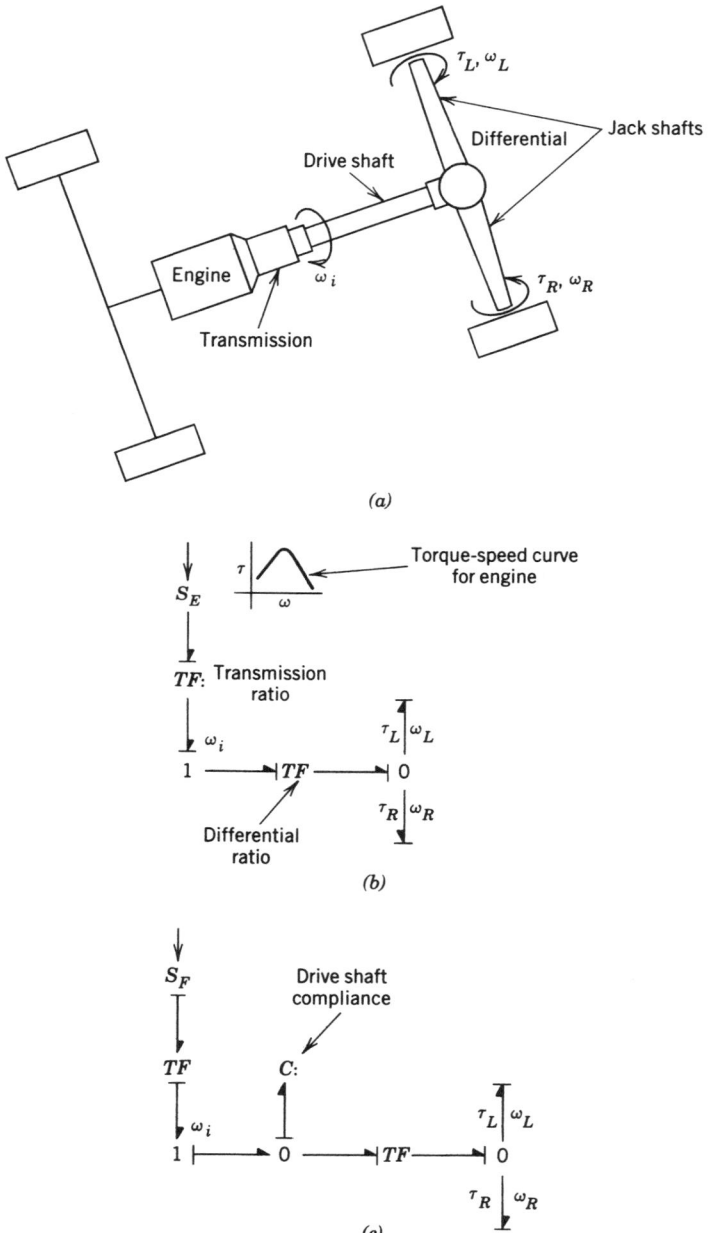

FIGURE 9.27. Drive train schematic and bond graph. (*a*) Schematic; (*b*) bond graph without drive shaft compliance; (*c*) bond graph with drive shaft compliance.

scribed causality must be used in order to allow the 3-D bond graph fragment to interact easily with a necessary coordinate transformation. Likewise, a complex coordinate transformation was presented which must be used with a prescribed causality for ease of coupling with other parts of an overall system model. The coordinate transformation shown here was the one particularly useful for vehicle dynamics.

The 3-D rigid-body development was demonstrated for vehicle dynamics applications. It was shown how wheel–tire models and drive train models could be developed separately from the 3-D mechanics and then made to interact in a computational model as long as prescribed causality was maintained. The complete model was not assembled here. However, it is hoped that sufficient detail was given to whet the reader's appetite so that uncontrollable hunger to read References [9], [10], and [11] results. These references use this approach for constructing very complicated vehicle system models for use in design and control synthesis.

REFERENCES

[1] Sir William Thomson and P. G. Tait, *Principles of Mechanics and Dynamics*, New York: Dover, 1962.

[2] C. Lanczos, *The Variational Principles of Mechanics*, Toronto: Univ. of Toronto Press, 1957.

[3] R. C. Rosenberg, "Multiport Models in Mechanics," *Trans. ASME, J. Dyn. Syst. Meas. Control*, **94**, Ser. G, No. 3, 206–212 (Sept. 1972).

[4] S. Crandall, D. Karnopp, E. Kurtz, and D. Pridmore-Brown, *Dynamics of Mechanical and Electromechanical Systems*, New York: McGraw-Hill, 1968.

[5] D. Karnopp, "Power-Conserving Transformations: Physical Interpretations and Applications Using Bond Graphs," *J. Franklin Inst.*, **288**, No. 3, 175–201 (Sept. 1969).

[6] R. C. Rosenberg, "State-Space Formulation for Bond Graph Models of Multiport Systems," *Trans. ASME, J. Dyn. Syst. Meas. Control*, **93**, Ser. G, No. 1, 35–40 (Mar. 1971).

[7] G. Martin, *Kinematics and Dynamics of Machines*, New York: McGraw-Hill, 1969, p. 7.

[8] H. B. Pacejka, "Introduction into the Lateral Dynamics of Road Vehicles," Third Seminar on Advanced Vehicle Dynamics, Amalfi, May 1986.

[9] D. C. Karnopp, "Bond Graphs for Vehicle Dynamics," *Vehicle Syst. Dynam.*, **5** (1976).

[10] D. L. Margolis and J. Asgari, "Sophisticated yet Insightful Models of Vehicle Dynamics Using Bond Graphs," ASME Symposium on Advanced Automotive Technologies, 89 ASME WAM, San Francisco, 1989.

[11] D. L. Margolis, "Bond Graphs for Vehicle Stability Analysis," *Int. J. Vehicle Design*, **5**, No. 4 (1984).

[12] B. Barasözen, P. Rentrop, and Y. Wagner, "Inverted *n*-Bar Model in Description and in State Space Form," *Math. Model. Syst.*, **1**, 272–285 (1995).

[13] D. Karnopp, "Lagrange's Equations for Complex Bond Graph Systems," *Trans. ASME J. Dynam. Syst. Meas. Control*, **99**, Ser. G, 300–306 (1977).

[14] D. Karnopp, "Understanding Multibody Dynamics Using Bond Graphs," *J. Franklin Inst.*, **334B**, No. 4, 631–642 (1997).

[15] D. Karnopp, "An Approach to Derivative Causality in Bond Graph Models of Mechanical Systems," *J. Franklin. Inst.*, **329**, No. 1, 65–75 (1992).

[16] D. Margolis and D. Karnopp, "Bond Graphs for Flexible Multibody Systems," *Trans. ASME, J. Dynam. Syst. Meas. Control*, **101**, 50–57 (1979).

[17] A. Zeid and C.-H. Chung, "Bond Graph Modeling of Multibody Systems: A Library of Three Dimensional Joints," *J. Franklin Inst.*, **329**, 605–636 (1992).

PROBLEMS

9-1 A particle can slide on a frictionless horizontal plane constrained by a linear spring pivoted at the origin of x, y and with spring constant k and free length l_0. Make a bond graph for this system by using r, θ for the vectors $\mathbf{q}_C$ and $\mathbf{q}_k$ and $\dot{x}$, $\dot{y}$ for the vector $\mathbf{v}_I$. Can the system accept all integral causality? Make another bond graph by using x, y for $\mathbf{q}_k$, $\dot{x}$, $\dot{y}$ for $\mathbf{v}_I$, and r for $\mathbf{q}_C$.

9-2 Three mass particles are attached to a spring-supported rigid massless bar that executes small vibratory motion. The vector $\mathbf{q}_C$ is x_1, x_2, the vector v_I is V_1, V_2, V_3, and several choices for $\mathbf{q}_k$ are possible. Show a bond graph for the system with the following choices for $\mathbf{q}_k$:

(a) $\mathbf{q}_k = x_1, x_2 = \mathbf{q}_C$;

(b) $\mathbf{q}_k = x_3, \theta$, where x_3 is the displacement of m_3 and θ is the angle of inclination of the bar; and

(c) $\mathbf{q}_k = x_1, x_3$.

Pick one bond graph, and write the equations of motion for the system using the graph.

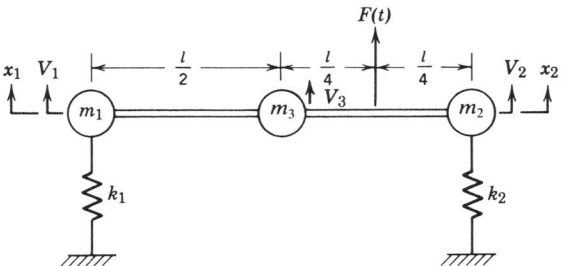

9-3 Set up bond graph representations of the system shown below using the three alternative *MTF* forms shown. Write *complete* state equations in each case, using integration-causality methods or Lagrange equations. Comment on any advantages or disadvantages you see in the alternative formulations.

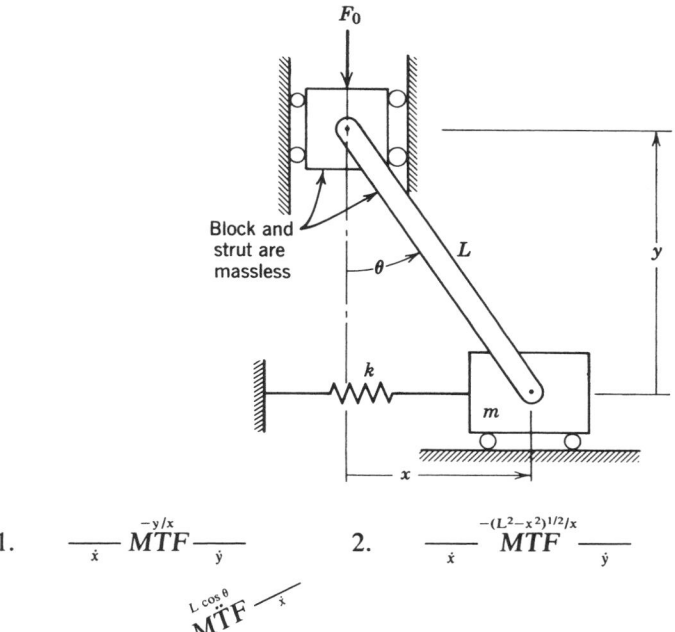

1. $\xrightarrow{\dot{x}} \overset{-y/x}{MTF} \xrightarrow{\dot{y}}$ 2. $\xrightarrow{\dot{x}} \overset{-(L^2-x^2)^{1/2}/x}{MTF} \xrightarrow{\dot{y}}$

3. $\xrightarrow{\dot{\theta}} 1 \underset{\overset{-L\sin\theta}{M\ddot{T}F} \xrightarrow{\ddot{y}}}{\overset{L\cos\theta}{M\ddot{T}F} \xrightarrow{\ddot{x}}}$

9-4   Consider the model of an off-road vehicle shown below. Note that the sur-

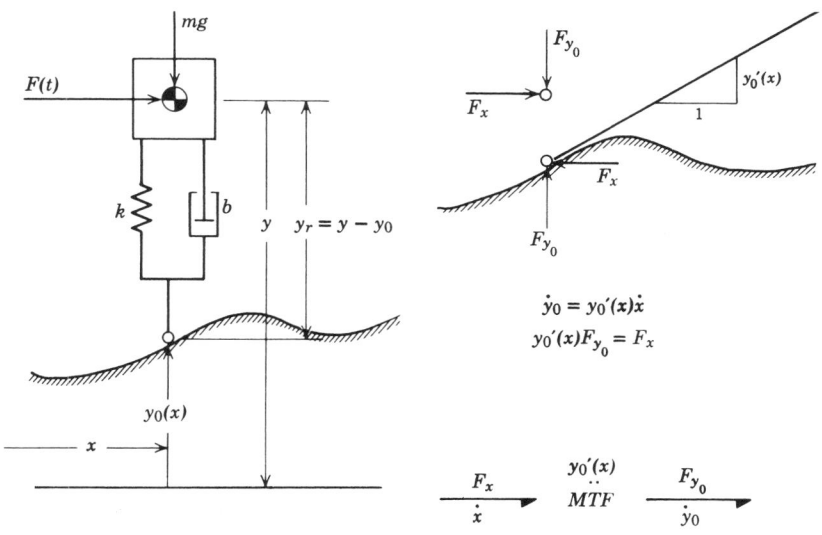

$$\dot{y}_0 = y_0'(x)\dot{x}$$
$$y_0'(x)F_{y_0} = F_x$$

$$\xrightarrow[\dot{x}]{F_x} \overset{y_0'(x)}{\underset{..}{MTF}} \xrightarrow[\dot{y}_0]{F_{y_0}}$$

face interaction may be represented by the equations and bond graph element shown. Develop a bond graph and state-space equations for this model.

9-5   Construct a bond graph for this two-degree-of-freedom vehicle model. Use two forms for the I-field representing the rigid body: one matrix representation using the $\dot{y}_1$, $\dot{y}_2$ port variables directly, the other using the auxiliary variables, y_c and θ, and a multiport TF.

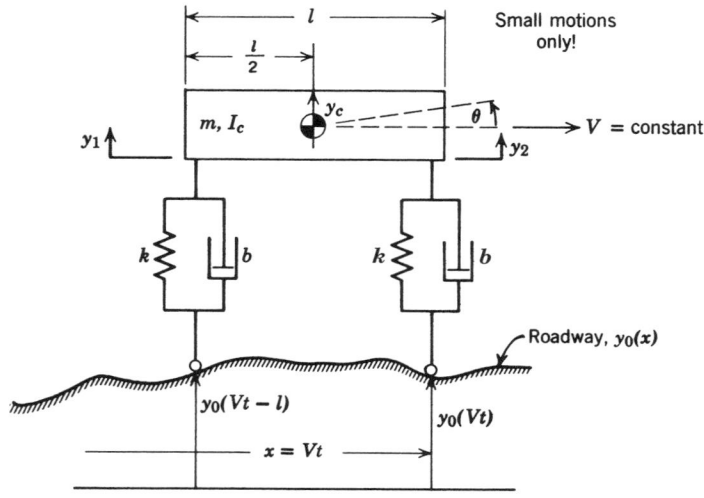

9-6   (In which the inherent nastiness of geometric nonlinearity is exhibited.)

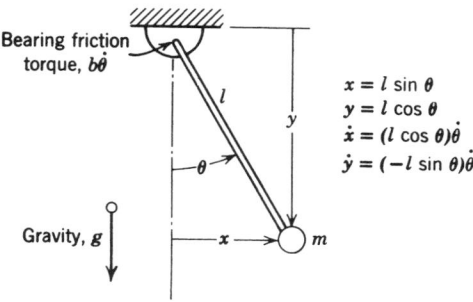

Consider the simple pendulum of length l, mass m, swinging through large angles under the influence of gravity and bearing friction.

(a)   Set up a modulated transformer representation of the relations between $\dot{x}$, $\dot{y}$, and $\dot{\theta}$ and the corresponding forces F_x, F_y and torque τ that are enforced by the rigid bar. Demonstrate that the multiport displacement-modulated transformer relations are power conservative.

(b) Show a bond graph for the system using your *MTF*. Set up sign conventions and a causal pattern in which integration causality is applied to the *x*-motion —*I*.

(c) Write the equations corresponding to your graph. Be sure that your state space is complete and in the standard form.

(d) Show that your state space is consistent with $ml^2\ddot{\theta} = -b\dot{\theta} - mgl\sin\theta$, which would be the result of many standard analyses.

9-7 Consider the double pendulum shown (see Problem 9-6).

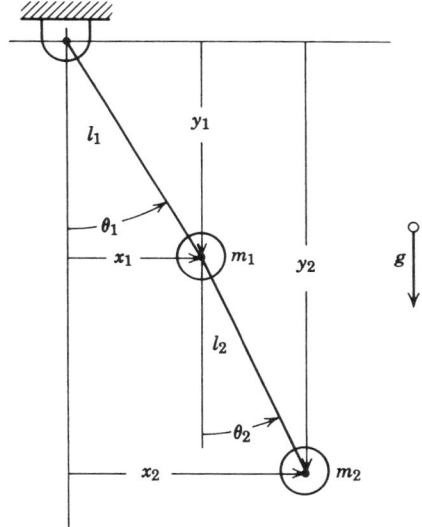

(a) Set up the multiport displacement-modulated transformer that relates $\dot{x}_1$, $\dot{y}_1$, $\dot{x}_2$, $\dot{y}_2$ to $\dot{\theta}_1$, θ_1, $\dot{\theta}_2$, θ_2, and exhibit both the velocity and force relations in matrix form.

(b) Using the *MTF*, show a system bond graph, and predict one possible set of state variables using the integration-causality method.

(c) Use the junction structure of (b) to compute $T(\theta_1, \dot{\theta}_1, \theta_2, \dot{\theta}_2)$—kinetic energy—and generalized torques so that Lagrange equations can be utilized as an alternative procedure to find state equations. What are the state variables if Lagrange equations are used?

9-8 The uniform thin bar shown below is pivoted at A. The pin slides freely in the vertical guideway. Represent the bar by three 1-port —*I*s representing the horizontal and vertical motion of the center of mass and rotation about the center of mass.

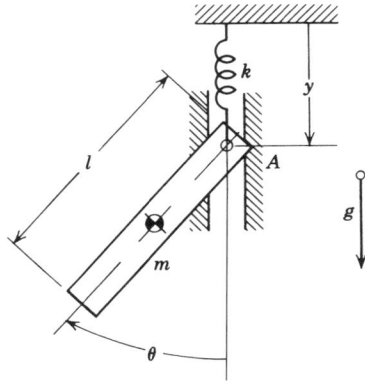

(a) Show that the junction structure of the multiport *MTF* involved may be rearranged to yield a bond graph for the system of the following form:

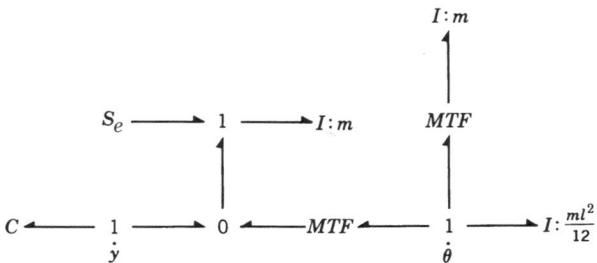

(b) Apply causality to the system, and write the equations of motion.

9-9 Let the pendulum have mass m and moment of inertia J. Assume $\theta \ll \pi$.

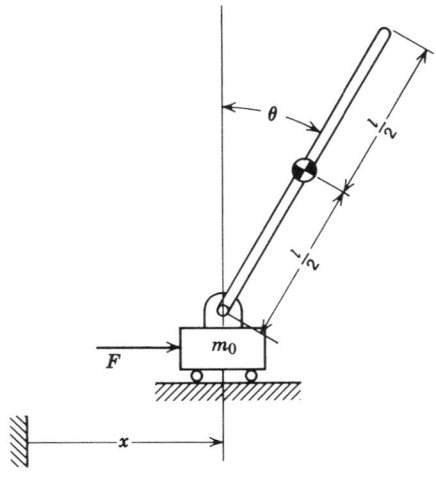

(a) Noting that the equilibrium point for the upside-down pendulum is $\theta = 0$, write expressions for kinetic energy and potential energy valid up to second order in θ and $\dot{\theta}$. Use the expressions to validate the linear bond graph shown below:

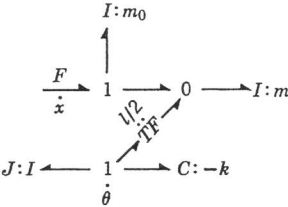

Find the negative spring constant for the $-C$ that models the gravity effect.

(b) Assuming F is an input, and that the $-I$s with parameters m_0 and J have integral causality, write state equations, and indicate the matrix that must be inverted because of the differentiation causality.

(c) Show that the system can be represented by

$$F \underset{\dot{x}}{\overset{=\dot{P}_x}{\longrightarrow}} I \underset{\dot{\theta}}{\overset{\dot{P}_\theta}{\longleftarrow}} C: +k,$$

and find the matrix representation of the I-field.

9-10 Following the pattern of analysis of Problem 9-9, find a bond graph for the double-inverted pendulum shown on page 343. Write state equations for the system.

9-11 The German-language automobile model shown below uses the subscripts, v, h, and R, for *vorn* (forward), *hintern* (rear), and *Rad* (wheel), respectively. Otherwise, it seems to be sufficiently international to be bond graphed.

Consider small pitch angles for the vehicle. Recall that the sum of the vertical forces is equal to the mass times the acceleration of the center of mass and that the sum of the torques about the center of mass is the centroidal moment of inertia times the angular acceleration.

(a) Construct a bond graph for the model in which the vehicle mass m and the centroidal moment of inertia appear as 1-port element parameters.

(b) Show that consistent causality may be assigned for your bond graph, and list the state variables. Discuss very briefly any difficulties you may find in assigning causality.

(c) Show that if the vehicle body is considered to be an I-field, the assignment of causality is simplified and the writing of state equations is easier.

(d) Either find the properties of the 2-port I-field describing the rigid body or outline a procedure to find them.

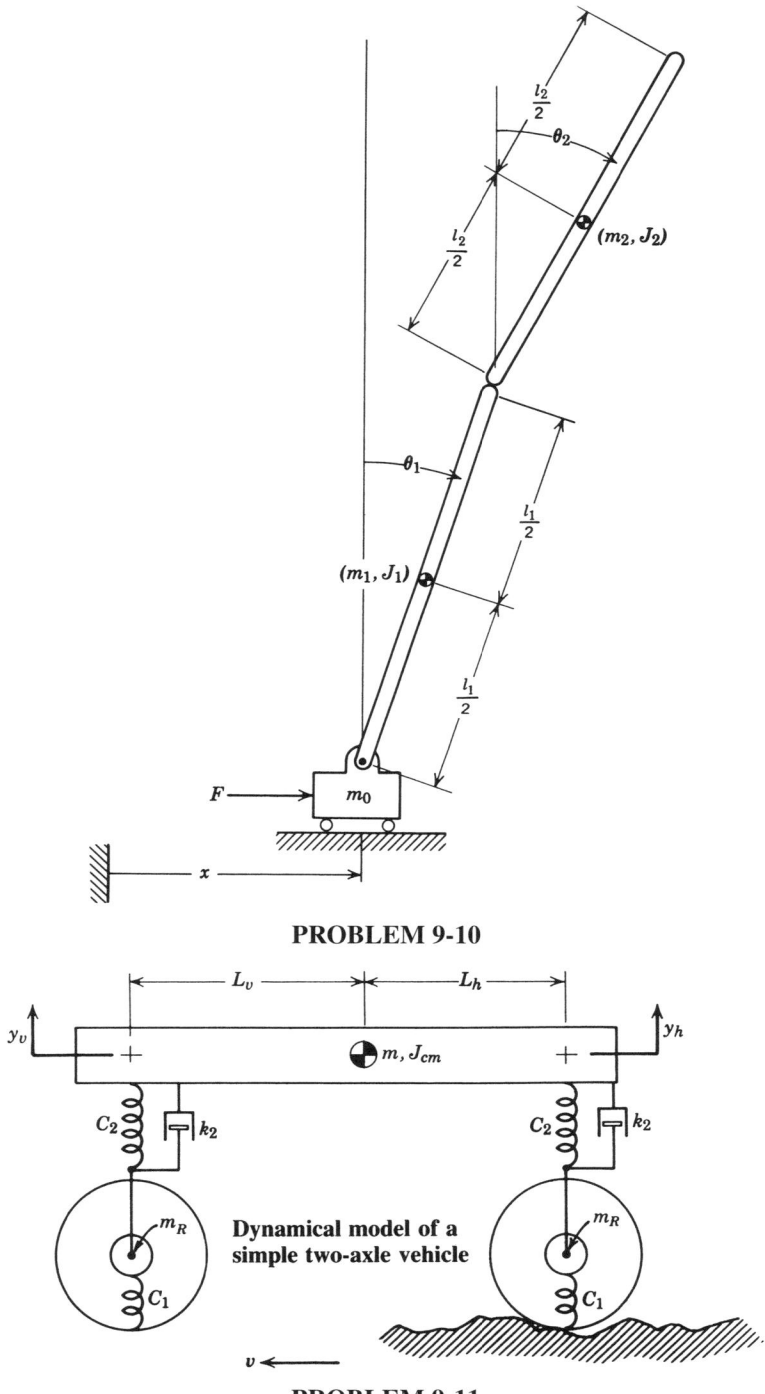

PROBLEM 9-10

Dynamical model of a
simple two-axle vehicle

PROBLEM 9-11

(e)   Can you think of a way to represent the nonlinear behavior of the wheels when they leave the ground?

9-12   The mass m is supported on a thin inextensible wire stretched across the frame with tension T.

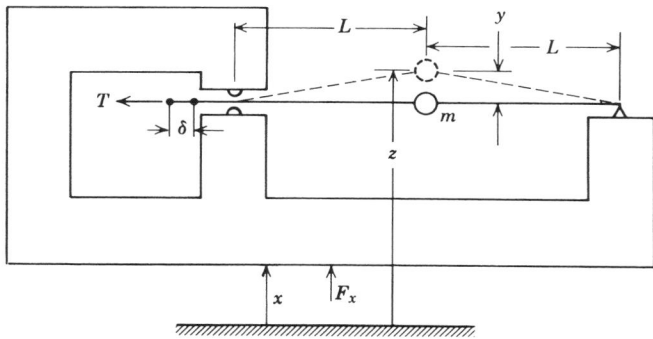

(a)   Show that if $T = $ const, and if $y/L \ll 1$, then the force on the mass is given approximately by $F = -(2y/L)T$, and a bond graph for the system (neglecting the inertia of the frame) is

$$k_{eq} = \frac{2T}{L} : C \leftarrow 1 \overset{F_y}{\underset{\dot{y}}{}} \overset{F_z \mid \dot{z}}{0} \overset{F_x}{\underset{\dot{x}}{}}$$

(b)   When T is allowed to vary by connecting a force source to the wire, the equivalent spring seems to simply have a "variable constant." This concept is not very profound, however, since a $-C$ element must be conservative, and with variable T the relation between F_y and y is not.

Show that if $y/L \ll 1$, the distance the end of the wire moves against the force source, δ, is $\delta = y^2/L$. Use this relation to show a bond graph that represents the effect of the wire by means of a force source and a displacement-modulated transformer. Note that this model shows a causal restriction that was not evident in the previous bond graph.

9-13   Consider the spherical pendulum shown in part a of the figure below. Verify the bond graph shown in part b by finding the constitutive laws for the MTF relating $\dot{\theta}_1, \dot{\theta}_2$ to $\dot{x}, \dot{y}, \dot{z}$. Write out the multiport MTF as a junction structure using 0- and 1-junctions and 2-port MTFs.

Now show that all three Is cannot accept integral causality, because the MTF cannot accept flow input causality on all three of the $\dot{x}, \dot{y}, \dot{z}$ bonds. Show that consistent complete causality can be achieved if only two of the Is have integral causality.

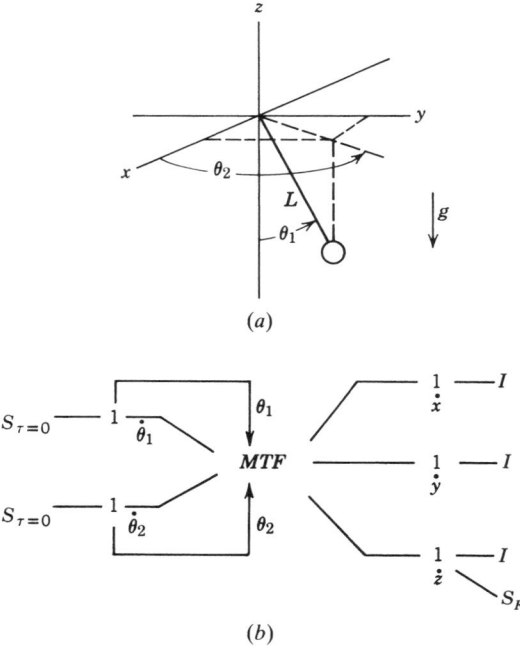

(a)

(b)

9-14 A brief outline of the relation between applied torque and angular momentum for general motion of a rigid body using components in a coordinate frame moving with the body and aligned with the principle axes is as follows: Following the notation of Reference [4], Section 4.4, a rigid body may be defined by a relation between angular velocity ω, angular momentum $\mathbf{H}$, and an inertia matrix $\mathbf{I}$, as follows:

$$\mathbf{I}\omega = \mathbf{H}. \tag{i}$$

In principal axis coordinates, this becomes

$$\begin{bmatrix} I_1 & 0 & 0 \\ 0 & I_2 & 0 \\ 0 & 0 & I_3 \end{bmatrix} \begin{bmatrix} \omega_1 \\ \omega_2 \\ \omega_3 \end{bmatrix} = \begin{bmatrix} H_1 \\ H_2 \\ H_3 \end{bmatrix}. \tag{ii}$$

As long as torques are computed about a fixed point or the center of mass of the body, the following is true:

$$\frac{d\mathbf{H}}{dt} = \tau, \tag{iii}$$

where $\mathbf{H}$ is the angular momentum vector and τ is the torque vector. When the components of $\mathbf{H}$ and τ are chosen in the moving coordinate frame, then

Eq. (iii) becomes

$$\frac{\partial \mathbf{H}}{\partial t_{\text{rel}}} + \omega \times \mathbf{H} = \tau, \tag{iv}$$

or

$$\begin{bmatrix} \dot{H}_1 \\ \dot{H}_2 \\ \dot{H}_3 \end{bmatrix} + \begin{bmatrix} \omega_2 H_3 - \omega_3 H_2 \\ \omega_3 H_1 - \omega_1 H_3 \\ \omega_1 H_2 - \omega_2 H_1 \end{bmatrix} = \begin{bmatrix} \tau_1 \\ \tau_2 \\ \tau_3 \end{bmatrix}, \tag{v}$$

in which the $\omega \times \mathbf{H}$ term corrects the terms representing change relative to the moving frame to correctly portray the total change in $\mathbf{H}$ relative to the inertial frame.

Verify that the modulated gyrator-ring structure correctly represents Euler's equations (v):

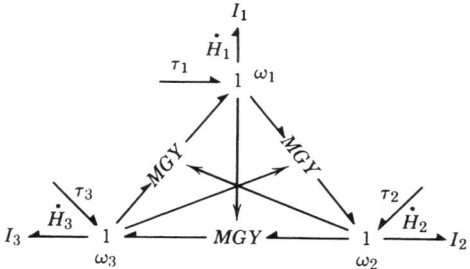

Select one of the *MGY*s, and write the equations that it represents. Verify power conservation in the form

$$\dot{H}_1\omega_1 + \dot{H}_2\omega_2 + \dot{H}_3\omega_3 = \tau_1\omega_1 + \tau_2\omega_2 + \tau_3\omega_3.$$

9-15 The mechanical system shown in part *a* of the figure below contains one rigid body that rotates about a fixed axis and one that moves in more general motion about a single fixed point. Using the results of Problem 9-14, verify the bond graph representation shown in part *b*. By computing ω_1, ω_2, ω_3 in terms of θ, ψ, find the constitutive laws for the multiport *MTF*.

If θ and ψ were used as generalized coordinates, a nonlinear, fourth-order set of state equations could readily be found by applying Lagrange's equation to $T-V$. If all dynamic elements could accept integral causality, a fifth-order state space would result. This suggests that derivative causality may be necessary for some of this system. See if this is true by expanding the multiport *MTF* into a junction structure of 0- and 1-junctions and 2-port *MTF*s and applying causality.

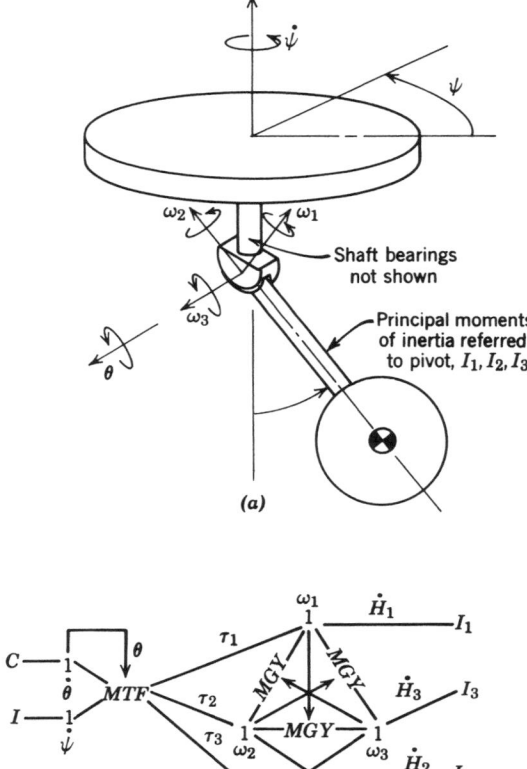

(a)

(b)

9-16   The coordinate transformation of Figure 9.6 takes the body-fixed center-of-mass velocity and angular velocity components and outputs the inertial an-

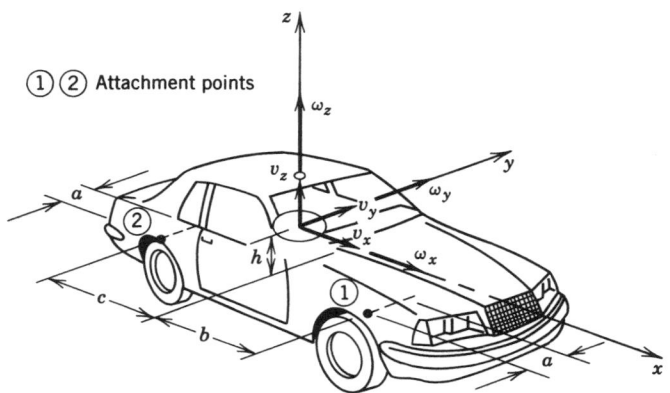

①② Attachment points

gular velocity of the body and the inertial velocity components of the center of mass. We frequently need the velocity components at attachment points on the rigid body. Using the figure shown, derive the coordinate transformation between center-of-mass body-fixed components and the body-fixed components at the attachment points shown.

9-17 The figure shows two rigid bodies connected by a frictionless spherical joint. Two body-fixed coordinate frames are shown. The spherical joint is at attachment point a on body 1 and attachment point b on body 2. Using a shorthand notation similar to that of Figure 9.6, construct a bond graph model for this interaction, assign causality, and discuss the problems associated with deriving a computational model.

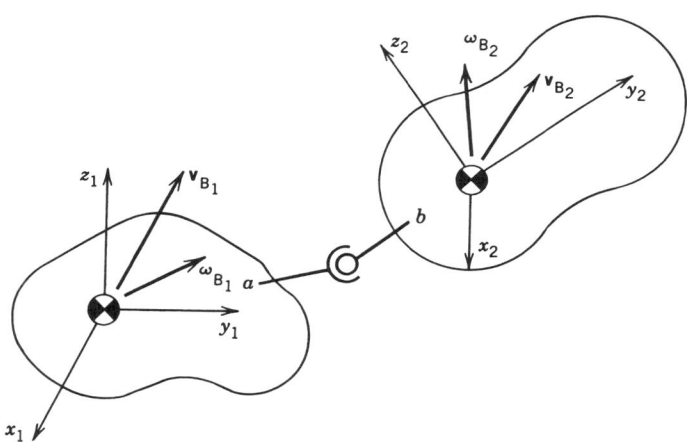

9-18 The figure on page 349 shows a rigid body with a spring–damper suspension at each corner. A body-fixed coordinate frame is attached. A constraint system, not shown, ensures that ω_z (the angular velocity about the moving z-axis) remains zero. Using Figure 9.2, show the simplified bond graph fragments that will represent the rigid-body part of this system.

9-19 For small displacements it is reasonable to assume that the suspension units in Problem 9-18 generate only vertical forces and that the body-fixed z-axis remains virtually vertical. This assumption eliminates the need for the coordinate transformation from body-fixed components at the attachment points to inertial components. Construct a bond graph model of the system under these assumptions. It is still convenient to use matrix transformations between center-of-mass motion and individual attachment points.

9-20 Planar motion of a mass particle is shown in the figure with respect to both a body-fixed frame and an inertial frame. The polar coordinates r, θ are used for the inertial description. A reduced bond graph fragment for planar motion

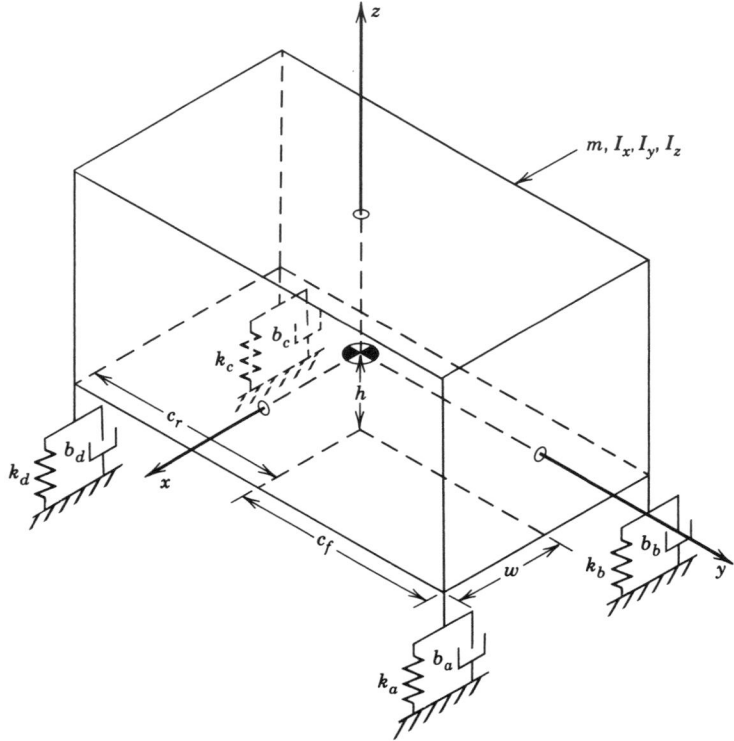

PROBLEM 9-18

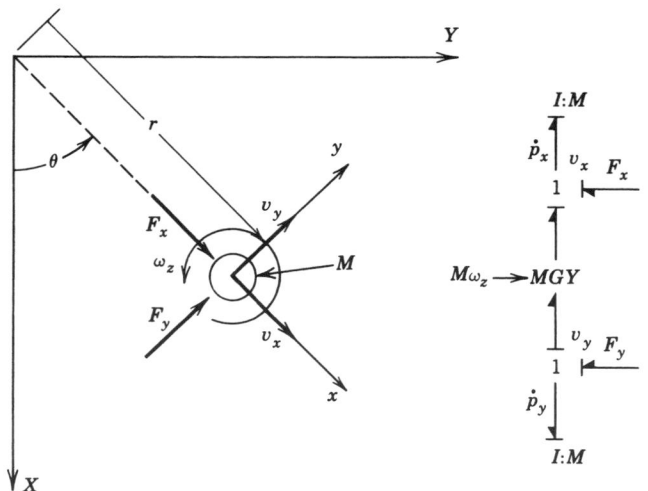

PROBLEM 9-20

is also shown in the figure. From classical dynamics, the absolute acceleration in r and θ directions are

$$a_r = \ddot{r} - r\dot{\theta}^2, \qquad a_\theta = r\ddot{\theta} + 2\dot{r}\dot{\theta}.$$

Using the bond graph for planar motion, show that the absolute acceleration is the same as for polar coordinates.

9-21 A classical simplified vehicle model, known as the *bicycle* model, is shown in a top view in the figure. The vehicle travels in the x, y plane and has no width. There is also no suspension, and only cornering forces are considered. The front wheel can be steered through the angle δ. The only angular velocity considered is that about the body-fixed z (vertical) axis, ω_z. There is no wheel mass, nor any bushings to consider. Using Figures 9.2, 9.25, and 9.26, construct a complete bond graph model of this vehicle and assign causality.

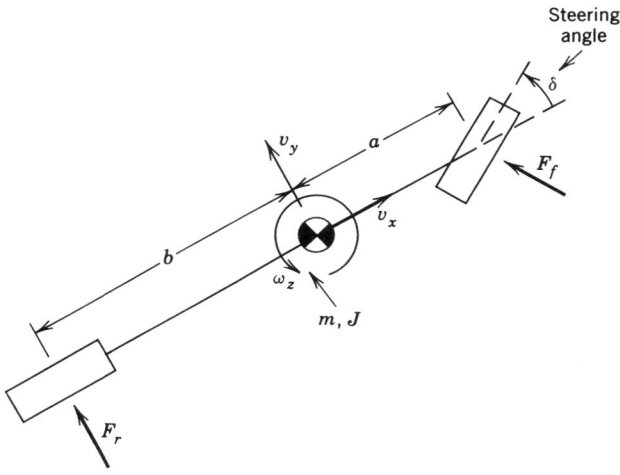

9-22 One of the uses of the bicycle model of a ground vehicle is for stability considerations. For this purpose it is assumed that v_x is large compared to v_y, and v_x is virtually constant. Also, the steering angle δ is small, and the slip angles front and rear are small enough that the cornering forces are related to their respective slip angles by the linear relationships

$$F_f = C_f \alpha_f, \qquad F_r = C_r \alpha_r,$$

where

$$\alpha_f = \frac{\text{transverse velocity at front}}{v_x},$$

$$\alpha_r = \frac{\text{transverse velocity at rear}}{v_x}.$$

Modify the bond graph from Problem 9-21 to reflect these additional simplifications. Assign causality and identify state variables.

9-23 Derive the state equations for the system of Problem 9-22. Put them into standard matrix form for linear equations, remembering that v_x is constant for
this model.

9-24 The equations from Problem 9-23 are

$$\frac{d}{dt}\begin{bmatrix} p_y \\ p_J \end{bmatrix} = \begin{bmatrix} \dfrac{-(C_r + C_f)}{mv_x} & \dfrac{bC_r - aC_f}{Jv_x} - \dfrac{mv_x}{J} \\ \dfrac{bC_r - aC_f}{mv_x} & \dfrac{-(a^2C_f + b^2C_r)}{Jv_x} \end{bmatrix}\begin{bmatrix} p_y \\ p_J \end{bmatrix} + \begin{bmatrix} C_f \\ aC_f \end{bmatrix}\delta.$$

Perform an eigenvalue analysis and determine the conditions for stability of
the vehicle.

9-25 It is well known that pneumatic tires have sidewall compliance that can affect
stability. The figure shows the bicycle model with springs, k_f and k_r, between
the wheel and the contact patch. It is assumed that the springs are always in
line with their respective axles as the vehicle moves.

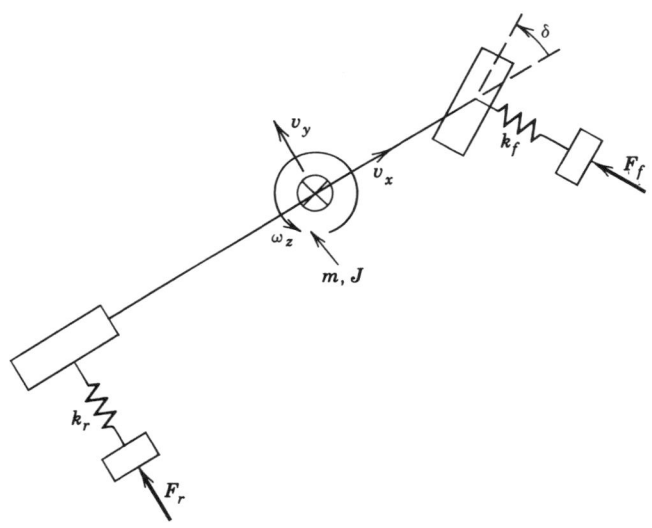

Construct a bond graph of the new model.

10

DISTRIBUTED-PARAMETER SYSTEMS

Distributed-parameter systems are, analytically, those represented by partial rather than total differential equations. Physically, distributed systems are engineering components and devices that cannot be accurately approximated by the "lumped" assumptions used thus far in the text. Remember that *all* engineering models are approximations and are made up by the modeler to answer specific questions concerning system understanding, design, control, and so on. As modelers we are always trying to use the simplest model that will lead to understanding of the actual system. As Einstein once said, "Everything should be made as simple as possible, but not simpler."

So far in the text we have said that engineering systems are built from components exhibiting inertia-like, compliance-like, or resistance-like effects. A mass has been assumed rigid with no finite stiffness, and springs have been compliant with no finite mass. Of course, all materials exhibit inertial and compliance effects, and it is our modeling assumptions that justify separating these effects into exclusive "lumps."

There are real engineering systems in which this lumping process is far from obvious. An automobile for which we are interested in suspension design can be represented as a rigid body and characterized by its mass and moments of inertia. A tractor–trailer rig for which we are interested in suspension design cannot be represented as rigid bodies if we are to accurately model the actual system dynamics. This is because the frames of the tractor–trailer are flexible, and their bending motion is essential to understanding the system. Automobile frames bend also, but their bending frequencies are significantly higher than those of the truck and are typically out of the range of interest. Thus we can lump-model the car, but not the truck.

Multistory buildings, elevated roadways, flexible structures for outer space, and long hydraulic fluid lines are all examples of engineering systems that cannot be modeled without some consideration of distributed dynamic effects. This chapter introduces the idea of distributed system modeling within the framework of bond graphs. Simple lumping techniques are discussed first, along with their limitations. Finite-mode modeling is then shown to produce the most accurate low-order representation possible for distributed systems. Combining lumped and distributed systems becomes obvious using this approach, and causality still dictates the computational ease of the resulting system model. The authors think that you will find this chapter most interesting and useful.

10.1   SIMPLE LUMPING TECHNIQUES FOR DISTRIBUTED SYSTEMS

In the course of deriving the distributed representation of a dynamic component or subsystem, it is typical to start with spatially distributed finite lumps and then take the limit as the lumps become infinitesimal in size. The lumped-parameter model of a distributed system is motivated by this approach to deriving the continuous equations. We simply stop with the finite lumps and perform no limiting process. This approach feels good and is physically understandable; however, even though the continuous equations are approached as the number of finite lumps becomes infinite, it is difficult to predict the rate of convergence. Thus, a large number of lumps may be required to obtain accuracy at low frequencies. In addition, each new lump, while improving low-frequency prediction, introduces new, totally inaccurate high frequencies. This approach to distributed system representation must be used cautiously, and with awareness on the part of the modeler. Reference [1] surveys the different approaches to distributed system modeling, of which this finite-lumping technique is one. We shall require the continuous representation for what follows, so we first demonstrate the finite-lump approach.

Longitudinal Motions of a Bar   Figure 10.1a shows a continuous representation of a uniform bar of length L, cross-sectional area A, and mass density ρ. It is fixed at the left end and acted upon by an external force $F(t)$ at the right. The continuous variable x locates any section of the bar, and the variable $\xi(x, t)$ represents the displacement of any cross section from an initially undisplaced state. Figure 10.1b shows a finite element of mass, $\rho A \, \Delta x$, with an imbalance in the normal stress σ across it. Newton's law yields

$$A\sigma(x + \Delta x) - A\sigma(x) = \rho A \, \Delta x \frac{\partial^2 \xi}{\partial t^2}. \tag{10.1}$$

Figure 10.1c shows the same section subject to simple strain, yielding the relationship

$$\sigma(x) = \frac{E\left[\xi(x) - \xi(x - \Delta x)\right]}{\Delta x}, \tag{10.2}$$

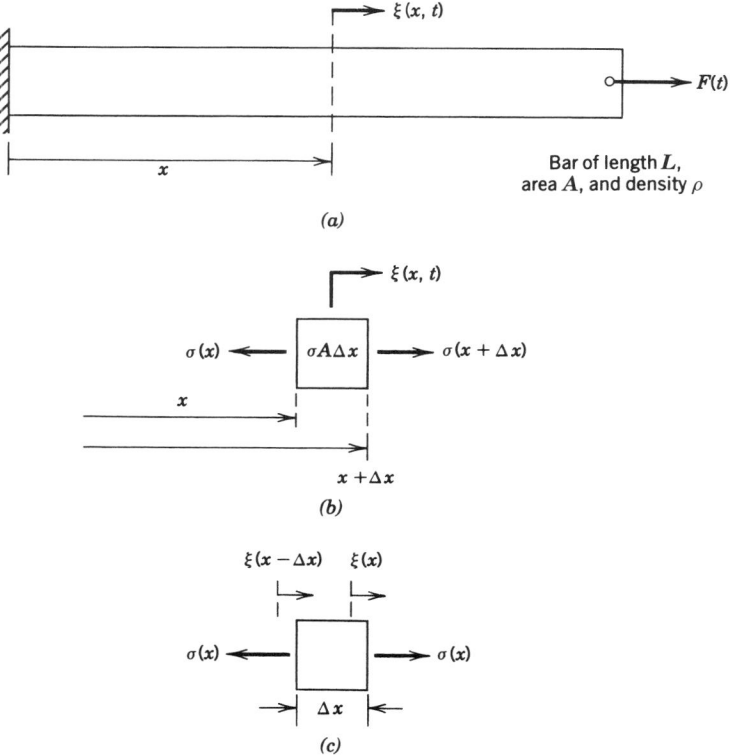

FIGURE 10.1. Vibrating bar.

where E is Young's modulus. Notice that Eqs. (10.1) and (10.2) are written as though the lumps occupied different points in space, when in fact, as Δx becomes infinitesimal, the mass lump and stiffness lump are at the same spatial location.

We will soon let $\Delta x \to 0$, but first we interpret (10.1) and (10.2) as finite lumps, and construct a lumped approximation to the system. Let $A\sigma = F$, and write (10.1) as

$$F_{i+1} - F_i = \frac{d}{dt} p_i \tag{10.3}$$

and (10.2) as

$$F_i = \frac{EA}{\Delta x} q_i, \tag{10.4}$$

where

$$p_i = \rho A \, \Delta x \dot{\xi}_i \tag{10.5}$$

is the momentum of the ith lump and

$$q_i = \xi_i - \xi_{i-1} \tag{10.6}$$

is the relative displacement between the ith and the $(i-1)$st lump. This yields the equations

$$\frac{d}{dt} p_i = \frac{EA}{\Delta x}(q_{i+1} - q_i), \tag{10.7}$$

$$\frac{d}{dt} q_i = \frac{p_i - p_{i-1}}{\rho A \, \Delta x}. \tag{10.8}$$

These equations are those generated by the internal elements in the bond graph model of Figure 10.2. The first element and the nth element must be treated separately, since they are boundary elements.

From the bond graph of Figure 10.2, the first element yields

$$\frac{dq_i}{dt} = \frac{p_i}{\rho A \, \Delta x}, \tag{10.9}$$

and the nth element yields

$$\frac{dp_n}{dt} = F(t) - \frac{EA}{\Delta x} q_n. \tag{10.10}$$

To use this approach for an actual system, we would select the number of lumps we wish to include and let Δx equal the bar length divided by the number of lumps. This choice will then set the I and C parameters. We would like to choose the number of included lumps based upon some accuracy criteria; however, as mentioned previously, it is not possible to determine, a priori, the convergence to the continuous model for general uses of this lumped model as part of an overall system. In fact, this approach to representing inherently distributed systems is usually used in configura-

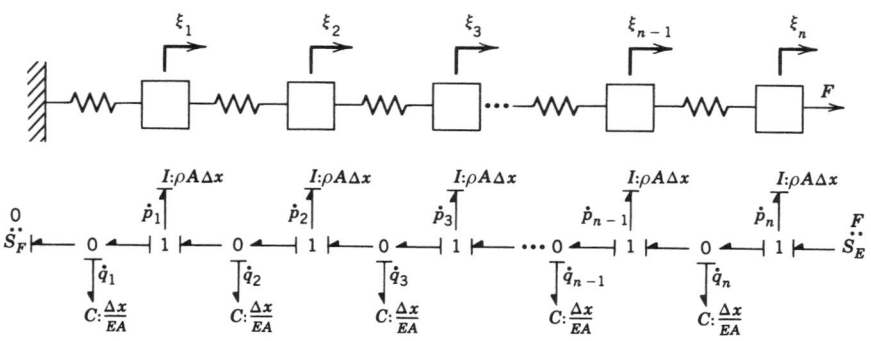

FIGURE 10.2. Bond graph finite-lump model of vibrating bar.

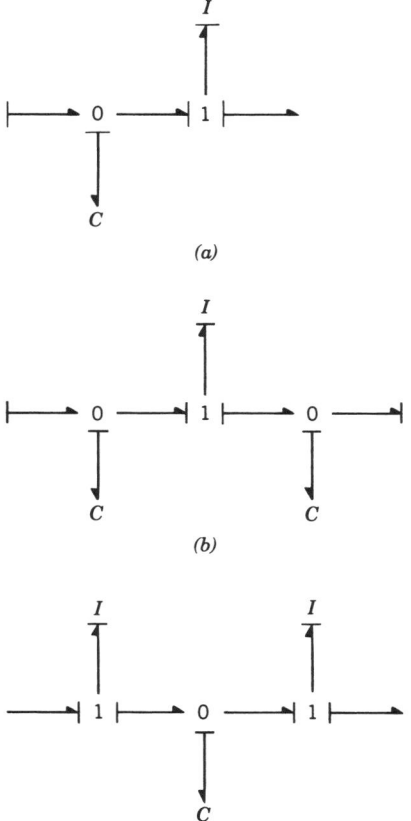

FIGURE 10.3. Typical configurations for finite-lump model of a bar.

tions as shown in Figure 10.3, the choice being dictated by causal considerations of the entire system.

Returning to Eqs. (10.1) and (10.2), the continuous model can be derived by letting $\Delta x \rightarrow 0$ while letting small changes in variables become differentials. Thus (10.1) becomes

$$A\frac{\partial \sigma}{\partial x} = \rho A \frac{\partial^2 \xi}{\partial t^2},\qquad (10.11)$$

for the internal domain of the bar, and (10.2) becomes

$$\sigma = E\frac{\partial \xi}{\partial x}.\qquad (10.12)$$

Combining (10.11) and (10.12) yields the continuous representation for longitudinal vibration of a bar with no external forcing:

$$E \frac{\partial^2 \xi}{\partial x^2} = \rho \frac{\partial^2 \xi}{dt^2}, \tag{10.13}$$

which should be recognized as the simple wave equation.

When dealing with continuum elements, external forcing, such as $F(t)$ in Figure 10.1, is considered to be distributed over the spatial domain of the element so that the force acting on any incremental element, $F(x, t)$, would be

$$F(x, t) = f(x, t) \Delta x, \tag{10.14}$$

where $f(x, t)$ is the force per unit length and could be distributed over the spatial dimension x of the rod in this example.

For a point force, such as $F(t)$ acting at $x = L$ in Figure 10.1, the distributed force per unit length is mathematically expressed through use of the delta function as

$$f(x, t) = F(t)\delta(x - L). \tag{10.15}$$

The total external force acting on the rod is obtained through integration with respect to x over the length, and turns out to be $F(t)$ as expected, since the area under the delta function is unity. The delta function correctly locates the force, since the delta function is zero except where its argument is zero. We shall make extensive use of the point force representation in some of the following development. For our rod example, Eq. (10.13) becomes

$$\frac{F(t)}{A}\delta(x - L) + E \frac{\partial^2 \xi}{\partial x^2} = \rho \frac{\partial^2 \xi}{\partial t^2}, \tag{10.16}$$

when the external forcing at the rod end is included.

We now have two types of representations for the bar. The first is a finite-lump representation, which will generate a high-order state space when many lumps are used. This lumped representation is viscerally satisfying because it is exactly the type of modeling we have been doing all along. We can imagine springs being compressed and stretched, causing the inertial elements in between to move. It is easy to "see" the physics of the system. Unfortunately, this finite-lump approach may not be particularly good as the number of included lumps increases. The state space grows very quickly while the accuracy does not.

The second representation for the bar is the linear partial differential equation (10.16). This recognizes that the bar is actually of infinite order, and as long as the simple wave equation is assumed to be valid, quite a lot can be said about the continuous behavior of the bar. However, if dissipation or nonlinear effects are included (which is more often the case than not), then we are generally forced to solve the continuum equations numerically. This always means backing up into a finite representation of one form or another. We shall see that the continuous representation yields a much more accurate and computationally efficient finite-order model than does the lumping process described above, even (especially!) as the continuum is approached by including more and more lumps.

Before proceeding to the use of the continuous model, transverse motion of a beam is discussed.

Transverse Beam Motion We assume here that the beam is long and slender and that it is of uniform cross-sectional area A, mass density ρ, Young's modulus E, shear modulus G, area moment of inertia I, and length L. Figure 10.4 shows the beam displaced at some instant of time. The spatial variable x defines a position along the beam, and $w(x, t)$ is the transverse displacement of the position x at the time t.

Also shown in Figure 10.4 is a finite element of the beam with the various forces and moments acting on it. The angle θ is the rotation of the neutral axis with respect to a horizontal reference, and the angle ϕ is the rotation of a plane cross section with respect to a vertical reference. If $\phi = \theta$, then we say there is no shear deformation. In general $\phi \neq \theta$, and we define the shear angle γ as

$$\gamma = \theta - \phi. \tag{10.17}$$

We shall derive the equations of motion for a Timoshenko model of the beam. Reference [2] discusses this model thoroughly. The Timoshenko model includes shear

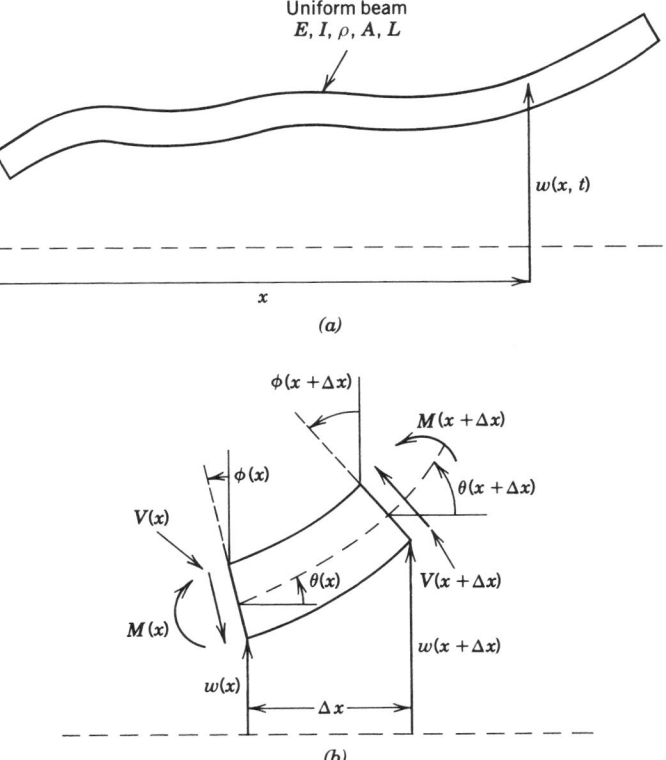

FIGURE 10.4. Uniform beam in transverse motion.

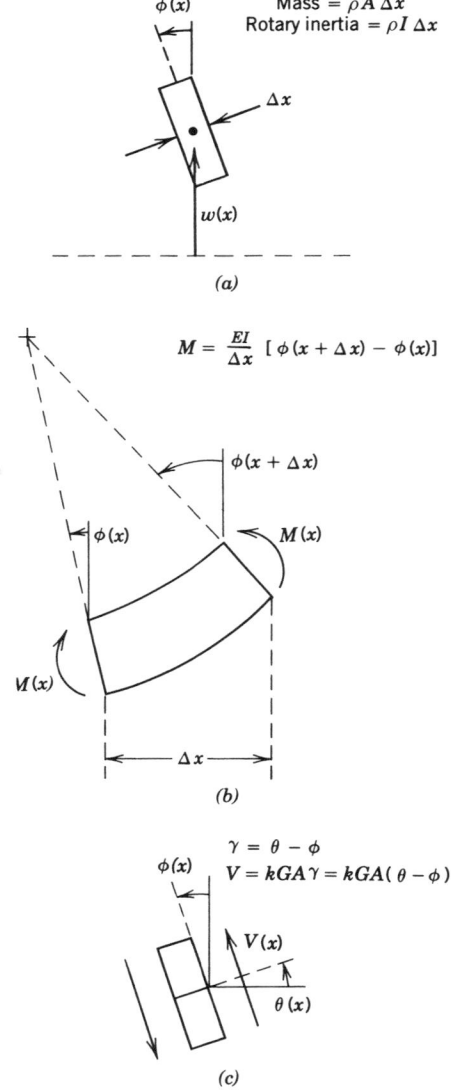

FIGURE 10.5. Beam elements for use in Timoshenko model.

deformation and the rotary inertia of each cross section. We shall later neglect these effects to reduce our model to the perhaps more familiar Bernoulli–Euler model of the beam.

From Figures 10.4*b* and 10.5 we can use Newton's laws for the element as

$$V(x + \Delta x) - V(x) = \rho A \, \Delta x \frac{\partial^2 w}{\partial t^2}, \qquad (10.18)$$

and

$$M(x + \Delta x) - M(x) + V(x + \Delta x)\,\Delta x = \rho I\,\Delta x \frac{\partial^2 \phi}{\partial t^2}, \qquad (10.19)$$

where it should be noted from Figure 10.5 that the inertia properties and compliant properties of the beam are assumed to be spatially separate.

From Figure 10.5b, we assume that the bending moment can be calculated as though pure bending took place. Thus,

$$M(x) = \frac{EI}{\Delta x}\left[\phi(x + \Delta x) - \phi(x)\right]. \qquad (10.20)$$

From Figure 10.5c, we assume that the shear $V(x)$ can be related to the shear angle by

$$V(x) = kGA(\theta - \phi), \qquad (10.21)$$

where

$$\theta(x) = \frac{w(x + \Delta x) - w(x)}{\Delta x}, \qquad (10.22)$$

the slope of the neutral axis. In Eq. (10.21), k is a parameter that accounts for the actual nonuniform shear distribution across a cross section.

Equations (10.18)–(10.22) suggest the lumped representation shown in Figure 10.6 and as a bond graph in Figure 10.7. Each element has mass $\rho A\,\Delta x$ and rotary inertia $\rho I\,\Delta x$. The spring elements k_b provide the bending moment and react to the difference in ϕ between successive elements, as indicated in Eq. (10.20). The spring elements k_s are the shear springs, and they react to the difference between θ_i

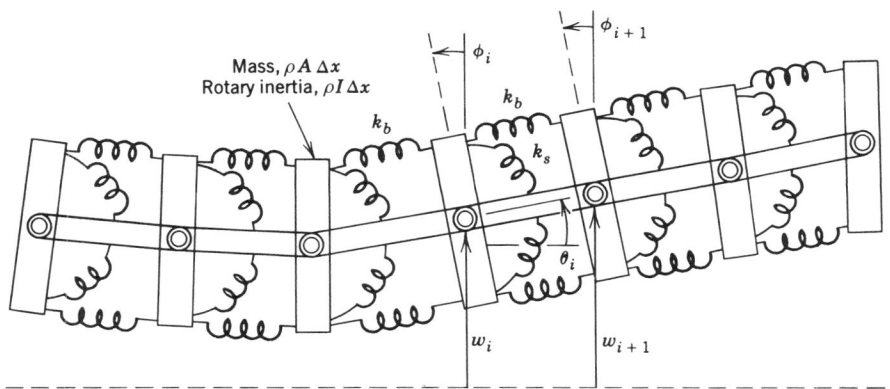

FIGURE 10.6. Lumped representation of Timoshenko beam.

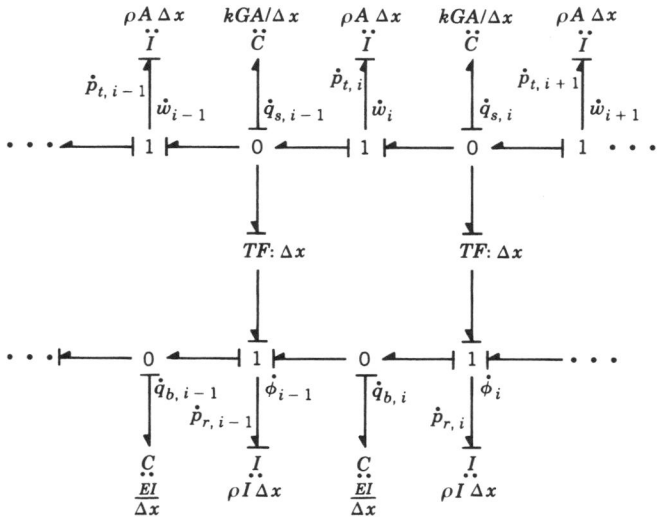

FIGURE 10.7. Bond graph finite-lump model of Timoshenko beam.

and ϕ_i, at any location i, as indicated in Eq. (10.21), where

$$\theta_i \cong \frac{w_{i+1} - w_i}{\Delta x}, \tag{10.23}$$

as indicated by Eq. (10.22).

The bond graph of Figure 10.7 shows the lumped model of the Timoshenko beam. The parameters would be chosen similarly to those for the bar done earlier. We would hope to base the number of included lumps upon some use criteria for the model, then simply select Δx to be the beam length divided by the number of included lumps.

As used in Figure 10.7, the bond graph possesses all integral causality. Thus state equations can be straightforwardly derived using the procedures described in an earlier chapter. Care must be exercised when using this representation as part of an overall model, to terminate the lumping process in a manner that preserves integral causality. This will greatly facilitate computation.

We are now in a position to derive the continuous representation of the beam. Before doing this, it is interesting to reduce our lumped model to a Bernoulli–Euler beam model. This model neglects rotary inertia and shear deformation. In Figure 10.7, we simply remove the lower inertial elements and the upper shear compliances. This results in the lumped model of Figure 10.8. Notice that integral causality still exists, although the shear forces associated with the upper 0-junctions are no longer set by the shear stiffness elements, but now are algebraically set by the bending stiffness.

The Timoshenko model can be reduced to a continuum representation by letting $\Delta x \to 0$ and letting small differences divided by Δx become derivatives. Thus, from

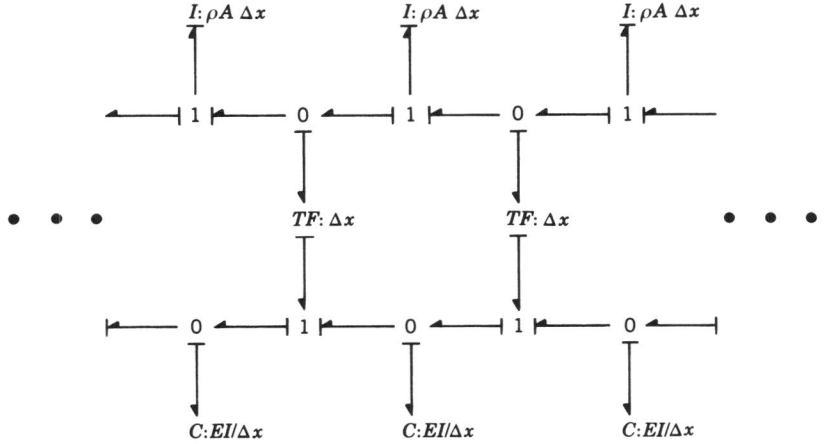

FIGURE 10.8. Bond graph finite-lump model for Bernoulli–Euler beam.

Eq. (10.18),

$$\lim_{\Delta x \to 0} \frac{V(x + \Delta x) - V(x)}{\Delta x} = \frac{\partial V}{\partial x} \tag{10.24}$$

and

$$\frac{\partial V}{\partial x} = \rho A \frac{\partial^2 w}{\partial t^2}. \tag{10.25}$$

Similarly, from Eq. (10.19)

$$\frac{\partial M}{\partial x} + V = \rho I \frac{\partial^2 \phi}{\partial t^2}, \tag{10.26}$$

and from (10.20),

$$M = E I \frac{\partial \phi}{\partial x}. \tag{10.27}$$

From Eqs. (10.21) and (10.22)

$$\theta = \frac{\partial w}{\partial x}, \tag{10.28}$$

and

$$V = k G A \left(\frac{\partial w}{\partial x} - \phi \right). \tag{10.29}$$

We can now combine Eqs. (10.25) and (10.29) to obtain

$$kGA\frac{\partial}{\partial x}\left(\frac{\partial w}{\partial x} - \phi\right) = \rho A\frac{\partial^2 w}{\partial t^2}, \tag{10.30}$$

and we can combine Eqs. (10.26), (10.27), and (10.29) to obtain

$$EI\frac{\partial^2 \phi}{\partial x^2} + kGA\left(\frac{\partial w}{\partial x} - \phi\right) = \rho I\frac{\partial^2 \phi}{\partial t^2}. \tag{10.31}$$

Equations (10.30) and (10.31) are the two partial differential equations in the two unknowns, $w(x, t)$ and $\phi(x, t)$, for the continuous uniform Timoshenko beam. These can be reduced to the Bernoulli–Euler model by neglecting rotary inertia in (10.31) and letting the shear stiffness become infinite, so that

$$\theta = \phi = \frac{\partial w}{\partial x}. \tag{10.32}$$

Then, substituting into Eq. (10.30) yields

$$-\frac{\partial}{\partial x}EI\frac{\partial^2 \phi}{\partial x^2} = -EI\frac{\partial^4 w}{\partial x^4} = \rho A\frac{\partial^2 w}{\partial t^2}, \tag{10.33}$$

or

$$EI\frac{\partial^4 w}{\partial x^4} + \rho A\frac{\partial^2 w}{\partial t^2} = 0. \tag{10.34}$$

If point forces $F_1(t)$ and $F_2(t)$ were acting on the Bernoulli–Euler beam at locations x_1 and x_2, respectively, along the beam, Eq. (10.34) would become

$$F_1\delta(x - x_1) + F_2\delta(x - x_2) - EI\frac{\partial^4 w}{\partial x^4} = \rho A\frac{\partial^2 w}{\partial t^2}. \tag{10.35}$$

While the Timoshenko model is a more realistic representation in that the high-frequency behavior is more accurately predicted, the solution is too cumbersome to pursue here. The interested reader should see Reference [2]. We will deal exclusively with the Bernoulli–Euler model.

Before continuing into the main emphasis of this chapter, it is perhaps useful to summarize the salient points that have been demonstrated here by example.

1. A lumped-parameter model can usually be found for a distributed system by following the derivation for the continuous equations up to the point where a typical element of the continuum is allowed to become infinitesimal and then simply retaining a finite size for the element. Such a lumped model may not be very accurate without retaining a large number of small elements.

2. The lumped-parameter representations have the advantage that they can be straightforwardly combined into an overall system model. Also, nonlinear effects pose no particular problems. The disadvantage to the lumped-parameter

approach described so far is that the model will generate a large state space and can cause severe computational problems due to large disparities in the time scales of the distributed portion of the model and the remainder of the system.

3. True continuum models can yield insight into system behavior when analytical solutions are available, which typically restricts us to the linear case. However, it is usually difficult to incorporate a continuum model into an overall system model with interactions with complex lumped systems (or other continuous systems) external to the continuum. When this is attempted, it becomes a problem of solving the partial differential equations for the continuum part, subject to complicated boundary conditions for the attached parts. This is really not a practical approach for obtaining an insightful representation of an overall system. It is much better to find an accurate, low-order, lumped-parameter representation for the continuous part of the system which can interact easily with external dynamics. We do this next.

10.2 LUMPED MODELS OF CONTINUA THROUGH SEPARATION OF VARIABLES

Many physical elements from which engineering systems are constructed possess a continuous distribution of inertia and compliance properties. The rod and beam presented previously are examples of such elements, as are stretched cables, membranes, plates, and shells. An aircraft fuselage is an example of a complex system composed of distributed elements. When it is justified, and when we as modelers choose to do so, these distributed elements can be represented by linear partial differential equations in space and time.

The actual physical dynamics of these distributed elements and systems are composed of propagating waves which reflect from boundaries and add together to produce what is actually observed at any point and time. There is mathematical development that we could pursue to demonstrate that continuous systems with distributed inertia and compliance properties possess wavelike behavior. However, this will not be carried out here, and the interested reader should see Reference [3].

Instead, we will pursue separation of variables, an alternative approach to representing the motion time history of linear, distributed wavelike systems. As will be seen, this approach yields an elegant lumped-parameter bond graph representation of distributed elements which can easily be coupled with bond graph models of attached dynamic systems to yield an extremely accurate overall system model possessing all the formulational and computational virtues that bond graphs exude. It is possible to pursue separation of variables in a very general way and demonstrate some very useful characteristics of the approach that apply to any appropriate continuum. Again, Reference [3] is recommended for that development. Here, we will demonstrate the approach through example.

The Bar Revisited The equation for forced longitudinal vibration of a bar is Eq. (10.16), repeated here as

$$\rho \frac{\partial^2 \xi}{\partial t^2} - E \frac{\partial^2 \xi}{\partial x^2} = \frac{F(t)}{A} \delta(x - L). \tag{10.36}$$

Separation of variables begins by temporarily setting the force $F(t)$ to zero and assuming that the displacement at any point and time, $\xi(x, t)$, can be separated into a product of a function of x only, $Y(x)$, and a function of time only, $f(t)$:

$$\xi(x, t) = Y(x) f(t). \tag{10.37}$$

Substitution into the homogeneous form of (10.36) yields

$$\rho Y \frac{d^2 f}{dt^2} - E f \frac{d^2 Y}{dx^2} = 0. \tag{10.38}$$

Dividing by $\rho Y f$ yields

$$\frac{1}{f} \frac{d^2 f}{dt^2} = \frac{E}{\rho} \frac{1}{Y} \frac{d^2 Y}{dx^2}. \tag{10.39}$$

The collection of terms on the left of (10.39) is, by assumption, dependent on time only, while the collection of terms on the right is dependent on x only. For this to be true for all x and t, each collection of terms must be equal to the same constant. By convention we let

$$\frac{1}{f} \frac{d^2 f}{dt^2} = -\omega^2, \tag{10.40}$$

with the result

$$\frac{d^2 f}{dt^2} + \omega^2 f = 0, \tag{10.41}$$

$$\frac{d^2 Y}{dx^2} + \frac{\rho}{E} \omega^2 Y = 0. \tag{10.42}$$

In order for a separated solution to exist, the two total differential equations (10.41) and (10.42) must be satisfied.

It is Eq. (10.42), for the spatial function $Y(x)$, that is of great interest. For this simple example, involving the simple wave equation, the spatial differential equation is straightforwardly solved:

$$Y(x) = A \cos kx + B \sin kx, \tag{10.43}$$

where

$$k^2 = \frac{\rho}{E} \omega^2. \tag{10.44}$$

To complete the solution for $Y(x)$, we must now specify the boundary conditions.

For the bar of Figure 10.1, the displacement at the left end, $\xi(0, t)$, is zero. At the right end, we have a choice of ways to express the boundary condition. We can include the force $F(t)$ in the boundary condition by balancing it with the normal stress at $x = L$. Thus

$$\sigma(L, t)A = F(t), \qquad (10.45)$$

or

$$EA\frac{\partial \xi}{\partial x}(L, t) = F(t). \qquad (10.46)$$

We can also think of the force as being applied a very small distance upstream of the right end, so that the right end itself is stress free. Thus,

$$\sigma(L, t) = E\frac{\partial \xi}{\partial x}(L, t) = 0. \qquad (10.47)$$

This replacement of the boundary forcing with a limiting form of forcing in the bar interior is analytically convenient, and it causes no problems, since we are only interested in solutions to within a finite scale of fineness. The boundary conditions for our example become

$$\xi(0, t) = Y(0)f(t) = 0,$$

or

$$Y(0) = 0, \qquad (10.48)$$

and

$$\frac{\partial \xi}{\partial x}(L, t) = \frac{dY}{dx}(L) \cdot f(t) = 0,$$

or

$$\frac{dY}{dx}(L) = 0. \qquad (10.49)$$

Application of (10.48) and (10.49) to (10.43) yields

$$A = 0, \qquad (10.50)$$

and

$$Bk \cos kL = 0. \qquad (10.51)$$

Equation (10.51) is called the frequency equation, and it is analogous to the characteristic equation for the eigenvalues of a lumped system.

The solution continues by recognizing that if B or k is zero, then $Y(x) = 0$ and $\xi(x, t) = 0$, a genuine but uninteresting solution. Instead, we let

$$\cos kL = 0, \tag{10.52}$$

with the result

$$k_n L = (2n - 1)\frac{\pi}{2}, \qquad n = 1, 2, 3, \ldots, \tag{10.53}$$

and, from Eq. (10.44),

$$\omega_n^2 = \frac{E}{\rho}k_n^2 = \frac{E}{\rho}\frac{(k_n L)^2}{L^2},$$

or

$$\omega_n = \sqrt{\frac{E}{\rho}}\frac{(2n - 1)}{L}\frac{\pi}{2}, \qquad n = 1, 2, 3, \ldots. \tag{10.54}$$

Accompanying each ω_n is a special shape function, $Y_n(x)$, from Eq. (10.43):

$$Y_n(x) = B_n \sin\left(k_n L\frac{x}{L}\right) = B_n \sin\left((2n - 1)\frac{\pi}{2}\frac{x}{L}\right), \qquad n = 1, 2, 3, \ldots. \tag{10.55}$$

The constants B_n are arbitrary, and it is convenient here to set them equal to unity.

The shape functions $Y_n(x)$ are frequently called eigenfunctions, mode shapes, or normal modes. Corresponding to each mode shape is a natural frequency ω_n. Figure 10.9 shows a few mode shapes for a bar fixed at the left end and free at the right. One interpretation of a mode shape and its frequency is that, if the rod is initially displaced in the shape of a mode and released, it will oscillate harmonically in that modal configuration at the corresponding modal frequency. Another fact about these modes and frequencies is that any motion time history that the bar executes, from any initial conditions, is a linear combination of the mode shapes oscillating at their respective natural frequencies. And another most useful fact about these mode shapes is that they are orthogonal, in that

$$\int_0^L Y_n(x)Y_m(x)\, dx = 0, \qquad n \neq m, \tag{10.56}$$

or

$$\int_0^L \sin\left((2n - 1)\frac{\pi}{2}\frac{x}{L}\right)\sin\left((2m - 1)\frac{\pi}{2}\frac{x}{L}\right) dx = \begin{cases} 0, & n \neq m, \\ L/2, & n = m, \end{cases} \tag{10.57}$$

for the bar example. We shall make use of this fact soon.

We are interested in the forced response of the bar, but so far we have dealt only with the unforced response. We now make use of the fact [3] that the response of the bar, forced or unforced, can always be expressed as a linear combination of the mode

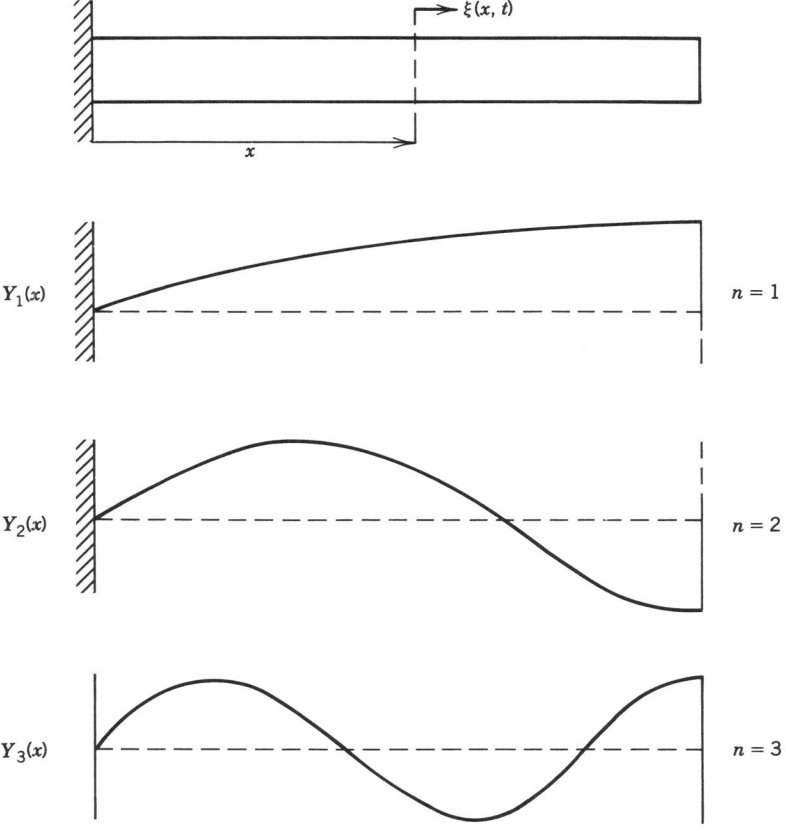

FIGURE 10.9. Mode shapes for the bar.

shapes. We write

$$\xi(x, t) = \sum_{n=1}^{\infty} Y_n(x)\eta_n(t) \tag{10.58}$$

and use this assumption in the forced equation (10.36). Thus

$$\sum_n \rho A Y_n \ddot{\eta}_n - \sum_n A E \frac{d^2 Y_n}{dx^2}\eta_n = F(t)\delta(x - L). \tag{10.59}$$

Now, multiply each term by the mth mode shape $Y_m(x)$, and integrate term by term over the bar length:

$$\sum_n \left(\int_0^L \rho A Y_n Y_m \, dx \right) \ddot{\eta}_n + \sum_n \left(\int_0^L \rho A Y_n Y_m \, dx \right) \omega_n^2 \eta_n$$

$$= \int_0^L F(t) \delta(x - L) Y_m \, dx, \tag{10.60}$$

where Eq. (10.42) has been used in the second term by substituting

$$\frac{d^2 Y_n}{dx^2} = -\frac{\rho}{E} \omega_n^2 Y_n. \tag{10.61}$$

As the summation is carried out for $n = 1, 2, \ldots$, each integral on the left-hand side of (10.60) is zero except for $n = m$, leaving only one equation,

$$\left[\int_0^L \rho A Y_m^2 \, dx \right] \ddot{\eta}_m + \left[\int_0^L \rho A Y_m^2 \, dx \right] \omega_m^2 \eta_m = \int_0^L F(t) \delta(x - L) Y_m \, dx. \tag{10.62}$$

The integral on the left side is called the modal mass m_m:

$$m_m = \int_0^L \rho A Y_m^2 \, dx, \qquad m = 1, 2, \ldots. \tag{10.63}$$

For the bar, it is simply

$$m_m = \frac{\rho A L}{2}, \tag{10.64}$$

independent of the mode number.

The integral on the right side of (10.62) is called the modal forcing function and is particularly simple to evaluate for point forcing. The delta function assures that no contribution to the integral occurs until $x = L$, at which point $Y_m(x)$ is virtually constant as we sweep across the delta function and equals $Y_m(L)$. Thus the integral becomes

$$\int_0^L F(t) \delta(x - L) Y_m(x) \, dx = F(t) Y_m(L) \int_0^L \delta(x - L) \, dx = F(t) Y_m(L). \tag{10.65}$$

Thus, Eq. (10.62) becomes

$$m_m \ddot{\eta}_m + k_m \eta_m = F(t) Y_m(L), \tag{10.66}$$

where m_m is the modal mass

$$m_m = \int_0^L \rho A Y_m^2(x) \, dx = \frac{\rho A L}{2}, \tag{10.67}$$

k_m is the modal stiffness

$$k_m = m_m \omega_m^2, \tag{10.68}$$

and $Y_m(x)$ and ω_m are known from the unforced analysis for the mode shapes.

The beauty of this approach to distributed systems is that, due to mode orthogonality, Eq. (10.66) is decoupled, and each $\eta_m(t)$ can be solved for separately and then combined with the mode shapes $Y_m(x)$ in Eq. (10.58) to produce the actual response,

$$\xi(x, t) = \sum_{m=1}^{\infty} Y_m(x) \eta_m(t). \tag{10.69}$$

Obviously, only a finite number of modes can be retained in the solution.

This development is becoming long, so we move quickly to a very interesting result. Define the modal momentum p_m as

$$p_m = m_m \dot{\eta}_m \tag{10.70}$$

and the modal displacement q_m as

$$q_m = \eta_m. \tag{10.71}$$

Then Eq. (10.66) can be written as

$$\frac{d}{dt} p_m = -k_m q_m + F(t) Y_m(L), \tag{10.72}$$

and

$$\frac{dq_m}{dt} = \frac{p_m}{m_m}. \tag{10.73}$$

We have reduced the mth modal equation to two first-order state equations in terms of bond graph energy variables, p_m and q_m. Figure 10.10 shows a bond graph that would duplicate these modal equations for $m = 1, 2, 3, \ldots$. Notice that integral causality exists for all m included modes and that the external force F properly excites each mode according to Eqs. (10.72) and (10.73). The transformers are simply the mode shapes from Eq. (10.55) evaluated at the location of the force. The bond graph also enforces that the flow on the external force bond, which must be $\partial \xi(L, t)/\partial t$, is the summation of the modal flows multiplied by appropriate transformer moduli:

$$\frac{\partial \xi}{\partial t}(L, t) = Y_1(L) \frac{p_1}{m_1} + Y_2(L) \frac{p_2}{m_2} + \cdots + Y_m(L) \frac{p_m}{m_m}, \tag{10.74}$$

which is exactly Eq. (10.69) evaluated for the velocity at $x = L$. In fact, from Eq. (10.69),

$$F(t) \frac{\partial \xi}{\partial t}(L, t) = \sum_{m=1}^{\infty} F(t) Y_m(L) \dot{\eta}_m(t), \tag{10.75}$$

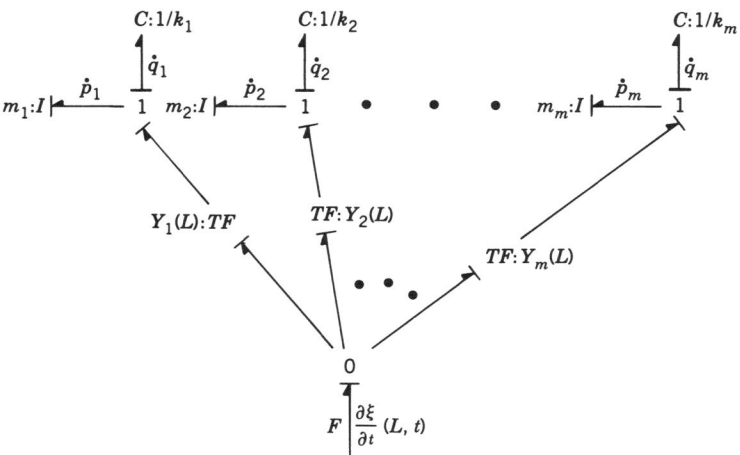

FIGURE 10.10. Bond graph for modal vibration of a bar forced at $x = L$.

but $F(t)Y_m(L)$ is the mth modal force, and $\dot{\eta}_m(t)$ is the mth modal flow. Thus the external force times the velocity at the input port equals the summation of the modal forces times corresponding modal flows. This is a statement of a power-conserving transformation, and the transformer fan of Figure 10.10 represents this transformation from physical variables to modal variables and back.

A good question at this point is: Why construct a bond graph when we have already tediously derived the equations (10.66) subject to (10.58)? The answer becomes obvious when we realize that the external force $F(t)$ is probably not a specified input, but rather resulted from interaction with some external system. For instance, if the end of the rod had an attached transducer which we decided to represent by a simple mass–spring–damper system, the bond graph of the interaction would be as shown in Figure 10.11. The modal dynamics are identical to Figure 10.10, and the external dynamic system simply attaches to the external port, where the end forcing $F(t)$ is now an internal force, as shown in Figure 10.11. Since integral causality still exists, equations for the entire interacting system can be straightforwardly derived. Of course, only a finite number of modes can be retained for the distributed part of the system.

If the rod end were terminated with a compliant member and then attached to another distributed rod, the model would be as shown in Figure 10.12, where n modes have been retained for each rod. Again, integral causality exists, and equation derivation is straightforward.

The reader who is gaining some familiarity with bond graph procedures should be starting to appreciate the utility and elegance of constructing models using this finite-mode approach. We will allow this good feeling to persist a little longer before discussing some of the pitfalls of this approach.

Before leaving the rod example, it should be noted that any location on the rod can be used as an output. We simply consider any desired output point as a location

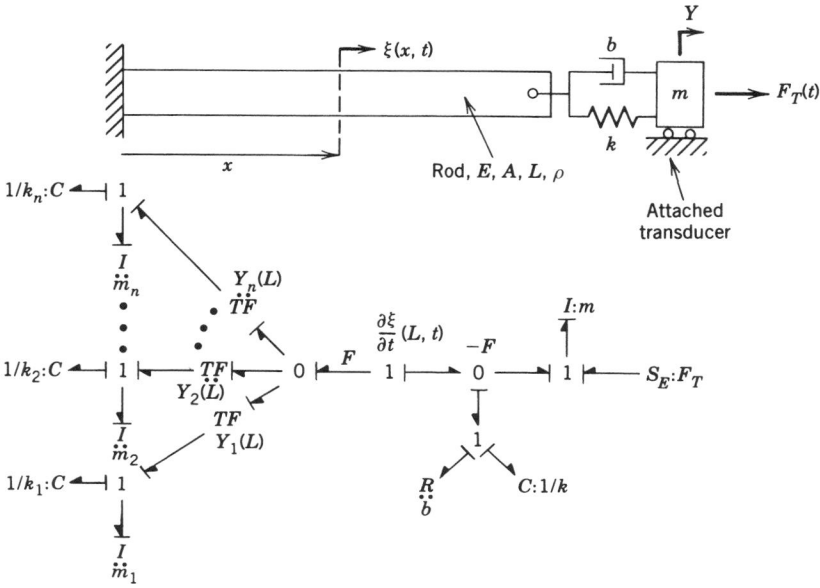

FIGURE 10.11. Finite-mode model of a rod with a lumped system attached at $x = L$.

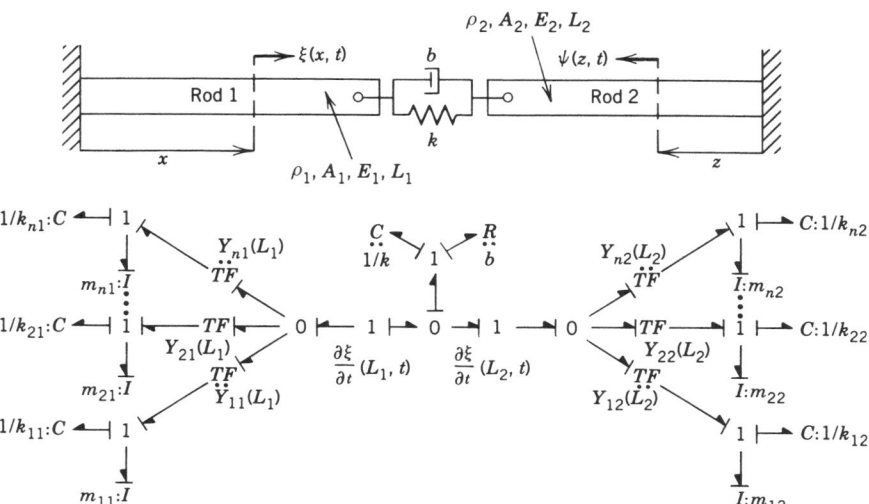

FIGURE 10.12. Two distributed rods interacting through a compliant dissipative element.

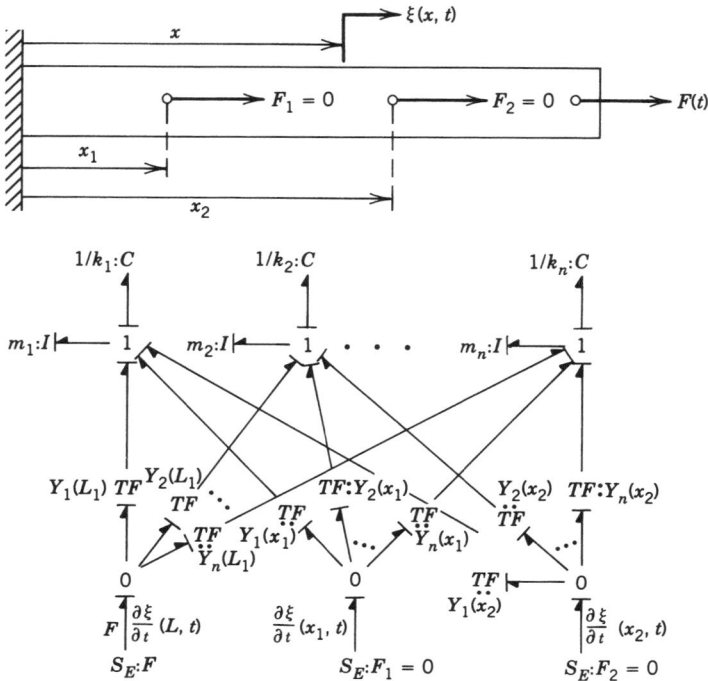

FIGURE 10.13. Bond graph showing how to obtain output at any point along the rod.

for a force input equal to zero. The bond graph for two such output points, x_1 and x_2, is shown in Figure 10.13. As can be seen, drawing every bond in the bond graph can become very messy. In practice we do not include output points on the bond graph. Instead we use Eq. (10.69) and recognize that

$$\frac{\partial \xi}{\partial t}(x_i, t) = Y_1(x_i)\frac{p_1}{m_1} + Y_2(x_i)\frac{p_2}{m_2} + \cdots + Y_n(x_i)\frac{p_n}{m_n}. \tag{10.76}$$

Some excellent questions that may have arisen by now are:

1. Is this approach only good for longitudinal vibration of rods fixed at one end and free at the other?
2. How many retained modes are enough?
3. What happens when causal flow inputs to the modes exist?

We will answer these questions in the following sections.

Bernoulli–Euler Beam Revisited One criticism of the finite-mode approach for representing distributed system dynamics is that the problem must be reformulated any time the boundary conditions change. Also, the boundary conditions in physical systems are ill-defined. For instance, if we decide to represent a truck frame as a beam, there are no definite boundary conditions. There are only additional physical elements, such as suspension components, engine mounts, and cab mounts, attached at various points. The frame is not pinned or built in, so what modes should we use?

It turns out that the most general modes to use for finite-mode modeling of distributed systems are those modes associated with force-free boundaries. A heuristic argument is that modes associated with fixed boundaries, such as the modes for our bar example evaluated at the fixed end, can never produce a response at the fixed end other than zero, regardless of how hard we might push on that location. The modes are all zero at the fixed end, and from Eq. (10.58), the motion will always be zero there. Thus, if we decide that some previously assumed fixed boundary was not really fixed, but actually has a mass attached, then we must reformulate the problem and use different mode shapes.

On the other hand, if we had started with force-free modes and later decided that a boundary was attached to ground through a stiff spring, the force-free modes are perfectly capable of adding up, according to Eq. (10.58), to a very small motion. The motion can even be zero if we decide to fix a previously force-free boundary. Thus, when using the finite-mode approach, it is best to use force-free modes unless the boundary conditions are well known and not subject to change.

The use of force-free modes adds a variation to the bond graph structure of Figure 10.13. This will be presented through the example of the Bernoulli–Euler beam described in Section 10.1. Unfortunately, the force-free modes are more difficult to derive than are modes for some other boundary conditions. However, once obtained (analytically, approximately, or even experimentally), they become part of our modeling arsenal, to be stored in our computer files, for use over and over again.

Figure 10.14 shows the Bernoulli–Euler beam acted upon by two external forces, $F_1(t)$, and $F_2(t)$. The equation of motion is repeated here from Eq. (10.35):

$$EI\frac{\partial^4 w}{\partial x^4} + \rho A\frac{d^2 w}{\partial t^2} = F_1\delta(x - x_1) + F_2\delta(x - x_2). \qquad (10.77)$$

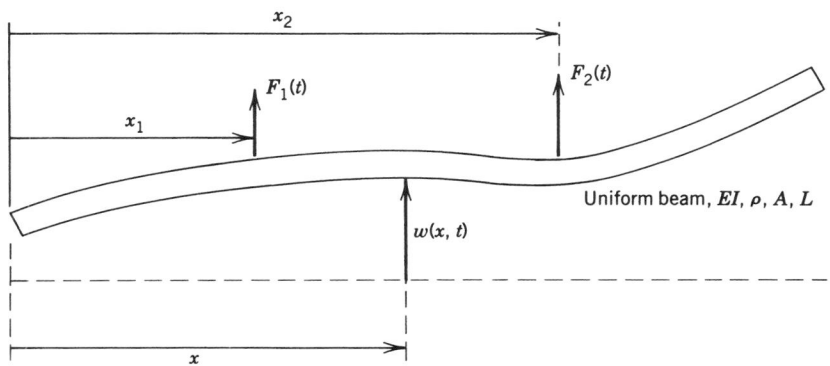

FIGURE 10.14. Uniform Bernoulli–Euler beam with point forcing.

The force-free boundary conditions are that no shear force and no moment exist at $x = 0$ and $x = L$. From Eqs. (10.26), (10.27), and (10.28), these conditions can be written in terms of $w(x, t)$ as

$$\frac{\partial^2 w}{\partial x^2}(0, t) = \frac{\partial^2 w}{\partial x^2}(L, t) = 0 \tag{10.78}$$

for the zero-moment constraint and

$$\frac{\partial^3 w}{\partial x^3}(0, t) = \frac{\partial^3 w}{\partial x^3}(L, t) = 0 \tag{10.79}$$

for the zero-shear constraint.

Proceeding as we did in the bar example, we try a separated solution for the homogenous form of (10.77) and assume

$$w(x, t) = Y(x)f(t). \tag{10.80}$$

Substituting into (10.77) yields

$$EI\frac{d^4 Y}{dx^4}f + \rho AY\frac{d^2 f}{dt^2} = 0. \tag{10.81}$$

Dividing each term by ρAYf yields

$$\frac{EI}{\rho A}\frac{1}{Y}\frac{d^4 Y}{dx^4} + \frac{1}{f}\frac{d^2 f}{dt^2} = 0. \tag{10.82}$$

This equation can be satisfied at all x and all t only if both terms equal the same constant. By convention we let the second term equal $-\omega^2$, yielding

$$\frac{d^2 f}{dt^2} + \omega^2 f = 0 \tag{10.83}$$

and

$$\frac{d^4 Y}{dx^4} - \frac{\rho A}{EI}\omega^2 Y = 0, \tag{10.84}$$

or

$$\frac{d^4 Y}{dx^4} - k^4 Y = 0, \tag{10.85}$$

where

$$k^4 = \frac{\rho A}{EI}\omega^2. \tag{10.86}$$

Equation (10.85) is a total differential equation which, when solved subject to the boundary conditions (10.78) and (10.79), will yield the mode shapes and associated mode frequencies. Using Eq. (10.80) in the boundary conditions yields

$$\frac{d^2Y}{dx^2}(0) = \frac{d^2Y}{dx^2}(L) = \frac{d^3Y}{dx^3}(0) = \frac{d^3Y}{dx^3}(L) = 0. \tag{10.87}$$

The spatial equation (10.85) has the general solution [3]

$$Y(x) = A\cosh kx + B\sinh kx + C\cos kx + D\sin kx. \tag{10.88}$$

We will not cover all the algebra here; however, using (10.87) in (10.88) yields the frequency equation

$$\cosh k_n L \cos k_n L = 1, \tag{10.89}$$

and the mode shape functions

$$Y_n(x) = (\cos k_n L - \cosh k_n L)(\sin k_n x + \sinh k_n x)$$
$$- (\sin k_n L - \sinh k_n L)(\cos k_n x + \cosh k_n x). \tag{10.90}$$

We now solve (10.89) for the special values of $k_n L$, and use these in (10.86) to obtain the mode frequencies as

$$\omega_n^2 = \frac{EI}{\rho A} \frac{(k_n L)^4}{L^4}. \tag{10.91}$$

At this point we must recognize that the force-free boundary conditions permit $\omega_n = k_n = 0$ to be a mode frequency. Using this information yields

$$\frac{d^4Y}{dx^4} = 0, \tag{10.92}$$

or

$$Y = c_1 x^3 + c_2 x^2 + c_3 x + c_4. \tag{10.93}$$

There are two possible solutions to (10.93) that satisfy the boundary conditions. They are $Y = \text{const}$ and $Y = ax + b$. These are called rigid-body modes, and it is convenient to think of them as rigid-body vertical translation of the entire beam and rigid-body rotation of the beam about the centrally located center of mass. Thus

$$Y_{00} = 1, \tag{10.94a}$$

and

$$Y_0 = x - \frac{L}{2}. \tag{10.94b}$$

The bending modes are given by (10.90) and are rather complicated functions. We will find that the rigid-body modes are orthogonal in the same sense as previously presented,

$$\int_0^L \rho A Y_n(x) Y_m(x)\, dx = 0, \qquad n \neq m, \tag{10.95}$$

for $n, m = 00, 0, 1, 2, 3, \ldots$.

The calculation of the forced response proceeds in a similar fashion to the bar example. We assume that the forced solution has the form

$$w(x, t) = \sum_{n=0}^{\infty} Y_n(x) \eta_n(t) \tag{10.96}$$

and use this in the forced equation (10.77). We then multiply each term by $Y_m(x)$ and integrate with respect to x from $x = 0$ to $x = L$. We then use the orthogonality property of the modes and obtain

$$\left(\int_0^L \rho A Y_n^2\, dx \right) \ddot{\eta}_n + \left(\int_0^L \rho A Y_n^2\, dx \right) \omega_n^2 \eta_n = F_1 Y_n(x_1) + F_2 Y_n(x_2), \tag{10.97}$$

which has exactly the same form as Eq. (10.66) for the bar example. The first zero-frequency mode yields

$$\left[\int_o^L \rho A(1)^2 dx \right] \ddot{\eta}_{00} = F_1 + F_2, \tag{10.98}$$

or

$$m \ddot{\eta}_{00} = F_1 + F_2, \tag{10.99}$$

which simply states that the external forces accelerate the center of mass of the beam. The other zero-frequency mode yields

$$\left[\int_0^L \rho A \left(x - \frac{L}{2} \right)^2 dx \right] \ddot{\eta}_0 = F_1 \left(x_1 - \frac{L}{2} \right) + F_2 \left(x_2 - \frac{L}{2} \right), \tag{10.100}$$

or

$$J_g \ddot{\eta}_0 = F_1 \left(x_1 - \frac{L}{2} \right) + F_2 \left(x_2 - \frac{L}{2} \right), \tag{10.101}$$

which simply states that the moment of the external forces about the center of mass produces angular acceleration $\ddot{\eta}_0$, where J_g is the centroidal moment of inertia of the beam.

The bond graph for a beam with force-free boundary conditions is shown in Figure 10.15. The rigid-body modes, which will be present whenever force-free boundaries are assumed, appear simply as inertia elements with no associated modal stiffness. The first inertia parameter is beam mass m, and the second is the beam moment

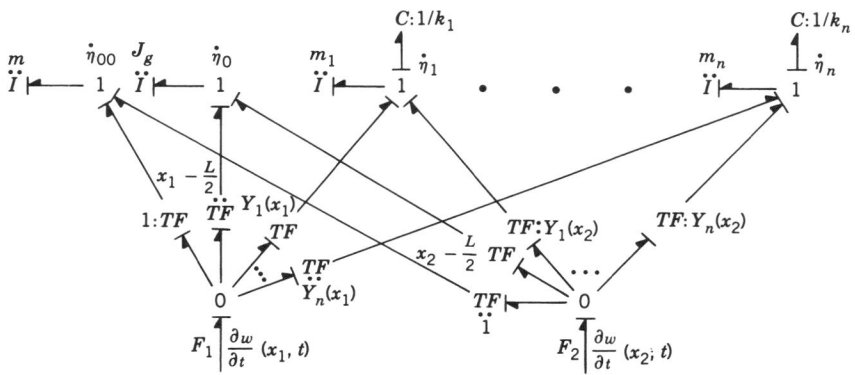

FIGURE 10.15. Bond graph for Bernoulli–Euler beam with force-free boundaries.

of inertia J_g. The —TF— elements connected to the rigid-body modes correctly apply the forces and moments to these elements. The rest of the structure is identical to the bar example. The modal masses are still, symbolically,

$$m_n = \int_0^L \rho A Y_n^2 dx, \qquad n = 1, 2, \ldots, \tag{10.102}$$

and the modal stiffnesses are still

$$k_n = m_n \omega_n^2.$$

Only this time the modes Y_n are given by Eq. (10.90), and the frequencies ω_n come from solving Eq. (10.89). Nobody said life would be easy, but the integration required in (10.102) need only be carried out once in your life in the nondimensional form

$$\frac{m_n}{m} = \int_0^1 Y_n^2 \left(\frac{x}{L}\right) d\frac{x}{L} \tag{10.103}$$

and then stored away for use whenever needed. The integration would probably be carried out numerically anyway.

The true virtue and elegance of this modeling procedure comes when we recognize that the external forces, F_1 and F_2 in this example, probably result from some attached dynamic system and can be appended to Figure 10.15 in straightforward bond graph fashion. For instance, the beam in Figure 10.16 has suspension components at the two ends and a one-degree-of-freedom system attached at $x = x_F$. Perhaps this is an optical bench with a laser turning mirror at $x = x_F$, and we are trying to design the suspension to isolate the mirror from ground motion. The finite-mode bond graph is also shown in Figure 10.16. We simply treat all attached components as external forcing and show these with 0-junctions for each force location. Each 0-junction is attached to each 1-junction with the modal components, through —TF— elements with moduli equal to appropriate mode functions evaluated at the location

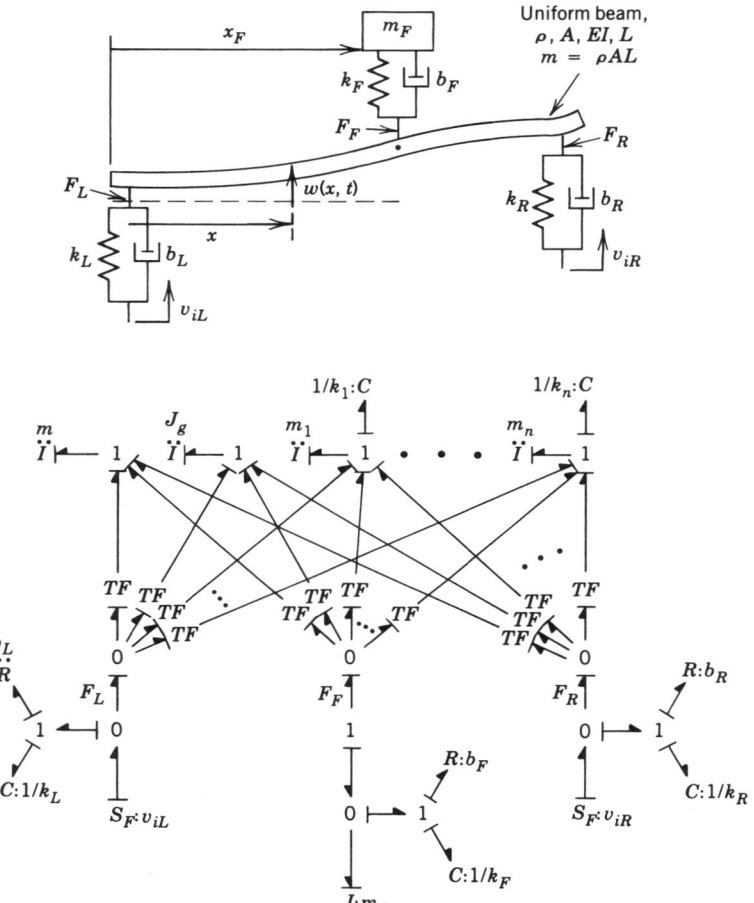

FIGURE 10.16. An example of a distributed system interacting with attached lumped systems.

of the force. The attached systems are then appropriately appended to the external 0-junction bonds to form a complete, low-order, very accurate system model.

While we have not been as general in our development as is possible, it should be realized now that the structure of Figure 10.15 is entirely general and represents *all* distributed structures that exhibit normal-mode behavior. The mode shapes change and the modal masses and stiffnesses change from system to system. But the bond graph structure is invariant. The modes can be obtained analytically, as was done here for the bar and the beam, or they can be obtained approximately, as described in Reference [4] for vibrating plates. They can be obtained by finite-element methods and then used in the bond graph structure to construct low-order, understandable models. And the modes can even be obtained experimentally using structural dynamic testing devices and programs. No matter how the modes are obtained, their finite-mode representation is as shown in Figure 10.15, and they are ready to interact with any external dynamic elements, even nonlinear ones.

Also, while not derived here, external moments can also be applied to the structure of Figure 10.15. If a moment is applied instead of a force, the associated —TF— elements, coming from the moment 0-junction, have the mode shape slopes, (dY_n/dx), evaluated at the location of the moment as their moduli instead of the mode shapes themselves.

10.3   GENERAL CONSIDERATIONS OF FINITE-MODE BOND GRAPHS

The reader who has begun to appreciate the virtues of bond graph modeling with regard to ease in putting all types of systems together, ease in identifying physical state variables, ease in deriving state equations, and ease in obtaining computer solutions through use of some very automated software, has certainly begun to appreciate the ease with which distributed systems can be made to interact with dynamic subsystems. Some questions still remain in the practical application of the finite-mode concept, and some of the most prominent are answered in this section.

How Many Modes to Retain   When a distributed component is part of an overall physical dynamic system model, we often have some idea of the frequency band of interest. This comes from our engineering judgment, and unfortunately, there is no substitute for exercising this judgment when we construct any system model, regardless of the modeling approach. If we were constructing a model of a long-haul truck interacting with the roadway unevenness, and decided to model the frame as a beam with the wheels attached through their suspension elements, we would make use of our knowledge that input frequencies from the road rarely exceed 20 Hz at normal driving speeds, wheel hop in trucks is around 15 Hz, and suspension frequencies are usually less than 2 Hz. It would then be obvious that very high-frequency beam modes would contribute very little to the response in the 20-Hz frequency range of interest.

We can quantify this thought somewhat by considering how a lightly damped mode responds to a harmonic input. Figure 10.17 shows a general finite-mode representation with one input location. The equation governing the contribution of the ith mode is

$$m_i \ddot{\eta}_i + k_i \eta_i = F Y_i. \tag{10.104}$$

If there is some damping associated with the mode, then (10.104) can be written as

$$\ddot{\eta}_i + 2\xi_i \omega_i \dot{\eta}_i + \omega_i^2 \eta_i = \frac{F}{m_i} Y_i, \tag{10.105}$$

where ξ_i is the mode damping ratio.

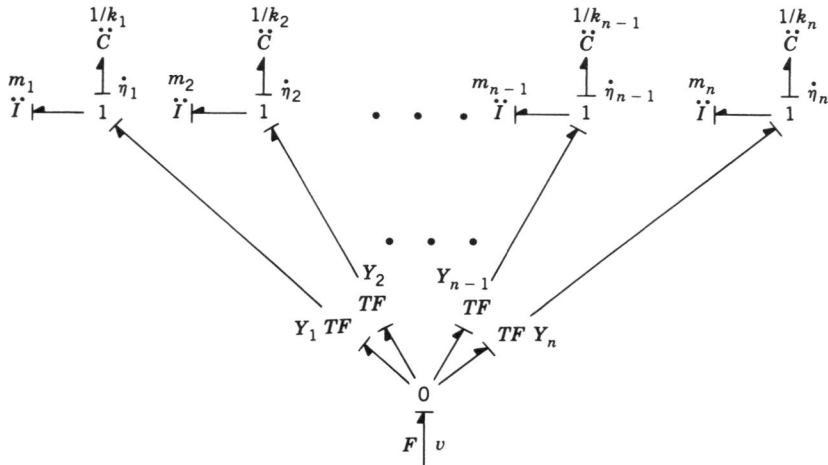

FIGURE 10.17. General finite-mode bond graph.

For F harmonic, the mode frequency response for $\dot{\eta}_i$ has the magnitude

$$\left| \frac{\dot{\eta}_i}{\frac{F}{m_i} Y_i} \right| = \frac{\frac{\omega}{\omega_i} \frac{1}{\omega_i}}{\left[\left(1 - \frac{\omega^2}{\omega_i^2}\right) + \left(2\xi_i \frac{\omega}{\omega_i}\right)^2 \right]^{1/2}}. \tag{10.106}$$

This response is sketched in Figure 10.18. At low excitation frequency, (10.106) becomes

$$\dot{\eta}_i|_{\omega/\omega_i \ll 1} \sim Y_i F \frac{1}{m_i \omega_i^2} \omega = Y_i \frac{F}{k_i} \omega, \tag{10.107}$$

and the response is governed by the modal stiffness, k. We call this low-frequency range the *stiffness-controlled region* in Figure 10.18. At frequencies near ω_i, the response is

$$\dot{\eta}_i|_{\omega/\omega_i \cong 1} \cong \frac{F}{m_i} Y_i \frac{1}{2\xi_i \omega_i} = F Y_i \frac{1}{b_i}, \tag{10.108}$$

where b_i is the mode damping constant, $b_i = 2\xi_i \omega_i m_i$. We call this frequency range near $\omega/\omega_i = 1$ the *resistance-controlled region*. Finally, at frequencies high compared to ω_i, we get

$$\dot{\eta}_i|_{\omega/\omega_i \gg 1} \sim \frac{F}{m_i} Y_i \frac{1}{\omega}, \tag{10.109}$$

and we see that the response is governed by the modal mass only. This high-frequency range is called the *mass-controlled region*.

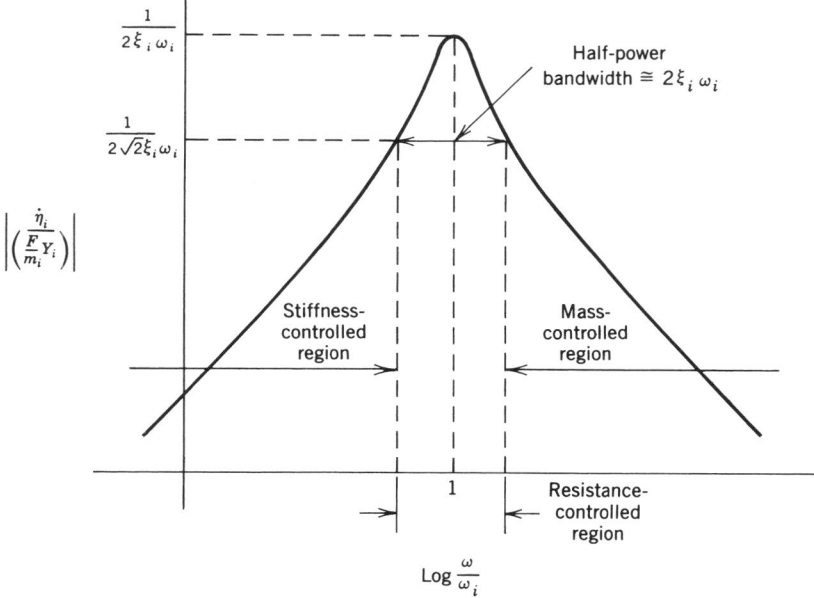

FIGURE 10.18. Frequency response of a single mode.

Assuming that we have reasonable knowledge of input frequency content, then we shall know which modes will be excited significantly below their mode frequencies and will therefore be in the stiffness-controlled region. For those infinity of modes in the stiffness-controlled region, their inertia and resistance are of no consequence and can be ignored. At this point we can either prune these high-frequency modes altogether or choose to retain their modal stiffness to improve the static stiffness representation of the distributed system.

In fact, for distributed systems with only one input location, the residual compliance due to the infinity of stiffness-controlled modes can be calculated straightforwardly. Consider Figure 10.19, where n dynamic modes have been retained, and n has been selected to extend well beyond the frequency range of interest. The remaining modes are represented by their modal compliance only. Notice that all the stiffness-controlled modes are in derivative causality.

The contribution of the modal flows $(\dot{q}_{n+1}, \dot{q}_{n+2}, \ldots)$ of the stiffness-controlled modes to the output velocity v we will call the *residual velocity*, v_R. From the bond graph,

$$v_R = Y_{n+1}\dot{\eta}_{n+1} + Y_{n+2}\dot{\eta}_{n+2} + \cdots, \qquad (10.110)$$

but

$$\dot{\eta}_{n+1} = \frac{1}{k_{n+1}}\dot{F}_{n+1}, \qquad \dot{\eta}_{n+2} = \frac{1}{k_{n+2}}\dot{F}_{n+2}, \qquad \text{etc.} \qquad (10.111)$$

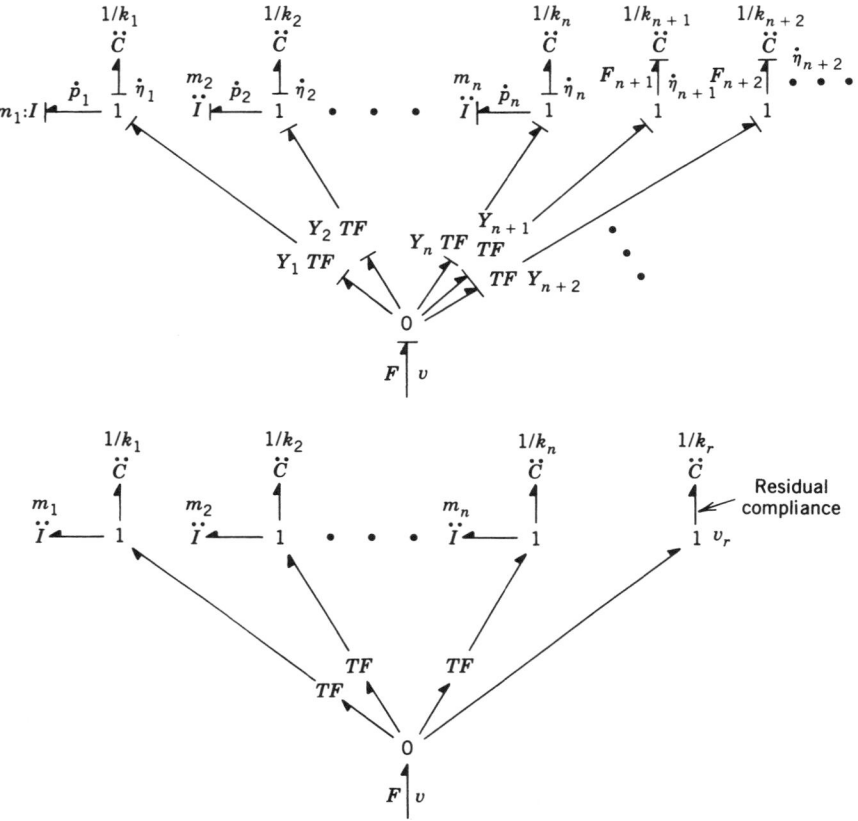

FIGURE 10.19. Finite-mode bond graph including residual compliance.

and

$$F_{n+1} = Y_{n+1}F, \qquad F_{n+2} = Y_{n+2}F, \qquad \text{etc.} \qquad (10.112)$$

Thus,

$$v_R = \frac{Y_{n+1}^2}{k_{n+1}}\dot{F} + \frac{Y_{n+2}^2}{k_{n+2}}\dot{F} + \cdots, \qquad (10.113)$$

or

$$v_R = \left(\sum_{j=1}^{\infty} \frac{Y_{n+j}^2}{k_{n+j}}\right)\frac{d}{dt}F. \qquad (10.114)$$

This same contribution to the output velocity would occur on introducing the single residual stiffness element k_r, where

$$\frac{1}{k_r} = \sum_{j=1}^{\infty} \frac{Y_{n+j}^2}{k_{n+j}},\qquad(10.115)$$

as shown also in Figure 10.19. The summation of Eq. (10.105) converges very quickly and is easily performed numerically.

It should be pointed out that, when more than one input is acting on the distributed system, the residual compliance is not calculated from the simple formula (10.115). Instead, a C-field representation is needed for the stiffness-controlled modes. The field constitutive properties can be determined, but frequently it is easier to simply eliminate the pruned modes altogether.

As a rule of thumb, it is recommended that modes be retained up to a frequency at least a factor of 2, but no more than a factor of 5, higher than the highest frequency of interest.

How to Include Damping The normal-mode representation for distributed system dynamics has been formulated without damping. And, strictly speaking, separation of variables and the general bond graph structure of Figures 10.15 and 10.17 are only correct in the absence of damping altogether or the presence of damping in a very special form. In general, distributed elements that we want to include in overall system models are lightly damped. We cannot identity exactly from where on the structure the damping arises. We only know for certain that, once excited, the energy will be dissipated. The mechanisms are complex; some dissipation is due to minute plastic deformations of the material, and some to the radiation of energy from material surfaces. One thing that is certain is that there are no little dampers attached to the structure that one can point to and therefore include in a model in order to represent the damping of the distributed elements.

It is customary to not attempt to perform detailed modeling of the damping mechanisms, but instead to include damping functionally by incorporating it into the individual modes. This is accomplished by simply appending R-elements to the mode oscillators of the modal bond graph. The general structure of Figure 10.17 will then become as shown in Figure 10.20. Treating each mode oscillator as an individual single-degree-of-freedom system, the resistance parameter R_i is set by

$$R_i = 2\xi_i \omega_i m_i,\qquad(10.116)$$

where ξ_i is the modal damping ratio. Typically structures are lightly damped, with ξ_i in the range of 0.01–0.1, where the high value would perhaps be representative of some composite structures. If experimental data were available, we could improve upon this seemingly crude inclusion of damping. However, usually the inclusion of modal damping as described above produces very accurate and useful models.

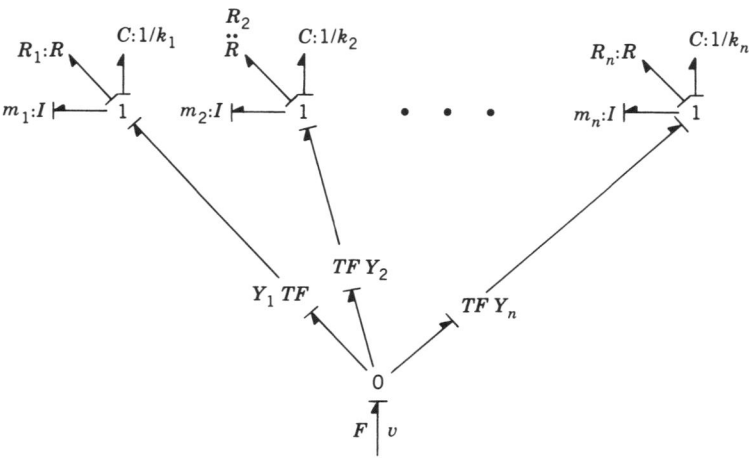

FIGURE 10.20. General finite-mode bond graph with modal damping.

Causality Consideration for Modal Bond Graphs From Chapter 2, the reader should be familiar with the concept of causality and with the incredible information it supplies with regard to equation formulation and the computability of one's model. We know that integral causality throughout a model indicates that formulation will be straightforward. If any derivative causality exists or some arbitrary assignment of causality is required, then we know that algebraic problems exist in the model, and formulation of equations can be difficult, if not impossible, when nonlinear elements are involved.

Another implication of causality is with regard to the general unification of some part of a system model with the models of other parts of the system. For instance, the bond graph modal representation of Figure 10.20 is not a complete model, because we intend to connect the external bond, carrying F and v, with some other dynamical elements. In Figure 10.20, the force F is a causal input to the modes, and the velocity v is a causal output. The bond graph possesses all integral causality. We thus could straightforwardly formulate the state equations for this bond graph fragment, and, as long as we agreed to use this fragment only in its present causality, these equations would never change. We would simply need to know the input F and the equations would deliver the output v. Of course, any attached dynamical model will have some causal restrictions, since it must have F as an output and accept v as an input. Such is the construction of computable models.

As we have seen in every finite-mode model presented so far, integral causality exists for all I- and C-elements, and the effort variable has been the causal input on all external bonds. Consider Figure 10.21, where a general n-mode representation is shown along with two external input locations. The causal effort F_1 is an input at one input location, but the causal flow v_2 is an input at the other location. The reader should assign causality to show that, with a causal flow input, one I-element in the modes must be in derivative causality. For this example, the nth modal inertia has

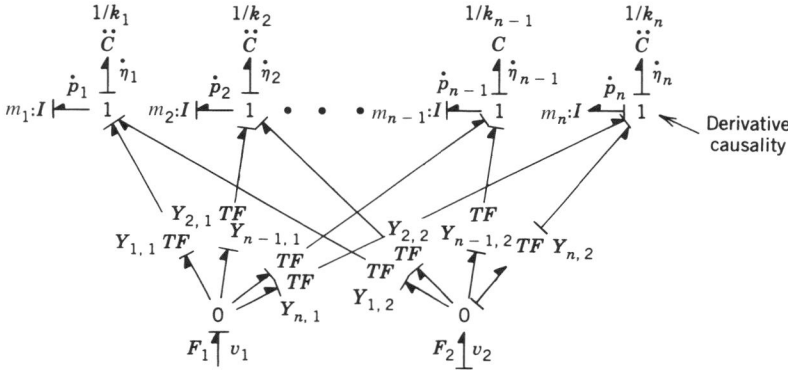

FIGURE 10.21. Finite-mode bond graph with causal flow input at one input location.

been selected to be put into derivative causality. In general, it can be shown that for each additional causal flow input to the modes, one additional modal inertia will be forced into derivative causality.

One result of the derivative causality is that a state variable is lost, since p_n is no longer a state variable, and the accuracy of the n-mode description is somewhat less. This really presents no problem, since we can always include additional modes and in fact have probably already included more than enough modes if recommended procedures have been followed in formulating the original model.

The real disadvantage of allowing derivative causality is in the formulation of system equations for analysis or simulation. If we attempt to use procedures described in Chapter 5 to deal directly with the derivative causality, we write

$$p_n = m_n \dot{\eta}_n \qquad (10.117)$$

and follow the causality for $\dot{\eta}_n$, obtaining

$$p_n = m_n \frac{1}{Y_{n,2}} \left(v_2 - Y_{n-1,2} \frac{p_{n-1}}{m_{n-1}} - \cdots - Y_{2,2} \frac{p_2}{m_2} - Y_{1,2} \frac{p_1}{m_1} \right). \qquad (10.118)$$

We see that p_n depends upon all $n-1$ other modal momenta in the mode description, and it also depends upon the external input, v_2. If we attempt to write the state equations for I-elements in integral causality, we shall find that $\dot{p}_1, \dot{p}_2, \ldots, \dot{p}_{n-1}$ will all require $\dot{p}_n$ in their equations, and $\dot{p}_n$, from (10.118), is

$$\dot{p}_n = m_n \frac{1}{Y_{n,2}} \left(\dot{v}_2 - Y_{n-1,2} \frac{\dot{p}_{n-1}}{m_{n-1}} - \cdots - Y_{2,2} \frac{\dot{p}_2}{m_2} - Y_{1,2} \frac{\dot{p}_1}{m_1} \right). \qquad (10.119)$$

Thus, all the modal momenta are coupled, and obtaining explicit state equations becomes very difficult. To make matters worse, the derivative of the input velocity, $\dot{v}_2$, is also required. The velocity v_2 probably is causally determined by some inertial elements external to the modes; thus additional momentum variables will couple with the modal momenta, making the algebraic problem worse. In fact, this coupling of dynamic elements external to the mode structure practically precludes the derivation of equations for the modes as separate entities to be used over and over with different external dynamic systems.

A matter of computational importance is that when derivative causality is permitted, we invariably end up with a mode shape factor (transformer modulus) in the denominator of the algebraic expression (10.119). In Eq. (10.119), $Y_{n,2}$ (mode shape n evaluated at position 2) is in the denominator of the right-hand side. If it should happen that $Y_{n,2}$ is zero or very small (position 2 is at or very near a node of mode n), then computational problems will result. Reference [5] discusses this problem further. The bottom line of this discussion is that it would be very nice to avoid derivative causality when using finite-mode models for distributed systems.

It turns out that the derivative causality can be avoided while still permitting causal flow inputs to the mode. We simply include an additional modal compliance in the model *without* the associated modal inertia. And we include as many additional modal compliances as there are external inputs with flow-in causality. This addition of modal C-elements without their associated I-elements will pose no accuracy problems, since the original choice of n modes was based upon ensuring that the higher frequency modes were already stiffness controlled. Thus additional modes are certainly stiffness controlled, and their modal inertia is of no consequence to the system response.

Figure 10.22 shows the general n-mode structure of Figure 10.21 repeated with the addition of the $(n + 1)$st modal compliance. Notice that v_2 is still a causal flow input to the modes, but no derivative causality exists. Thus, no algebraic problems exist, and equation formulation is straightforward.

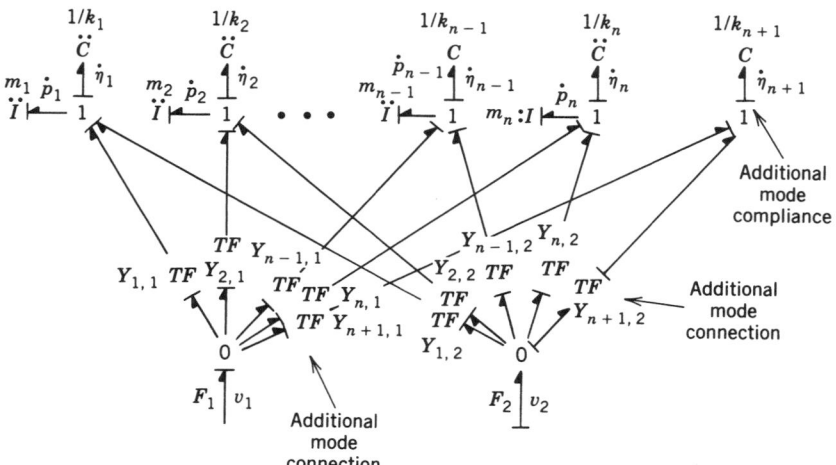

FIGURE 10.22. Finite-mode bond graph with causal flow input plus an additional modal compliance.

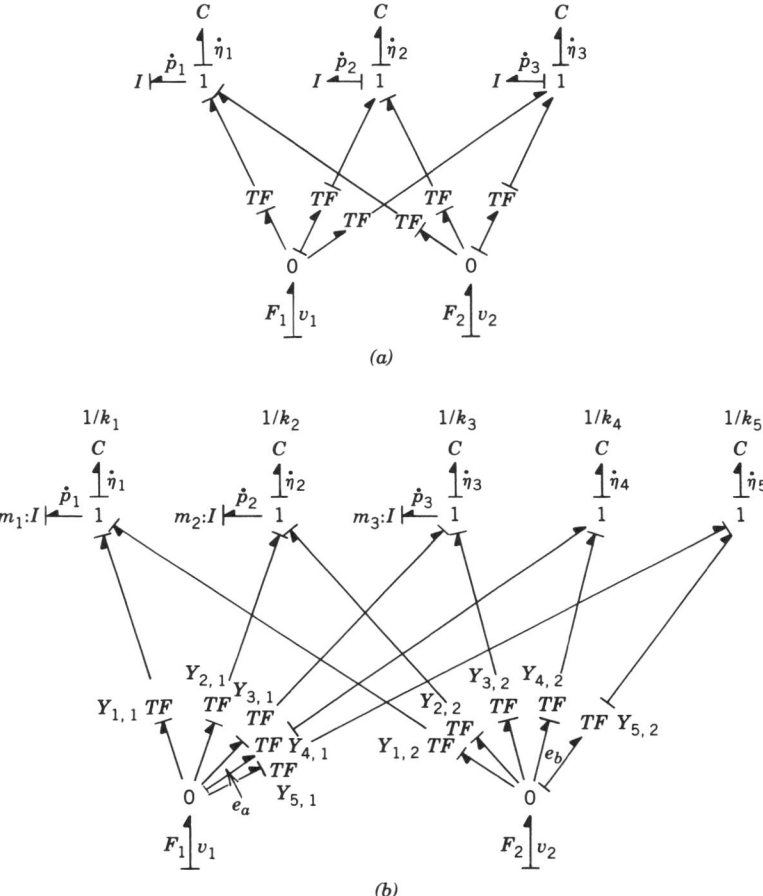

FIGURE 10.23. Finite-mode bond graph with two causal flow inputs.

Before leaving this section it must be admitted that the act of appending additional modal compliances to avoid causality problems does present an algebraic problem when more than one causal flow input is present. Figure 10.23 exposes this problem. In part a, a three-mode model is shown with two causal flow inputs. Readers should convince themselves that the only permissible causality forces modal inertias 2 and 3 into derivative causality. Figure 10.23b shows the same model after using the trick of appending two additional modal compliances without their associated inertias. Readers should assign causality to this model and show that, after assigning integral causality to all I- and C-elements, there are still some unassigned bonds. This occurrence always indicates the presence of an algebraic loop. The effort e_a has been chosen arbitrarily to set the effort on the 0-junction indicated, and as a result, the causal picture is complete. The effort e_b has also been labeled in Figure 10.23b.

Unlike the algebraic problem when derivative causality is permitted, the present algebraic loop can be solved, once and for all, internally to the bond graph structure. It is straightforward to derive

$$e_a = \frac{1}{Y_{4,1}} \left(k_4 \eta_4 - Y_{4,2} e_b \right) \tag{10.120}$$

and

$$e_b = \frac{1}{Y_{5,2}} (k_5 \eta_5 - Y_{5,1} e_a). \tag{10.121}$$

This can obviously be solved explicitly for e_a and e_b and then used in the state-space equation formulation. The algebraic problem created by appending extra modal compliances in order to avoid derivative causality is very manageable, while dealing with the derivative causality is generally not.

10.4   ASSEMBLING OVERALL SYSTEM MODELS

In the previous sections a procedure was developed for representing all types of distributed continua as finite-mode bond graph models. Only beams and bars were introduced through example; nevertheless, multidimensional elements can be represented with the same bond graph structure as was developed for beams and bars. Only the mode shapes, modal masses, and modal frequencies differ from continuum to continuum. This statement is made somewhat tongue in cheek, since obtaining modal information for multidimensional structures can be virtually impossible.

If an overall structure is composed of many (more or less conventional) elements, then an overall structural model can be constructed by appropriately connecting finite-mode models of the elements. Figure 10.24 shows an A-frame for a ground vehicle composed of three uniform beams. In Figure 10.24b the frame is shown with internal forces and moments exposed. The forces F_1, F_2, and F_3 are external forces which ultimately will represent suspension forces coming through the force locations. We are assuming that beams 1 and 2 do not twist due to the motion of beam 3.

In Figure 10.25 a first cut at an assembled model is shown. The words "Modes of beam 1" and so forth are used to represent the 1-junctions with attached modal oscillators and the two rigid-body inertias. The forces and moments are shown with 0-junctions, and the fan of bonds emanating from the 0-junctions must connect to each 1-junction in the appropriate mode set through transformers whose moduli are force-free beam mode shapes evaluated at the location of the forces. Thus V_1 acts on beams 2 and 1, M_1 acts on beams 2 and 1, V_2 acts on beams 2 and 3, and V_3 acts on beams 3 and 1. Causality considerations tell us that derivative causality will exist because beam contact interfaces will share common velocities and angular velocities. Thus, our second and final model will have additional modal compliances to rid us of the derivative causality. Reference [6] shows this example fully developed.

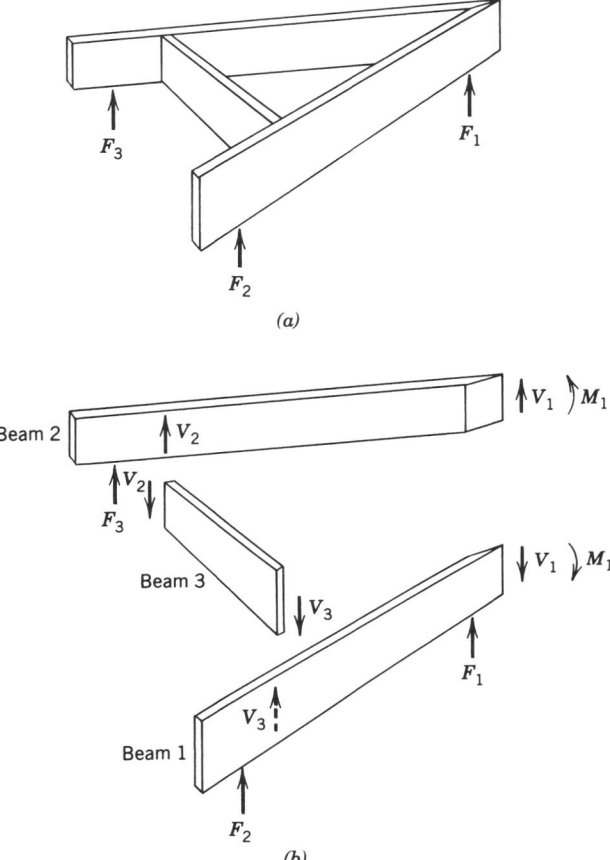

FIGURE 10.24. An A-frame composed of three uniform beams.

As a final example (Figure 10.26), consider a longitudinal bar attached at $z = z_1$ to a uniform beam. A spring k is attached to the beam at one end, and a velocity excitation $v_i(t)$ exists at the left end of the bar. Frequency considerations have led us to a two-mode (plus the rigid-body mode) representation for the bar and a one-mode (plus two rigid-body modes) representation for the beam. Force-free modes are used for both distributed elements in order to have the most general representation possible. The final bond graph is also shown in Figure 10.26. Notice that the bar has one causal flow input; thus one additional modal compliance is included for the bar. The beam also has one causal flow input; thus the beam modes also require an additional modal compliance. The overall model has all integral causality and is ready for straightforward formulation.

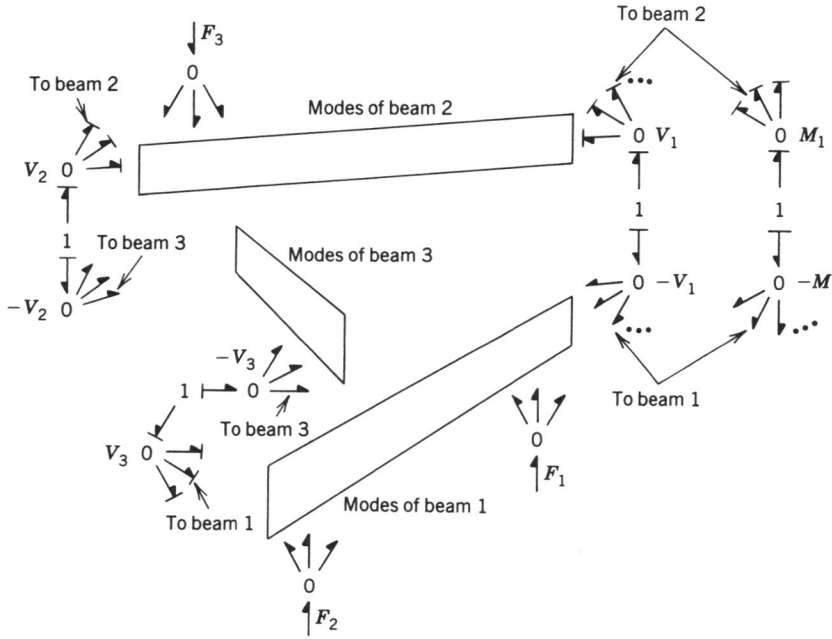

FIGURE 10.25. Schematic of the finite-mode bond graph for the A-frame.

10.5   SUMMARY

This chapter has dealt with the modeling of inherently distributed systems so that the distributed elements could be included in an overall system model in a reasonable manner. The first modeling approach described dealt with the use of finite lumps, contiguously aligned, and retained in sufficient numbers that low-frequency accuracy would result. This approach is all right for very crude approximations, using one or two lumps, but not practical when many lumps are required.

The main emphasis of this chapter is the finite-mode approach for obtaining very accurate, low-order, overall system models incorporating all types of distributed and lumped dynamic elements. It is hoped that the reader appreciates the elegance and utility of this approach. If not, it is suggested that References [6], [7], and [8] be read. These show applications of finite-mode modeling to very realistic systems, such as the assembled frame of a ground vehicle, the dynamics of the structure holding the world's largest laser, and the dynamics of the human lung.

REFERENCES

[1] D. L. Margolis, "A Survey of Bond Graph Modeling for Interacting Lumped and Distributed Systems," *J. Franklin Inst.*, **19**, No. 1/2 (Jan. 1985).

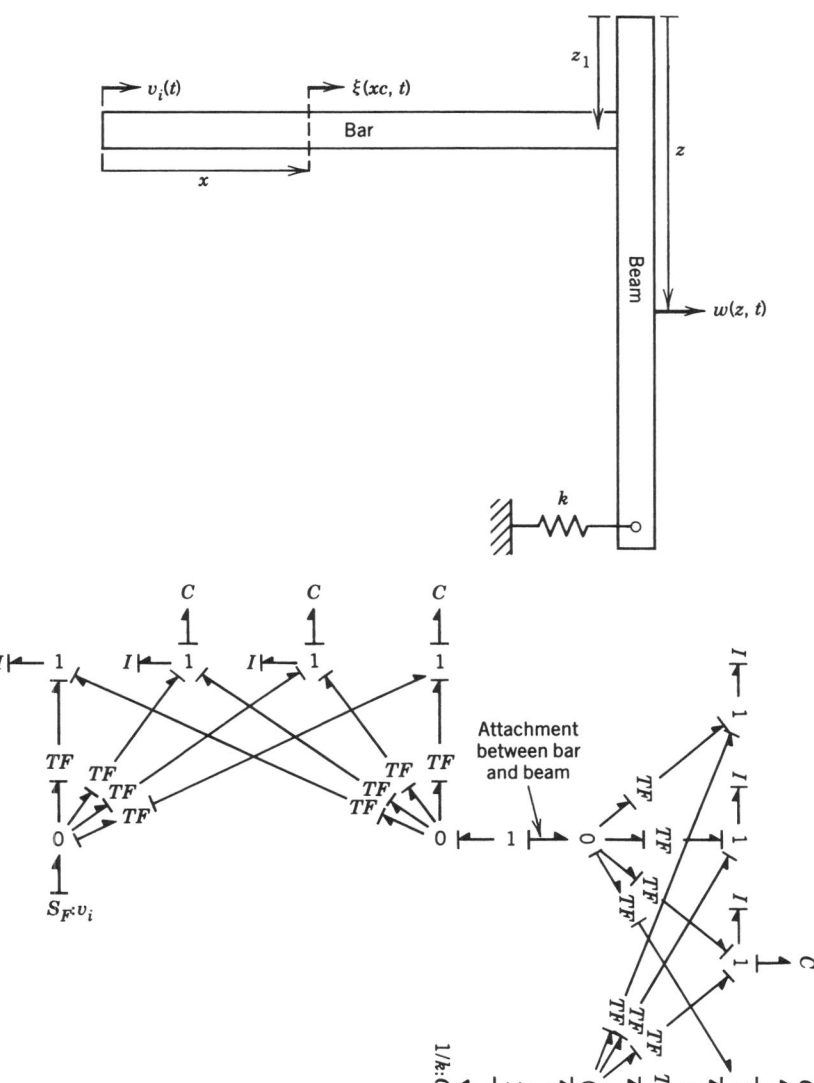

FIGURE 10.26. A bar interacting with a beam.

[2] H. Crandall et al., *Dynamics of Mechanical and Electromechanical Systems*, New York: McGraw-Hill, 1968.

[3] L. Meirovitch, *Analytical Methods in Vibrations*, New York: Macmillan, 1967.

[4] A. W. Leissa, "Vibration of Plates," NASA Report SP-160, 1969.

[5] D. L. Margolis, "Bond Graphs for Distributed System Models Admitting Mixed Causal Inputs," *ASME J. Dyn. Syst. Meas. Control*, **102**, No. 2, 94–100 (June 1980).

[6] D. L. Margolis, "Bond Graphs, Normal Modes, and Vehicular Structures," *Vehicle Syst. Dyn.*, **7**, No. 1, 49–63 (1978).

[7] D. L. Margolis, "Dynamical Models for Multidimensional Structures Using Bond Graphs," *ASME J. Dyn. Syst. Meas. Control*, **102**, No. 3, 180–187 (Sept. 1980).

[8] D. L. Margolis and M. Tabrizi, "Acoustic Modeling of Lung Dynamics Using Bond Graphs," *J. Biomech. Eng.*, **105**, No. 1, 84–91 (Feb. 1983).

PROBLEMS

10-1   Using the simple lumping technique of Section 10.1, construct as low-order a model as possible for the bars shown in the Figure such that integral causality exists for all elements.

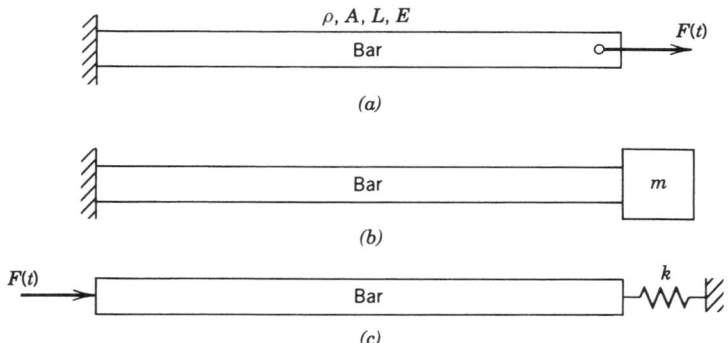

10-2   For the systems of Problem 10-1, assign causality and derive state equations. Write down, using symbols from Problem 10-1, reasonable parameters for each system.

10-3   A uniform Bernoulli–Euler beam is suspended at its two ends and has a dynamic system attached at its center. Construct a low-order, finite-lump model using lumps similar to those in Figure 10.8.

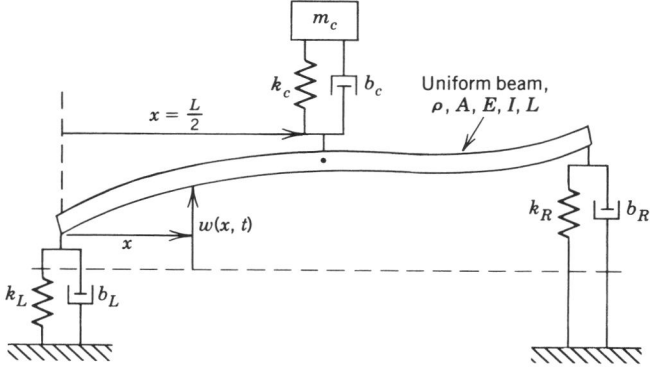

10-4 For the system of Problem 10-3, derive a state representation.

10-5 The transverse motion of a tightly stretched string is governed by the simple wave equation, identically to the bar in longitudinal motion,

$$T\frac{\partial^2 w}{\partial x^2} + F\delta(x - x_1) = \rho\frac{\partial^2 w}{\partial t^2}.$$

In the figure the boundaries are force free because the attachment points are massless. The boundary conditions are

$$\frac{\partial w}{\partial x}(0, t) = \frac{\partial w}{\partial x}(L, t) = 0.$$

Assume a separated solution

$$w(x, t) = Y(x)f(t),$$

and derive the frequency equation and mode shapes, including the rigid-body mode.

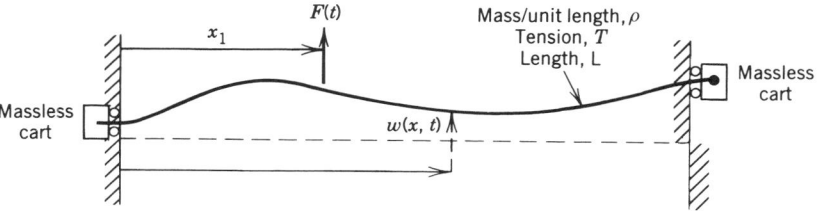

10-6 For Problem 10-5, assume the forced response is given by

$$w(x, t) = \sum_{n=0}^{\infty} Y_n(x)\eta_n(t),$$

and derive expressions for the modal masses and stiffnesses. Convince yourself that a bond graph structure similar to Figure 10.17 (only including the rigid-body mode) is appropriate for this system.

10-7 Two tightly stretched strings are connected as shown in the figure. Construct a finite-mode model for this system including two modes for each string. Since the string ends are fixed, use the mode shapes for fixed ends, and use the symbol $Y_n(x)$ to represent these modes.

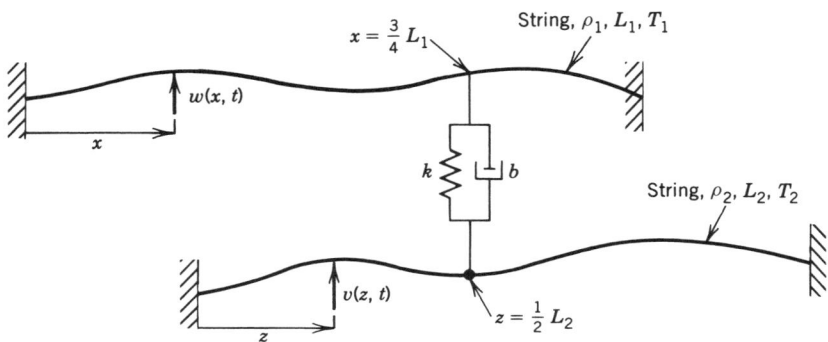

10-8 A tightly stretched string has dynamic elements on its boundaries as shown in the figure. Construct a bond graph for this system including two dynamic modes for the string (plus the rigid-body mode, of course). Assign causality, and convince yourself that the derivative causality exists. Use symbols $Y_n(x)$ for the mode shapes and modal parameters.

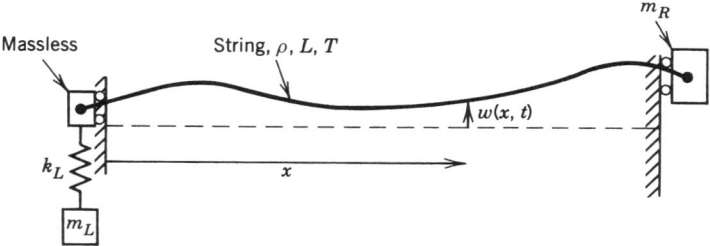

10-9 For the system of Problem 10-8, add an additional modal compliance, and show that no derivative causality exists. Derive the state equations.

10-10 The figure shows a bar, fixed at its left end, with three different right-end boundary conditions. In part a of the figure, the right end is fixed, in b it is attached to a spring, and in c it is attached to a velocity source. For reference, the mode shapes and frequencies for the fixed–fixed bar are

$$\omega_n^2 = \frac{E}{\rho}\left(\frac{n\pi}{L}\right)^2, \quad Y_n(x) = \sin n\pi\frac{x}{L}, \qquad n = 1, 2, \ldots,$$

and the mode shapes and frequencies for a bar fixed at $x = 0$ and free at $x = L$ are

$$\omega_n^2 = \frac{E}{\rho}\left((2n-1)\frac{\pi}{2L}\right)^2, \quad Y_n(x) = \sin(2n-1)\frac{\pi}{2}\frac{x}{L}, \qquad n = 1, 2, \ldots.$$

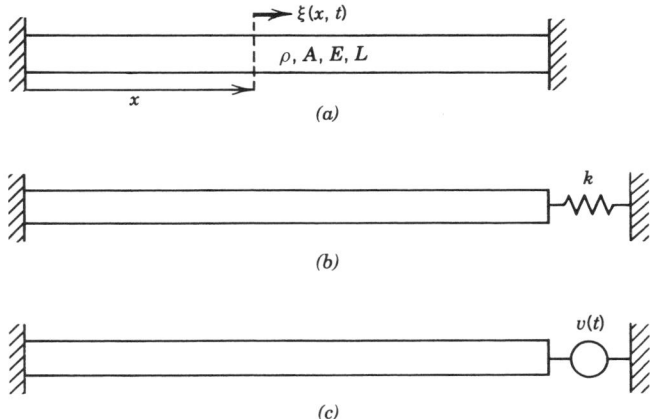

(a)

(b)

(c)

(a) Construct a two-mode bond graph model for the bar in part *b* using the fixed–free mode shapes and properly attaching the spring k. Assign causality, derive equations, and derive the characteristic equation. Let $k \to \infty$, effectively fixing the right end, and determine the natural frequencies of the constrained system. Compare with the frequencies for a fixed–fixed bar.

(b) Construct a two-mode model for the bar in part *c*, again using the fixed–free modes. Assign causality, and notice the derivative causality caused by the velocity input $v(t)$. Let $v(t) = 0$, effectively fixing the right end; deal with the derivative causality algebraically; and derive the characteristic equation for this system. Compare the calculated frequency with the frequency for the fixed–fixed bar.

(c) Finally, rid yourself of derivative causality by appending an additional modal compliance. Again derive the characteristic equation, and compare the calculated frequencies with those for a fixed–fixed bar.

The reader should be impressed at how well the force-free modes converge to fixed-boundary modes even when only two modes are used.

10-11 A truck frame is modeled as a uniform beam in the figure. We would like to design a control system, using actuators at the suspension locations, that can provide damping for the bending motion of the frame. Construct a one-mode model (two rigid-body modes plus one dynamic mode) for the frame, and properly attach the suspension elements. Derive a set of state equations from which we can launch our control studies. If derivative causality exists, modify your model to fix it, and then derive equations.

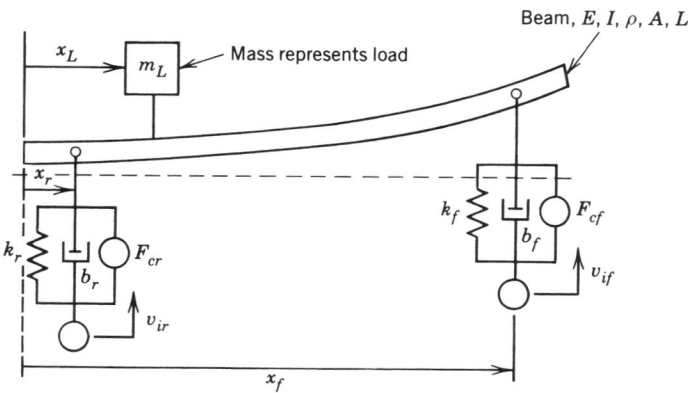

10-12 The figure shows a simple structure composed of two vertical beams and a horizontal bar. Construct a finite-mode model of this system, assuming the beams execute only transverse motion and the bar only longitudinal motion. Use symbols for modes and modal parameters, but state what modes you are using. Predict any formulation problems, and tell how you would modify the model to avoid these problems.

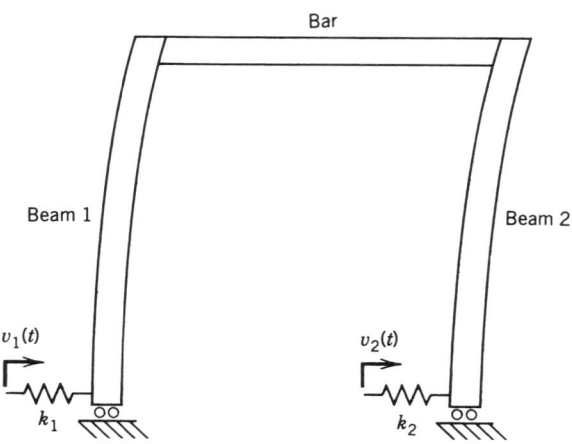

10-13   A tightly stretched string is fixed at both ends, as in Problem 10-7. Perform a modal analysis of a tightly stretched string and convince yourself that the mode shapes and frequencies are

$$\omega_n^2 = \sqrt{\frac{T}{\rho}} \frac{n\pi}{L}, \qquad n = 1, 2, \ldots,$$

$$Y_n(x) = \sin n\pi \frac{x}{L},$$

where T is the string tension and ρ is the mass per unit length. Derive expressions for the modal masses, m_n, and modal stiffnesses, k_n.

10-14   The tightly stretched string from Problem 10-13 now has a mass, m, attached at $x = \frac{1}{4}L$:

Treat the string lengths $\frac{1}{4}L$ to the left and $\frac{3}{4}L$ to the right of the mass as nonmodal tension springs and derive an expression for the natural frequency of the system.

10-15   Model the system in Problem 10-14 using the finite-mode bond graph. Include the first two modes for the string and, of course, attach the mass. Assign causality, and you will discover that derivative causality exists. Use formulation methods from Chapter 5 and derive the state equations. Perform a linear analysis and derive an expression for the system natural frequencies. Compare this result to that from Problem 10-14.

10-16   Add an additional modal compliance to the model in Problem 10-15 and show that no derivative causality exists. Derive state equations.

10-17   The tightly stretched string in Problem 10-13 now has two masses, m_1 and m_2, attached at positions $x = \frac{1}{4}L$ and $x = \frac{3}{4}L$. Formulate a two-mode bond graph model. Assign causality starting with the modal masses and show

that both attached masses are in derivative causality. Append one additional modal compliance and show that one attached mass is in derivative causality. Append an additional modal compliance and show that all integral causality exists but there are unassigned bonds indicating an algebraic loop. Perform the algebra using procedures from Chapter 5 and derive state equations.

11

MAGNETIC CIRCUITS AND DEVICES

Many useful electrical and electromechanical devices contain magnetic circuits. Multiport models for some of these devices have already been studied in previous chapters, but here the magnetic flux paths will be modeled in detail. A detailed model is required, of course, if one is to design a motor, solenoid, transformer, or the like, although a system analyst may be content with an overall multiport model that adequately predicts only the external port behavior.

11.1 MAGNETIC EFFORT AND FLOW VARIABLES

The toroidal coil shown in Figure 11.1 can serve as an ideal configuration for defining magnetic variables and establishing bond graph representations for magnetic circuits. If the toroid is made of soft iron, we expect the device to behave at the single electrical port as an inductance with a linear relation between current i and flux linkage λ, at least for moderate values of i. This model shows no physical effects inside the coil, however.

Inside the iron core, magnetic *flux* is induced whenever current flows through the coil. Let the total flux be called φ; it is measured in the SI system in *webers*. (In this unit system one weber is equivalent to one volt-second. There are several units for magnetic variables in use, but for simplicity, only the units appropriate to the SI system will be presented here.) The magnetic field may be described by the *magnetic flux density vector* **B** with the units of *teslas*, or *webers per square meter*. The length of **B** corresponds to the amount of flux passing through an area element perpendicular to the flux lines, and the direction of **B** is along the flux lines. The **B**-vector points in the direction the north pole of a compass needle would point if it were free to line itself up with the magnetic field. The **B**-vector is also often called the *magnetic induction*.

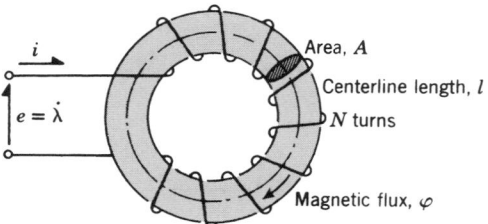

FIGURE 11.1. Toroidal coil.

In Figure 11.1, if we can assume that each of the N turns of conductor around the toroidal core links all the flux φ in the toroid, then the flux linkage λ is related to φ by

$$\lambda = N\varphi. \tag{11.1}$$

In practice, when many layers of wire are wound around on a core, some turns of wire do not link all of the flux lines. In this case, sometimes described by saying that some flux "leaks" out of the coil, Eq. (11.1) can be still valid except that N is a nondimensional number representing an *effective number of turns*, rather than an actual number of turns. In what follows, we shall refer to N as a number of turns without necessarily including the adjective "effective."

By differentiating Eq. (11.1) with respect to time, the relation between the port voltage e and the magnetic variable $\dot\varphi$ is found:

$$e = \dot\lambda = N\dot\varphi, \tag{11.2}$$

which is just Faraday's law applied to the coil.

The driving force which tends to set up φ in the core is the *magnetomotive force*, M, which is proportional both to the number of turns of conductor and to the current flowing in the coil:

$$M = Ni. \tag{11.3}$$

The magnetomotive force, which is analogous to the electromotive force, has the units of current, amperes (although it is conventionally given in ampere-turns, despite the fact that the number of turns is really dimensionless). In the ideal, dissipationless case, M and φ for the coil are related by a nonlinear or linear relation. This relation is sketched in Figure 11.2.

In Figure 11.2a, a typical relation for a ferromagnetic material core is sketched. A *soft* ferromagnetic material is one that is easily magnetized and demagnetized by means of a current-carrying coil, that is, a material for which M and φ are related by a single valued curve so that $\varphi = 0$ when $M = 0$. *Hard* ferromagnetic materials, used for permanent magnets, show a hysteresis loop when M is cycled, and φ can remain high even when M is zero. For now, we shall concentrate on soft ferromagnetic

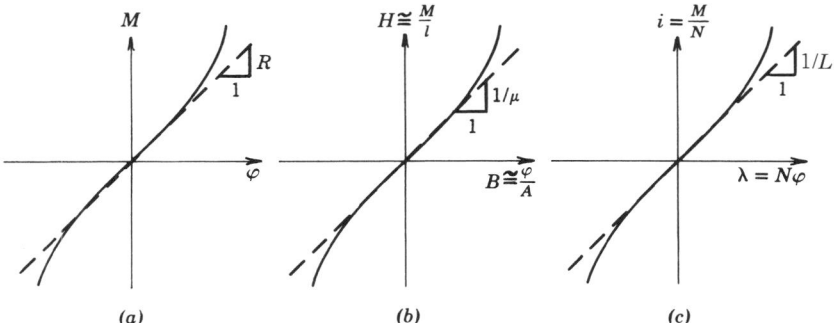

FIGURE 11.2. Core properties. (*a*) Magnetomotive force versus flux; (*b*) field strength versus flux density; (*c*) current versus flux linkage.

materials. Such materials exhibit a saturation effect as shown. The increase in φ as M is increased slows down, and, practically, there is a limiting value of flux that the core can attain when M is very large. (Most steels saturate at B-values of less than 2 T.)

In order to characterize a core material, it is convenient to exhibit a curve independent of any particular core configuration. For this purpose the so-called B–H curve is frequently used. In Figure 11.2*b* the magnitude of the **B**-vector, B, is shown as φ divided by the core cross-sectional area A. This way of computing B would be exact if the flux were uniformly distributed throughout the core. The *magnetizing force*, or *field strength* H, is the magnetomotive force per unit length, with units of ampere-turns per meter. For the toroidal coil, H is approximately M divided by the centerline length of the coil, l. When only a single B–H curve is given, it is implied that the material is *isotropic*, that is, that the relation between B and H is the same no matter how the field is oriented in the material. For certain crystal structures, this is not the case, and several different B–H curves for different orientations of **B** can be found.

In Figure 11.2*c*, the port characteristics of the inductor are sketched by modifying the M–φ or B–H curve using N. We see that the saturation effect corresponds to a nonlinear region of the —I element representing the inductance. Since most B–H curves exhibit saturation, electrical chokes must be nonlinear, though in normal operation of these devices a linear inductance model is often sufficiently accurate.

It is of interest to compute the inductance for the coil. Assuming that the curves of Figure 11.2 can be approximately represented by a straight line near the origin, L may be found as follows:

$$L = \frac{\lambda}{i} = \frac{N\varphi}{M/N} = \frac{N^2 A B}{l H}. \tag{11.4}$$

The initial slope of the B–H curve is given the symbol μ and called the *permeability*; it has units of tesla-meters per ampere

$$\mu = \frac{B}{H},$$ (11.5)

so that $L = N^2 A \mu / l$. (Free space itself has a permeability, μ_0, and sometimes μ is expressed as a relative permeability, that is, the value of B/H is given by $\mu \mu_0$ rather than just μ.)

For the entire coil, one may express the slope of the M–φ curve in the linear region in two ways. The *permeance*, **P**, is given by the following formulas:

$$\mathbf{P} = \frac{\varphi}{M} = \frac{\mu A}{l}.$$ (11.6)

The *reluctance*, **R**, is given by

$$\mathbf{R} = \frac{1}{\mathbf{P}} = \frac{M}{\varphi} = \frac{l}{\mu A}.$$ (11.7)

As a mnemonic device, one may say that a long small-area flux path in a material of small permeability is reluctant to permit the establishment of much flux. The parameters L, **R**, and μ are shown as slopes in the sketches in Figure 11.2. Table 11.1 summarizes the magnetic variables and parameters discussed, together with their SI units and some remarks concerning typical values.

In order to proceed further with the analysis of magnetic circuits and devices, it is useful to begin classifying variables such as M and φ and parameters such as **P** and **R**. Historically, the reluctance **R** was sometimes thought of as analogous to electrical resistance, so that flux was thought of as analogous to current and the magnetomotive force as analogous to the electromotive force. Although this analogy can be used, it is not satisfying as a basis for a bond graph, since an electrical resistor dissipates power while a coil exhibiting reluctance stores energy. In fact, the reluctance and the permeance are linear parameters of a —C or —I element. From a bond graph point of view it seems reasonable that M, the magnetomotive force, should be an effort quantity, but the flow should not be flux, as in the more traditional analogy, but rather the time rate of change of flux, $\dot{\varphi}$.

The deficiencies of the reluctance–resistance analogy are apparent when one attempts to study a dynamic system containing electrical and mechanical elements as well as a magnetic current (see Reference [1]). It was only after the gyrator had been accepted as a useful network element, however, that the analogy most natural for bond graphs could be proposed [2].

Suppose we return to the basic equations, (11.2) and (11.3), and identify $\dot{\varphi}$ as a flow variable and M as an effort variable. Then N is clearly a gyrator parameter. In addition, since φ is a displacement variable, then Eq. (11.6) shows that **P** is a capacitance parameter and **R**, the reluctance, is the inverse capacitance or stiffness parameter. All the relations are elegantly summed up in the bond graph of Figure 11.3. Of course, the combination $\rightarrow GY \rightarrow C$ does behave like $\rightarrow I$ at the external port, as it must. The gyrator is necessary if we wish to consider both electromotive force and magnetomotive force as effort variables.

TABLE 11.1. Summary of Magnetic Variables and Parameters

Symbol	Name	SI Unit	Equivalent Units, Remarks
φ	Total magnetic flux	Weber (Wb)	Volt-second (V-s)
$\dot{\varphi}$	Flux rate	Wb/s	V
B	Magnetic flux density, magnetic induction	Tesla (T)	Weber per square meter, (Wb/m^2); magnitude of a vector quantity
M	Magnetomotive force, MMF	Ampere (-turn) (A)	Number of turns of a conductor is dimensionless, but is included if $M = Ni$ is used where N = turns of wire carrying current i
H	Magnetic field strength	Ampere (-turn) per meter (A/m)	Magnitude of a vector quantity
μ_0	Permeability of free space, $4\pi \times 10^{-7} \cong 1/800,000$	Henry per meter (H/m)	Tesla-meter per ampere (T-m/A); relates B to H in air: $B = \mu_0 H$ or $H \cong 800,000 B$
μ	Permeability of material	H/m	T-m/A; relates B to H in isotropic material: $B = \mu H$
μ_r	Relative permeability of material	Dimensionless	$\mu = \mu_r \mu_0$; for ferromagnetic materials, μ_r varies from about 400 to 400,000 for such materials, $2B < H < 2000B$
P	Permeance of circuit element	Henry (H)	Weber per ampere (Wb/A), used in formula $\varphi = \mathbf{P}M$
R	Reluctance of circuit element	H^{-1}	A/Wb; $\mathbf{R} = 1/\mathbf{P}$ used in formula $M = \mathbf{R}\varphi$
E	Energy	Joule (J)	$\mathbf{E} = \int M\, d\varphi$ (A-Wb = A-V-s)

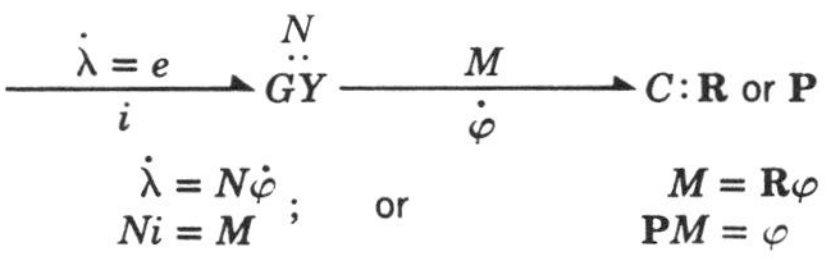

$$\xrightarrow[\;i\;]{\dot{\lambda}=e} \overset{N}{\underset{}{GY}} \xrightarrow[\dot{\varphi}]{M} C : \mathbf{R} \text{ or } \mathbf{P}$$

$$\begin{matrix} \dot{\lambda} = N\dot{\varphi} \\ Ni = M \end{matrix} ; \quad \text{or} \quad \begin{matrix} M = \mathbf{R}\varphi \\ \mathbf{P}M = \varphi \end{matrix}$$

FIGURE 11.3. Bond graph for coil of Figure 11.1.

TABLE 11.2. Mechanical, Electrical, and Magnetic Bond-Graph Variable and Parameters

General	Mechanical	Electrical	Magnetic
Effort	Force, F	Electromotive force, e	Magnetomotive force, M
Flow	Velocity, V	Current, i	Flux rate, $\dot{\varphi}$
Displacement	Distance, X	Charge, q	Flux, φ
Capacitance parameter	Compliance, C	Capacitance, C	Permeance, $\mathbf{P}$
Stiffness parameter	Stiffness, k	$1/C$	Reluctance, $\mathbf{R} = 1/\mathbf{P}$

The classification of variables which will be used subsequently is listed in Table 11.2 for convenience. Note that the capacitance parameter only is to be applied in the linear case. The magnetic circuit generally involves nonlinear C-elements with characteristics as sketched in Figure 11.2.

11.2   MAGNETIC ENERGY STORAGE AND LOSS

Figure 11.2 showed several ways to exhibit the properties of a slightly nonlinear but lossless magnetic core. In this section the energy storage and loss will be investigated in more detail.

Figure 11.4 shows two small pieces of material of physical volume Al. On the left, a uniform stress σ is associated with a force F and causes an extension δ and strain ϵ in the isotropic material. The total energy stored would be

$$\mathbf{E} = \int F \, d\delta = \int \sigma \, Al \, d\epsilon = Al \int \sigma \, d\epsilon. \tag{11.8}$$

Thus the energy per unit volume is the area shown shaded in the σ-ϵ diagram when the material is deformed from a state of zero stress and strain. When the material is in the linear elastic range, $\sigma = E\epsilon$ and the energy is $E\epsilon^2/2$, but if σ versus ϵ is *any* single-valued curve, the material is obviously energy conservative.

On the other hand, if the material yields, the stress and strain follow a hysteresis loop such as the one shown dashed in the sketch, and the area enclosed by the loop represents the energy loss per cycle per unit volume. Mechanical elements which remain elastic are represented by C-elements, but when yielding occurs, some kind of R-element must be used to model the energy loss.

The material on the right of Figure 11.4 behaves under the action of a magnetic field in a manner similar to the material on the left under the action of a mechanical stress field. Assuming that the flux φ has a uniform density B and that the material is isotropic, the total stored energy would be computed according to the classifications of variables in Tables 11.1 and 11.2 as follows:

$$E = \int M \, d\varphi = \int lHA \, dB = Al \int H \, dB. \tag{11.9}$$

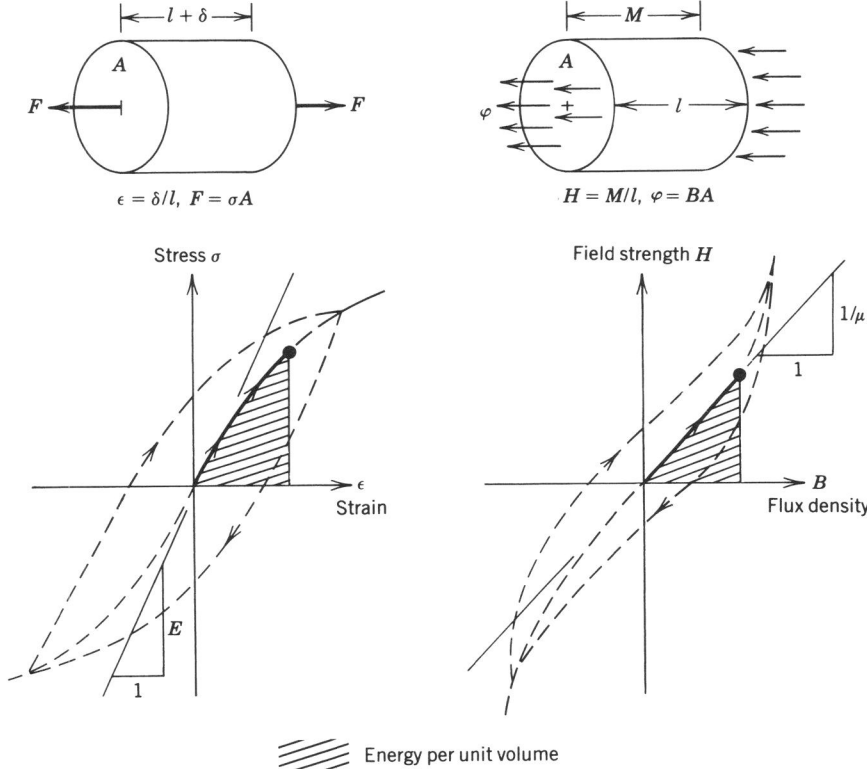

$\epsilon = \delta/l, \; F = \sigma A$

$H = M/l, \; \varphi = BA$

FIGURE 11.4. Mechanical and magnetic energy densities.

Thus the magnetic energy per unit volume is represented by the hatched area in the B–H plot on the right in Figure 11.4. In the linear range, $H = B/\mu$ and the energy is $B^2/2\mu$, but if B and H lie on *any* single-valued curve, energy is conserved.

When B and H are cycled, a hysteresis loop indicated by the dashed loop forms, and the area inside the loop represents the energy loss per cycle per unit volume. For soft magnetic materials, the loops are quite narrow and a C-element with a curved constitutive law can be used to model the saturation effect without loss. The hysteresis effect requires the use of R-elements, since energy loss is involved.

The detailed modeling of hysteresis is generally quite complicated, since it involves a sort of memory effect. One way to handle this is to replace the single C-element, which has a single displacement state variable, with a combination of C- and R-elements, which then serves to remember aspects of the loading of the hysteretic system through the values of a number of hidden state variables. The procedure is discussed in Reference [3].

Permanent magnets are made of hard magnetic materials with very large hysteresis loops. These materials resemble elastic–plastic mechanical materials which can retain a large strain after having been yielded. A permanent magnet retains a large **B**-

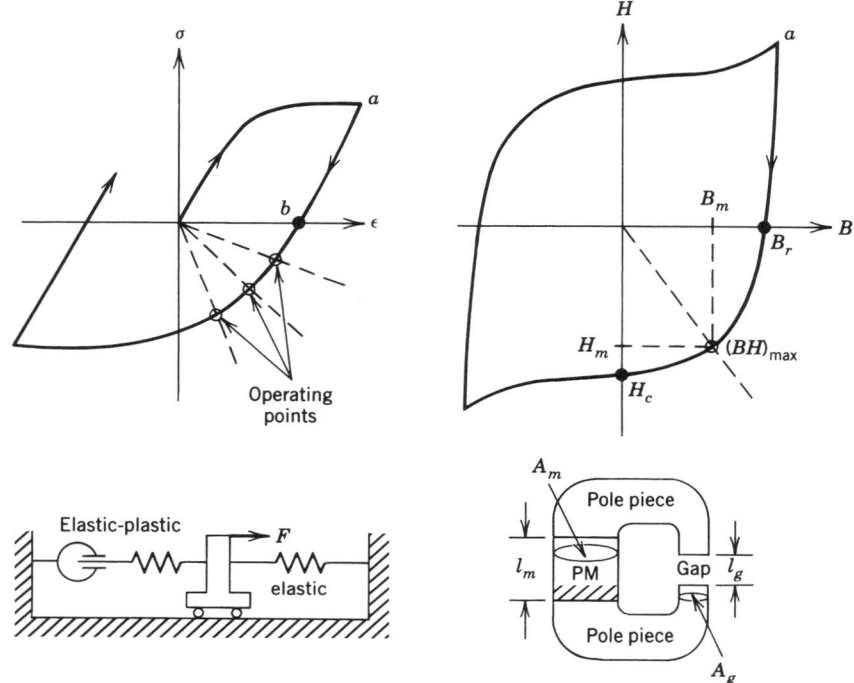

FIGURE 11.5. Hysteresis loops for an elastic–plastic material and a permanent-magnet material.

field after magnetization even in the presence of an **H** or MMF in a direction which would tend to demagnetize the material.

Figure 11.5 shows the stress–strain plot of an elastic–plastic material and the B–H plot for a typical permanent magnet. In most plots the B and H axes are interchanged from the way they are presented here. When this is done, the hysteresis loops are traversed in the opposite direction and the areas representing stored energy change places. The unconventional H–B plot is required to keep the displacement variable φ or B on the horizontal axis and the effort variable M or H on the vertical axis, as has been done consistently for all energy domains in this book.

On the left of Figure 11.5 is a mechanical system which will help explain how permanent magnets can be used to produce flux in air gaps for use in motors, gyrators, electrodynamic loudspeakers, and other electromechanical transducers instead of through the use of current-carrying coils. In the stress–strain diagram, if the material is strained beyond its yield point to point a, when the stress is relaxed, the material returns to point b with a permanent strain. A similar phenomenon occurs for the magnetic material represented on the right. After being magnetized to point a on the H–B hysteresis loop, the material would retain a *B-value of* B_r, the *remanence* or *remanent field*, if H were returned to zero. To achieve $B = 0$, one would have to apply a negative field strength of magnitude H_c, the *coercivity*. The parameters B_r and H_c are often used to describe the properties of permanent magnets.

Now suppose that some elastic–plastic material is connected in mechanical series to some elastic material as shown schematically in the lower left of Figure 11.5. If the force F is used to stretch the elastic–plastic material beyond its yield point and then removed, the system will not return to point b on the stress–strain diagram, because the elastic element will tend to recompress the yielded element. The stress will be negative, and the strain will be less than the strain at point b. Several possible operating points are shown, depending upon how stiff the elastic element happens to be.

An analogous situation arises when a permanent magnet is used in a magnetic circuit with an air gap as shown in the lower right of Figure 11.5. If we assume for simplicity that the pole pieces have negligible reluctance and we neglect any leakage flux, then all the flux from the magnet will also traverse the gap, and the total MMF from the magnet will appear across the gap. The magnet will have to work against a negative H caused by the MMF at the gap necessary to force the flux through the gap. We now investigate a means for choosing optimum magnet and gap dimensions.

A common principle is that a given magnet should be used so that the total energy in the air gap is maximum. The MMF–φ relation in the air gap is linear,

$$M_g = R_g \varphi_g, \tag{11.10}$$

so the total energy is

$$E = R_g \varphi_g^2 / 2 = M_g \varphi_g / 2, \tag{11.11}$$

and we wish to maximize $M_g \varphi_g$. The assumptions mean that the magnet flux is equal to the gap flux and the magnet MMF is just the negative of the gap MMF:

$$\varphi_m = \varphi_g, \qquad M_m = -M_g. \tag{11.12}$$

Then, assuming a uniform flux density and field strength and using the magnet length l_m and area A_m, we can relate $M_g \varphi$ to the B and H values in the magnet:

$$\left| M_g \varphi_g \right| = |M_m \varphi_m| = (A_m B)(l_m H) = (A_m l_m)(BH), \tag{11.13}$$

where only the magnitude of H is considered, although H is actually negative. The point is that the gap energy is maximized by operating the magnet at that point in the B–H plane where the product BH is maximum. Let us now define the point B_m, H_m on the hysteresis curve to be the point at which BH is maximum, $(BH)_{\max}$. We then dimension the gap in such a way that the magnet operates at B_m, H_m.

Three simple equations apply: (1) the magnet and gap fluxes are equal,

$$B_m A_m = B_g A_g; \tag{11.14}$$

(2) the magnitudes of the magnet and gap MMFs are equal,

$$H_m l_m = H_g l_g; \tag{11.15}$$

and (3) in the air gap

$$B_g = \mu_0 H_g. \tag{11.16}$$

Elimination of B_g and H_g yields the final result:

$$\frac{A_g l_m}{A_m l_g} = \frac{B_m}{\mu_0 H_m}. \tag{11.17}$$

The term on the right in Eq. (11.17) can be thought of as a normalized slope of the line passing through the $(BH)_{\max}$ point. Its value is often tabulated for magnet materials. The value of $(BH)_{\max}$ is a measure of the strength of the magnet material.

Note that by adjusting the magnet and gap areas, the B-value in the gap can be greater than in the magnet. If $A_g = A_m$, then the ratio of l_m to l_g is just $B_m/\mu_0 H_m$. For older metallic magnets this ratio was usually greater than unity, leading to designs with magnets much longer than the air gap length, but recent high-strength rare earth magnets have a ratio of about unity, which means that the magnet length is about the length of the air gap unless $B \gg B_m$ is required.

For more details on the design of magnetic circuits to produce constant fields either with permanent magnets or with current-carrying coils see References [4] and [5]. We now proceed to model circuits in which fields may vary dynamically.

11.3 MAGNETIC CIRCUIT ELEMENTS

Using the effort–flow identifications of Table 11.2, it is possible to begin a study of lumped-parameter elements for magnetic circuits. In Figure 11.6a, a coil around a length of core is shown contributing an increase in MMF. In this model, the length of core has no reluctance. In Figure 11.6b, an MMF drop is shown related to the flux φ. When two flux paths are joined as shown in Figure 11.6c, the result can be modeled with a 0-junction, since the three flux rates sum to zero and there is a single MMF. Finally, in Figure 11.6d, an air gap is modeled essentially in the same way as a length of core material. The air gap has high reluctance for its length or low permeability (essentially the permeability of free space, μ_0) and does not exhibit saturation as iron cores do. In computing reluctance or permeance parameters for —C elements of flux paths, we use Eqs. (11.6) and (11.7), in which areas and lengths for the flux path elements must be estimated judiciously, with consideration given to the actual paths of the flux lines.

Using the elements of Figure 11.6, a bond graph for a simple magnetic circuit such as the one shown in Figure 11.7 may be constructed. The MMF can be treated just like voltage in an electric circuit. In Figure 11.7b the MMF variables have been each assigned to a 0-junction, and drops in MMF across —C elements and a rise due

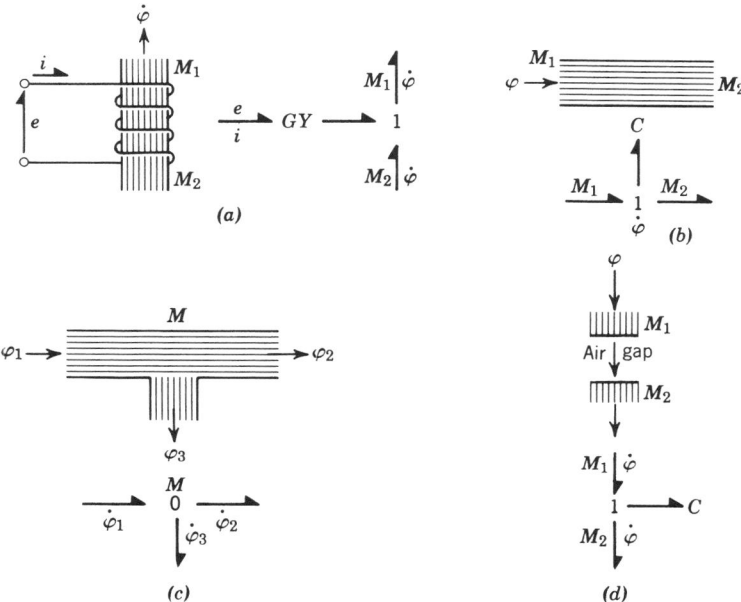

FIGURE 11.6. Bond graphs for magnetic circuit elements. (*a*) Driver coil; (*b*) piece of ferromagnetic core; (*c*) core junction; (*d*) air gap.

to the coil have been indicated using 1-junctions. Since only differences in MMF are significant ultimately, M_d is chosen as a zero MMF point, and the bond graph can be simplified. It may be helpful to consider the previously studied case of the analogous electric circuit shown in Figure 11.7*c*.

After the ground node has been eliminated and the 2-port 0-junctions removed, it is found that three —*C* elements are joined to a single 1-junction. This means that all *C*s have the same flow variable and the effort drops add. In magnetic terms, this means that for the linear case, a single *C*-element with a reluctance equal to the sum of the three reluctances of the original *C*s can be substituted as shown in Figure 11.7*d*. In the nonlinear case, an equivalent —*C* relation can be found by simply adding MMF drops for the three original *C*s as the common flux is varied over the range of interest.

A more complicated magnetic circuit appears in Figure 11.8. The flux path associated with φ_3 could be physically present, or it could be a lumped representation of the path taken by leakage flux that bypasses the air gap. In fact, there always is some flux that escapes from the main path and spreads into the surrounding space. Although the real leakage paths are distributed in space, it is often sufficient to include a single equivalent leakage path and leakage flux. Since the leakage path reluctance is usually dominated by the high reluctance of the air portion, it is common to assume that the leakage path can be represented by a linear, high-reluctance —*C*. The basic bond graph of Figure 11.8*b* is simplified by the choice of a zero MMF point, and

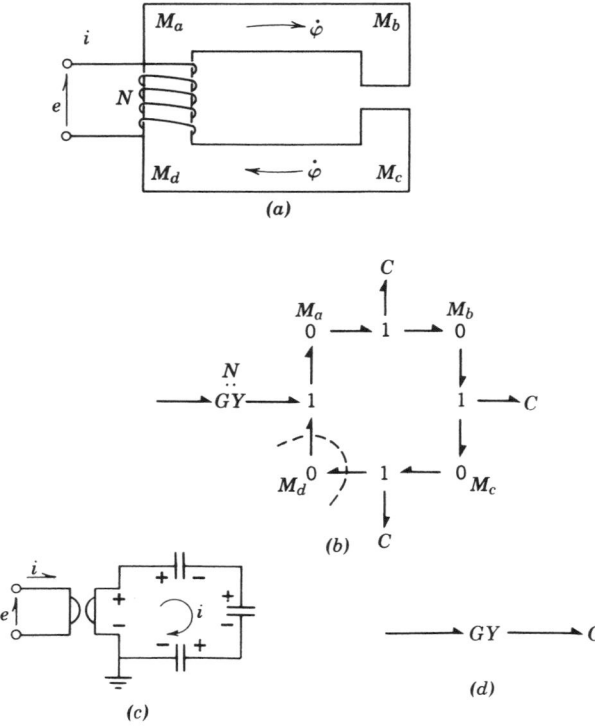

FIGURE 11.7. Bond graph model of magnetic circuit with air gap. (*a*) Sketch of device; (*b*) bond graph; (*c*) analogous electric circuit; (*d*) simplified bond graph.

the bond graph of Figure 11.8*c* results. The loop may be eliminated using the bond graph identity for the 0–1–0–1 ring, since the signs are proper for the reduction. The bond graphs of Figure 11.8*d* and *e* result. In the last form, it is clear that *C*2, *C*4, and *C*6 could be combined, as could *C*1 and *C*5. If these *C*s were linear, their reluctances would be summed. Ultimately, a single equivalent *C* could be found for the entire collection of *C*s, 0s, and 1s, but then, of course, none of the internal MMFs or fluxes could be found.

When no loss mechanisms are included in magnetic circuits, then they appear internally as networks of *C*s, and it is often convenient to reduce the system by finding equivalences among the subfields. At a more detailed level, however, there are energy losses associated with eddy currents in the core and other effects. Laminations in the core of a magnetic circuit help in reducing the loss associated with eddy currents, but accurate models require the insertion of loss elements. Analysis of a laminated core can show that the *R–C* transmission line model for the core may be used for the linear case [2]. A study of such detailed models would take too long for present purposes, and in practice a simpler approach often suffices. It is often sufficient to append a magnetic resistance to the 1-junction of Figure 11.6*b*. This implies an extra MMF

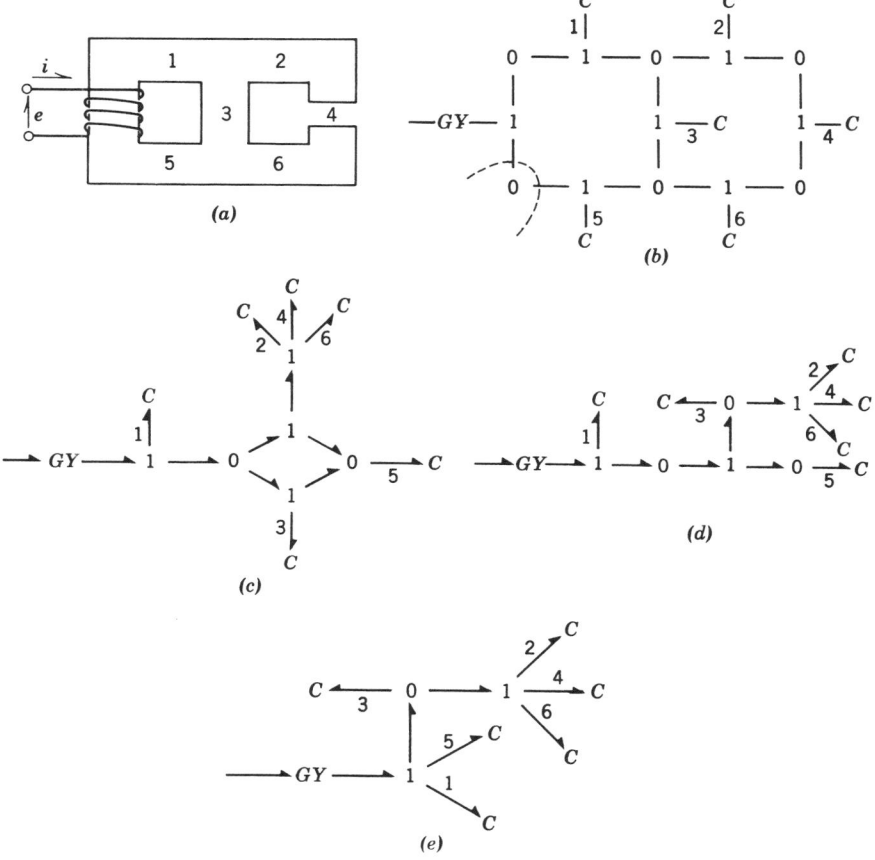

FIGURE 11.8. Circuit with extra flux path to model leakage flux. (a) Sketch; (b) basic bond graph; (c)–(e) simplified bond graphs.

drop associated with a length of core that is a function of $\dot{\varphi}$, in addition to the drop in MMF that is associated with φ. As might be expected, it is not easy to predict the magnitude of the resistance in the absence of experimental evidence. On the other hand, particularly for a limited frequency range, it is often possible to adjust a linear resistance relation to provide a quite accurate representation of losses in the core.

11.4 MAGNETOMECHANICAL ELEMENTS

Before exhibiting magnetic circuits containing loss elements, it is important to show a method of handling magnetomechanical transducers. (The solenoid studied in Chapter 8 was treated in multiport fashion without analyzing the magnetic circuit in detail.) The basic idea behind many transducers is that mechanical motion can alter

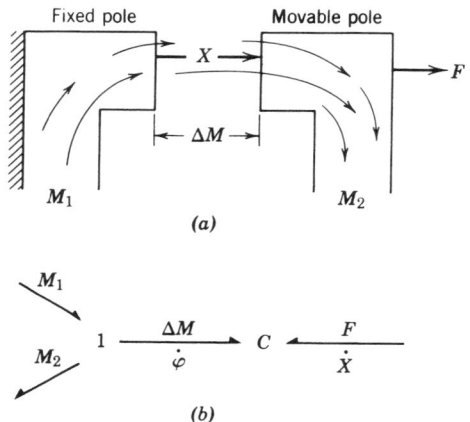

FIGURE 11.9. Magnetomechanical transduction. (*a*) Sketch; (*b*) bond graph representation.

the relation between flux and MMF in a flux path. This type of transducer is often called a "variable-reluctance" transducer.

Consider the device of Figure 11.9, in which a force F is associated with a movable pole piece that communicates with a fixed pole through an air gap of variable length X. A flux φ passes through the gap and is associated with an MMF drop ΔM. The device can be represented by a C-field, as shown in Figure 11.9b, and is analogous to the movable-plate capacitor of Chapter 8.

Since this particular device uses an air gap, it is reasonable to assume that the device is linear, but that the reluctance $\mathbf{R}$ is a function of X, that is,

$$\mathbf{R}(X)\varphi = \Delta M. \tag{11.18}$$

Using arguments similar to those in Chapter 7, we may write the energy $\mathbf{E}$ as

$$\mathbf{E}(X, \varphi) = \frac{1}{2}\mathbf{R}(X)\varphi^2. \tag{11.19}$$

Then,

$$\Delta M = \frac{\partial E}{\partial \varphi} = R(X)\varphi \tag{11.20a}$$

and

$$F = \frac{\partial E}{\partial X} = \frac{1}{2}\frac{dR}{dX}\varphi^2. \tag{11.20b}$$

From Eq. (11.7), we might expect

$$\mathbf{R}(X) \cong \frac{X}{\mu_0 A}. \tag{11.21}$$

So

$$F \cong \frac{\varphi^2}{2\mu_0 A}, \tag{11.22}$$

indicating the F would be roughly constant for small values of X if φ were held constant. In most cases, however, φ is more nearly inversely proportional to $\mathbf{R}$ (when an MMF is held constant somewhere in the circuit), so F tends to vary as X^{-2} for small X. For very large values of X, the circuit model itself breaks down, and the flux does not follow the paths assumed. Any movement of core pieces will be associated with changes in stored energy and hence with forces or torques. The bond graph of Figure 11.9*b* may be used even when the magnetic relations are nonlinear due to saturation, but the computation of the energy is more complicated.

Figure 11.10 shows a ferromagnetic rotor which is subjected to an aligning torque τ when a field is induced in the stator. There are several ways to compute the torque, but perhaps the easiest involves energy considerations. We can again use Eq. (11.19) but with $\mathbf{R}$ a function of θ instead of X. Assuming that the flux is concentrated in the narrow gap of length l_g, Eq. (11.7) can be used with the area of the gap as a function of θ rather than the length. Letting the axial length of the rotor be l, the effective area of the two air gaps is $l(l_0 - r\theta)$, so the reluctance is

$$\mathbf{R} = \frac{2l_g}{\mu_0 l(l_0 - r\theta)}. \tag{11.23}$$

Then

$$\mathbf{E}(\theta, \varphi) = \frac{l_g \varphi^2}{\mu_0 l(l_0 - r\theta)} \tag{11.24}$$

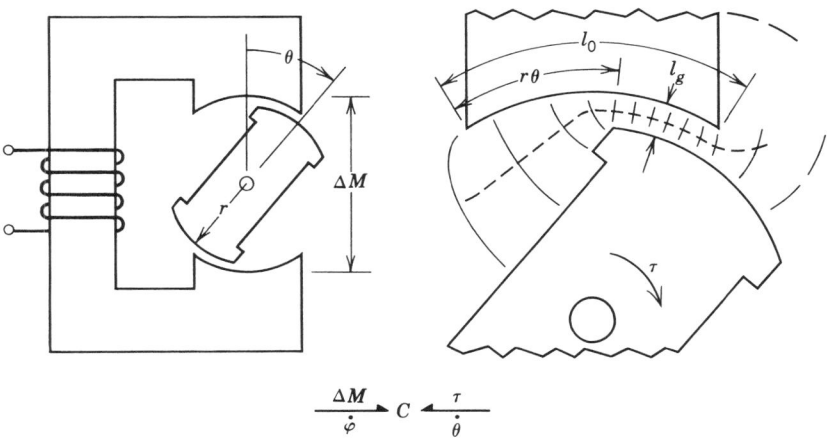

FIGURE 11.10. Rotary, variable-reluctance transducer.

and

$$\tau = \frac{\partial E}{\partial \theta} = \frac{l_g r \varphi^2}{\mu_0 l (l_0 - r\theta)^2}, \tag{11.25}$$

$$\Delta M = \frac{\partial E}{\partial \varphi} = \frac{2 l_g \varphi}{\mu_0 l (l_0 - r\theta)}. \tag{11.26}$$

The last two relations describe the 2-port C-field in integral causality.

Other equivalent expressions may not be so useful for bond graph dynamic models. For example, defining

$$B = \frac{\varphi}{l (l_0 - r\theta)}, \tag{11.27}$$

Eq. (11.25) can be written

$$\tau = \frac{l r l_g B^2}{\mu_0}. \tag{11.28}$$

While correct, Eq. (11.28) may obscure the fact that B varies with θ and φ as the rotor turns and the excitation changes.

11.5 DEVICE MODELS

As an example of the utility of the bond graph models for magnetic circuits developed above, consider the relay of Figure 11.11. Bond graphs are an aid in the study of such systems, since three energy domains are involved. From the schematic diagram of the device, Figure 11.11a, one may begin assembling a bond graph model by indicating MMF values and then representing MMF drops with 1-junctions and C- and R-elements. At this point, some judgment is required, since it is not clear how many flux paths should be represented, nor is it clear where the eddy current losses will be most severe. When the bond graph is simplified after choosing M_0 as the zero MMF, as in Figure 11.11c, it becomes clear that several C- and R-elements can be combined.

Discounting the loss elements, the transduction from electrical energy to mechanical energy is accomplished by a C-field addressed at one port through a gyrator or, in other words, by an IC-field transducer. This would have been predicted by the methods used in Chapter 8. However, the present approach using the detailed model of the magnetic circuit allows a designer to study the flux paths in detail and to model internal losses. From a designer's point of view, a single relay is a rather complex dynamic system. Conventional descriptions of models of such systems using equations are not particularly easy to follow. For an example of a real device modeled at about the level of complexity shown in Figure 11.11, see Reference [6]. The reader who takes the trouble to read this paper carefully may become convinced that the

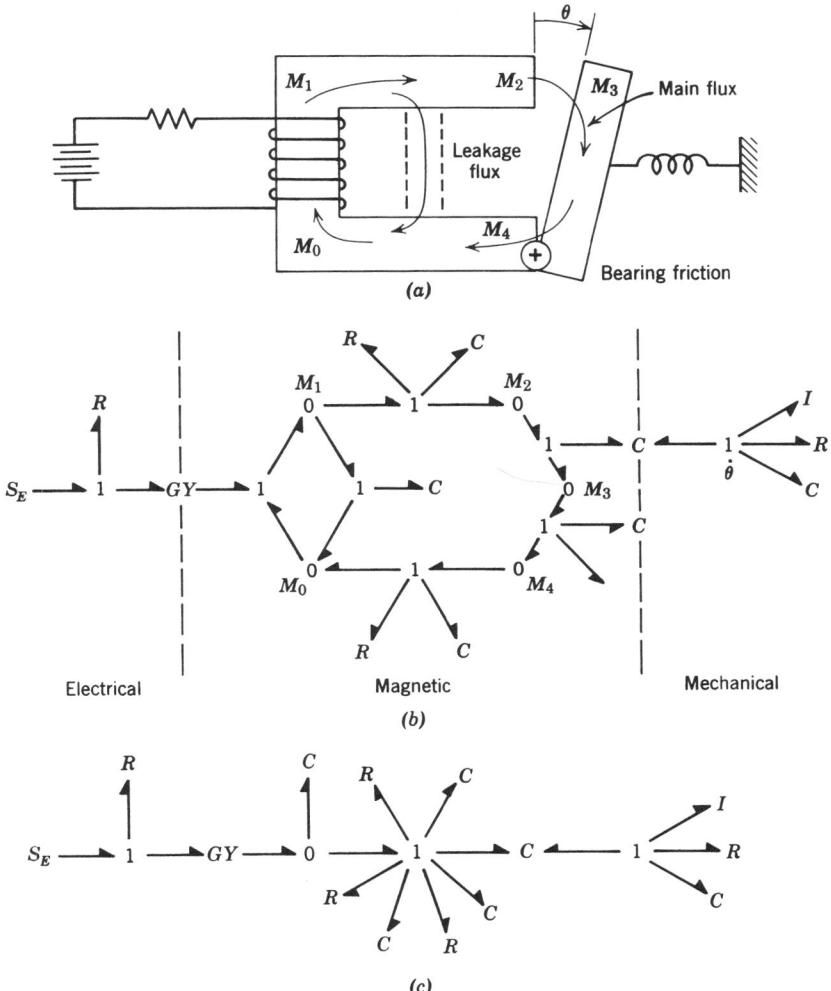

FIGURE 11.11. Relay. (*a*) Schematic diagram; (*b*) basic bond graph; (*c*) simplified bond graph.

bond graph representation yields a compact and insightful means of displaying the physical assumptions used in creating a multiple-energy-domain model.

It is possible to create a very general bond graph model which describes a wide variety of electromagnetic–mechanical devices, including all the power-conserving and energy-storing devices discussed previously. The model is developed in Reference [7], and the devices are described in more detail in References [8–10].

We consider four devices and three forms of force or torque law which will be incorporated into one general bond graph model. The derivation of the fundamental

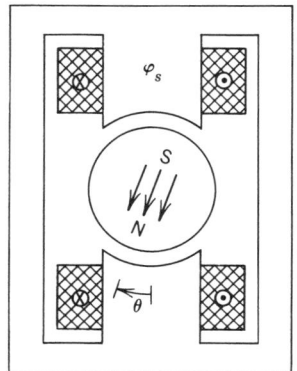

FIGURE 11.12. Synchronous motor with salient poles and permanent-magnet rotor discussed in Reference [8].

form of the force law appears in Reference [8] as applied to synchronous motors such as the salient pole version shown in Figure 11.12.

The flux linked by the stator windings is called φ_s in Reference [8] and is imagined split into a component due to the coil current i_s and a part due to the permanent magnet rotor position θ:

$$\varphi_s = L_s(\theta)i_s + \varphi_{sr}(\theta), \tag{11.29}$$

in which $L_s(\theta)$ is a position-dependent inductance. The model is therefore electrically linear, which means that saturation in the magnetic circuit is assumed not to occur. The magnetic energy $\mathbf{W}_m$ is also split into a part having to do with current in the stator windings and a part having to do with the permanent magnet position, $\mathbf{W}_{mr}$:

$$\mathbf{W}_m = \frac{1}{2}L_s(\theta)i_s^2 + W_{mr}(\theta). \tag{11.30}$$

An energy argument leads to the following expression for the electromagnetic torque T_e:

$$T_e = i_s\frac{d\varphi_{sr}}{d\theta} - \frac{dW_{mr}(\theta)}{d\theta} + \frac{1}{2}i_s^2\frac{dL_s(\theta)}{d\theta}. \tag{11.31}$$

This expression, although derived for a fairly specific type of rotary device, actually can be interpreted as showing the correct form for the three basic types of forces present in electromagnetic actuators.

In Figure 11.13, two fundamental types of linear actuators discussed in Reference [9] are sketched. In previous bond graph literature, case a would be modeled either by an energy-conservative IC-field or by a C-field if the magnetic circuit variables were explicitly included, and case b would be modeled by a gyrator.

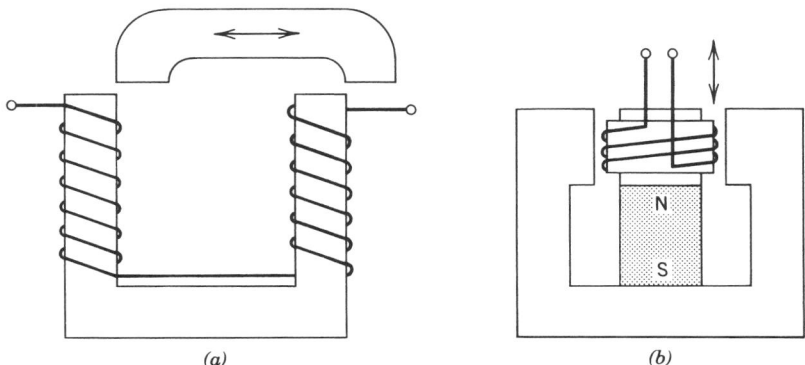

FIGURE 11.13. Linear motion actuators discussed in Reference [9]. (*a*) Variable-reluctance device; (*b*) voice coil device.

In Reference [9], however, a general force law is stated which applies to both devices:

$$F_e = i\frac{d\varphi_0}{dx} - \frac{dW_0(x)}{dx} + \frac{\partial}{\partial x}\int_0^i \varphi_i(x, i')\, di'. \qquad (11.32)$$

The terms in Eq. (11.32) are strictly analogous to those in Eq. (11.31). The rotor angular position θ is replaced by the linear position x of the moving element. The symbol φ_0 corresponds to φ_{sr}, and W_0 to W_{mr}. The integral term has to do with self-inductance and could also be represented as in Eq. (11.30) using a variable-inductance coefficient.

Finally, in Figure 11.14, a very different type of linear actuator is analyzed in which permanent magnets (PMs) form the moving element. The three iron cores are

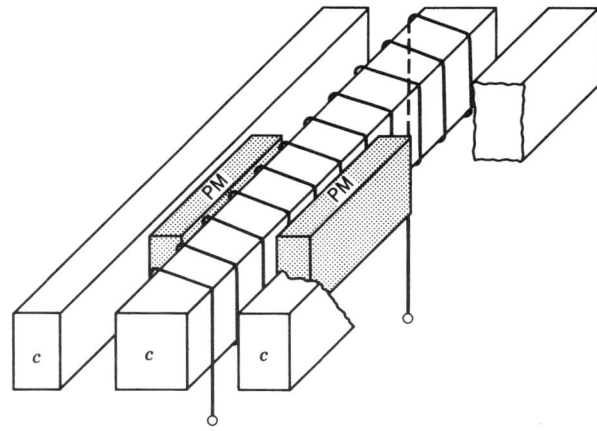

FIGURE 11.14. Linear motor with moving permanents as discussed in Reference [10].

stationary with a simple winding distributed along the middle one. The force law used in Reference [10] is

$$F = I\frac{d\varphi_{cm}}{dp} - \frac{dW_m}{dp} + \frac{1}{2}I^2\frac{dL_c}{dp}. \tag{11.33}$$

The similarities to Eqs. (11.31) and (11.32) are evident if one notes that here p represents the magnet position.

We now create a bond graph capable of representing Eqs. (11.31)–(11.33) and thus capable of modeling any of the four devices in Figures 11.12–11.14, as well as many others. The bond graph will also predict the induced voltage effects consistent with the forces and will allow the simple addition of inertia, friction, and resistance effects. For simulation purposes, the forces will actually be expressed in terms of flux or flux linkage state variables rather than as functions of current.

The bond graph to be developed for the actuators is shown in Figure 11.15. The mechanical variables F and $\dot{x}$ stand for force and linear velocity, but for rotary actuators they could be interpreted as torque and angular velocity, respectively. The force law corresponding to Eqs. (11.31)–(11.33) is

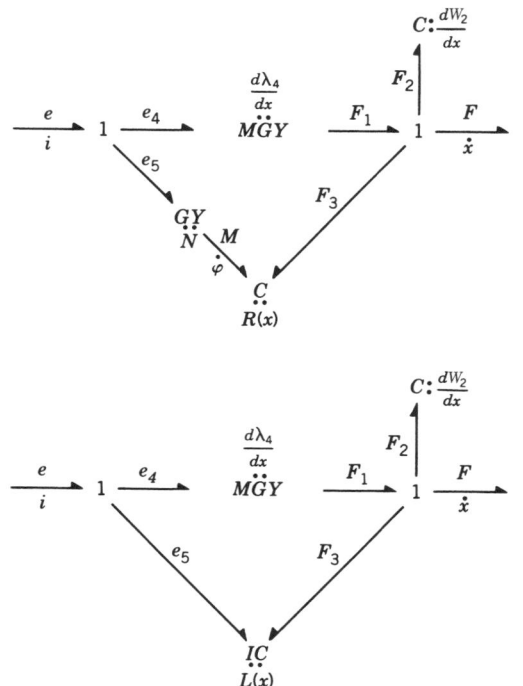

FIGURE 11.15. General bond graph for actuators in C-field form using magnetic circuit variables M and $\dot{\varphi}$ and in IC-field form using current and voltage variables.

$$F = i\frac{d\lambda_4}{dx} - \frac{dW_2}{dx} + \frac{1}{2}i^2\frac{dL}{dx}, \tag{11.34a}$$

$$F = F_1 - F_2 - F_3, \tag{11.34b}$$

where i is the coil current, λ_4 is the flux linkage due to the permanent magnet, W_2 is the magnetic energy in the magnetic circuit due to the permanent magnet, and the C-field or the IC-field represents the energy stored due to the coil self-inductance effect. Using a gyrator for the coil of N turns, we can conveniently relate the voltage component e_5 and current i to the rate of change of flux, $\dot\varphi$, in the coil and to the magnetomotive force M. Then a 2-port C-field can represent the magnetomechanical energy storage. For the electrically linear case, the position-dependent reluctance $R(x)$ is useful to express the stored energy in terms of x and φ. An alternative description uses as electric variables the current i and the voltage component $e_5 = \lambda_5$ and uses an IC-field. In this case, the position-dependent inductance $L(x)$ can be used to express the stored energy as a function of x and the self-flux linkage of the coil, λ_5. See Chapter 8.

The three force components in Eqs. (11.34)–(11.34a) should be compared directly with the corresponding terms in Eqs. (11.31)–(11.33). The components F_1, F_2, and F_3 will be discussed separately.

The force component F_1 can sometimes be computed directly from the Lorentz force. This would be the case for the voice coil of Figure 11.13b. A more general approach is to start from a flux linkage expression. Suppose $\lambda_4(x)$ is the coil flux linkage due to the permanent-magnet field (if one is present). In Figure 11.13b, the flux linkage varies with the position of the coil with respect to the magnetic circuit. In Figures 11.12 and 11.14 the configuration of the magnetic circuit changes as the magnet moves, and hence the flux linkage changes in the coils.

The voltage component e_4 is found from the rate of change of the flux linkage λ_4:

$$e_4 = \dot\lambda_4 = \frac{d\lambda_4}{dx}\frac{dx}{dt}. \tag{11.35}$$

By power conservation, we find

$$F_1 = \frac{d\lambda_4}{dx}i, \tag{11.36}$$

which is represented by the modulated gyrator in Figure 11.15. For some devices, such as voice coils with modest excursions and, roughly, d-c motors, $d\lambda_4/dx$ is nearly constant and the modulated gyrator can be replaced by a constant-parameter gyrator. On the other hand, if the field is produced not by a permanent magnet but rather by an externally excited field, the modulated gyrator parameter might vary with time or field current as well as with position.

The force component F_2 arises when a permanent magnet moves in a magnetic circuit. The magnet will have generally preferred positions corresponding to positions of minimum reluctance. Even without coil current there can be magnetic forces depending on position. Such forces are of the same nature as mechanical spring forces. If the energy in the magnetic circuit due to the permanent magnet, $\mathbf{W}_2(x)$, is given as a function of position, then energy conservation requires

$$F_2 = \frac{d\mathbf{W}_2(x)}{dx}. \tag{11.37}$$

Cases in which there is no permanent magnet (Figure 11.13a) or in which x does not influence $\mathbf{W}_2$ (Figures 11.13b and 11.14) have no force component F_2, so the C-element representing F_2 can be eliminated.

The final force component F_3 is due to the self-inductance effects of the coil current. For the electrically linear case, the energy $\mathbf{W}_3$ may be expressed as a function of the flux φ in the coil and the position variable x by using the reluctance $\mathbf{R}(x)$ in the form

$$W_3(x, \varphi) = \frac{\mathbf{R}(x)\varphi^2}{2}. \tag{11.38}$$

Differentiation of this expression yields the 2-port C-field laws for magnetomotive force M and force component F_2 in conservative form implied in the upper bond graph of Figure 11.15, that is,

$$M(x, \varphi) = \frac{\partial \mathbf{W}_3}{\partial \varphi} = \mathbf{R}(x)\varphi, \tag{11.39}$$

$$F_3(x, \varphi) = \frac{\partial \mathbf{W}_3}{\partial x} = \frac{\varphi^2}{2} \frac{d\mathbf{R}(x)}{dx}. \tag{11.40}$$

The gyrator with parameter N relates the voltage component e_5 to the rate of change of flux $\dot\varphi$ and the current i to the magnetomotive force M, yielding

$$e_5 = N\dot\varphi, \tag{11.41}$$

$$Ni = M. \tag{11.42}$$

Alternatively, the IC-field bond graph in the lower part of Figure 11.15 can be described by the flux linkage associated with the coil current λ_5 as the position-dependent self-inductance $L(x)$. The energy $\mathbf{W}_3$ is then

$$\mathbf{W}_3(x, \lambda_5) = \frac{\lambda_5^2}{2L(x)}, \tag{11.43}$$

which yields the IC-field laws for current force by differentiation:

$$i = \frac{\partial \mathbf{W}_3}{\partial \lambda_5} = \frac{\lambda_5}{L(x)}, \tag{11.44}$$

$$F_3(x, \lambda_5) = \frac{\partial \mathbf{W}_3}{\partial x} = -\frac{\lambda_5^2}{2} \frac{1}{L^2(x)} \frac{dL(x)}{dx}, \tag{11.45}$$

where

$$\frac{d\lambda_5}{dt} = e_5. \tag{11.46}$$

The reluctance and the inductance are related by the expression

$$L(x) = \frac{N^2}{\mathbf{R}(x)}. \tag{11.47}$$

The basic bond graph expressions for F_3 involve the state-variable flux φ or flux linkage λ_5, while the corresponding expressions in Eq. (11.34) and the analogous expressions in Eqs. (11.31)–(11.33) use the current. To see that these expressions are exactly equivalent, we merely use Eq. (11.44) to eliminate λ_5 in terms of i in Eq. (11.45), obtaining

$$F_3(x, i) = -\frac{i^2}{2}\frac{dL(x)}{dx}. \tag{11.48}$$

The C-field laws (11.39), (11.40) and the IC-field laws (11.44), (11.45) are useful in simulation because they are in integral causality form.

The bond graph incorporates the force laws but has the additional advantage that it also automatically incorporates the voltage law for the electromagnetically induced voltage e in the form

$$e = e_4 + e_5 \tag{11.49}$$

with e_4 from Eq. (11.35) and e_5 from Eqs. (11.39)–(11.41) or from Eqs. (11.44), (11.45). Thus, the complete actuator description is included in the bond graph in a power- and energy-consistent form.

Figure 11.16 shows the lower bond graph of Figure 11.15 supplemented by electrical resistance, mechanical inertia, and mechanical friction elements. Complete integral causality is shown: The input signals to the device are an external voltage e_{ext} and an external force F_{ext}, and the model responds with the current i and veloc-

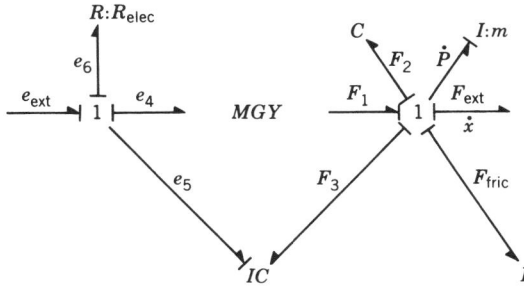

FIGURE 11.16. General bond graph with coil resistance and mechanical mass effects added. Integral causality shown.

ity $\dot{x}$. Certain interconnections to other system components could, of course, induce derivative causality.

This bond graph can be simplified to reduce to known models of specific devices. A permanent-magnet d-c motor, for example, does not need the IC-field, because the inductance is not a strong function of x or θ. Thus, a simple constant inductance I attached to bond 5 suffices. Also, the MGY can usually be simplified to a constant-parameter GY, and the C-element on bond 2 can be eliminated if the "cogging" torque is neglected. Analogous simplifications can be made for the devices of Figures 11.13b, 11.14. The device of Figure 11.13a has no permanent magnet, and if residual magnetic effects are neglected, both the MGY and the C on bond 2 may be removed. Then, only the energy-storing C- or IC-field remains, as in the models for relays or movable-core solenoids. Finally, there are cases in which the electrical time constants are short compared to the system time scales or the mechanical inertia and friction can be neglected. In such cases, the model may be drastically simplified and only very limited causal patterns may be possible.

Although this brief introduction to magnetic system bond graphs by no means exhausts the subject, the basic ideas have been set forth, and it is hoped that the interested reader can extend bond graph techniques to more complicated situations in magnetic systems.

REFERENCES

[1] E. Colin Cherry, "The Duality between Interlinked Electric and Magnetic Circuits and the Formation of Transformer Equivalent Circuits," *Phys. Soc. London Proc.*, **62**, 101–111 (1949).

[2] R. W. Buntenbach, "Improved Circuit Models for Inductors Wound on Dissipative Magnet Cores," *1968 Conference Record of Second Asilomar Conference on Circuits and Systems*, 68C64-ASIL, New York: IEEE, 1969, pp. 229–236.

[3] D. Karnopp, "Computer Models of Hysteresis in Mechanical and Magnetic Components," *J. Franklin Inst.*, **316**, No. 5, 405–415 (1983).

[4] J. K. Watson, *Applications of Magnetism*, New York: Wiley, 1980.

[5] M. G. Say and E. O. Taylor, *Direct Current Machines*, 2nd ed., London: Pitman, 1986.

[6] P. G. Stohler and H. R. Christy, "Simulation and Optimization of an Electromechanical Transducer," *Simulation*, **13**, No. 4, 202–210 (1969).

[7] D. Karnopp, "Bond Graph Models for Electromagnetic Actuators," *J. Franklin Inst.*, **319**, No. 1/2, 173–181 (1985).

[8] E. M. H. Kamerbeek, "Electric Motors," *Philips Tech. Rev.*, **33**, No. 8/9, 215–234 (1973).

[9] J. Timmerman, "Two Electromagnetic Vibrators," *Philips Tech. Rev.*, **33**, No. 8/9, 249–259 (1973).

[10] L. Honds and K. H. Meyer, "A Linear D.C. Motor with Permanent Magnets," *Philips Tech. Rev.*, **40**, No. 11/12, 329–337 (1982).

PROBLEMS

11-1 Consider the device of Figure 11.7. Suppose you are given the permeability of the core material and of air, the pertinent physical dimensions of the core, and the number of turns, N. Estimate the inductance of the device if the core material remains in the linear range of the B–H curve.

11-2 In Figure 11.8, find an expression for the capacitance parameters for the sections of core material in terms of permeability and physical dimensions in the linear case. Combine the capacitance parameters into a single equivalent capacitance, and show how to estimate the inductance using this equivalent capacitance.

11-3 In the transducer sketched below, a slug of magnetic material slides partly in and out of the flux path.

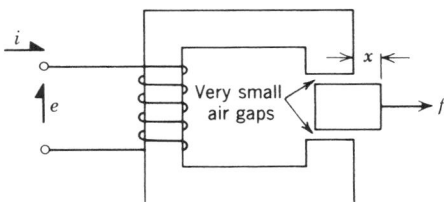

Make a simple bond graph model of the device, neglecting all loss effects and all leakage paths. By imagining how the reluctance at the slug would vary with displacement x, discuss qualitatively the difference you might expect between the f-versus-x relation at constant current for this device and for the device of Figure 11.11.

11-4 The figure shows typical demagnetization curves for several permanent-magnet materials. (This is a conventional replotting of the lower right hand quadrant of the hysteresis curve in Figure 11.5.) The heavy curve represents a typical high-energy-product rare earth material. The $(BH)_{\max}$ point for this material has $B_m/\mu_0 H_m \cong 1$, and one can see that the Alnico materials have corresponding $B_m/\mu_0 H_m$ values greater than 1.

Consider the design of a loudspeaker voice coil as shown schematically in Figure 11.13b. The cylindrical coil is to have a diameter of 30 mm. The air gap is 2 mm with an axial length of 10 mm. The magnet is also cylindrical, with a diameter of 30 mm and unknown length. Assuming that the rare earth magnet material is represented by the heavy curve on the B–H plot, find the optimum length of the magnet, I_n, as well as the flux density in the gap, B_g. Neglect any reluctance in the pole pieces which lead the flux to the radial gap.

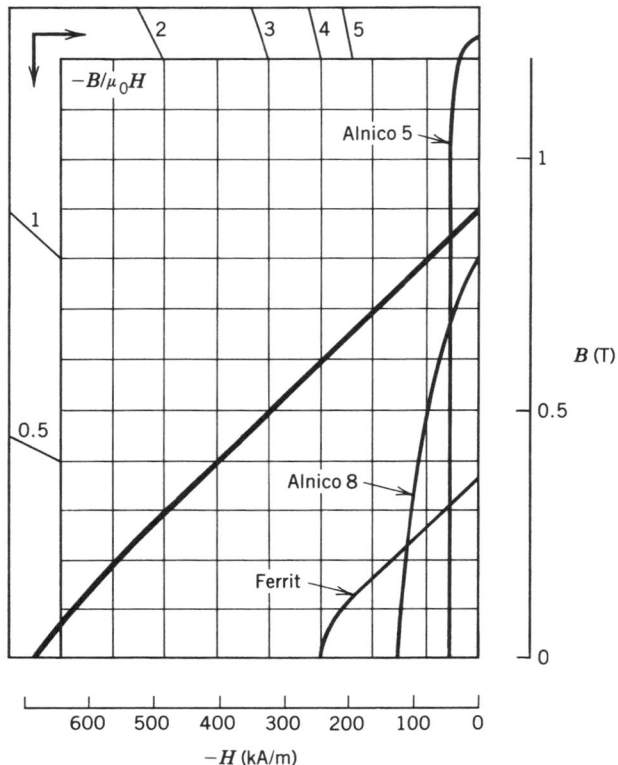

11-5   The bond graph below represents a simplified version of the relay model of Figure 11.11:

$$S_e \xrightarrow[i]{V_0(t)} 1 \xrightarrow[N]{} GY \xrightarrow[\varphi]{M} C \xrightarrow[\dot{x}]{F} 1 \xrightarrow{} TF \xrightarrow[\dot{\theta}]{\dot{p}_\theta} I{:}J_0$$

with $\mathbf{R}:\dot{R}$ over the first 1, $R_0 + \frac{x}{\mu_0 A}$ over C, $C{:}k$, and $1{:}\dot{x}$.

The core loss elements have been eliminated, and the total core reluctance $\mathbf{R}_0$ has been incorporated in the 2-port C-field constitutive law. The air gap effective length x is related to the angle θ by the length l. The mechanical spring force is $k(x - x_0)$, which applies as long as the relay has not encountered a stop. The moment of inertia about the pivot is J_0, and the C-field reluctance is $R_0 + x/\mu_0 A$. [See Eqs. (11.18)–(11.22).] Apply causality and write the state equations for φ, x, and p_θ.

11-6   Consider the force associated with a flux φ across an air gap as illustrated in Figure 11.9. Consider only the gap reluctance approximated by Eq. (11.21) and that the MMF ΔM is generated by a coil of N turns carrying a constant current I. Express the force F as a function of x and I.

11-7 The bond graphs of Figure 11.15 apply to the devices of Figures 11.10, 11.12, 11.13a,b, and 11.14. However, in several cases parts of the general bond graph are not needed, because they represent effects which are either totally absent or negligible in normal operation. Show simplified bond graphs for the five devices sketched, and note why some elements can be eliminated.

11-8 Suppose a permanent-magnet lifting system were capable of inducing a B-value of 0.5 T across a very narrow air gap between the magnet surface and a piece of steel. Develop an expression for the force per unit area in terms of B and μ_0, and find the force developed for an area 1 cm $\times$ 1 cm. How many kilograms could be lifted by this magnet?

11-9 Consider the design of a voice coil actuator of the type shown in Figure 11.13b. The design parameters are to be

$$B = 0.5\ T, \qquad R = 8\Omega, \qquad V_{\max} = 10\ V,$$
$$\text{Allowable current density} = 20 \times 10^6 \text{A/m}^2 = J,$$
$$\text{Resistivity of copper wire} = 1.72 \times 10^{-8}\Omega \cdot m = \rho_{\text{Cu}}.$$

(The resistance of the coil is $R = \rho_{\text{Cu}}l / A_w$, where l is the wire length and A_w is the wire area.) What would the maximum force of this device be?

11-10 Consider a device similar to that shown in Figure 11.10 but with a permanent magnet replacing the coil. Suppose that the total reluctance of the magnetic circuit is $\mathbf{R}(\theta)$. [Equation (11.23) indicates how $\mathbf{R}(\theta)$ for the gap might vary for small angles, but $\mathbf{R}(\theta)$ would be valid for all θ and would include core reluctance.] The bond graph below shows a 2-port C based on $\mathbf{R}(\theta)$ as well

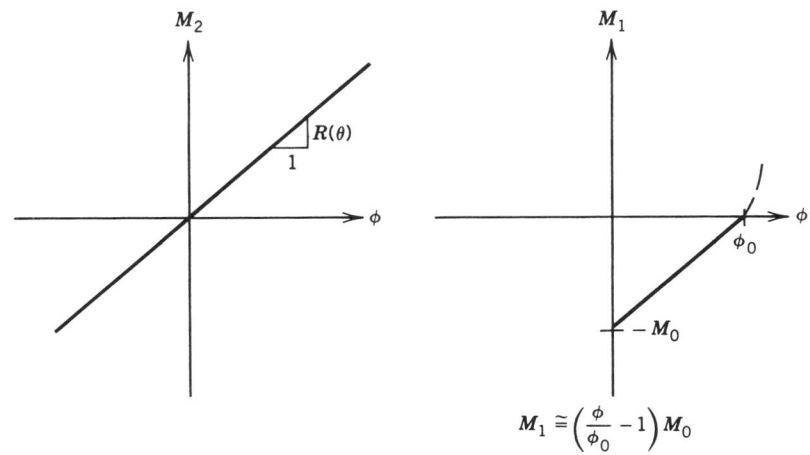

$$M_1 \equiv \left(\frac{\phi}{\phi_0} - 1\right) M_0$$

as a 1-port C representing the demagnetization curve for the permanent magnet. The relevant parts of the constitutive laws are also sketched. (The $M_1 - \varphi$ sketch can be related to the lower right hand quadrant of the $B-H$ hysteresis loop shown in Figure 11.5.)

$$P.M.: C \xrightarrow[\varphi]{M_1} 1 \xrightarrow[\varphi]{M_2} \overset{R(\theta)}{C} \xrightarrow[\theta]{\tau}$$

Noting that $M_2 = -M_1$, the flux φ can be eliminated from the constitutive laws so that the "cogging torque" τ can be expressed in terms of the parameters M_0, φ_0, and the reluctance $R(\theta)$. Find the law which relates τ to θ.

12

THERMOFLUID SYSTEMS

12.1 BASIC THERMODYNAMICS IN BOND GRAPH FORM

Thermodynamics is essentially the universal science of physical processes. All of the models of physical systems that have been made so far may be studied from the point of view of thermodynamics, and indeed, when C- and I-fields were constrained to conserve energy or R-fields were arranged to dissipate power, thermodynamic arguments were used. On the other hand, since heat flow and temperature dependence have not yet been discussed, most of our dynamic models can be characterized as *isothermal models*; that is, we treated our multiports as if they were immersed in an infinite reservoir that maintains constant temperature even when energy is lost or supplied. In fact, we know that all elements change their constitutive laws with temperature to greater or lesser extent, and the transfer of heat and change in temperature is the special province of thermodynamics.

Much of engineering thermodynamics is not very dynamic at all. It deals with changes between equilibrium states but often is not concerned with the process of change itself. In many cases, useful information may be extracted from a knowledge of the constraints that end-point equilibrium states of a system must satisfy. But if we are really interested in the path a system will follow in state space, then we must study what has come to be called *irreversible thermodynamics*, which involves nonequilibrium systems.

In recent years, there has been an upsurge of interest in making dynamic models for thermodynamic systems in which a variety of electrical, chemical, mechanical, fluid dynamic, and heat transfer effects are important. Here we can only give an introductory account in which the model systems discussed previously are shown to have thermodynamic implications. We do this by the device of considering systems of such physical extents and time scales that local pseudoequilibrium conditions exist. In this way, we may modify some familiar relations from thermostatics into a slowly varying form of thermodynamics.

Another restriction for the present is to consider only closed systems, that is, systems in which no mass passes through the system boundaries. This restriction to a Lagrangian point of view will be lifted in the section on fluid mechanics, but the control volume or Eulerian point of view, which is often used for certain types of systems, introduces a set of modeling difficulties that are best left for later discussion.

Consider first, then, a fixed amount of a pure substance which is in at least pseudoequilibrium, so that all parts of the substance are at essentially identical conditions of pressure, temperature, density, and so forth.

We consider below a unit mass of the substance. Extensive quantities such as volume, internal energy, and entropy for a unit mass will be designated by the lowercase letters v, u, s. When a mass of the substance, m, is involved, the corresponding variables will be designated by uppercase letters: $V = mv$, $U = mu$, and $S = ms$. The substance is supposed to be of such small extent and the disturbances on such a slow time scale that wave motion, turbulence, and so forth are negligible. Then in the absence of motion and electromagnetic or surface-tension forces, we may assume that such a pure substance has only two independent properties. All other properties are related to any two independent properties by equations of state for the substance.

A useful, and logical, way to describe the constitutive relations for a pure substance begins with the statement that the internal energy per unit mass, u, of the substance is a function of the volume v and entropy s per unit mass:

$$u = u(s, v). \tag{12.1}$$

The well-known Gibbs equation relates changes in u to changes in v and s:

$$du = T \, ds - p \, dv, \tag{12.2}$$

where T is the thermodynamic temperature and p is the pressure. (In all that follows, T will indicate thermodynamic or absolute temperature, a positive quantity.)

The relation (12.2) involves energy or work. If the substance changes state slowly enough, we may treat it as if it were a lumped-parameter multiport with the relation

$$\frac{du}{dt} = T \frac{ds}{dt} - p \frac{dv}{dt}. \tag{12.3}$$

The term $T \, ds/dt$ is associated with a flow of heat or, sometimes, with dissipative work such as might be done by a paddle wheel slowly stirring a fluid substance. The term $p \, dv/dt$ represents the sort of reversible power with which we have been dealing frequently in dynamic models.

Since

$$du = \frac{\partial u}{\partial s} ds + \frac{\partial u}{\partial v} dv,$$

it is clear that

$$T = \frac{\partial u}{\partial s} \quad and \quad -p = \frac{\partial u}{\partial v}. \tag{12.4}$$

Also

$$\left.\frac{\partial T}{\partial v}\right|_s = \frac{\partial^2 u}{\partial v \, \partial s} = \left.\frac{\partial(-p)}{\partial s}\right|_v, \tag{12.5}$$

which we may recognize as a reciprocal relation similar to those found for other energy-storing fields. In thermodynamics, Eq. (12.5) is called a *Maxwell reciprocal relation*.

The constitutive laws of the pure substance may be represented by a C-field; thus,

$$\xrightarrow[\dot{s}]{T} C \xrightarrow[\dot{v}]{P}$$

if we are willing to call T an effort, $\dot{s}$ a flow, and s a displacement. Integral causality,

$$\underset{\dot{s}}{\overset{T}{\vdash}} C \underset{\dot{v}}{\overset{P}{\dashv}}$$

implies

$$T = T(s, v), \qquad p = p(s, v), \tag{12.6}$$

which are also implied by Eqs. (12.1) and (12.2). The only unusual feature of the pure-substance C-field is the sign convention on the $p\dot{v}$ port, which is negative in the traditional form of the Gibbs equation (12.2).

The internal energy is associated with all integral causality and with constitutive laws in the form of (12.6). For mixed- and all-derivative causality, the constitutive laws are switched around. In thermodynamics this switching around of independent and dependent variables is often accomplished using the Legendre transformations of u. The enthalpy h, Helmholtz free energy f, and Gibbs free energy ϕ are all Legendre transformations of the internal energy u and correspond to different causal patterns. The enthalpy is

$$h(s, p) \equiv u + pv, \tag{12.7}$$

and its derivatives are

$$\frac{\partial h}{\partial s} = T(s, p), \tag{12.8}$$

$$\frac{\partial h}{\partial p} = v(s, p). \tag{12.9}$$

Using $h(s, p)$, another Maxwell relation can be found:

$$\left.\frac{\partial T}{\partial p}\right|_s = \frac{\partial^2 h}{\partial s \, \partial p} = \left.\frac{\partial v}{\partial s}\right|_p. \tag{12.10}$$

The causal pattern for the C-field corresponding to h is shown below:

$$h = h(s, p) \quad \Leftrightarrow \quad \overset{T}{\underset{s}{\longmapsto}} C \overset{P}{\underset{v}{\longmapsto}}$$

The Helmholtz free energy is defined by the transformation

$$f(T, v) \equiv u - Ts. \tag{12.11}$$

Its derivatives are

$$\frac{\partial f}{\partial T} = -s(T, v), \tag{12.12}$$

$$\frac{\partial f}{\partial v} = -p(T, v). \tag{12.13}$$

And yet another Maxwell relation is

$$\left.\frac{\partial(-s)}{\partial v}\right|_T = \frac{\partial^2 f}{\partial T \, \partial v} = \left.\frac{\partial(-p)}{\partial T}\right|_v. \tag{12.14}$$

The causal pattern corresponding to f is

$$f = f(T, v) \quad \Leftrightarrow \quad \overset{T}{\underset{s}{\dashv}} C \overset{P}{\underset{v}{\dashv}}$$

Finally, the Gibbs free energy is defined by a double Legendre transformation on u:

$$\phi(T, p) = u + pv - Ts. \tag{12.15}$$

The derivatives of ϕ are

$$\frac{\partial \phi}{\partial T} = -s(T, p), \tag{12.16}$$

$$\frac{\partial \phi}{\partial p} = +v(T, p), \tag{12.17}$$

and the Maxwell relation is

$$\left.\frac{\partial(-s)}{\partial p}\right|_T = \frac{\partial^2 \phi}{\partial T \, \partial p} = \left.\frac{\partial v}{\partial T}\right|_v. \tag{12.18}$$

The Gibbs free energy corresponds to all-derivative causality for the C-field:

$$\phi = \phi(T, p) \quad \Leftrightarrow \quad \overset{T}{\underset{s}{\dashv}} C \overset{P}{\underset{v}{\longmapsto}}.$$

The energy function u and the co-energy functions h, f, and ϕ, which are Legendre transformations of u, have been expressed in terms of their own natural variables.

When this is done, the constitutive functions for the substance may be found by differentiation, and such constitutive laws automatically ensure that the C-field will be conservative. On the other hand, one could write out constitutive laws in any form without the use of state functions such as u, h, f, or ϕ; the only difficulty is that arbitrary constitutive laws would not, in general, allow the existence of an internal energy function. In a cycle, a substance with arbitrary constitutive laws might allow net energy production and thus the construction of a perpetual motion machine of the first kind. In a sense, deriving constitutive laws from u, h, f, or ϕ is safer than not using state functions, since conservation of energy will be built into the relations.

A classical example of constitutive laws not stated in terms of state function derivatives is the perfect gas. The equation of state for a perfect gas is

$$pv = RT, \tag{12.19}$$

in which R is a constant. It is often stated, in addition, that u is a function of temperature only. This means that if the constitutive laws for the gas were used to express $u(s, v)$ in terms of, say, T and v, then u would be a function only of T and not of v. Let us derive this fact using the state functions.

If Eq. (12.19) is solved for p in terms of T and v, the constitutive law in the form of Eq. (12.13) is obtained:

$$\left. \frac{\partial f}{\partial v} \right|_T = -p(T, v) = \frac{-RT}{v}. \tag{12.20}$$

Using this result, the Helmholtz free energy can be found by integration:

$$f(T, v) = -RT \ln v + \psi(T), \tag{12.21}$$

where $\psi(T)$ is some function of temperature alone. Then, using Eq. (12.12),

$$-s(T, v) = \frac{\partial f}{\partial T} = -R \ln v + \frac{d\psi(T)}{dT}. \tag{12.22}$$

Finally, substituting Eq. (12.22) into Eq. (12.11) and solving for u,

$$u(T, v) = f(T, v) + Ts = -RT \ln v + \psi(T) + RT \ln v - T \frac{d\psi(T)}{dT}$$

$$= \psi(T) - T \frac{d\psi(T)}{dT}, \tag{12.23}$$

which demonstrates that u is only a function of T.

Although u is a function of T, we need u in terms of s and v in order to evaluate the complete equations of state. There are two complete equations of state because the gas is a 2-port C-field. We need more information than just Eq. (12.19). [We could solve for the function ψ in Eq. (12.21) for example.] It is more common to assume that the two so-called specific heats are constant. The specific heat at a constant pressure, c_p, is defined as

$$c_p = \left.\frac{\partial h}{\partial T}\right|_p,$$ (12.24)

and the specific heat at constant volume, c_v, is

$$c_v = \left.\frac{\partial u}{\partial T}\right|_v.$$ (12.25)

By substituting Eq. (12.19) into Eq. (12.7) and differentiating as in Eq. (12.24), we find that

$$c_p = c_v + R.$$ (12.26)

If one now assumes that c_v is constant, then Eq. (12.25) may be integrated to yield

$$u = c_v(T - T_0),$$ (12.27)

and, noting that Eqs. (12.7) and (12.19) imply that h is also a function of T alone, Eq. (12.24) may be integrated similarly to yield

$$h = c_p(T - T_0),$$ (12.28)

in which the subscript 0 stands for a state in which it is assumed that $u = h = s = 0$.

Rearranging the basic equation (12.2) and using Eq. (12.27), one can find s:

$$ds = \frac{du}{T} + p\frac{dv}{T} = c_v\frac{dT}{T} + R\frac{dv}{v},$$

or

$$s = c_v \ln\frac{T}{T_0} + R \ln\frac{v}{v_0},$$

which may be solved for $T(s, v)$:

$$T = T_0 e^{s/c_v} \left(\frac{v}{v_0}\right)^{-R/c_v}.$$ (12.29)

Further manipulations of Eq. (12.2) yield

$$ds = c_p\frac{dv}{v} + c_v\frac{dp}{p},$$

$$s = c_p \ln\frac{v}{v_0} + c_v \ln\frac{p}{p_0},$$

which yields $p(s, v)$,

$$p = p_0 e^{s/c_v} \left(\frac{v}{v_0}\right)^{-c_p/c_v}.$$ (12.30)

Equations (12.29) and (12.30) are the complete relations for the C-field in the form (12.4). It is desirable, however, to show that these equations do indeed derive from $u(s, v)$. Using Eq. (12.27) and (12.29), we have

$$u(s, v) = c_v T_0 \left[e^{s/c_v} \left(\frac{v}{v_0} \right)^{-R/c_v} - 1 \right], \tag{12.31}$$

from which we find

$$T = \frac{\partial u}{\partial s} = T_0 e^{s/c_v} \left(\frac{v}{v_0} \right)^{-R/c_v},$$

which agrees with Eq. (12.29), and

$$-p = \frac{\partial u}{\partial v} = c_v \frac{T_0}{v_0} e^{s/c_v} \left(\frac{-R}{c_v} \right) \left(\frac{v}{v_0} \right)^{(-R/c_v)-1} = \frac{-RT_0}{v_0} e^{s/c_v} \left(\frac{v}{v_0} \right)^{-(R+c_v)/c_v},$$

$$\tag{12.32}$$

which agrees with Eq. (12.30) upon using Eqs. (12.19) and (12.26).

The reader probably feels that all of the above manipulations of the perfect gas law have only produced some complicated laws in Eqs. (12.29) and (12.30) in place of the simple law $pv = RT$. But it is important to remember that (1) $pv = RT$ is not a complete characterization of the C-field of the gas and (2) $pv = RT$ is in a form that only allows differential causality on the $T\dot{S}$ port. Equations (12.29) and (12.30) are complete and in integral causality form at both C-field ports and hence are useful in making a dynamic model of a quantity of gas as a thermomechanical transducer.

12.2 HEAT TRANSFER IN TRUE BOND GRAPHS AND PSEUDO–BOND GRAPHS

The $T\dot{S}$ port for the pure-substance C-field may be used to model all types of power flow that are associated with entropy increase, or, in other words, irreversible effects. Changes in entropy of a pure substance can be accomplished by a variety of dissipative effects such as stirring with a paddle wheel or heating with an electrical resistor immersed in a fluid. In each case, one may identify the power dissipated with $T\dot{S}$ in order to find the change in state of the substance. An important use of $T\dot{S}$ ports is in modeling the effects of heat flow. Using $\dot{Q}$ to stand for the rate of transfer of heat in power units, we may often identify $\dot{Q}$ with $T\dot{S}$.

Consider first the simple case of conduction heat transfer shown in Figure 12.1a. The idea is that two reservoirs of thermal energy at absolute temperatures T_1 and T_2 are allowed to communicate through a thermal resistance but in no other way. Generally, a heat flow $\dot{Q}$ will be set up between the two reservoirs. It is common experience that heat flows from higher toward lower temperature, and, indeed, this observation is behind the second law of thermodynamics. In Figure 12.1b, possible

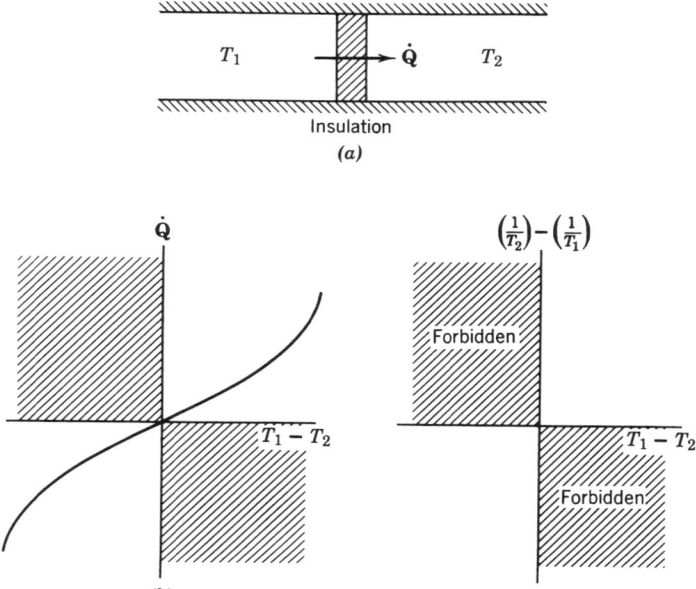

FIGURE 12.1. Conduction heat transfer. (a) Thermal resistance; (b) allowed relations between Q and $T_1 - T_2$; (c) forbidden regions for $1/1T_2 - 1/1T_1$.

relations between $\dot{Q}$ and $T_1 - T_2$ are sketched for the thermal resistance. All we need to assume is that $\dot{Q}$ is related to $T_1 - T_2$ in such a way that any values of $\dot{Q}$ and $T_1 - T_2$ would plot in the first and third quadrants of the $\dot{Q}$ versus $T_1 - T_2$ plane. (Although it is common to assume that $\dot{Q}$ is a function of $T_1 - T_2$ as sketched, the argument is true if $\dot{Q}$ is positive when $T_1 - T_2$ is positive, negative when $T_1 - T_2$ is negative, and zero when $T_1 - T_2$ is zero.

If we now write

$$T_1 \dot{S}_1 = \dot{Q} = T_2 \dot{S}_2, \tag{12.33}$$

which implies that the heat flow out of one body is instantaneously equal to the heat flow into the other body, then the net entropy flow rate, $\dot{S}_2 - \dot{S}_1$, may be found:

$$\dot{S}_2 - \dot{S}_1 = \frac{\dot{Q}}{T_2} - \frac{\dot{Q}}{T_1} = \dot{Q}\left(\frac{1}{T_2} - \frac{1}{T_1}\right) = \dot{Q}\frac{T_1 - T_2}{T_1 T_2}. \tag{12.34}$$

In Figure 12.1c, it is shown that $1/T_2 - 1/T_1$ is positive, negative, or zero according as $T_1 - T_2$ is positive, negative, or zero. This is true because the absolute temperatures are inherently positive quantities. Thus, the net entropy production rate, which is the product of $\dot{Q}$ and $1/T_2 - 1/T_1$, is positive for any finite value of $T_1 - T_2$ and only vanishes when $\dot{\mathbf{Q}}$ vanishes. The thermal resistor may be represented by the 2-port field shown below:

$$\overset{T_1}{\underset{\dot{S}_1}{\rightharpoonup}} R \overset{T_2}{\underset{\dot{S}_2}{\rightharpoonup}} .$$

Although the 2-port resistive field is power conservative [Eq. (12.33)], it is neither a transformer nor a gyrator. It has the peculiar property that, with the sign convention shown above,

$$\dot{S}_2 - \dot{S}_1 \geq 0,$$

which implies that the entropy flow leaving the field is greater than the entropy flow entering the field no matter which way the heat is flowing. When the thermal resistor is part of a system, it will tend to increase the entropy of the system whenever any heat flows through the resistor.

A possible constitutive law for the resistor is

$$\dot{\mathbf{Q}} = H(T_1 - T_2),$$

or

$$\dot{S}_1 = \frac{H(T_1 - T_2)}{T_1}, \qquad \dot{S}_2 = \frac{H(T_1 - T_2)}{T_2}, \qquad (12.35)$$

where the heat transfer coefficient H is assumed to be constant or a slowly varying function of the average temperature, $(T_1 + T_2)/2$. As a check, the net entropy production rate is

$$\dot{S}_2 - \dot{S}_1 = \frac{H(T_1 - T_2)}{T_2} - \frac{H(T_1 - T_2)}{T_1} = \frac{H(T_1 - T_2)^2}{T_1 T_2} > 0. \qquad (12.36)$$

Note that Eq. (12.35) is written in the causal form,

$$\overset{T_1}{\underset{\dot{S}_1}{\rightharpoonup}} R | \overset{T_2}{\underset{\dot{S}_2}{\rightharpoonup}} .$$

Given any two temperatures $T_1 > 0$, $T_2 > 0$, it is possible to solve for $\dot{\mathbf{Q}}$, $\dot{S}_1$, and $\dot{S}_2$. The remaining causal forms are not so useful. For example,

$$\overset{T_1}{\underset{\dot{S}_1}{\rightharpoonup}} R \overset{T_2}{\underset{\dot{S}_2}{\rightharpoonup}} $$

implies that given any $\dot{S}_1$ and $\dot{S}_2$, the resistance should provide T_1 and T_2. But we cannot actually impose arbitrary variables $\dot{S}_1$ and $\dot{S}_2$, since Eq. (12.36) must be obeyed. This causality is therefore not useful for dynamic systems. Even mixed causalities such as

$$\begin{array}{c} \dfrac{T_1}{\dot{S}_1} \; R \; \dfrac{T_2}{\dot{S}_2} \end{array}$$

are fraught with difficulties. Inverting the last equation of (12.35), we find

$$T_2 = \frac{H T_1}{\dot{S}_2 + H},$$

which seems to imply that T_2 can be negative if $\dot{S}_2 < -H$. Actually, $\dot{S}_2$ is limited, given T_1, by the requirement that T_2 must be positive. Thus, the values $\dot{S}_2$ may be assigned are limited by the choice of T_1. In this sense, the mixed causal patterns are less useful than the causality of Eq. (12.35) in which the (positive) temperatures T_1 and T_2 can be assigned arbitrarily. In what follows, the thermal resistor will be assumed to accept only the causality $\dashv R \vdash$.

12.2.1   A Simple Example

Figure 12.2 shows a system incorporating two chambers containing compressible fluids. One chamber is of constant volume, so that the internal energy of the fluid contained in the chamber varies only because of heat flow through the partition, which is modeled as a thermal resistance. The volume of the other chamber is changed as a piston moves back and forth.

The bond graph for the system shown in Figure 12.2b incorporates two capacitance elements for the fluids and an R-field for the thermal resistance. The 0-junction is used simply to achieve inward sign conventions for both the $T_1 \dot{S}_1$ port of the left-hand C and the resistive field, so that the sign conventions assumed previously are reflected in the model. The 0-junction assures that $T_1 = T_2$ but $\dot{S}_1 = -\dot{S}_2$.

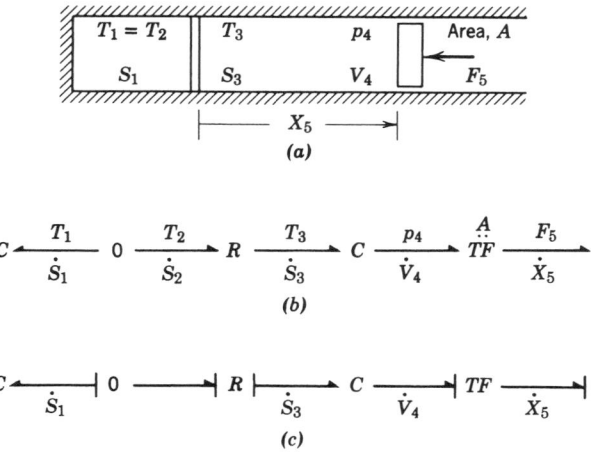

FIGURE 12.2. Example system. (a) Schematic diagram; (b) bond graph; (c) bond graph with integral causality.

In Figure 12.2c integral causality is shown for the system. The preferred R-field causality is compatible with integral causality on both C-elements. The state variables are S_1 (the entropy of the fixed-volume fluid), S_3 (the entropy of the variable-volume fluid), and V_4 (the volume of the variable chamber). The velocity $\dot{X}_5(t)$ plays the role of an input variable.

The constitutive laws for the fluids could be found by starting with the internal energy per unit mass, $u(s, v)$, a function of entropy per unit mass and specific volume, and then converting to the total energy U as a function of total entropy S and volume V. This would require a knowledge of the total mass and the initial state of the fluids. The pressure and temperature constitutive relations could then be found by differentiating U with respect to S and V. The constitutive relations for the thermal resistance have been discussed previously.

Clearly, the model for the system represented by the bond graph can only be appropriate for slow changes. We assume, for example, that the fluids are in pseudoequilibrium, so that the temperature and pressure are essentially uniform throughout the volumes at every instant. This would not be the case if one needed to worry about acoustic waves or the details of heat transmission in the fluids themselves. Also, although the entire system is power conservative, the entropy of the two C-elements will increase if any heat flows. From the point of view of the external port, this irreversibility makes it appear that energy is lost. For example, if X_5 is taken through a cycle starting from an equilibrium state at which $T_2 = T_3$, the fluid temperature in the variable volume will change, heat will flow, and, during the cycle, a net energy loss will be observed at the external port. What the bond graph shows is that energy is not lost, but rather converted back into thermal energy that cannot be completely converted back into mechanical energy except in the limiting case in which X_5 moves so slowly that T_2 and T_3 are virtually identical and the net entropy production almost vanishes.

In the model of Figure 12.2 no irreversible phenomena other than heat transfer have been included. It would not be hard to include other dissipative effects. For example, if the piston had Coulomb friction and if all the mechanical energy lost in friction were converted into thermal energy, a simple model would involve the creation of another equivalent entropy flow to the fluid C-field equal to the dissipated power divided by the fluid temperature. A bond graph for this case is shown in Figure 12.3.

The Coulomb friction resistor may be described by a relation such as

$$F_6 = A \operatorname{sgn} \dot{X}_5,$$

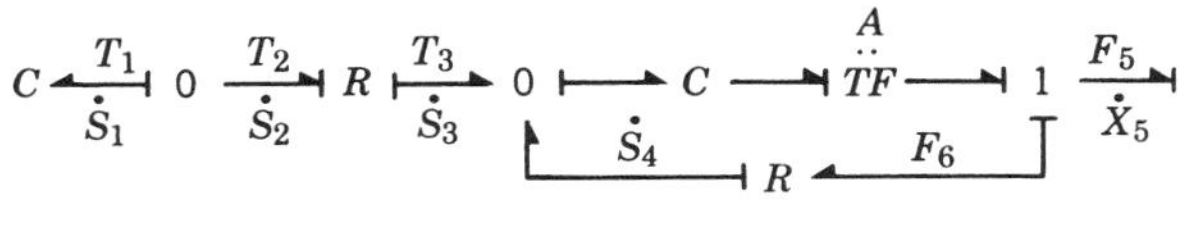

FIGURE 12.3. System of Figure 12.2 with piston friction.

so that the power dissipated is

$$F_6 \dot{X}_5 = A \dot{X}_5 \operatorname{sgn} \dot{X}_5 \geq 0.$$

The entropy flow $\dot{S}_4$ is then just

$$\dot{S}_4 = \frac{1}{T_3} A \dot{X}_5 \operatorname{sgn} \dot{X}_5 \geq 0.$$

The 2-port R-field version of a mechanical resistor can accept either causality at the mechanical port but only the causality shown in Figure 12.3 at the thermal port. Actually, the frictional energy may not reach the fluid instantaneously, as assumed in the model: It may heat the piston and wall material first and then warm the fluid. A more complex model with other thermal capacitances could model such effects.

12.2.2 An Electrothermal Resistor

All dissipation results in thermal effects, which can sometimes be neglected or treated separately. For example, electrical resistors heat up as electrical power is dissipated, but in many cases the heating does not change the characteristics of the device very much if a means of cooling the resistor is provided. In such cases, ordinary circuit models assume that the circuit components remain essentially at a constant temperature. Here, we construct a model that contains both electrical and thermal effects.

The resistor is sketched in Figure 12.4a. The body of the resistor is assumed to have a fairly uniform temperature T and to be immersed in an atmosphere at temperature T_0. The resistor is assumed to dissipate electrical power:

$$ei \geq 0. \tag{12.37}$$

As long as Eq. (12.37) is obeyed, we may assume that the relation between e and i for the resistor has some temperature dependence.

The power lost electrically is often spoken of as being converted to heat, so we will assume that

$$ei = \dot{Q} = T \dot{S}, \tag{12.38}$$

in which it is a moot point whether $\dot{Q}$ should be regarded as a heat flow in the usual sense. Two causal forms for an R-field representation are useful:

$$\overset{e}{\underset{i}{\rightharpoonup}} R \overset{T}{\underset{\dot{S}}{\rightharpoonup}}$$

$$e = e(i, T), \qquad \dot{S} = \frac{i e(i, T)}{T}, \tag{12.39}$$

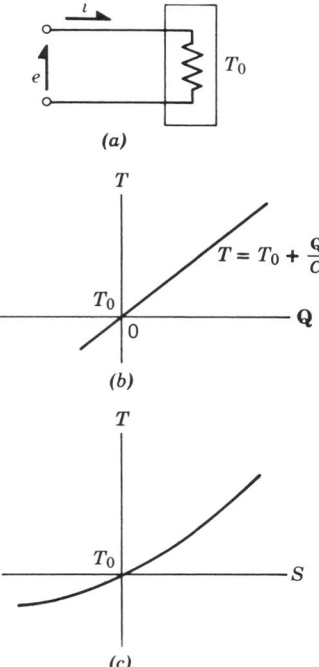

FIGURE 12.4. The electrothermal resistor. (a) Schematic diagram; (b) definition of thermal capacity; (c) capacity relationship in terms of entropy.

and

$$\frac{e}{i} \dashv R \vdash \frac{T}{\dot{S}},$$

$$i = i(e, T), \qquad \dot{S} = \frac{ei(e, T)}{T}. \tag{12.40}$$

As in previous examples, there is a preferred causality at the thermal port.

The temperature of the resistor body depends on how much thermal energy has been stored up in the resistor. In what follows, we neglect any work done against the pressure of the atmosphere by expansion of the resistor, that is, the assumption is that no significant power flows at the $P\dot{V}$ port. The temperature, then, will depend on the amount of energy in the form $T\dot{S}$ that has been absorbed.

It is common to assume that T is a function of $\mathbf{Q}$, but for consistency we express T as a function of S. These two points of view are readily reconciled, since $\dot{\mathbf{Q}}$ and S are related by Eq. (12.38). Suppose, for example, that a thermal capacity C (approximately constant) has been defined as in Figure 12.4b and in the equation

$$T = T_0 + \frac{\mathbf{Q}}{C}, \tag{12.41}$$

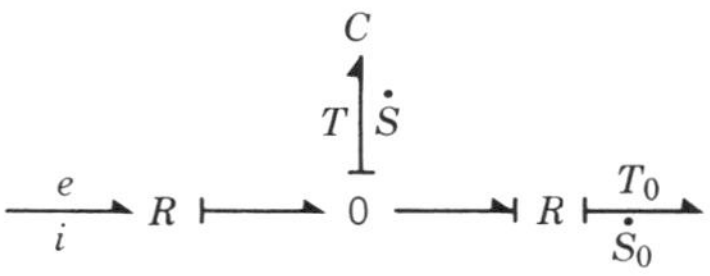

FIGURE 12.5. Bond graph for electrothermal resistor.

in which we have integrated $\dot{Q}$ to find Q at a time when $T = T_0$. Using Eqs. (12.38) and (12.41), S may be found:

$$S = \int_0^S dS = \int_0^Q \frac{dQ}{T_0 + Q/C} = \frac{C \ln(T_0 + Q/C)}{T_0}, \tag{12.42}$$

where S and Q are both assumed to vanish at the initial time. Solving Eq. (12.42) for Q in terms of S and then substituting the result into Eq. (12.41), we find

$$T = T_0 e^{S/C}, \tag{12.43}$$

which is the relationship for the thermal capacitance when S is used instead of C. [In reality, since T is an absolute temperature, $Q \cong T_0 S$ and $T \cong T_0(1 + S/C)$ for modest excursions of T from T_0.]

The bond graph of Figure 12.5 shows the complete model for the resistor. The entire system is conservative, but power is lost from the electrical port, since the power on some heat flow bonds cannot reverse. Both the electrothermal and heat transfer R-fields represent irreversible effects and generate entropy. As long as the temperature dependence of the electrical resistance is not strong, one may avoid the complicated model of Figure 12.5 in favor of a simple 1-port R. However, in principle, the electrical and thermal systems are always coupled bilaterally, so that models of this type are required.

The reader will have noticed in this brief introduction to bond graph models for thermodynamic systems that only closed systems were modeled; that is, no mass crossed the boundary of the system. Much of the science of thermodynamics is concerned with flow processes in which not only power, but also mass, flows into and out of the system. While such systems can often be modeled with bond graphs, the convection of energy as mass moves through the boundary of a control volume complicates the modeling process. Some flow-process models, in which the distinction between Lagrangian and Eulerian descriptions of fluid motion parallels that between closed and open thermodynamic systems, will be discussed in the following sections.

12.3   FLUID DYNAMIC SYSTEMS

We have already established effort, flow, displacement, and momentum variables for fluid systems of the closed-circuit hydraulic type, in which the static pressure times the volume flow rate represents most of the transmitted power. We now take a deeper look at fluid systems in general.

Much of fluid mechanics work is concerned with field problems in which a flow field in two or three dimensions is to be determined. Such problems are usually described by partial differential equations. Unless an analytic solution to such a problem happens to be known, one must resort to finite-difference or finite-element techniques. Such techniques represent the continuum with a large number of similar lumps. Although bond graphs can be made for the "microlumps" involved in partial differential equations or their finite approximations, there often is little to be gained by such a representation. The microlumps are all very similar, and they interact with their fellows only in standardized ways.

In this section, the main concern will be with the gross type of lumping commonly done when a fluid system is a small part of a larger system and therefore cannot practically be represented in great detail. Typical examples of such cases occur when hydraulic or pneumatic elements form part of a control system. In what follows, the main concern will be with interior flows and with fluid dynamic interaction with solid mechanical elements through the influence of forces, motions, pressures, and volume flow rates.

As a simple example of the sort of approximations often made in the analysis of fluid systems, consider the line element shown in Figure 12.6. The model, which in

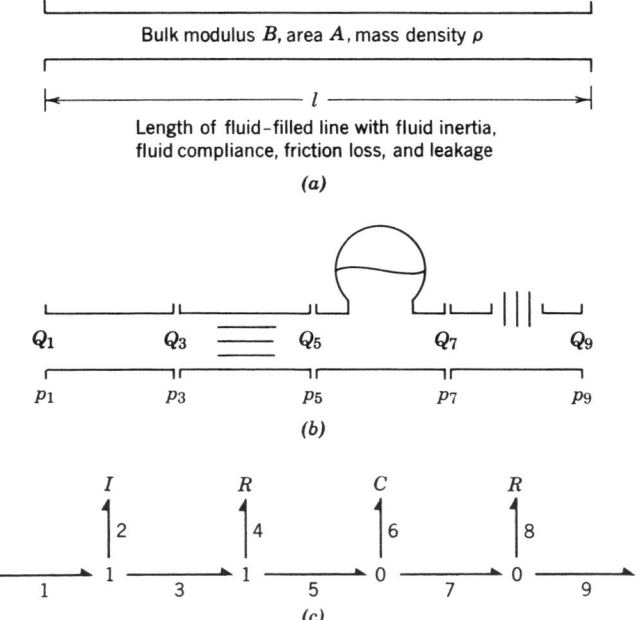

FIGURE 12.6. Lumped representation of fluid-filled lines. (*a*) Sketch of line with some parameters; (*b*) schematic diagram; (*c*) bond graph.

the linear case is analogous to electrical-transmission-line models, attempts to treat a variety of effects clearly present in real lines. The model is plausible for short segments of line and for linear elements—yet it is often used for long lengths l and for nonlinear elements. As will be seen, there are some philosophical snags with this model if it is used without careful thought.

First of all, it is clear that the model is one-dimensional. The volume flow rate past a unit cross-sectional area, Q, and the pressure p clearly must be thought of as averaged quantities over a field of velocities and pressures that vary over the area. Since the velocity profile of the flow may vary widely during transient conditions, it is clear that the variables p and Q and the parameters of the $—R$, $—C$, and $—I$ elements in Figure 12.6 cannot be evaluated exactly unless the nature of the flow over the cross-sectional area is known in advance. Nevertheless, in system design studies, one rarely is sure in advance how the flows in various parts of the system will behave, so that one must make at least a preliminary estimate of the system parameters before the dynamics of the flow can even be estimated.

Consider first the problem of estimating the inertia of the fluid in the line. An elementary derivation of the inertia coefficient (See Reference [1], for example) proceeds as follows from Figure 12.6: The mass of the fluid in the line is ρAl, the force tending to accelerate the fluid is $A(p_1 - p_3)$, and the velocity of the fluid is Q_2/A. Thus

$$\rho Al \frac{d}{dt} \frac{Q_2}{A} = A(p_1 - p_3), \qquad \text{or} \quad Q_2 = \frac{A}{\rho l} p_{p_2}, \tag{12.44}$$

where p_{p_2} is the pressure momentum of bond 2 (or the time integral of p_2). This result shows that the inertia coefficient expressed as a "mass" is $\rho l/A$ when p, Q variables are used. That the inertia of small-area tubes is larger than the inertia of larger area tubes comes as something of a surprise.

There are several problems with the simple derivation given above. First, in keeping with the assumption of one-dimensionality, the fluid in the pipe was treated as if it moved as a rigid body. It is difficult to improve on this assumption until it is known how the velocity profile of the fluid changes in space and time. The types of flows being considered here are often called "quasi-one-dimensional flows," and in the steady state, when a well-developed velocity profile is known at each cross section, one may identify Q/A as an *average* velocity. In later calculations, the average of the square of the velocity over the cross section will be needed, and this quantity may be taken as $\beta Q^2/A^2$, where the correction factor $\beta \geq 1$ can be calculated if the velocity profile is known (see Reference [2]). The factor β is unity only for a uniform velocity distribution that can occur for frictionless, irrotational flow. For transient conditions the velocity distribution is often very hard to estimate, so in the remainder of this section we will assume a uniform distribution even in cases in which this cannot strictly be true. The accuracy of some results may be slightly improved by the introduction of factors such as β, although the estimation of the factors in the absence of experimental data may be difficult.

Second, unless one is willing to assume that the fluid is incompressible, the proper value of the density ρ is open to question. If $l = dx \rightarrow 0$, then $\rho(x, t)$ might represent an instantaneous value of ρ at position x, but if l is finite, then existence of capacitance in Figure 12.6 is not compatible with using ρ as a constant. If ρ varies, then, of course, the two ends of the slug of fluid do not move with the same velocity. Later on, a more sophisticated look at the problem will be taken. For now, let it simply be noted that for hydraulic systems in which the density changes are small, an average density for ρ is often sufficient, despite the contradiction implied by the use of a compliance together with a constant inertia.

What is really wrong with this elementary derivation is that the slug of fluid is treated as a rigid body, even though it is clear that a control volume comprising a length of pipe through which mass and momentum flow is being considered. As will be demonstrated, the derivation happens to be essentially correct when the two ends of the pipe have identical cross-sectional areas, since the momentum flow terms at the ends then cancel. The elementary derivation does not generalize readily when pipes of varying area are encountered, and it is surprising that few authors of elementary system dynamics texts even mention the control volume basis for their fluid mechanical system dynamics.

The element $R4$ in Figure 12.6 represents a loss in pressure beyond that required to accelerate the fluid. The relation between p_4 and Q_4 would be easy to specify as a generally nonlinear relationship if one could use the data for flow in pipes that has been determined experimentally. But almost all the data on friction factors is for fully developed steady flow and would not apply to any but the slowest transient conditions. For this reason, the analyst must be prepared to experiment with the friction loss law until the model response matches experimental data sufficiently well.

The next element in Figure 12.6 is intended to model the compliance of the fluid and pipe walls. The flow Q_6, which is the difference between Q_5 and Q_7, represents a loss in flow between the ends of the pipe. The pressure p_6 (which equals p_5 and p_7) can be determined from the integral of Q_6 when one can define a bulk modulus B:

$$p_6(t) = p_6(0) + \frac{B}{V_0} \int_0^t Q_6 \, dt = p_6(0) + \frac{B}{V_0} V_6(t), \qquad (12.45)$$

where V_0 is the volume of the fluid in the pipe at $t = 0$ when the pressure is $p_6(0)$. One might think that, simply by replacing the linear relation (12.45) with some nonlinear relation between p_6 and V_6 to model compressibility effects of a gas, the same model would serve in the case in which large density changes occur, but the situation is not quite so simple. As is demonstrated below, when density variations are significant, the thermodynamics of the situation must be studied.

Finally, R_8 represents a loss of flow in the pipe section due to leakage. Clearly, when the ps represent absolute pressure rather than gage pressure, R_8 should react to the difference between p_8 and the pressure external to the pipe, rather than just to p_8 itself. This requires that R_8 be attached to a 1-junction that computes the difference between the internal and external pressure and imposes the condition that flow out of the pipe is the same as flow into the external atmosphere.

From the brief introduction to the lumped line elements given above, it is clear that there are some fundamental difficulties in justifying the lumping process. Many difficulties disappear when only a linearized lumped model is desired, since it then becomes plausible to neglect small changes in quantities that have finite mean values. In order to see more clearly the nature of some of the more commonly used fluid dynamic models, one needs to consider the partial differential equations of fluid mechanics and nonlinear effects.

12.3.1   One-Dimensional Incompressible Flow

In order to illuminate the connection between standard techniques in fluid mechanics and a lumped representation using bond graph elements, a derivation of Bernoulli's equation is given first. Consider the problem of Figure 12.7a. Let s represent distance

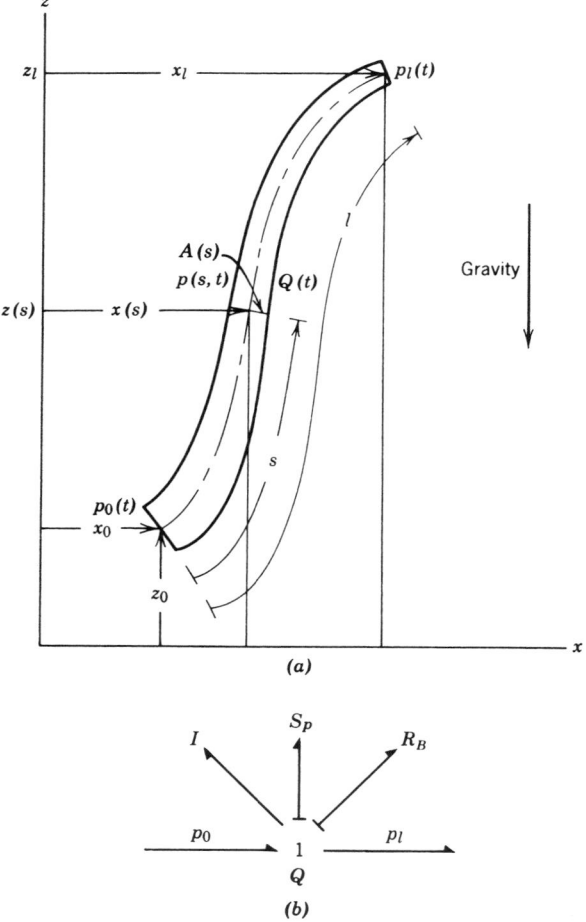

FIGURE 12.7. Constrained motion of an incompressible fluid. (a) Sketch of system; (b) bond graph.

along the centerline of a curved rigid pipe of length l. Then $A(s)$ is the pipe cross-sectional area, $x(s)$ and $z(s)$ describe the horizontal and vertical positions of the pipe centerline, respectively, $v(s, t)$ represents the (average) fluid velocity at position s and time t, and $p(s, t)$ represents the pressure.

Newton's law yields

$$\rho \frac{Dv}{Dt} = \rho \frac{\partial v}{\partial t} + \rho v \frac{\partial v}{\partial s} = -\frac{\partial p}{\partial s} - \rho g \frac{dz}{ds}. \tag{12.46}$$

Noting that, because of incompressibility, the volume flow rate $Q(t)$ is independent of s, we see that v is given by

$$v(s, t) = \frac{Q(t)}{A(s)}. \tag{12.47}$$

Substituting Eq. (12.47) into Eq. (12.46), we have

$$\rho \left(\frac{\dot{Q}(t)}{A(s)} + \frac{Q^2(t)}{A(s)} \frac{\partial 1/A(s)}{\partial s} \right) = -\frac{\partial p}{\partial s} - \rho g \frac{dz}{ds}, \tag{12.48}$$

which may be integrated in s from $s = 0$ to $s = l$:

$$I \dot{Q} + \frac{\rho Q^2}{2} \left(\frac{1}{A_l^2} - \frac{1}{A_0^2} \right) = p_0(t) - p_l(t) - \rho g (z_l - z_0), \tag{12.49}$$

where

$$I = \rho \int_o^l \frac{ds}{A(s)}, \tag{12.50}$$

and one could substitute $\beta Q^2/A^2$ for the Q^2/A^2 terms as discussed previously.

It is interesting to note that Bernoulli's equation [essentially Eq. (12.49)] can be represented exactly by the lumped elements shown in Figure 12.7b. The linear inertia coefficient defined in Eq. (12.50) reduced to that found by the nonrigorous method of Eq. (12.44) if $A(s)$ is constant, but several other terms appear in Eq. (12.49) that did not appear in Eq. (12.44). The gravity term that is represented by a constant-pressure source in Figure 12.7b is easily understood, but the term involving Q^2 requires explanation.

First of all, if $A(s)$ is constant, then $A_1 = A_0$ and the Q^2 term vanishes, thus showing that Eq. (12.44) is essentially correct for the constant-area case. However, in deriving Eq. (12.44), the flow of fluid through the ends of the pipe section was not properly considered. To use Newton's law in its simplest form, one must follow the flow of the fluid; that is, one must use a Lagrangian description. However, one really wants to treat the pipe as a control volume through which fluid passes, that is, with an Eulerian description. The Q^2 term in Eq. (12.50) may be thought of as a dynamic pressure-correction term which was fortuitously absent from the case of Eq. (12.44).

Another way to interpret the result of Eq. (12.50) is to note that the power flow $p(t)Q(t)$ does not represent the total flow of energy past a stationary point of the pipe. Terms of the form $\rho Q^2/2A^2$ represent dynamic pressure associated with the kinetic energy of the fluid. In the system under study, p and Q contain all the information required to find the dynamic pressure, since the velocity is Q/A. The Q^2 term in Eq. (12.49) may be represented as a resistance in Figure 12.7b, since it a relation between Q and a pressure (the dynamic pressure). When the resistance characteristic is plotted as in Figure 12.8, it is clear that if one keeps track only of the pQ power, then the dynamic pressure may be converted back and forth into static pressures as in nozzles and diffusers.

The resistance required to represent Bernoulli's equation is unusual in two ways. First, no matter whether $A_0 > A_1$ or $A_0 < A_1$, there are regimes of operation in which the resistance supplies power. Physically, this means that dynamic pressure is being partially converted to static pressure and hence to an apparent power flow in the form pQ. Second, although the pressure in the resistance is uniquely given for any flow, if one gives the pressure, then there is either no corresponding flow or two of them. Thus, there is a very strong causal preference for this resistance, as shown in Figure 12.7b.

The latter feature is readily explained by the observation that the basis for our derivation was the assumption that the fluid filled the pipe, and hence $v = Q/A$. From some pressure conditions at the ends of the pipe, this is not true—a jet can form in the pipe. Also, the equations derived may be valid only for one direction of flow. In studying the flow of water through a nozzle into the atmosphere, for example, it may make no sense to consider reverse flow, because the nozzle would fill with air and the

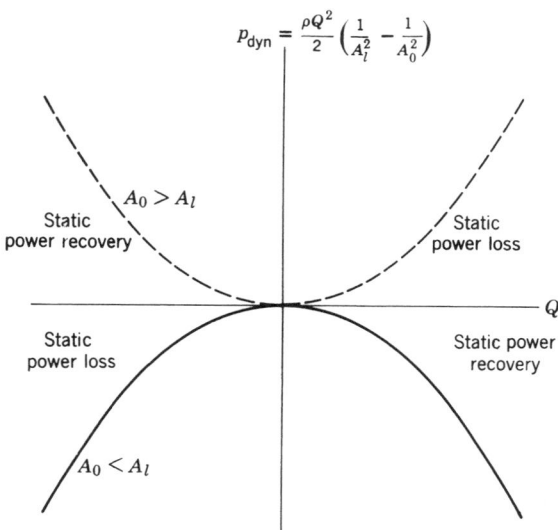

FIGURE 12.8. Constitutive law for the dynamic pressure resistor.

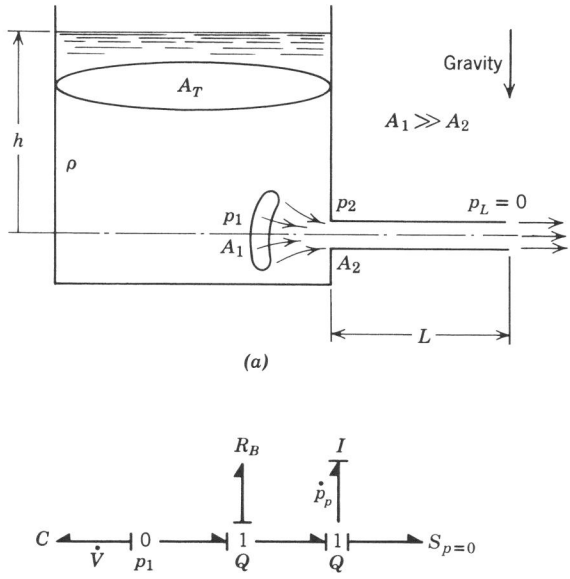

FIGURE 12.9. Tank-emptying problem. (*a*) Sketch of system; (*b*) bond graph.

equations would no longer be meaningful. Suffice it to say that in any application of Bernoulli's equation including the bond graph representation shown in Figure 12.7, one must use care that the proper branch of the resistance relation is being used, to avoid nonsensical results. Such difficulties with resistance relations are rare in other types of physical systems.

As an illustration of the utility of the bond graph representation, consider the classical elementary problem of estimating the time required for a tank to empty through a pipe. The system is shown in Figure 12.9. For simplicity, no friction losses will be considered, but the Bernoulli resistor may be thought of as indicating the loss in kinetic energy of the fluid that leaves the system. The tank has a capacitance of $A_T/\rho g$, where A_T, the tank area, is large compared to the pipe area. As in the classical analysis, we imagine the fluid entering the end of the pipe from a large area, A_1, at pressure p_1 and essentially zero velocity. The Bernoulli resistor then gives a pressure $\rho Q^2/2A_2^2$, where Q is related to the pressure momentum by a relation of the form of Eq. (12.44). Using the notation shown in Figure 12.9 and noting from the bond graph that two state variables—$V(t)$, the volume, and $p_p(t)$, the pressure momentum—are required, we find

$$\dot{V} = -\frac{A_2}{\rho L} p_p, \tag{12.51}$$

$$\dot{p}_p = \frac{-\rho}{2A_2^2} \left(\frac{A_2}{\rho L} p_p\right)^2 + \frac{\rho g V}{A_T}, \tag{12.52}$$

or

$$-\frac{L}{A_2}\ddot{V} = \frac{-\rho}{2A_2^2}(\dot{V})^2 + \frac{\rho g}{A_T}V \tag{12.53}$$

as long as $V \geq 0$. Note that, although no static power flows into the zero-pressure source representing the atmosphere, there is a flow, and with a diffuser one could recover power from this flow.

Control volume problems can become quite complex if the control volume itself is not fixed in inertial space. Bond graph methods can clarify the modeling of such systems, but space limitations prohibit a full discussion here. An example of such a system appears in Reference [6]. We now go on to discuss the implications of dropping the incompressibility assumption, and we find that further difficulties await us and that more approximations are involved in the hydraulic system bond graphs used previously than first might have been imagined.

12.3.2 Representation of Compressibility Effects

For small changes in density, it is easy to see that compressibility effects are readily modeled by means of linear capacitors, as discussed previously. For large changes in density, however, it is best to begin with a study of the thermodynamics of a pure substance.

As we have seen in the previous section, the internal energy per unit mass of a fluid, u, depends on two independent properties: the density ρ and the entropy s. Changes in these quantities are related by the Gibbs equation:

$$du = T\,ds - p\,d\left(\frac{1}{\rho}\right), \tag{12.3}$$

where T is the thermodynamic temperature, p is the pressure, $1/\rho$ is the specific volume, and

$$u = u\left(s, \frac{1}{\rho}\right). \tag{12.1}$$

The characteristics of the fluid are conveniently summed up by Eq. (12.1), in which we have used $1/\rho$ for the specific volume, since we will use v for velocity in this section. The Gibbs equation for a unit mass or a fixed amount of matter can be represented by a C-field, as we have seen.

Using the C-field representation, it is straightforward to model systems in which fixed amounts of a fluid are compressed and expanded and heated. When the fluid passes into and out of a control volume, however, the situation is somewhat more complex. In Figure 12.10, for example, fluid is compressed in a fixed volume by allowing fluid to pass slowly in and out of a port. In this case, the total energy contained in the volume V_0, denoted by U, depends on the mass m contained in the volume and the internal energy per unit mass, u. Changes in U occur not only because of the flow

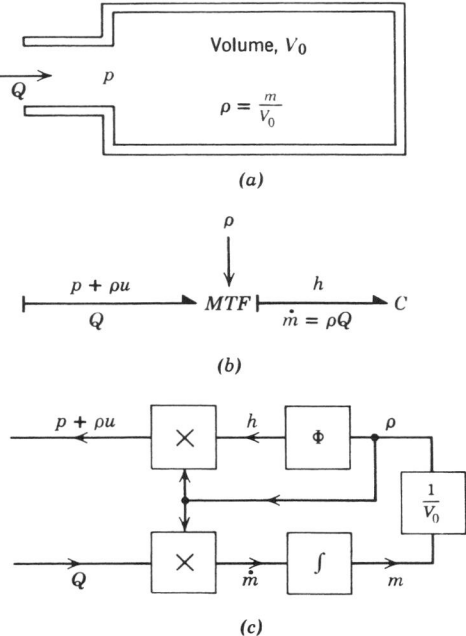

FIGURE 12.10. Isentropic compression of a fluid. (*a*) Fluid flow into a rigid volume; (*b*) bond graph; (*c*) block diagram.

work pQ but also because of the convection of energy. If one uses the mass flow rate $\dot{m} = \rho Q$ instead of Q as a flow variable, then

$$dU = u\, dm + \frac{p}{\rho} dm \qquad (12.54)$$

in the isentropic case. Defining the enthalpy h by the relation

$$h = u + \frac{p}{\rho}, \qquad (12.55)$$

it now appears that the power flow past the port is given by

$$h\dot{m} = u\dot{m} + \frac{p}{\rho}\dot{m} = u\rho Q + \frac{p}{\rho}\rho Q = (\rho u + p)Q, \qquad (12.56)$$

which shows that pQ is only the hydrostatic power. As in the previous section, in which it was found necessary to supplement the static pressure with a dynamic pressure term to account properly for real power flow, so here it is necessary to supplement p with an extra term, ρu, to account for convected internal energy. Figure 12.10*b* shows a bond graph based on Eqs. (12.55) and (12.56), and a block diagram corresponding to integral causality is given in Figure 12.10*c*. Note that the dynamic pressure term has not been incorporated, so this model is only valid for low flow rates (as indeed is the C-field representation for the Gibbs equation). Also, note that the pressure and internal energy depend on density alone in the isentropic case, so the enthalpy also is determined by conditions inside the vessel. In more complicated cases, the enthalpy at the port is determined by conditions inside the vessel for outflow, $\dot{m} < 0$, but is determined by external conditions for inflow, $\dot{m} > 0$. This kind of causal switching with changes in direction of flow is known to occur in heat exchanger systems in which the flow can reverse but has not received much study in terms of bond graphs or any other means of system analysis.

Figure 12.11 shows how the bond graph is modified when isentropic compression occurs partly due to variable volume and partly due to inflow and outflow, as in the power cylinder of a pneumatic servomechanism. The C-field in Figure 12.11b has one $h\dot{m}$ port and one $p\dot{V}$ port. The implication is that the total energy U depends on m and V with

$$h = \frac{\partial U}{\partial m}, \qquad -p = \frac{\partial U}{\partial V}. \tag{12.57}$$

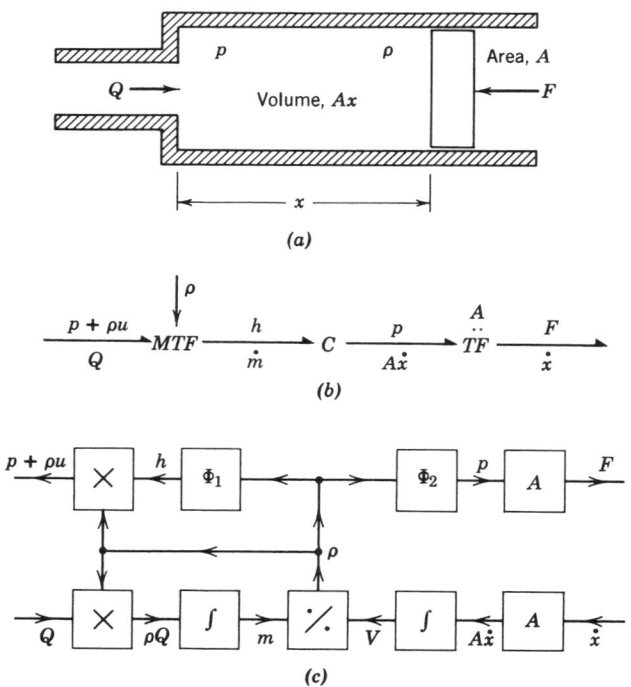

FIGURE 12.11. Compression in a ram. (a) Sketch of system; (b) bond graph; (c) block diagram.

It may be worthwhile to verify that the C-field is really conservative by checking the validity of the Maxwell reciprocal relation

$$\frac{\partial h}{\partial V} = \frac{\partial (-p)}{\partial m}. \tag{12.58}$$

With the isentropic assumption, the internal energy per unit mass is a function of specific volume only:

$$u = u\left(\frac{1}{\rho}\right), \qquad -p = \frac{du(1/\rho)}{d(1/\rho)}. \tag{12.59}$$

Expressing the specific volume as

$$\frac{1}{\rho} = \frac{V}{m}, \tag{12.60}$$

we have

$$p = p\left(\frac{V}{m}\right), \qquad u = u\left(\frac{V}{m}\right), \qquad h = u + p\frac{V}{m}; \tag{12.61}$$

then

$$\frac{\partial h}{\partial V} = u'\frac{1}{m} + p'\frac{V}{m^2} + p\frac{1}{m} = p'\frac{V}{m^2}, \tag{12.62}$$

where the primes denote the derivatives of the functions u and p and Eq. (12.4) has been used to cancel two terms in Eq. (12.62). Differentiating p with respect to m yields

$$\frac{\partial p}{\partial m} = \frac{-V}{m^2}p', \tag{12.63}$$

which, upon comparison with Eq. (12.62), validates Eq. (12.58).

The bond graph of Figure 12.11 and the equations it represents are a sophisticated version of models of compressibility effects in fluid servomechanisms in which the volume of fluid concerned is variable. In practice, a linearized compressibility parameter may simply be incorporated into the system equations, but, as is often the case, it is virtually impossible to proceed from a linearized model of an effect to a more accurate nonlinear model without starting again from the basic physics of the situation.

12.3.3   Inertial and Compressibility Effects in One-Dimensional Flow

The simple model of Figure 12.6, neglecting for the moment the loss elements, is a lumped approximation of the inertial effects and compressibility effects in a length of line. When a large number of such models with parameters appropriate to a length Δl

are cascaded and when $\Delta l \rightarrow 0$, the state equations for the model form an approximation to the partial differential equation called the *one-dimensional wave equation*. In the sections above, models for the inertia effect in incompressible flow and the compliance effect when no inertia is considered were developed. The question now arises whether one can simply cascade a series of lumps that alternately account for inertia and compressibility effects in the manner of Figure 12.6 and thereby construct a model of a distributed line.

The answer is in the affirmative in several important cases, namely:

1. *The Acoustic Approximation.* In this case, small deviations in pressure and density are modeled with linear equations. The inertial and compliance coefficients are calculated at the mean pressure and density. See Reference [5]. Such a model is useful in studying compressibility effects in oil hydraulics and water-hammer problems, as well as in acoustics.

2. *Lagrangian Descriptions.* When one is willing to follow the motion of a group of particles, then the inertia of the group is constant. As long as Newtonian mechanics is used, the system of equations can be nonlinear only due to nonlinear elastic or dissipative effects. For the vibration of bars and strings, for example, a Lagrangian description is natural, since the particles never move very far anyway. For beams and plates, the differential equations are more complex than the wave equation, but the Lagrangian description still allows one to use finite models with Is and Cs representing inertia and compliance effects. The junction structure is more complex than the simple 0–1–0–1 string of Figure 12.6, however. For fluid systems in which the particles move through fixed boundaries, the Lagrangian descriptions are used rather rarely.

For the cases above one may easily construct finite-element models from normal bond graph components that, when reduced to differential volumes, mirror the usual derivations of the partial differential equations of motion for the continuum model. Essentially, the inertial, compliance, and resistance aspects of the system may be treated separately in the finite elements, even though these aspects refer to a single point when the volume of the element is made to approach zero in the continuum description. As will be demonstrated, the Eulerian description common to fluid mechanics entangles the inertial and compliance aspects so thoroughly that it is difficult to construct a series of finite elements converging to the continuum description without using a large number of active bonds, except in the acoustic approximation.

Consider the isentropic, one-dimensional flow of fluid through a tube of unit cross-sectional area. Using s for the space coordinate, $v(s, t)$ for velocity, $\rho(s, t)$ for density, and $p(s, t)$ for pressure, the equations describing the system are

$$\rho \frac{\partial v}{\partial t} + \rho v \frac{\partial v}{\partial s} = -\frac{\partial p}{\partial s}, \tag{12.64}$$

$$\frac{\partial \rho}{\partial t} + \frac{\partial \rho v}{\partial s} = 0, \tag{12.65}$$

$$p = p(\rho). \tag{12.66}$$

Equation (12.64) is just Newton's law for a length of fluid with the acceleration expressed in Eulerian form. Equation (12.65) is a statement of conservation of mass that, with the constitutive law (12.66) for the gas, can be used to define the compressibility effect. The density, $\rho(s, t)$, enters both the inertia law (12.64) and the compressibility laws. In contrast, a Lagrangian description would express the inertia in terms of a constant mass of fluid in a reference state.

It is probably not sensible to attempt to construct a bond graph for Eqs. (12.64)–(12.66) unless ρ in Eq. (12.64) is nearly constant [in Eqs. (12.65) and (12.66) ρ *must* be allowed to vary] and $v\,\partial v/\partial s$ is a second-order small quantity. Basically, Eq. (12.64) is Newton's law following the flow infinitesimally. At each succeeding instant of time, the law refers to *different* infinitesimal slices of fluid. Thus, the time-varying inertial effect will not conserve energy as normal bond graph inertial elements do. Also, the $\partial v/\partial s$ term implies that the velocity is different at the two ends of the differential control volume. Thus, the inertial effect and the compressional effect both appear in Eq. (12.64). Finally, the factors appearing in these equations (pressure and velocity for volume flow, since we are considering unit area) do not multiply to give the true power flow at position s and time t. As we have seen, the convected internal energy has been left out of the equations. One should not, therefore, expect that these (correct) equations can be represented by a bond graph in which power conservation and energy conservation are built in. It seems that the alternatives, if one wishes to represent fluid dynamic lines in terms of bond graphs, are to (1) use a Lagrangian description; (2) use an Eulerian description, but with a restriction essentially to the acoustic approximation; and (3) consider only the incompressible or noninertial cases, as was done above. Thus, the intuitively appealing scheme of Figure 12.6 appears to be more restrictive than has been generally appreciated, even without consideration of shear stresses, boundary layers, and so forth, which also complicate the modeling of fluid transmission lines.

12.4   PSEUDO–BOND GRAPHS FOR COMPRESSIBLE GAS DYNAMICS

It has previously been seen that constructing bond graph models for fluid dynamic systems is straightforward when the Lagrangian approach of tracking a fixed packet of particles is used. The Lagrangian description is not convenient for computational purposes, and it is much better to describe fluid systems with respect to the Eulerian control volume, where observations are made at a fixed point in space as different fluid particles pass by. Unfortunately, some convective terms appear in the energy and momentum equations due to the fixed reference in the Eulerian approach, and the convective terms do not conveniently lend themselves to a bond graph representation in terms of power variables.

A major emphasis of this book is the construction of low-order, accurate, understandable models of all types of interacting physical engineering systems. Thermodynamic and nonlinear gas dynamic system parts are another type of interaction we would like to include in our unified approach. After all, the gas dynamics is only part of the physics of an internal combustion engine, a Wankel compressor, a helical rotor expander, or an air cushion vehicle. There are many other dynamic elements that must interact with the gas dynamics in order to produce an overall system model.

This section develops a pseudo–bond graph representation for gas dynamics. The bond graph fragment relies on causality in the usual way and connects to other bond graph fragments in the usual way. All the virtues of bond graph modeling are preserved, including the a priori identification of state variables, the ease of deriving state equations, and setting up a computational model. In the pseudo–bond graph fragment, the product of the bond effort and flow variables is not necessarily equal to power. Because of this, the concept of a power-conserving transformation is not operative, and care must be exercised when using transformers. This turns out to be a small price to pay in exchange for the enormous modeling flexibility the technique affords.

12.4.1   The Thermodynamic Accumulator

Consider the control volume in Figure 12.12 with the internal gas at instantaneous pressure P, absolute temperature T, density ρ, and volume V, moving at velocity v, and containing mass m and energy E.

The control volume can transport mass through the "in" port and the "out" port, but flow can go in either direction at either port. Finally, we can obtain work from the control volume by the volume expansion indicated in the figure. Notice that we assume that one pressure, one temperature, one density, etc. characterize the entire internal volume of the control volume.

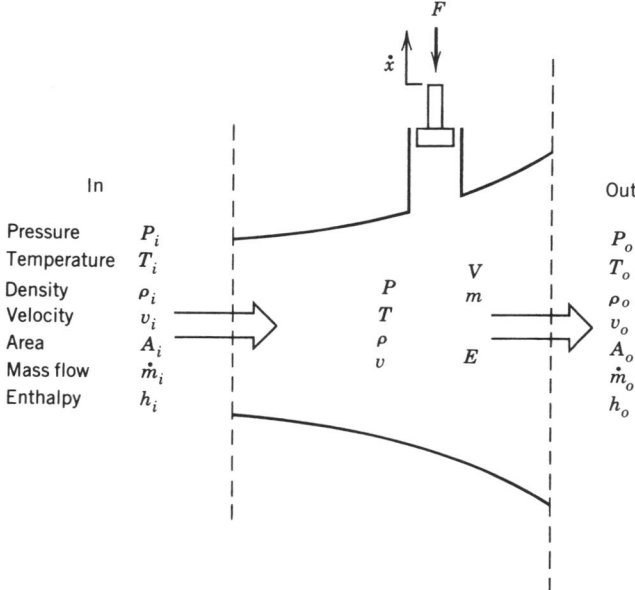

FIGURE 12.12. Control volume for gas dynamics.

The one-dimensional energy, mass, and momentum equations can be written in the following forms:

$$\frac{d}{dt}E = \left(h_i + \frac{v_i^2}{2}\right)\dot{m}_i - \left(h_0 + \frac{v_o^2}{2}\right)\dot{m}_o - P\frac{dV}{dt} \quad \text{(energy)} \quad (12.67)$$

$$\frac{d}{dt}m = \dot{m}_i - \dot{m}_o \quad \text{(mass)} \quad (12.68)$$

$$\frac{d}{dt}(mv) = v_i\dot{m}_i - v_o\dot{m}_o + P_iA_i - P_oA_o$$
$$+ \tfrac{1}{2}(P_i + P_o)(A_o - A_i) + \mathcal{R} \quad \text{(momentum)}. \quad (12.69)$$

Here E is the internal energy of the control volume,

$$E = mc_vT; \quad (12.70)$$

h is the enthalpy, where

$$h_i = c_pT_i, \qquad h_o = c_pT_o; \quad (12.71)$$

$\mathcal{R}$ is the resultant external force acting on the control volume; and we assume that the gas obeys the gas law,

$$PV = mRT. \quad (12.72)$$

For air, the gas parameters are

$$R \equiv \text{gas constant}$$
$$= 287 \text{ N-m/kg-K},$$
$$c_p \equiv \text{specific heat at constant pressure}$$
$$= 1005 \text{ N-m/kg-K}, \quad (12.73)$$
$$c_v \equiv \text{specific heat at constant volume}$$
$$= 718 \text{ N-m/kg-K}.$$

Also,

$$R = c_p - c_v, \quad (12.74)$$

and

$$\gamma = \frac{c_p}{c_v} = 1.4. \quad (12.75)$$

For now, we will not consider momentum variation within the control volume, and we will neglect the convected kinetic energy in the energy equation. This kinetic energy is typically small compared to the convected enthalpy, as becomes apparent if we consider that at room temperature of 20°C, or 293 K, the enthalpy is $h = c_p (293 \text{ K}) = 2.94 \times 10^5$ N-m/kg, while for a high gas velocity of 100 m/sec, the kinetic energy contribution is only $v^2/2 = 5 \times 10^3$ N-m/kg.

We will work with the equations

$$\frac{d}{dt} E = h_i \dot{m}_i - h_o \dot{m}_o - p \frac{dV}{dt}, \tag{12.76}$$

$$\frac{d}{dt} m = \dot{m}_i - \dot{m}_o, \tag{12.77}$$

$$\frac{d}{dt} V = \dot{V}, \tag{12.78}$$

$$E = m c_v T, \tag{12.79}$$

$$PV = mRT. \tag{12.80}$$

Equations (12.76), (12.77), and (12.78) look like first-order state equations, and they motivate the construction of the bond graph of Figure 12.13. The 3-port C-field in this figure has one true bond and two pseudobonds. The true bond has pressure P as its effort variable and volume rate $\dot{V}$ as its flow variable. The product, $P\dot{V}$, is power, and these variables have been used previously for lumped modeling of incompressible fluid systems.

One of the pseudobonds has energy flow $\dot{E}$ as the flow variable and temperature T as the effort. The other pseudobond uses pressure P as the effort and mass flow $\dot{m}$ as the flow. For both pseudobonds, the product of their effort and flow variables is not power. In fact, $\dot{E}$ is already power on the upper pseudobond.

Even though the C-field representation is not a true bond graph, nevertheless, it operates on associated effort and flow variables in a manner identical to a true C-field. As the causality indicates in Figure 12.13, the C-field possesses all integral causality and therefore accepts flow inputs on all three bonds ($\dot{E}$, $\dot{m}$, and $\dot{V}$). It then integrates these flows to produce the state variables E, m, and V. And finally the C-field operates

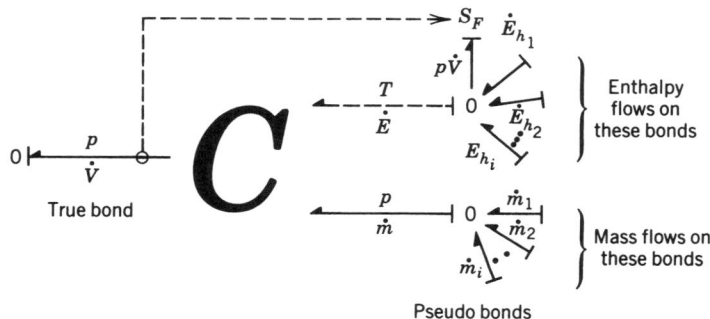

FIGURE 12.13. Bond graph of the thermodynamic accumulator.

on these state variables through appropriate constitutive laws to produce the outputs P and T. The constitutive relations come from Eqs. (12.79) and (12.80) and are

$$T = \frac{1}{c_v} \frac{E}{m} \qquad (12.81)$$

and

$$P = \frac{mRT}{V} = \frac{mR}{V} \frac{1}{c_v} \frac{E}{m} = \frac{R}{c_v} \left(\frac{E}{V} \right). \qquad (12.82)$$

Thus, if $\dot{E}$, $\dot{m}$, and $\dot{V}$ are prescribed (causally), then the thermodynamic accumulator will output T and P via the constitutive laws (12.81) and (12.82). By doing this we make compressible gas dynamics (subject to the operational assumptions that got us to this point) fit into our general unified approach for modeling systems.

The flow source, $SF{\leftarrow}$, associated with the T, $\dot{E}$ bond is necessary to properly account for the work done by the fluid in our control volume. The remaining external bonds associated with the T, $\dot{E}$ bond tell the accumulator the enthalpy flows $h_i \dot{m}_i$, from whatever system is attached to the accumulator, and, as the 0-junction enforces, the attached system learns of the temperature of the control volume. Similarly, the P, $\dot{m}$ pseudobond has external bonds which causally prescribe the mass flows from the attached system, while these external bonds also prescribe the control volume pressure as an input to the attached system.

We need to describe the origin of the transported energy flows (enthalpy flows)

$$\dot{E}_h = \sum_i \dot{E}_{h_i} = \sum_i h_i \dot{m}_i = \sum_i c_p T_i \dot{m}_i \qquad (12.83)$$

and the mass flows $\dot{m}_i$, where

$$\dot{m} = \sum_i \dot{m}_i, \qquad (12.84)$$

but first we demonstrate the use of the thermodynamic accumulator for a simple case.

Figure 12.14 shows a cylinder with a trapped mass of air under a moving piston with prescribed motion, $v_i(t)$. Without making any assumptions about the thermodynamic process, we now show that a model using the thermodynamic accumulator predicts a reversible adiabatic or isentropic process. The bond graph for this simple closed system is also shown in Figure 12.14.

The state equations are

$$\dot{E} = -P\dot{V},$$
$$\dot{V} = A_p v_i \quad \text{(prescribed)}, \qquad (12.85)$$
$$\dot{m} = 0 \quad \text{(closed system)}.$$

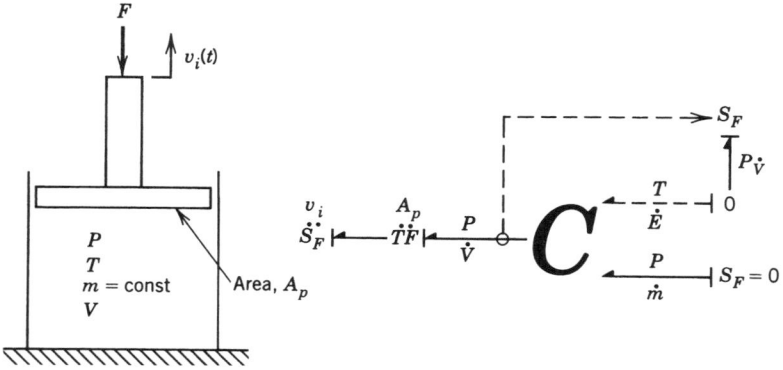

FIGURE 12.14. Nonlinear air spring and its bond graph model.

From the constitutive law of Eq. (12.82) we can write

$$\dot{E} = -\frac{R}{c_v}E\frac{\dot{V}}{V}. \tag{12.86}$$

If we were pursuing a numerical solution, we would numerically integrate this equation along with the equation for $\dot{V}$ and then output pressure and temperature from (12.81) and (12.82). Instead, we recognize that (12.86) has an analytical solution,

$$EV^{R/c_v} = \text{const.} \tag{12.87}$$

If we reintroduce the pressure from Eq. (12.82), then

$$\frac{c_v}{R}PV^{(R/c_v)+1} = \text{const,} \tag{12.88}$$

or

$$PV^{\gamma} = \text{const,} \tag{12.89}$$

using

$$\gamma = \frac{c_p}{c_v} = \frac{R}{c_v} + 1. \tag{12.90}$$

The reader should recognize that (12.89) is the P–V relationship for an isentropic process.

12.4.2 The Isentropic Nozzle

We now consider the energy flows $\dot{E}_{h_i}$ and mass flows $\dot{m}_i$ required as inputs to the thermodynamic accumulator. Figure 12.15 shows a nozzle with upstream pressure and temperature P_u, T_u; downstream pressure and temperature P_d, T_d; exit area A; and mass flow $\dot{m}$. If we assume that isentropic flow exists, then $\dot{m}$ depends upon the

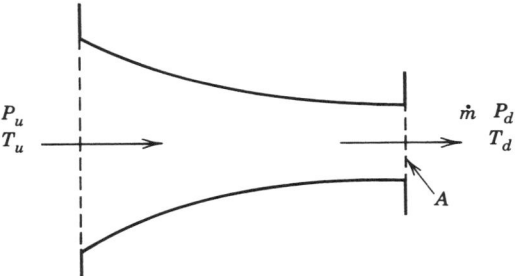

FIGURE 12.15. Isentropic nozzle.

pressure ratio

$$P_r = \frac{P_d}{P_u} \tag{12.91}$$

rather than the pressure drop $P_u - P_d$ across the nozzle. It can be derived [7] that $\dot{m}$ is given by

$$\dot{m} = A\frac{P_u}{\sqrt{T_u}}\sqrt{\frac{2\gamma}{R(\gamma - 1)}}\sqrt{P_r^{2/\gamma} - P_r^{(\gamma+1)/\gamma}}, \tag{12.92}$$

and for

$$P_r \leq P_{r\ \text{crit}} = \left(\frac{2}{\gamma + 1}\right)^{\gamma/(\gamma-1)} \tag{12.93}$$

the flow is "choked," and $\dot{m}$ is independent of the downstream pressure, P_d.

The transported energy $\dot{E}_h$ associated with this mass flow $\dot{m}$ is then

$$\dot{E}_h = c_P T_u \dot{m}. \tag{12.94}$$

We will consider the mass flows and energy transport flows into and out of the control volumes represented by the thermodynamic accumulator as isentropic nozzle flows given by Eqs. (12.92) and (12.94). The upstream side of the nozzle and the sign of $\dot{m}$ will be dictated by the sign of the pressure drop ΔP across the nozzle. If ΔP changes sign, then so will our assignment of the upstream end of the nozzle.

Consider the 4-port R-element of Figure 12.16, where all bonds have "effort in" causality. Although these are all pseudobonds, the 4-port R-element is functionally identical to any R-field using true power variables. This R-element operates on the input efforts P_a, P_b, T_a, T_b and delivers the output flows $\dot{m}_a$, $\dot{m}_b$, $\dot{E}_{h_a}$, $\dot{E}_{h_b}$. This can be expressed functionally as

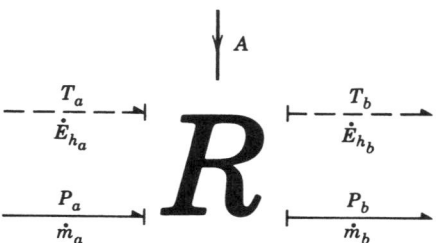

FIGURE 12.16. Bond graph for the isentropic nozzle.

$$\dot{m}_a = \dot{m}_a(P_a, P_b, T_a, T_b),$$
$$\dot{m}_b = \dot{m}_b(P_a, P_b, T_a, T_b),$$
$$\dot{E}_{h_a} = \dot{E}_{h_a}(P_a, P_b, T_a, T_b),$$
$$\dot{E}_{h_b} = \dot{E}_{h_b}(P_a, P_b, T_a, T_b).$$

(12.95)

We now define the computational procedures that allow the R-field of Figure 12.16 to represent the isentropic nozzle:

$$\text{if } P_a > P_b, \quad \text{then } P_u = P_a, \quad T_u = T_a, \quad P_d = P_b;$$
$$\text{if } P_a < P_b, \quad \text{then } P_u = P_b, \quad T_u = T_b, \quad P_d = P_a; \quad (12.96)$$

and

$$P_r = P_d/P_u. \tag{12.97}$$

Furthermore,

$$\text{if } P_r > P_{r\text{ crit}}, \quad \text{then } P_r = P_d/P_u;$$
$$\text{if } P_r \le P_{r\text{ crit}}, \quad \text{then } P_r = P_{r\text{ crit}}. \quad (12.98)$$

Calculate

$$\dot{m} = A\frac{P_u}{\sqrt{T_u}}\sqrt{\frac{2\gamma}{R(\gamma-1)}}\sqrt{P_r^{2/\gamma} - P_r^{(\gamma+1)/\gamma}}. \tag{12.99}$$

Now obtain outputs:

$$\text{if } P_a > P_b, \quad \text{then } \dot{m}_a = \dot{m}_b = \dot{m};$$
$$\text{if } P_a < P_b, \quad \text{then } \dot{m}_a = \dot{m}_b = -\dot{m}; \quad (12.100)$$

and in either case,

$$\dot{E}_{h_a} = \dot{E}_{h_b} = c_p T_u \dot{m}_a. \tag{12.101}$$

The relationships (12.96)–(12.101) fully represent the input–output behavior of the isentropic nozzle. And as long as we maintain the causality for which these relationships were derived, we can construct this computational procedure once and for all, to be used over and over when needed.

12.4.3 Constructing Models with the Thermodynamic Accumulator and Isentropic Nozzle

Figure 12.17 shows a cylinder with gas trapped above and below the piston. If no leakage exists between top and bottom chambers, then the model is simply two independent accumulators identical to the one in Figure 12.14. When leakage past the piston is included, the model is shown in Figure 12.17. Notice how the 4-port R-element interacts with the pseudobonds of the C-element. The C-element outputs, P_1, T_1, P_2, T_2 are the inputs to the R-element, which, through the computational procedure (12.96)–(12.101), outputs the mass and transported energy flows for use as inputs to the C-element. On the true bond side of the C-element, all is normal, in that the pressure outputs are transformed into forces on the piston, and the piston inertia, in turn, outputs the velocity $\dot{x}$, which then is transformed into volume velocities of the two chambers. Figure 12.17 demonstrates a very elegant coupling of mechanical and thermodynamic energy domains. Since integral causality exists, setting up state equations for computational solution is quite straightforward.

Figure 12.18 shows a two-stroke, piston-ported internal combustion engine and its bond graph model. This model has been used to aid in the understanding of the

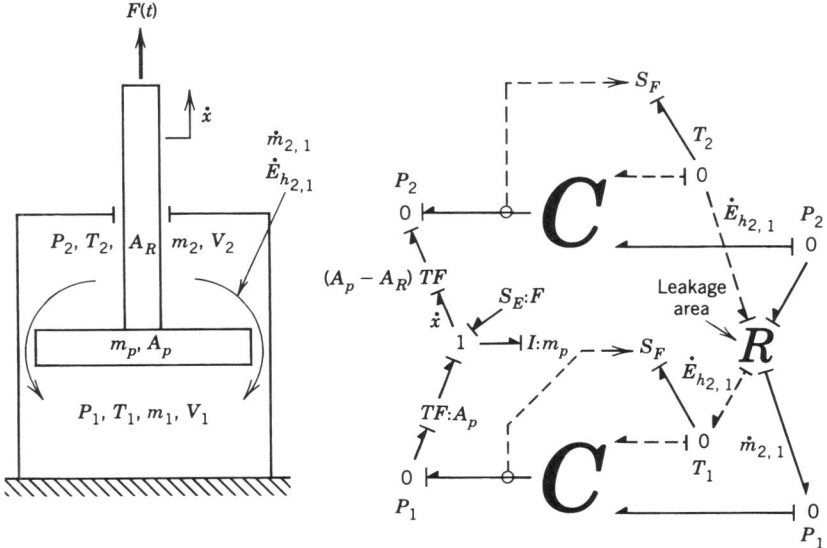

FIGURE 12.17. Two-sided air cylinder with leakage.

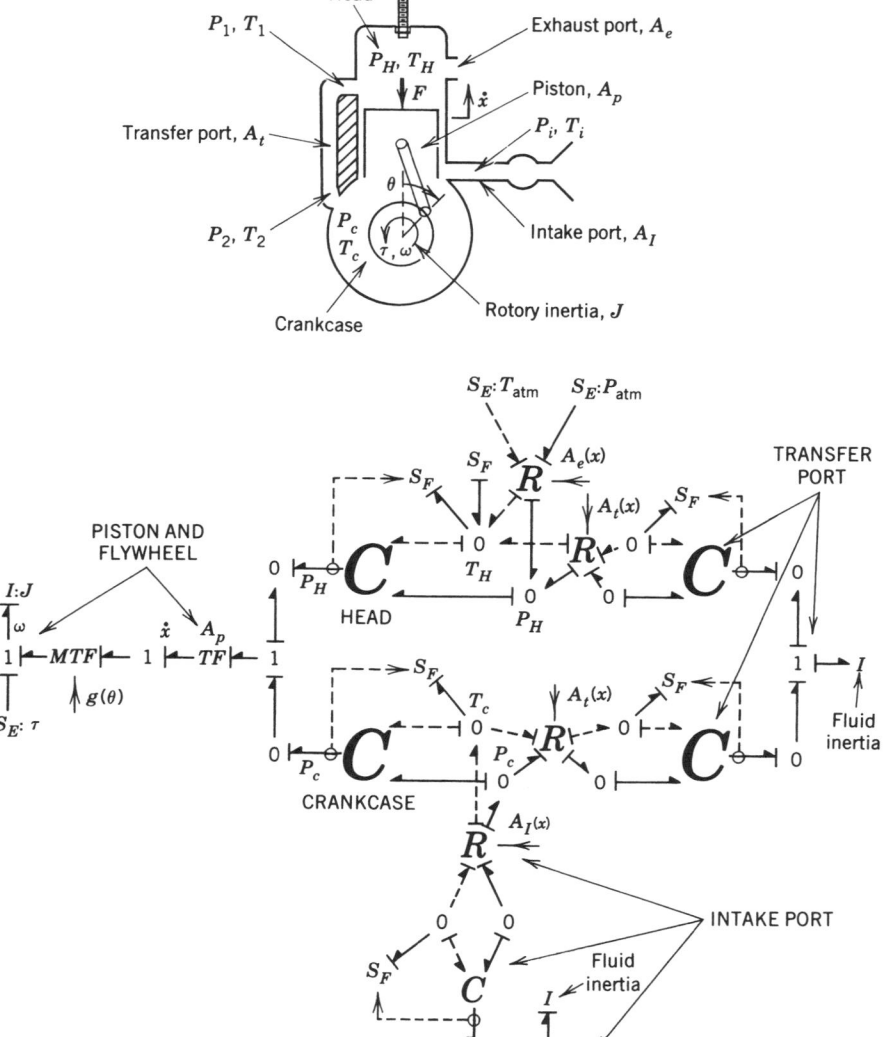

FIGURE 12.18. Bond graph for a two-stroke, piston-ported engine.

interacting dynamics of two-stroke engines, and it is shown here as another example of using the 3-port C-element and 4-port R-element to construct rather complex thermodynamic models. Readers should see reference [8] if they are interested in IC engines.

The two-stroke engine relies heavily on fluid dynamics for its operation. As the piston is driven downward from the pressure due to combustion of an air–fuel charge, it first opens the exhaust port as the piston moves past the exhaust opening. As exhaust gases escape into the exhaust system (not shown), the piston uncovers the transfer port, and the compressed air–fuel in the crankcase moves through the transfer passage to supply the head with the next charge for combustion. As the piston moves upward after passing through bottom dead center, the intake port is opened, and fresh air–fuel enters the expanding crankcase, ready for the next crankcase compression. The air–fuel in the head is compressed and ignited, initiating another cycle.

The inertia of the fluid in the intake and transfer passages, as well as the dynamics of the exhaust system, is essential to the high-performance operation of the engine. Once fluid is moving in the intake and transfer passages, its momentum continues to charge the crankcase and cylinder head against adverse pressure gradients even as the piston is closing the ports. It is this dynamic behavior that makes the engine so interesting and dynamic modeling so essential for design purposes.

The inertia effects of the transfer and intake are handled in a functional manner, and the transported momentum from Eq. (12.69) is neglected. We can, in fact, construct a pseudobond graph for the momentum equation, but this is beyond the scope of the present presentation. For the model shown here, it is assumed that the fluid inertia of the intake and transfer passages behaves like a trapped mass of fluid in the passages. The density of the fluid is set by the accumulators at the ends of the passage, depending upon the instantaneous flow direction. This has been shown experimentally to work well. The complete momentum equation must be used for the exhaust system.

Notice that integral causality exists throughout Figure 12.18; thus state-variable selection, equation formulation, and computer coding will be straightforward.

As a final example, consider Figure 12.19, which shows a Wankel compressor and its bond graph. The reader is urged to study this model and appreciate the elegance and utility of the representation.

12.4.4 Summary

In order to include compressible gas dynamic elements in an overall system model in a computationally convenient form, it was necessary to introduce the thermodynamic accumulator and associated pseudobonds. The 3-port C-element possesses all the characteristics of a true representation, and through causal considerations, the accumulator is a great aid in putting models together.

The isentropic nozzle and its 4-port R-element representation are required to interact with the accumulator to account for transported energy. As long as the causality presented in this section is not violated, then including even complex, nonlinear thermodynamics and gas dynamics in overall system models follows the same process as constructing any bond graph model for interactive components crossing many energy domains.

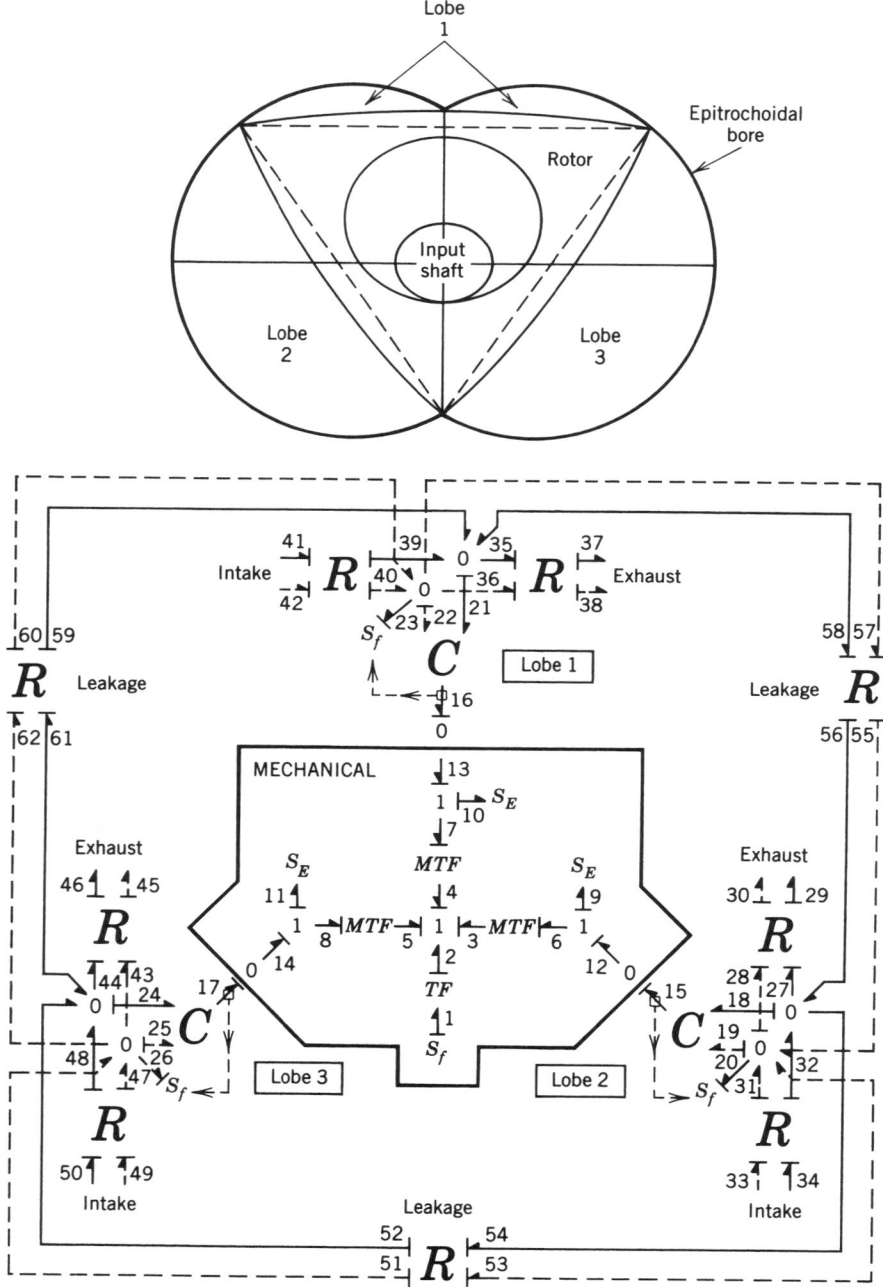

FIGURE 12.19. Wankel compressor and its bond graph.

REFERENCES

[1] R. H. Cannon, *Dynamics of Physical Systems*, New York: McGraw-Hill, 1967.

[2] W. M. Swanson, *Fluid Mechanics*, New York: Holt, Rinehart and Winston, 1970.

[3] C. J. Radcliffe and D. Karnopp, "Simulation of Nonlinear Air Cushion Vehicle Dynamics Using Bond Graph Techniques," *Proceedings 1971 Summar Computer Simulation Conference*, Boston, MA, July 19–21, 1971.

[4] L. Tisza, *Generalized Thermodynamics*, Cambridge, MA: MIT Press, 1966.

[5] P. M. Morse and K. U. Ingard, *Theoretical Acoustics*, New York: McGraw-Hill, 1968.

[6] D. C. Karnopp, "Bond Graph Models for Fluid Dynamics Systems," *Trans. ASME J. Dyn. Syst. Meas. Control*, **94**, Ser. G, No. 3, 222–229 (Sept. 1973).

[7] R. H. Sabersky, A. J. Acosta, and E. Hauptmann, *Fluid Flow*, New York: Macmillan, 1971.

[8] D. L. Margolis, "Modeling of Two-Stroke ICE Dynamics Using the Bond Graph Technique," *SAE Trans.*, pp. 2263–2275 (Sept. 1975).

[9] D. L. Margolis, "Bond Graph Fluid Line Models for Inclusion with Dynamic Systems Simulation," *J. Franklin Inst.*, **308**, No. 3, 255–268 (Sept. 1979).

PROBLEMS

12-1 Two bodies that do not expand or contract are in thermal contact through a thermal resistance. Make two bond graphs for the system using both $T\dot{Q}$ and $T\dot{S}$ variables. Explain the different relations for the —R— and —C elements when the two different variable sets are used. Using the $T\dot{S}$ bond graph and Figure 12.1, show that if $T_2 \neq T_1$ initially, the entropy of the system can only increase.

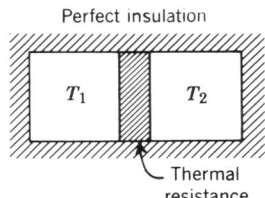

12-2 Consider the compression of a gas by means of a crank–piston arrangement. Let the cylinder have a single average temperature, and define thermal resistances between the gas temperature T and the cylinder temperature and between the cylinder temperature and the atmospheric temperature T_0. Make a bond graph that would allow you to predict the crank torque τ for low speeds of rotation, ω. (Note that you do not have enough information to evaluate all the system parameters.)

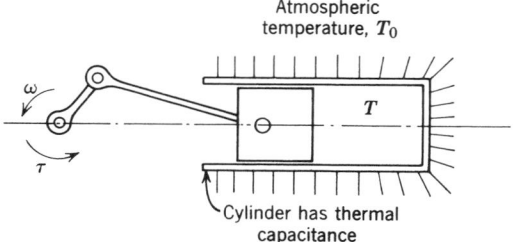

12-3   The dashpot in the suspension system has a force–velocity constitutive law that varies with the average temperature T of the dashpot, since it contains oil that changes viscosity with temperature. Make a simple model of the system that would predict how the dashpot heats up when the input base velocity $V(t)$ is given. Discuss your assumptions and how you might estimate the system parameters you need for the thermal part of your model.

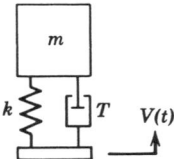

12-4   Suppose a shaft is connected to a paint stirrer such that, for low enough angular rates ω, all the power $\tau\omega$ goes to heating up the paint. Make a bond graph relating the mechanical power to an entropy flow $\dot{S}$ and including possible heat transfer to the atmosphere. Discuss the simplifying assumptions you have used.

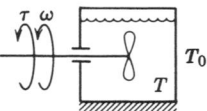

12-5   In the acoustic approximation, the bulk modulus B is given by

$$B = \rho_0 c^2,$$

where ρ_0 is the mean mass density of the fluid and c is the speed of sound. If Δp represents a small increase in pressure over the mean pressure p_0, then

$$\Delta p = \rho_0 c^2 \frac{\Delta \rho}{\rho_0},$$

where $\Delta \rho$ represents the change in density.

(a) Considering a fixed mass of fluid that occupies volume V when the pressure is p_0, show that

$$\frac{\Delta\rho}{\rho} = \frac{\Delta V}{V}, \qquad \Delta p = -B\frac{\Delta V}{V},$$

where $-\Delta V$ represents a *decrease* in volume.

(b) Evaluate the inertia and capacitance parameters for the length of pipe shown in Figure 12.6.

(c) Using $\lambda = c/f$, where λ is wavelength, f is frequency, and c is sound speed, relate the length l for the pipe segment to the highest circular frequency ω of interest, so that even the shortest wavelength will span several "lumps" if one uses Figure 12.16 to make a model of a long pipe by cascading many segments.

12-6 A high-speed hydraulic ram is forced by a pressure source, and we desire to predict how fast it can be stroked. Let the inlet pipe have length l and area A, and consider only inflow Q. The ram has mass m and area A_2 and a force $F(t)$ applied to it. To be conservative, assume that only the static pressure in the ram acts on the piston, that is, all the dynamic pressure is assumed to be lost. Make a bond graph for this system using a Bernoulli resistor to model the dynamic pressure loss. Apply causality and write equations of motion for the system.

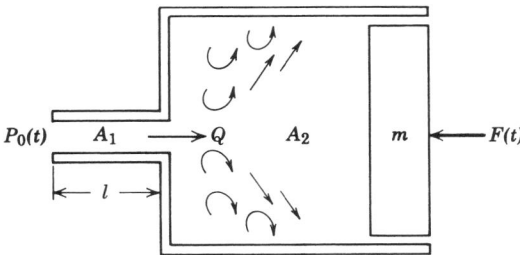

12-7 A schematic diagram of a muffler system employing two Helmholtz resonators and a resistive element is shown.

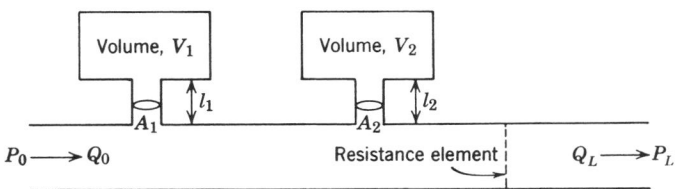

Assume all dimensions are less than one wavelength of the highest frequency of interest. Make a bond graph for the system in which the inertias of the necks of the resonators of effective lengths l_1, l_2, and areas A_1, A_2, the capacitances of the volumes V_1, V_2, and the resistance R are all represented. Using the results of Problem 12-5, list the capacitance and inertia parameters in terms of the density ρ_0 and the sound speed c.

12-8   Part of an automatic flow-metering system is shown. The main flow goes through a smooth venturi that reduces the area from A_m to A_v and in which we may assume that all dynamic pressure is fully recovered. The spring-mounted pistons deflect due to the pressure in the main pipe and at the throat. Orifices in the pipes connecting the piston chambers to the main pipe restrict the flows to the pistons to low values.

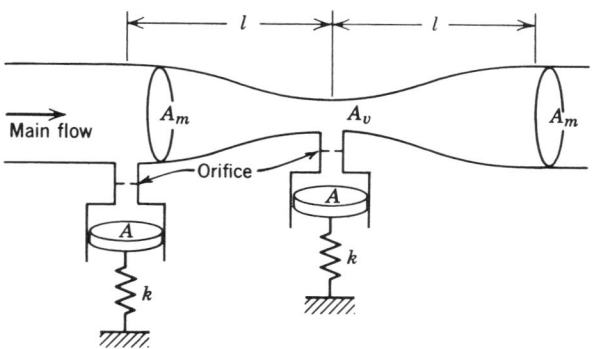

Construct a bond graph for this subsystem, and indicate inertia elements and Bernoulli resistors for the two area change sections of length l. Why should we expect the two pistons to deflect differing amounts? Would the difference between the two deflections serve to measure the main flow?

12-9   The figure shows an air cylinder that could be used as an actuator in a control system. The inlet ports at the top and bottom can be exposed to a supply at pressure P_s and temperature T_s or to exhaust at P_a and T_a. This switching would be accomplished through some valving, which is not shown.

   Construct a bond graph model of this component which is ready for installation into a system. Assign causality so we know what causality is acceptable from any attached system.

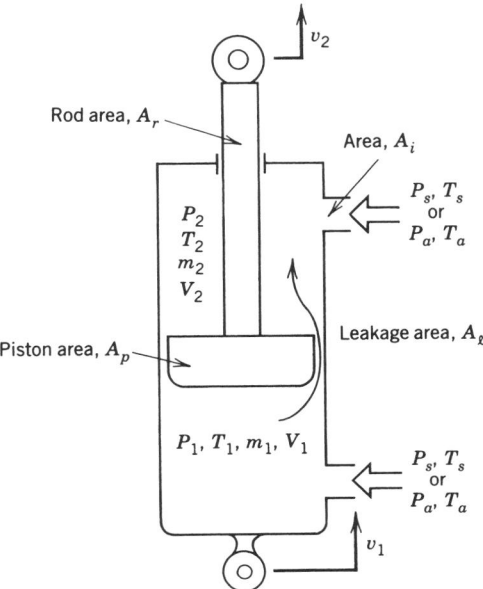

12-10   Install the actuator in the quarter car model shown. Using words and equations, set up the state equations so that you understand how a computational model will result.

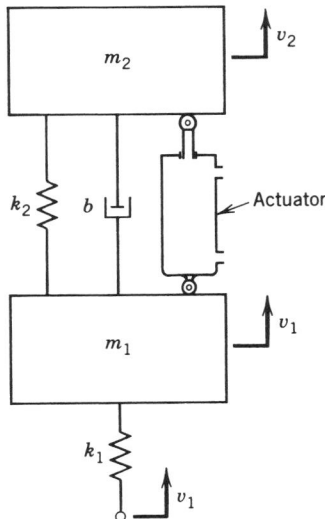

12-11   The figure shows the skirt of an air cushion vehicle. A supply at pressure P_s and temperature T_s fills the skirt volume, so that the vehicle "floats" on a cushion of air, while air escapes under the skirt to atmosphere. The height of the skirt above ground is called h and is time varying as the vehicle negotiates an uneven terrain. The ground input is the velocity $v_i(t)$

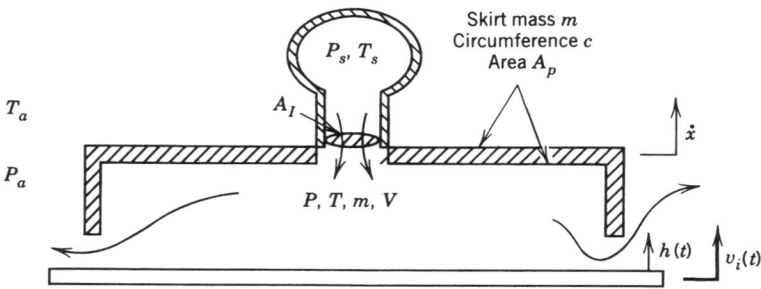

Construct a model of this system which, if solved, would predict the vertical motion time history of the vehicle. Since $h(t)$ is very small, it is all right to consider the relative motion between base and skirt as creating a volume change in the skirt.

12-12   The figure shows a schematic of a proposed variable-rate air spring capable of continuous variation. It is installed as a suspension element in the figure. It consists of a primary air cylinder coupled to a secondary cylinder with a motion-driven piston. If the piston is moved forward and mass is transferred from the front chamber to the rear chamber, then the vehicle mass will be suspended on a stiffer air spring, and the ride height will not have changed.

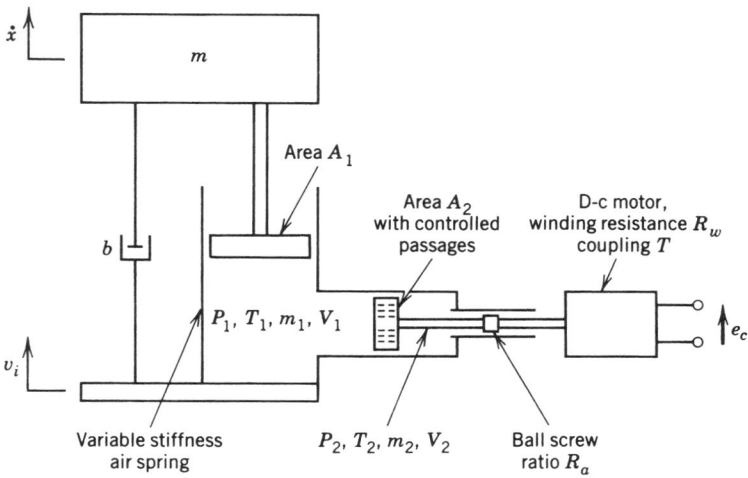

Construct a bond graph model of this system, including all the dynamics suggested in the figure. Assign causality to show that formulation will be straightforward. This model could be used to test control strategies for an adaptive suspension.

12-13   Consider the system from Problem 12-2. Neglect thermal capacitance of the piston walls but include leakage past the piston to the atmosphere. Using one thermodynamic accumulator and one 4-port isentropic nozzle, construct a bond graph model of the system. The crank radius is R, crank length is L, piston area is A_p, leakage area is A_L, and piston mass is m.

12-14   Include the thermal capacitance of the cylinder as well as convection from the cylinder surface to the atmosphere. The heat transfer coefficient for the cylinder inside wall is h_1, and for the outside wall it is h_2. The thermal capacitance of the wall is C_W. Modify the model from Problem 12-13 to include these effects.

12-15   The system from Figure 12.17 is an interesting one. Since work is done from the external port to drive gas through leakage area A_L, the gas must continuously get hotter to account for this input energy. If there were no leakage, then all work would be reversible and the gas would heat and cool as the piston cycled. Derive the equations for this system from the bond graph in Figure 12.17.

12-16   In Problem 12-15, assume the cylinder walls have thermal capacitance C_W and respective heat transfer coefficients h_1 and h_2 for the inside and outside cylinder wall. Account for convection to the cylinder wall on the inside and from the cylinder to the atmosphere, T_0, on the outside. Modify the bond graph and derive state equations for the system.

13

NONLINEAR SYSTEM SIMULATION

In Chapter 5, bond graph causality was shown to indicate, prior to any equation formulation, the state variables, the number of needed state equations, and any equation formulation problems. If causality was complete, and all energy storage elements possessed integral causality, then equation formulation is straightforward. If causality is either not complete and/or some energy storage elements are in derivative causality, then algebraic problems exist in the formulation of state equations.

In Chapter 4, some simulation of simple systems was demonstrated. These systems were linear, and no formulation problems existed. In general, there is little need to use numerical integration, from time step to time step, for linear systems. Analytical solutions are known for sets of first-order, linear state equations, and there are excellent commercial, linear analysis programs that can present solutions for linear systems without numerical integration.

In this chapter we look at the problems of generating solutions to complex, nonlinear systems where linear analysis tools are of little help. For most nonlinear systems, system response can be obtained only through numerical simulation. There are several excellent commercial equation solvers that will march out a step-by-step response for nonlinear systems. All these programs require equations to be in some particular form in order to use them. This chapter discusses the various forms that equations can take as a result of modeling decisions.

13.1 EXPLICIT FIRST-ORDER DIFFERENTIAL EQUATIONS

A bond graph model with all integral causality and no unassigned bonds after completion of the sequential assignment procedure (Section 5.2) yields explicit first-order differential equations of the general form

$$\dot{\mathbf{x}} = \mathbf{f}(\mathbf{x}, \mathbf{u}, \mathbf{t}), \tag{13.1}$$

where $\mathbf{x}$ is a vector of state variables, $\mathbf{u}$ is a vector of system inputs, and $\mathbf{f}(\cdot)$ is a vector of functions. Individual equations, as would be derived using methods from Chapter 5, would appear as

$$\dot{x}_1 = f_1(x_1, x_2, \ldots, x_n, u_1, u_2, \ldots, u_r, t),$$
$$\dot{x}_2 = f_2(x_1, x_2, \ldots, x_n, u_1, u_2, \ldots, u_r, t),$$
$$\vdots \tag{13.2}$$
$$\dot{x}_n = f_n(x_1, x_2, \ldots, x_n, u_1, u_2, \ldots, u_r, t).$$

Equations of this form are the most straightforward to solve computationally. The individual equations are written as difference equations, a reasonable time step, Δt, is selected, and the solution is marched out one time step at a time. There are countless algorithms available to solve explicit equations, but they all do basically the same thing [1].

The problem for the modeler is not so much involved with solving equations like (13.2), but much more involved with delivering the right-hand side of Eq. (13.2) to the computer equation solver. A nonlinear system may have nonlinearities that cannot be explicitly written as analytical functions. A 1-port resistance characterized by an effort–flow relationship might be derived from first principles as

$$e = gf^3, \tag{13.3}$$

where g is some constant. Such a relationship is easily incorporated into the right-hand side of Eqs. (13.2). But if one has only a table of e and f values perhaps generated experimentally, then the user has a bit more work to do to incorporate the data in Eqs. (13.2). Bounded inputs or displacements, or impact with solid barriers, present another challenge to the user to allow proper evaluation in Eqs. (13.2) at each time step. Thus, even though Eqs. (13.2) are the most convenient to solve computationally, there is still a lot of work to be done to properly incorporate system nonlinearities.

Fortunately for us, there are many excellent commercial packages available to march out solutions to explicit equations like (13.2). These software packages have many built-in functions that handle many commonly encountered nonlinearities. They also allow tables of data to be incorporated with automatic interpolation between data points. These packages have excellent plotting capabilities and make the job of the user almost pleasant.

As an example of reasonably straightforward, nonlinear simulation, consider the system in Figure 13.1. It consists of mass, m, in a gravity field, g, attached to an inertial frame with a spring and damper. The spring and damper both could be nonlinear so no spring constant or damping constant is shown in the figure. The pendulous motion is not restricted to small values of the angle, θ. Body-fixed coordinates x, y

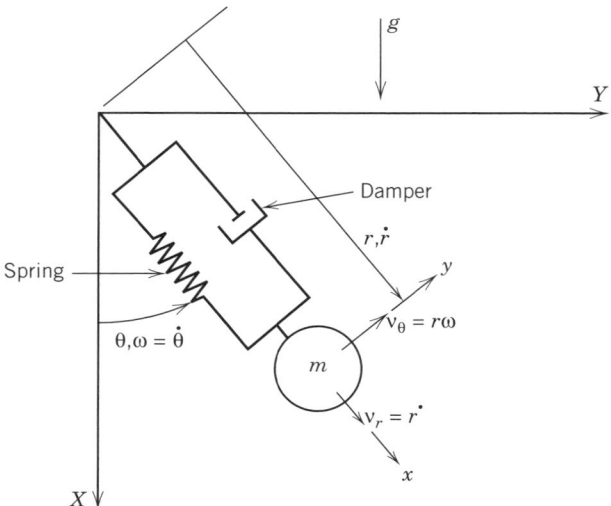

FIGURE 13.1. Nonlinear system exhibiting simultaneous pendulous and spring–mass oscillatory motions.

are used, and the radial and tangential velocities are indicated. The astute reader will recognize this example as a slightly modified version of Figure 9.8 in Chapter 9.

The bond graph for this system is shown in Figure 13.2. The modulated gyrator ($—MGY—$) is explained in Chapter 9 and is always required when using body-fixed coordinates. Causality has been assigned to the bond graph, and the causal-

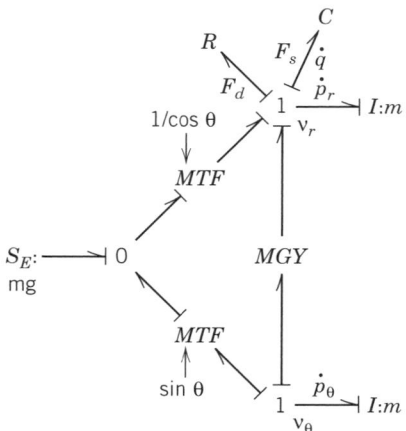

FIGURE 13.2. Bond graphs for the system from Figure 13.1.

ity completed with three elements in integral causality. The state variables are the r-direction momentum, p_r, the θ-direction momentum, p_θ, and the spring displacement, q. Thus, three state equations are needed to characterize the motion of this system. Note that the spring force, F_s, and damper force, F_d, have been indicated in the bond graph.

Automated equation derivation was demonstrated in Chapter 4 for some simple linear systems. This topic will be addressed again in this chapter. But, for now, the state equations are derived, using the procedure from Chapter 5, as

$$\dot{p}_r = -F_s - F_d + m\omega\frac{p_\theta}{m} + mg\cos\theta, \tag{13.4}$$

$$\dot{p}_\theta = -mg\sin\theta - m\omega\frac{p_r}{m}, \tag{13.5}$$

$$\dot{q} = \frac{p_r}{m}, \tag{13.6}$$

where

$$F_s = F_s(q), \tag{13.7}$$

with

$$F_d = F_d\left(\frac{p_r}{m}\right), \tag{13.8}$$

$$\omega = \frac{p_\theta}{mr}. \tag{13.9}$$

Equations (13.4)–(13.9) are explicit equations like (13.2). These equations are ready to be solved once the constitutive behaviors of the spring and damper are specified. Notice that the angle θ is needed but is not available from the equation set. Thus, although the bond graph has delivered explicit equations, additional thought is required before a solution can be pursued.

The nonlinear constitutive relationships for the spring and damper might be prescribed analytical functions or they may come from experimental data. These would be included straightforwardly in a simulation package. Since no energy storage depends on the angular displacement θ, the state space must be expanded by the user such that $\dot{\theta}$ and θ are available. This is done by writing the free integrator equations

$$\frac{d\theta}{dt} = \omega = \frac{p_\theta}{mr}, \tag{13.10}$$

$$\frac{dr}{dt} = v_r = \frac{p_r}{m}. \tag{13.11}$$

Note that r is needed even though the displacement q of the spring is available. This allows us to differentiate between the free length of the spring and the initial length of the pendulum.

These free-integrator equations are solved numerically along with the actual state equations, (13.4)–(13.6), and the solution can be straightforwardly generated. In Eq. (13.10) the user must be careful to avoid $r = 0$.

This expanding of the state space to make available modulating variables such as θ in this example is often necessary when dealing with mechanical systems that can move through large angular displacements. Fortunately, rates of change of these modulating variables are always available directly from the bond graph, so writing an extra state equation is no problem.

13.2   DIFFERENTIAL ALGEBRAIC EQUATIONS CAUSED BY ALGEBRAIC LOOPS

When causality does not complete on a bond graph after assigning causality by the rules of Chapter 5, an algebraic loop exists among some of the effort and flow variables of the model. This occurrence does not suggest anything about the quality of one's modeling decisions, but those decisions did lead to the algebraic problem. And this algebra must be dealt with prior to simulating the system.

When an algebraic loop exists, derivation of equations yields state equations like

$$\dot{\mathbf{x}} = \mathbf{f}(\mathbf{x}, \mathbf{u}, \mathbf{z}, t), \tag{13.12}$$

where

$$\mathbf{z} = \mathbf{g}(\mathbf{x}, \mathbf{u}, \mathbf{z}, t), \tag{13.13}$$

and $\mathbf{x}$ is the vector of state variables chosen from the energy storage elements in integral causality, and $\mathbf{u}$ is the vector of inputs. The vector $\mathbf{g}$ contains e and f variables that are part of the algebraic loop, and, unfortunately, these variables depend on themselves, as shown by the right-hand side of Eq. (13.13).

If it is possible to solve Eq. (13.13) for $\mathbf{z}$ as it depends on $\mathbf{x}$, $\mathbf{u}$, and t, then substitution into (13.12) yields explicit equations as described previously. This procedure was demonstrated in Section 5.4 for the case when the model is linear. When the model is nonlinear, it may not be possible to do the required algebra, especially if $\mathbf{z}$ contains several e and f variables.

Looking at Eqs. (13.12) and (13.13), we can imagine an iterative computational procedure that, starting with the initial values of $\mathbf{x}$ and $\mathbf{u}$, would manipulate the components of $\mathbf{z}$ until the right-hand side of Eq. (13.13) reproduces the values of the $\mathbf{z}$ variables to within some prescribed error bounds at which point substitution into Eq. (13.12) would allow one time step to be taken. This procedure would be repeated every time step and is computationally intensive. There are computer algorithms that do this procedure more or less automatically, and they will be discussed further in the section on automated simulation.

Another, and perhaps more satisfying, approach to dealing with the algebra is to revisit the modeling assumptions that led to the algebraic problem in the first place. By modifying one's assumptions to include some inertia or compliance at locations

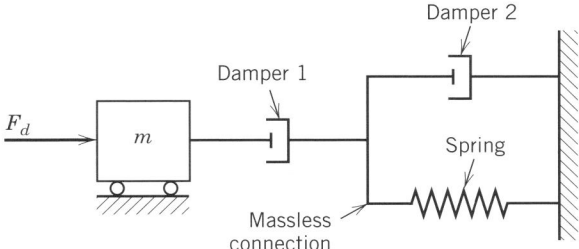

FIGURE 13.3. A simple system with an algebraic loop.

where it was previously ignored, the algebraic loops can be made to disappear, and explicit equations like (13.2) will result. This procedure was coined the *Karnopp–Margolis method* in Reference [2], demonstrated for a complex system in [3], and shown for a simple system here.

Figure 13.3 shows a mechanical schematic for a system consisting of one mass, one spring, and two damping elements. The spring and dampers are all nonlinear. Assigning causality by the rules of Chapter 5 yields the incomplete bond graph of Figure 13.4a. Arbitrarily assigning F_1 into the 0-junction results in the causally complete bond graph of Figure 13.4b. Since one arbitrary causal assignment was necessary, we know that one algebraic loop is present.

Assume the R- and C-elements are characterized functionally as

$$F_1 = \Phi_{R_1}(v_1), \tag{13.14}$$

$$v_2 = \Phi_{R_2}^{-1}(F_2), \tag{13.15}$$

$$F_s = \Phi_c(q_s). \tag{13.16}$$

Choosing F_1 as the auxiliary variable in the algebraic loop and using the procedure developed in Chapter 5, we can derive

$$F_1 = \Phi_{R_1}\left(\frac{p_m}{m} - v_2\right) \tag{13.17}$$

$$v_2 = \Phi_{R_2}^{-1}(F_1 - F_s). \tag{13.18}$$

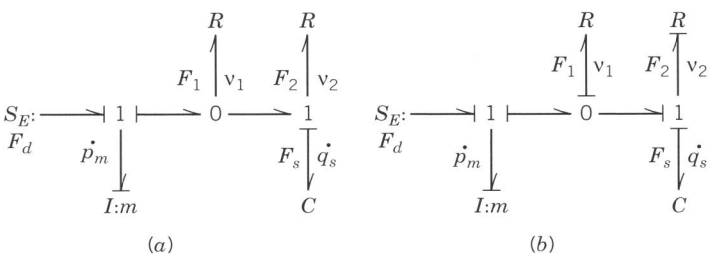

FIGURE 13.4. Bond graph for the system of Figure 13.3.

Combining (13.16), (13.17), and (13.18) yields

$$F_1 = \Phi_{R_1} \left\{ \frac{p_m}{m} - \Phi_{R_2}^{-1} [F_1 - \Phi_c(q_s)] \right\}. \tag{13.19}$$

This relationship is the example equivalent to Eq. (13.13) for the general case. Here we have F_1 related to state variables p_m and q_s and to itself.

The state equations are straightforwardly derived as

$$\dot{p} = F_d - F_1, \tag{13.20}$$

$$\dot{q}_s = \Phi_{R_2}^{-1} [F_1 - \Phi_c(q_s)]. \tag{13.21}$$

These equations are an example of the general Eqs. (13.12). Here the state derivatives depend on themselves and the auxiliary variable, F_1.

If the functions $\Phi_{R_1}(\cdot)$, $\Phi_{R_2}^{-1}(\cdot)$, and $\Phi_c(\cdot)$ are such that Eq. (13.19) could be solved for F_1, then that result can be used in Eqs. (13.20) and (13.21) to yield explicit equations, and the solution is straightforward. However, it should be appreciated that, even when only one variable is involved with the algebraic loop, the nonlinear functions may not allow solution of Eq. (13.19) for F_1, explicitly. If some of the functions are not analytical expressions, but rather tables of data, explicit solution of (13.19) would be impossible. Generally, if multiple auxiliary variables are involved in the algebra, explicit formulation is out of the question.

Returning to the original model of Figure 13.1, it was assumed that the connection in the middle is massless. This is a modeling decision based upon reasonable assessment of the requirements of the model. We know that the connection is not massless, just like we know the proposed constitutive behavior of the elements is not exact. In Figure 13.5, the massless connection is changed to a new assumption, namely, that the connection has some mass. Also shown in Figure 13.5 is the bond graph with causality assigned. This time the causality completes, and there is no algebraic loop. Derivation of state equations is guaranteed to be in explicit form.

Directly from the bond graph, these equations are

$$\dot{p}_m = F_d - F_1, \tag{13.22}$$

$$\dot{p}_m = F_1 - F_s - F_2, \tag{13.23}$$

$$\dot{q}_s = \frac{p_{m'}}{m'}, \tag{13.24}$$

where

$$F_1 = \Phi_{R_1} \left(\frac{p_m}{m} - \frac{p_{m'}}{m'} \right), \tag{13.25}$$

$$F_2 = \Phi_{R_2} \left(\frac{p_{m'}}{m'} \right), \tag{13.26}$$

$$F_s = \Phi_c(q_s). \tag{13.27}$$

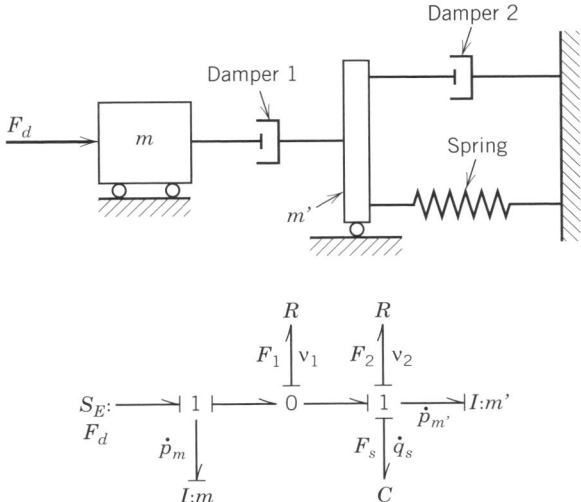

FIGURE 13.5. System from Figure 13.3 without the massless connection.

These equations are in explicit form, and, after the constitutive behavior of the non-linear elements is specified, they are ready for computer simulation.

Upon rethinking our original modeling assumptions in order to generate a causally complete bond graph and, thus, explicit state equations, two things of interest occurred. First, introduction of the originally unintended inertial element, m', has increased the order of the system from two state equations to three state equations. This is not significant in this simple example, but if the model was large and many algebraic loops existed, such that many additional state variables were required, the modeler may decide to deal with the algebra iteratively as mentioned previously.

The second result of introducing m' into the model is that, if m' is made very small, to simulate the originally intended system, a high-frequency component is introduced into the model. The inertial element, m', will interact with the spring and dampers to produce short-period and/or short-time-constant response components, and the computational time step must be appropriately shortened to capture these short time transients in the solution. In general, if the "parasitic elements" introduced to "fix" the causality of the original model are chosen so as to introduce dynamics about a factor of 10 faster than the original system, the simulation results will be imperceptibly different from the original system response. The increased time required for the simulation due to the shorter step size is about the same as the time needed to perform the algebraic iterations if the model is left in differential algebraic form.

Before leaving this section, the reader should realize that nonlinear systems do not officially have eigenvalues, time constants, or oscillation frequencies as defined in Chapter 6 for linear systems. So using these terms above while describing the effect of introducing parasitic elements to "fix" causality in nonlinear systems is not, strictly speaking, rigorous. However, I-elements interacting with C-elements will oscillate, I-elements interacting with R-elements will experience decaying motion over some period of time, and R-elements interacting with C-elements will dissipate energy over some time period. The engineer modeling a physical system will have no difficulty in interpreting the dynamics from introduced parasitic elements.

13.3   IMPLICIT EQUATIONS CAUSED BY DERIVATIVE CAUSALITY

In Chapter 5 we saw that, sometimes, after assigning causality to a bond graph, some I- or C-elements end up in derivative causality. As with incomplete causality from the previous section, the presence of derivative causality does not suggest anything about the quality of the model or one's modeling decisions. But it does indicate that an algebraic problem exists that must be dealt with prior to simulation of the system.

For linear systems, Chapter 5 demonstrated a procedure to handle the algebraic implications prior to deriving the state equations. For nonlinear systems, the procedure can still be used, but the algebraic manipulations may not be possible to perform.

In general, if derivative causality exists in a bond graph model, the state equations will have the general form

$$\dot{\mathbf{x}} = \mathbf{f}(\mathbf{x}, \mathbf{u}, t, \dot{\mathbf{x}}), \tag{13.28}$$

where $\mathbf{x}$ is the vector of state variables and $\mathbf{u}$ is vector of inputs. The right-hand side of Eq. (13.28) represents the nonlinear functions of the states, inputs, and state derivatives. These equations are in implicit form and cannot be solved easily numerically because a knowledge of input and state variables does not immediately produce values of the derivatives of the state variables.

One can imagine a computer algorithm that, starting from the initial values of the state variables, guesses initial values of the state derivatives, $\dot{\mathbf{x}}$, and then uses some iterative procedure until the right-hand side of Eq. (13.28) reproduces the left-hand side to within some defined accuracy. Once the $\dot{\mathbf{x}}$ are known, an integration step can be taken, and the process begins again. This solution of implicit equations is quite computer intensive. More will be said about this in the section on automated simulation.

Another approach to dealing with derivative causality is the same as dealing with algebraic loops. We look at the modeling decisions that led to derivative causality and make new modeling decisions that do not introduce the problem. We again include physically motivated parasitic elements that permit explicit equations while retaining the intended performance of the model.

Consider the example system of Figure 13.6. It consists of three nonlinear springs and a nonlinear damper. The springs' constitutive behaviors are

$$F_1 = \Phi_{c_1}(q_1), \tag{13.29}$$

$$F_2 = \Phi_{c_2}(q_2), \tag{13.30}$$

$$F_3 = \Phi_{c_3}(q_3). \tag{13.31}$$

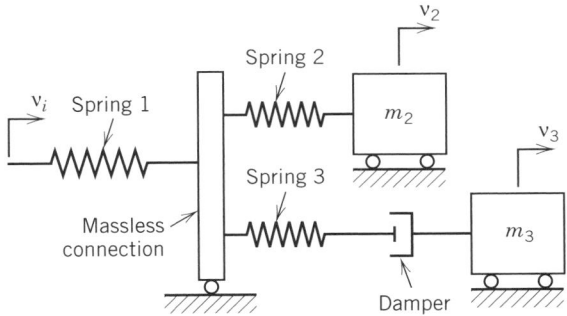

FIGURE 13.6. Nonlinear system exhibiting derivative causality.

The damper is characterized by

$$v_d = \Phi_R^{-1}(F_d). \qquad (13.32)$$

The bond graph for this system is shown in Figure 13.7, and causality assignment indicates that spring 3 is in derivative causality. Notice the causality on the damper dictates that the flow, v_d, is the output and the effort, F_d, is the input.

Procedures from Chapter 5, Section 5.5, show that state equation formulation will require the flow on that spring, $\dot{q}_3$, to be part of the equation derivation. From Eq. (13.31) we can write

$$q_3 = \Phi_{c_3}^{-1}(F_3), \qquad (13.33)$$

which says we must be able to invert the constitutive relationships associated with derivative causal energy storage elements. This may not be possible, but assuming it is, we continue. Causality tells us that

$$F_3 = F_1 - F_2, \qquad (13.34)$$

where F_1, F_2 are from Eqs. (13.29) and (13.30). Then,

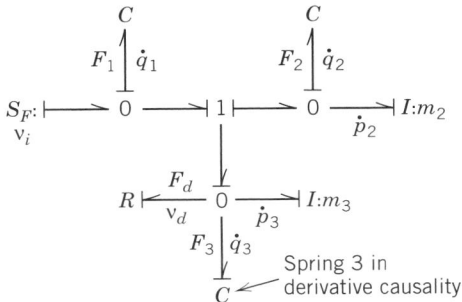

FIGURE 13.7. Bond graph for the system in Figure 13.6.

$$\dot{q}_3 = \frac{d\Phi_{c_3}^{-1}(F_3)}{dF_3}\dot{F}_3, \tag{13.35}$$

where

$$\dot{F}_3 = \frac{d\Phi_{c_1}}{dq_1}\dot{q}_1 - \frac{d\Phi_{c_2}}{dq_3}\dot{q}_3. \tag{13.36}$$

The state equations, directly from the bond graph, are

$$\dot{q}_1 = v_1 - v_d - \frac{p_3}{m_3} - \dot{q}_3, \tag{13.37}$$

$$\dot{q}_2 = -\frac{p_2}{m_2} + v_d + \frac{p_3}{m_3} + \dot{q}_3, \tag{13.38}$$

$$\dot{p}_2 = F_2, \tag{13.39}$$

$$\dot{p}_3 = F_1 - F_2, \tag{13.40}$$

where F_1, F_2 come from Eqs. (13.29) and (13.30) and v_d comes from Eq. (13.32), with

$$F_d = F_1 - F_2. \tag{13.41}$$

Notice that q_3 is not an independent state variable even though spring 3 stores energy. Because of the derivative causality, q_3 is algebraically related to the state variables q_1 and q_2. Using Eq. (13.36) in Eq. (13.35) and substituting into Eqs. (13.37) and (13.38), we end up with equations of the general form of Eq. (13.28), where state derivatives appear on the right-hand side of the state equations.

Presuming that the operations indicated in Eqs. (13.35) and (13.36) can be carried out, the iterative solution of these implicit equations could be done. It may also be possible to perform the algebra to get all state derivatives to the left-hand side of Eqs. (13.37) and (13.38) and manipulate until explicit equations result. In general, this will not be possible for nonlinear systems because the constitutive behavior of the elements will not necessarily be differentiable analytical functions but may be bounded functions with discontinuities and/or tables of data from testing real devices.

An alternative approach is to look back at our modeling assumptions and revise them a bit so we do not end up with derivative causality. Figure 13.8 shows the example system from Figure 13.6 without the assumption of a massless connection. This time a mass, m', has been included at the connection. The bond graph with m' included shows no derivative causality; thus explicit equations will result from the bond graph.

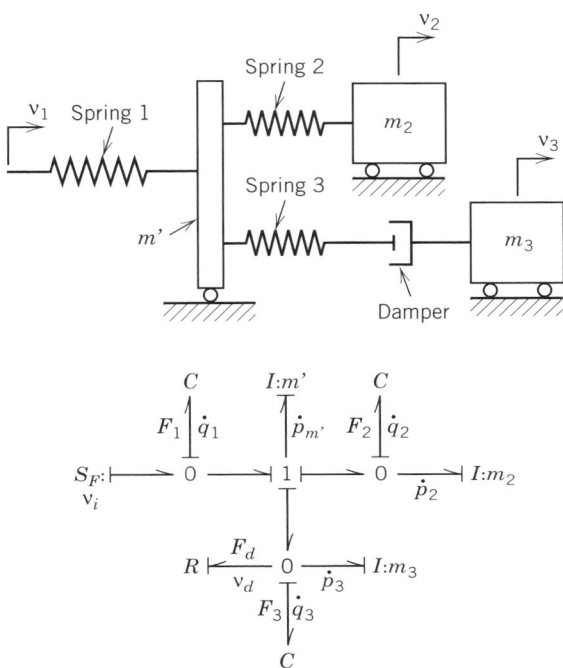

FIGURE 13.8. Revised system of Figure 13.6 without derivative causality.

This time the state equations are

$$\dot{q}_1 = v_i - \frac{p_{m'}}{m'}, \tag{13.42}$$

$$\dot{q}_2 = -\frac{p_2}{m_2} + \frac{p_{m'}}{m'}, \tag{13.43}$$

$$\dot{q}_3 = \frac{p_{m'}}{m'} - \frac{p_3}{m_3} - v_d, \tag{13.44}$$

$$\dot{p}_2 = F_2, \tag{13.45}$$

$$\dot{p}_3 = F_3, \tag{13.46}$$

$$\dot{p}_{m'} = F_1 - F_2 - F_3, \tag{13.47}$$

where F_1, F_2, and F_3 come from the constitutive relationships of Eqs. (13.29), (13.30), and (13.31) and v_d comes from Eq. (13.32) with $F_d = F_3$. Thus we have explicit, first-order, nonlinear state equations ready for computer solution.

For the case of derivative causality, when we revisited our modeling assumptions and introduced a "parasitic element," m', we ended up with two additional state equations. One is due to the introduced I-element, with mass m', and the other is due to switching the causality on spring 3 making q_3 an independent state variable. For low-order models requiring few parasitic element additions, the additional state equations will pose no computational problems. But for very large models, with many derivative causal elements, we may elect to deal with the problem by solving the implicit equations directly. The section on automated simulation will deal with this later.

Finally, when introducing parasitic elements to yield explicit equations, fast dynamic transients are introduced if the parasitic elements are sized to approximate the originally intended model. This was discussed in the previous section concerning algebraic loops. That discussion applies here as well.

13.4 AUTOMATED SIMULATION OF DYNAMIC SYSTEMS

Automated simulation here refers to the many commercial equation solvers that take equations of motion delivered in some prescribed format, march out solutions, and present results with virtually no computer programming required from the user. All of these packages accept equation formats as inputs, and some of them have graphical interfaces such that various graphical icons are interpreted as equations and delivered properly to the equation solver. The current software packages available throughout the world will not be specifically discussed in the text. Here we discuss some generalities of automated simulation.

There are several software packages available for very specific systems such as automobile dynamics, train dynamics, magnetic circuits, acoustics, plus others. These have specific models already internally built, and the user must only provide data to describe the particular system. These are very useful programs for the intended use but do not represent the system dynamic modeling emphasized in this text. Here we speak of system models that have been constructed by the modeler to develop insight into system behavior or control philosophies for prototyping. Here we discuss software packages that allow simulation of these models in the most straightforward way.

13.4.1 Sorting of Equations

For an equation solver to simulate one time step into the future, it requires the current values of the state-variable derivatives. These derivatives are calculated one at a time from the equations the user provides by automatically sorting the equations such that the updated information required to evaluate each derivative has already been computed. As a simple example, consider the equation

$$\frac{dx}{dt} = ax + b, \tag{13.48}$$

where

$$a = 6, \tag{13.49}$$

$$b = 8, \tag{13.50}$$

$$x = 2. \tag{13.51}$$

Without automatic sorting, dx/dt could not be evaluated unless Eqs. (13.49)–(13.51) have been previously evaluated so that a, b, and x have specific numerical values. With automatic sorting, the computer algorithm knows to evaluate a, b, and x prior to calculating dx/dt. This feature may appear trivial or obvious, but it frees the user from providing equations in proper sequence and greatly facilitates putting several submodels together without paying attention to order.

An equally important feature of sorting comes from the following simple example. Consider

$$\frac{dx}{dt} = ax + b, \tag{13.52}$$

where

$$b = c - b + x, \tag{13.53}$$

$$c = 6, \tag{13.54}$$

$$a = 8, \tag{13.55}$$

$$x = 2. \tag{13.56}$$

Here we have a problem sorting the equations. In Eq. (13.53), b depends on itself, and there is no way an automatic sorting routine can find a sequential evaluation of Eqs. (13.53)–(13.56) that will allow an evaluation of dx/dt in Eq. (13.52). An automatic sorting algorithm cannot sort these equations. In principle, one could simply solve Eq. (13.53) for $b = \frac{1}{2}c + \frac{1}{2}x$ and be done with it. A computer algorithm capable of handling differential algebraic equations* would take the values of c and x, then iterate on b until the right-hand side of Eq. (13.53) equaled the left-hand side. An explicit equation solver, given these equations, would respond with "can't sort" and quit. Thus, modern commercial software packages requiring explicit equations automatically inform when the model is *not* computable, but cannot simulate further.

As seen previously, bond graphs inform of algebraic formulation problems prior to deriving a single equation. This allows the modeler to address his or her modeling assumptions before attempting to simulate the model. This is a most useful virtue of bond graph modeling.

13.4.2   Implicit and Differential Algebraic Equation Solvers

When modeling assumptions lead to algebraic loops or derivative causality, explicit equation solvers cannot sort equations for simulation. If the algebra is not dealt with,

---

*There are software packages that will solve differential algebraic equations. These are not nearly as prevalent as explicit equation solvers.

then a solution can be marched out only by the iterative procedures described previously. The equations involved are differential algebraic equations (DAEs) or implicit differential equations.

As mentioned, there are commercial packages devoted to simulating very specific dynamic systems such as automobiles and trains. These models are in the area of dynamics called "multibody system dynamics," where the inertial bodies are all assumed to be rigid and no relative motion is permitted at attachment points among the bodies. Bond graph models of such systems would be full of derivative causality. For these specialty systems, DAE algorithms are used to solve system equations. Because the equations have a particular form due to the particular algebraic problem, the software can march out solutions with no programming from the user.

The modeling procedure developed in this text often results in algebraic problems that must be addressed by the modeler prior to computer solution. Something as common as Coulomb friction, if involved in an algebraic loop, cannot be handled by a DAE solver without significant user attention. While DAE solvers are powerful tools for automated simulation, in this text we emphasize models that yield explicit equations.

13.4.3 Icon-Based Automated Simulation

There are commercial software packages that allow graphical system descriptions which are automatically translated into equations of required format and given to the equation solver. These icon-based packages are predominantly associated with special applications like electric circuits, hydraulic circuits, mechanical spring–mass–damper systems, magnetic circuits, acoustic circuits, and more. Some of these programs allow "functional block" inputs, like block diagrams, and are less specific system intended than the aforementioned software.

There are several packages that translate bond graph descriptions into appropriate input files for equation solvers. Since bond graphs allow modeling of all types of energetic systems, linear or nonlinear, the graphical bond graph programs are the most appropriate for overall system dynamics coupling complex mechanical systems with actuators, sensors, and control.

If linear modeling assumptions are used in the construction of a model, then only one parameter is needed to describe each $-R$, $-I$, $-C$, $-TF-$, or $-GY-$ element. For multiport elements, the number of parameters increases but remains constrained. For nonlinear systems, there is an infinite variety of possible constitutive laws. For instance, if a bond graph had a C-element, a computer could interpret this symbol as having the linear constitutive law:

$$e = \frac{q}{C}, \qquad (13.57)$$

and the user would only have to supply the value of the compliance parameter, C. If that same C-element were nonlinear, such that

$$e = \Phi_c(q), \qquad (13.58)$$

the user would have to provide the computer the functional dependence through some user-provided function or tabular data. Some icon-based graphical bond graph software interprets the bond graph as though it were linear, then allows the user to modify the input file to reflect the intended nonlinearities or simply assign values to appropriate parameters if the element were actually linear.

The title of the section is "Automated Simulation of Dynamic Systems," but after reading all this description, the reader may feel there is nothing at all automated about simulation. It is true that low-order, linear systems can be simulated with virtually no effort from the user. But complex nonlinear systems do require significant user input.

Bond graphs allow organization of complex models into manageable parts, and they inform the user of computational problems through the assignment of causality. In the next section, a complex system simulation is demonstrated.

13.5   EXAMPLE NONLINEAR SIMULATION

In Chapter 12, thermodynamic modeling was developed, and a bond graph of a piston-ported, two-stroke engine was shown in Figure 12.18. The example model done here is a simplified version of a two-stroke engine called the "air motor."

The air motor is shown schematically in Figure 13.9. It consists of a slider–crank kinematic mechanism which drives a piston in a cylinder. Just after top dead center (TDC), the inlet value is opened exposing the piston to the supply pressure, P_s. The pressure, P, in the cylinder pushes the piston downward. As the piston passes the exhaust port, the piston modulates the flow area and effects the flow of air out of (or into) the cylinder. The idea of the simulation might be to determine the best location, x_e, of the exhaust port so as to extract the most useful work from the supply air. Our purpose here is just to demonstrate a nonlinear simulation. This particular example was chosen because all the components have been modeled previously in the text, but more importantly, there is no way that a linearized analysis will be helpful here. The only way to gain insight into the behavior of this system is through simulation.

It is assumed that a single pressure, P, temperature, T, mass, m, energy, E, and volume, V, characterize the instantaneous properties of the cylinder head. It is further assumed that the piston is massless and there is no leakage past the piston. Finally, it is assumed that the inlet area, A_i, is exposed instantly upon command, the cylinder head can be modeled as a single thermodynamic accumulator (Section 12.4.1), and the inlet and exhaust ports are each characterized as isentropic nozzles (Section 12.4.2).

The bond graph for the air motor is shown in Figure 13.10. Causality has been assigned and indicates all integral causality with no algebraic loops. Thus, although complex, formulation will yield explicit equations ready for simulation. The thermodynamic accumulator will contribute three state variables, and the flywheel inertia will contribute one state variable. Only four state equations are needed for this simulation.

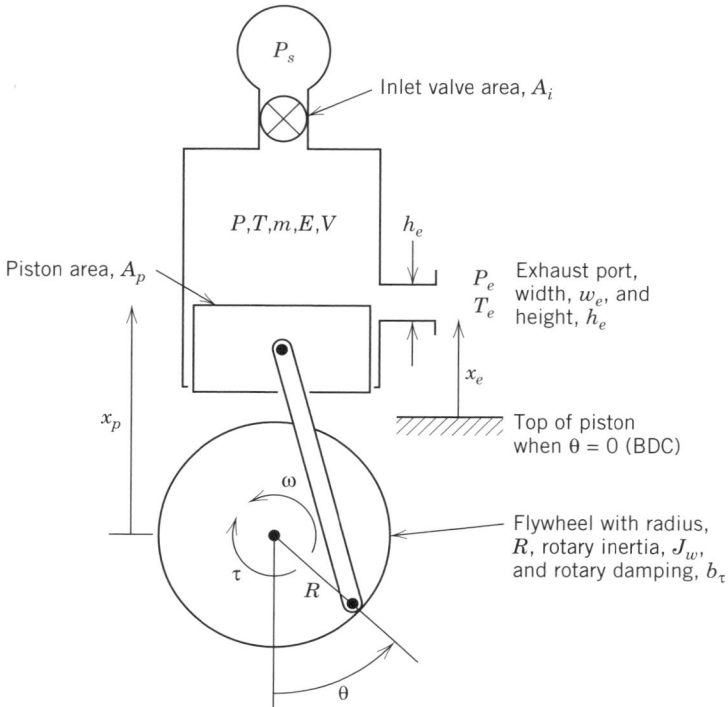

FIGURE 13.9. Schematic of an air motor.

The 4-port R-element for the inlet requires specification of the supply pressure, P_s, and supply temperature, T_s. It also needs to know the flow area, A_i, which will be specified through some logic statements provided by the user. It will be open when θ exceeds 180° and closed when the piston position, x_p, is less than the exhaust port location, x_e.

The 4-port R-element representing the exhaust port requires specification of the pressure, P_e, and temperature, T_e, just outside the port. The upstream conditions come from the causal outputs, P and T, from the 3-port, C-element. The exhaust port area, A_e, is modulated by the piston position, x_p, through some user-supplied logic. This is indicated by the ideal measurement of $\dot{x}_p$ followed by signal processing blocks.

The slider–crank relationship, $m(\theta)$, was derived in Section 8.1 and is repeated here,

$$m(\theta) = \left[R \sin\theta - \frac{R^2 \sin\theta \cos\theta}{(L^2 - R^2 \sin^2\theta)^{1/2}} \right], \tag{13.59}$$

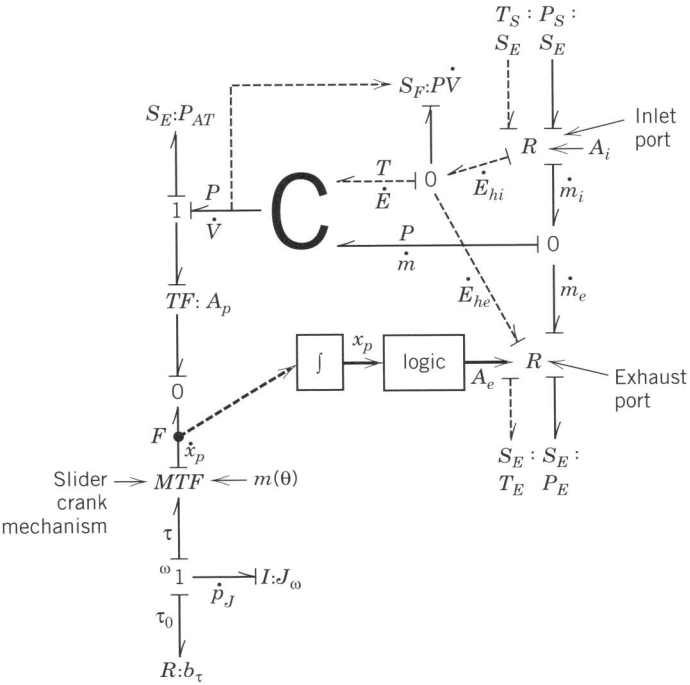

FIGURE 13.10. Bond graph of the air motor.

used such that

$$\dot{x}_p = m(\theta)\omega. \tag{13.60}$$

In Eq. (13.59) the reference for $\theta = 0$ is measured from vertically down (bottom dead center), which is different than the θ reference in Section 8.1. The state variables are E, m, and V, respectively the energy, mass, and volume of the accumulator, and p_J, the angular momentum of the flywheel.

Following the causal information, the equations can be derived as

$$\dot{E} = -P\dot{V} + \dot{E}_{hi} - \dot{E}_{he}, \tag{13.61}$$

$$\dot{m} = \dot{m}' - \dot{m}_e, \tag{13.62}$$

$$\dot{V} = -Ap\dot{x}_p \tag{13.63}$$

$$\dot{p}_J = -b_\tau \frac{p_J}{J_w} - \tau. \tag{13.64}$$

The enthalpy flow, $\dot{E}_{hi}$, and mass flow, $\dot{m}_i$, for the inlet port are outputs from the 4-port R-element described in Section 12.4.2. In that section a computational procedure is described by Eqs. (12.96)–(12.101). This will not be repeated here. We will just state here the causal implication of the 4-port; thus,

$$\dot{E}_{hi} = f_i(T_s, P_s, T, A_i), \tag{13.65}$$

$$\dot{m}_i = g_i(T_s, P_s, T, A_i). \tag{13.66}$$

Similarly, the energy and mass flows through the exhaust port can be expressed as

$$\dot{E}_{he} = f_e(T, P, T_e, P_e, A_e), \tag{13.67}$$

$$\dot{m}_e = g_e(T, P, T_e, P_e, A_e). \tag{13.68}$$

In these expressions, the outputs from the accumulator are related to the state variables by the "compliance" relationships,

$$T = \frac{1}{c_v}\frac{E}{m}, \tag{13.69}$$

and

$$P = \frac{R}{c_v}\frac{E}{V}. \tag{13.70}$$

These constitutive laws are repeated from Section 12.4.1.

In order to complete the formulation, we need $\dot{x}_p$ from Eq. (13.63) and the torque, τ, from Eq. (13.64). From the bond graph,

$$\dot{x}_p = m(\theta)\frac{P_J}{J_w} \tag{13.71}$$

$$\tau = m(\theta)Ap(P - P_{AT}) \tag{13.72}$$

Atmospheric pressure, P_{AT}, is included in the bond graph to convert from absolute pressure, needed on the thermodynamic side of the accumulator, to gauge pressure on the mechanical side.

We must recognize the need for θ and x_p as modulating variables. Here we expand the state space by writing

$$\dot{\theta} = \frac{P_J}{J_w}, \tag{13.73}$$

and θ will be solved along with the other state equations. The displacement, x_p, is available from solution of Eq. (13.71).

Finally, the area modulation A_i for the inlet and A_e for the exhaust are needed to complete the formulation. These would both be specified by logic supplied by the user. For example, for A_i,

$$\text{if } \theta = \,\geq 180° \quad \text{and} \quad x_p > x_e, \quad A_i = A_{i\text{MAX}}$$
$$\text{otherwise,} \qquad\qquad\qquad A_i = 0. \tag{13.74}$$

This will keep the inlet open until the exhaust port begins to open. The user must be a little careful here to make sure that θ gets reset to zero after each full rotation since θ will increase continuously from integration of Eq. (13.73) and Eq. (13.74) would not work after one rotation.

The exhaust area logic would be

$$A_e = 0;$$

$$\text{if } x_p \leq (x_e + h_e), \quad A_e = W_e[x_e + h_e + x_p]; \tag{13.75}$$

$$\text{if } x_p \leq x_e, \qquad\qquad A_e = w_e h_e.$$

These statements will modulate the area as the piston uncovers the port, then keeps the area at its maximum value when the port is uncovered, and sets the area to 0 when the port is covered. There may be other strategies that the user wants to try. These are just examples.

Eqs. (13.61)–(13.75) can be delivered to an explicit equation solver in any order desired, and the automatic sorting routines will evaluate each equation in its proper order to arrive at the state derivatives. An integrating algorithm will march the solution from one time step to the next. There is no need to attempt to put these equations in the functional explicit form,

$$\dot{\mathbf{x}} = \mathbf{f}(\mathbf{x}, \mathbf{u}, t), \tag{13.76}$$

since we know that this is a causal, computable, model and the sorting of equations can be carried out.

13.5.1   Some Simulation Results

The air engine parameters are,

$D_p = 80$ mm, piston diameter;

$R = 40$ mm, crank radius;

$L = 40$ mm, connecting rod length;

$V_{sq} = 40$ cc, volume in head at TDC;

$J_w = 1.64 \times 10^{-3}$ Kgm^2, flywheel rotary inertia;

$b_\tau =$ load resistance, to be varied to determine torque, speed, and power.

For the intake port,

$A_{i\text{MAX}} = 2.9$ cm^2, maximum inlet area;

$P_s = 200$ psig $= 14.3$ atm $= 14.3 \times 10^5$ N/m^2 supply pressure;

$T_s = 20°$C $= 293$ K, supply temperature.

For the exhaust port,

$h_e = 17$ mm, height of port;

$w_e = 17$ mm, width of port;

x_e is variable, placement of port relative to crank center.

User might try 0, 20 mm, 40 mm, ...

Initial Conditions The simulation is started at just past TDC with $\theta = 182°$. The pressure in the cylinder, P, is P_{atm}, and the temperature, T, is 20°C. The cylinder volume, V, is $V_{sq} = 40$ cc. From Eq. (13.70), we can calculate the initial energy, E, and from Eq. (13.69), we can calculate the initial mass, m. The initial angular momentum, p_J, is zero.

The simulation is initiated with these initial conditions, and simulation proceeds with the piston moving cyclically as energy and mass enter and exit the motor. The state equations were solved using a commercial software package and some representative responses are shown here.

Figure 13.11 shows the cylinder pressure, P, and exhaust and inlet areas as they depend on time. Figure 13.12 shows the cylinder temperature. The pressure and temperature both take several cycles to come to a more or less steady state. When the inlet area opens, the pressure spikes a bit due to the pressure difference between supply pressure and cylinder pressure and the fact that the piston is moving very slowly near TDC. The temperature shows the same behavior.

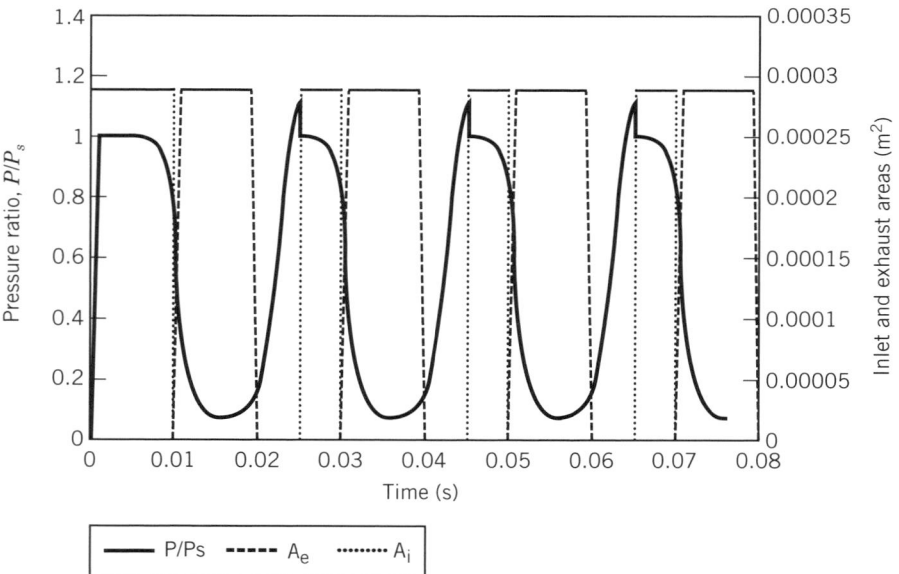

FIGURE 13.11. Pressure in the air engine cylinder.

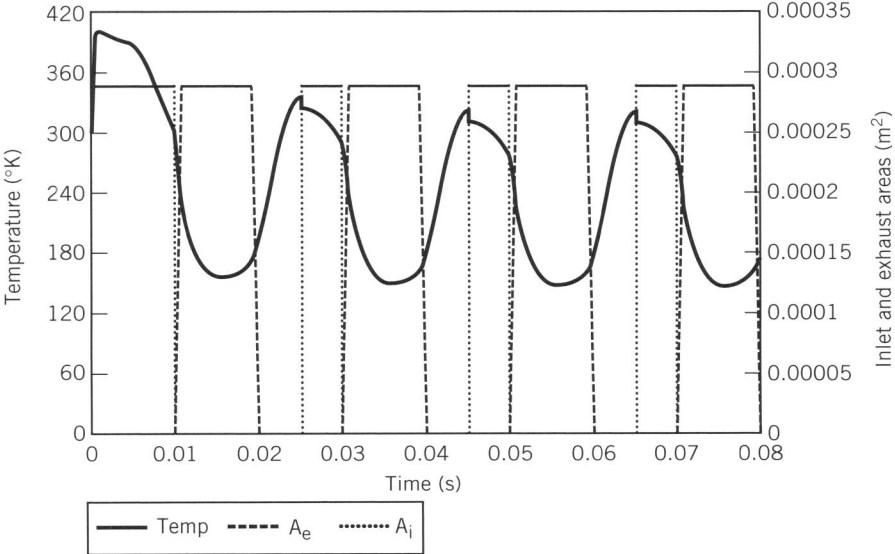

FIGURE 13.12. Temperature in the air engine cylinder.

Figure 13.13 shows the mass flow rate through the inlet port and exhaust port. Interestingly, the inlet port shows a very brief negative spike. This is due to the pressure in the cylinder being higher than the supply pressure when the inlet opens. There is

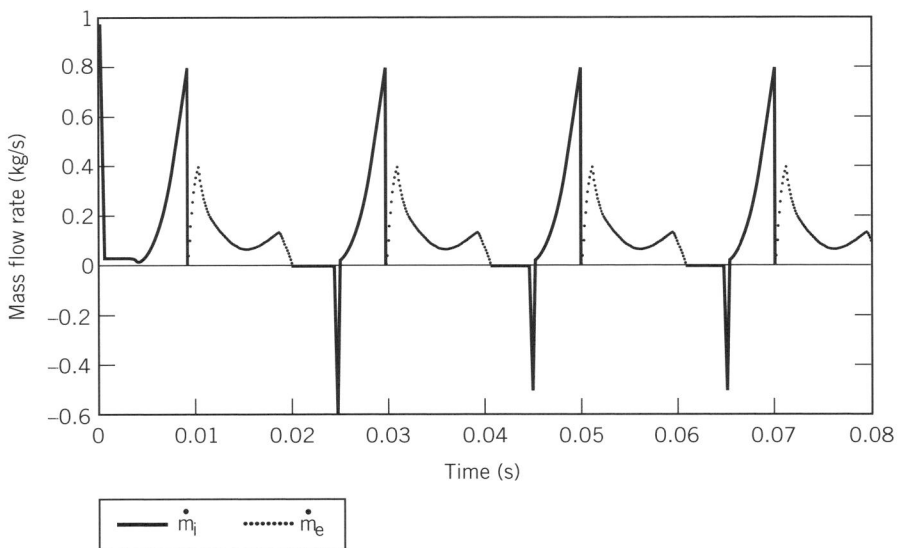

FIGURE 13.13. Mass flow rate through the inlet and exhaust ports.

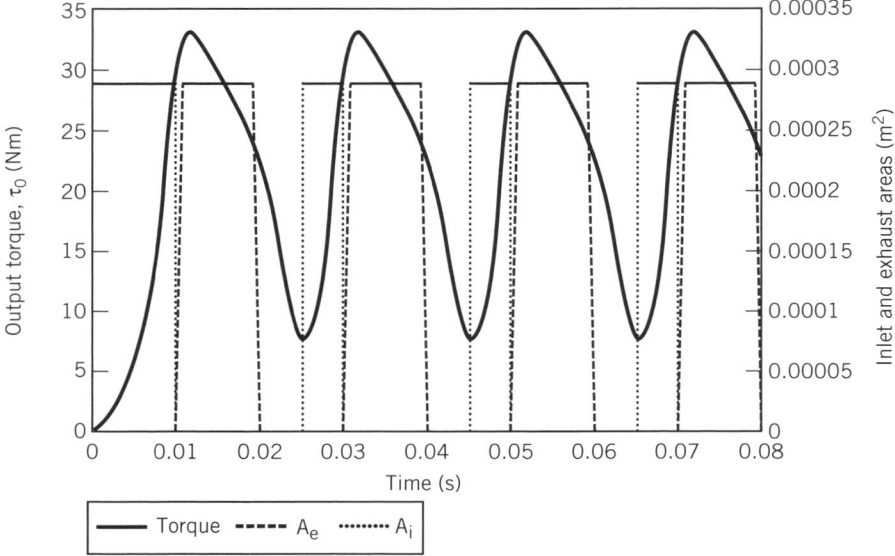

FIGURE 13.14. Output torque, τ, from the air engine.

a rapid readjustment followed by flow into the cylinder as the piston is driven downward. The exhaust flow increases rapidly as the exhaust port opens and then decreases as the cylinder pressure drops. The increase in mass flow just before the exhaust port closes is due to the piston having reached bottom dead center and moving upward. As the piston closes the exhaust area, it drives air out through the port.

Figure 13.14 shows the output torque of the engine. This is the torque measured on the torsional resistance, b_τ, from Figure 13.10. It is consistent with the pressure behavior. It is interesting that the torque peak does not coincide with the peak pressure. This is due to the inherent behavior of a slider–crank mechanism. Near TDC, where the pressure is highest, there is no moment arm to transmit torque to the flywheel. As rotation proceeds, the moment arm grows as the pressure decreases.

13.6   CONCLUSIONS

This chapter has attempted to expose the complexities of nonlinear simulation and the organization provided by using bond graphs to put complex systems together. Causality is a powerful tool in that formulation problems are exposed before equation formulation, providing the user choices in how to proceed toward a computable model. Causality also dictates the input and output variables for the nonlinear constitutive laws for the elements. This allows a priori determination of whether a particular relationship must be inverted if required by the formulation demands.

It was shown here and in Chapter 9 that nonlinear geometry often requires modulation variables that are not algebraically available from the state variables. In this case, the state space must be expanded to include "free integrators," which are solved along with the state variables. Bond graphs provide these free-integrator relationships.

Finally, an example was developed to demonstrate a complex simulation. It is hoped that the reader appreciates the usefulness of the bond graph in developing the simulation model.

REFERENCES

[1] B. Carnahan, H. A. Luther, and J. O. Wilkes, *Applied Numerical Methods*, New York: Wiley, 1969.

[2] A. Zeid and C.-H. Chung, "Bond Graph Modeling of Multibody Systems: A Library of Three-Dimensional Joints," *J. Franklin Inst.*, **329**, 605–636 (1992).

[3] D. Margolis and D. Karnopp "Analysis and Simulation of Planar Mechanisms Using Bond Graphs," *ASME J. Mechan. Des.*, **101**, No. 2, 187–191 (1979).

PROBLEMS

13-1   The system shown below has a nonlinear spring characterized by

$$F_s = g_s \delta_s^3$$

and a nonlinear damper characterized by

$$F_d = g_d v_d^3.$$

Construct a bond graph model, derive equations, and put into a form for computer simulation.

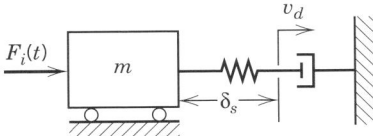

13-2   For the system of Problem 13-1, what changes must be made in the computer model if the spring and damper are characterized by square laws rather than cubic laws?

13-3   For the system of Problem 13-1, include friction between the mass and ground and reformulate the bond graph. Derive equations and put into a form ready for computer simulation. Describe the program statements that would be necessary in order to include friction.

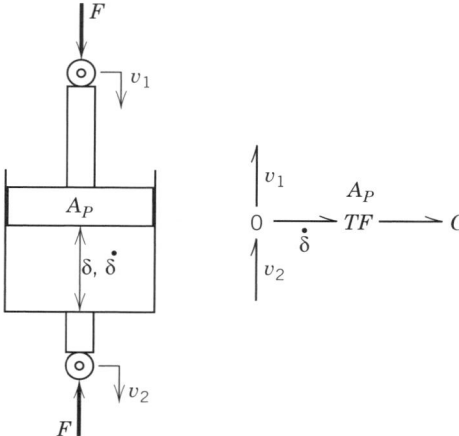

13-4   The device shown above is an air spring which is assumed to behave isentropically. The constitutive relationship for this device has been determined to be

$$F = P_0 A_P \left[\frac{1 - \left(\dfrac{A_P \delta}{V_0} \right)^{\gamma}}{\left(\dfrac{A_P \delta}{V_0} \right)^{\gamma}} \right],$$

where P_0 is atmospheric pressure and V_0 is the chamber volume when $A_P \delta / V_0 = 1$. A bond graph for the device is shown with attachment points exposed.

Include this air spring in the quarter car model shown below, construct a bond graph model, and derive a complete set of state equations. Organize

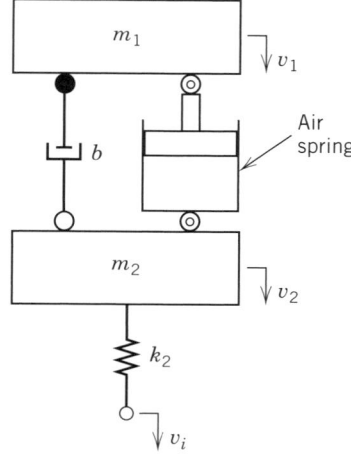

these equations into a form suitable for computer simulation. Explain any program statements needed to handle the special case when $A p \delta / V_0 \to 0$.

13-5 The system shown has two nonlinear dissipation elements with constitutive behavior,

$$F_1 = g_1 v_1^3, \qquad v_1 = v_m - v_2,$$
$$F_2 = g_2 |v_2| v_2.$$

Construct a bond graph model and determine that an algebraic loop exists. Attempt to derive state equations using the procedure from Chapter 5, Section 5.4. Now add a parasitic element to eliminate the algebraic loop. Derive state equations and organize them into a form suitable for computer simulation.

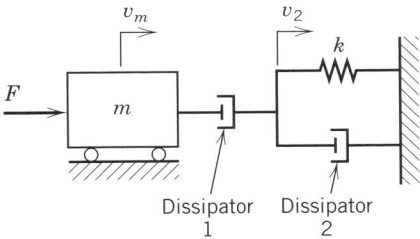

13-6 The system shown is a d-c motor with a flexible shaft and attached inertial load. The shaft is nonlinear and behaves according to

$$\tau = g(\theta_2 - \theta_1)^3.$$

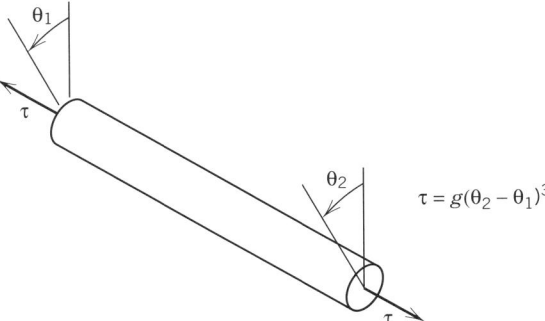

Construct a bond graph model, assign causality and convince yourself that derivative causality exists. Attempt to derive a computable model using the methods from Chapter 5, Section 5.5. Append a parasitic element to your model that eliminates derivative causality and assign a physical interpreta-

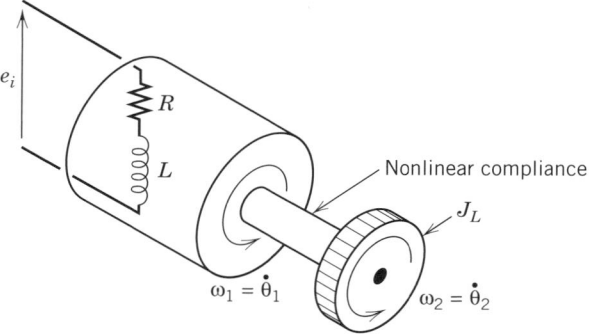

tion to this element. Derive the resulting explicit equations and put into a form ready for computer simulation.

13-7 Figure 12.17 shows a thermodynamic system and its causal bond graph. Derive the state equations for this system and show they are appropriate for computer simulation.

13-8 The system shown couples the air spring from Problem 13-4 and the slider–crank device from Eq. (13.59). Construct a bond graph model of the system and derive state equations. Expand the state space as necessary and set up your equations for computer simulation.

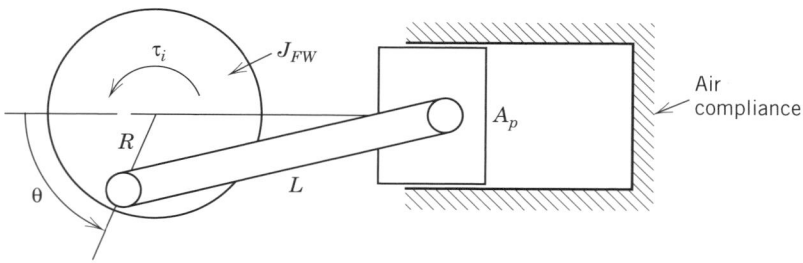

13-9 The mass, m, bounces off the rigid wall with no loss of energy.

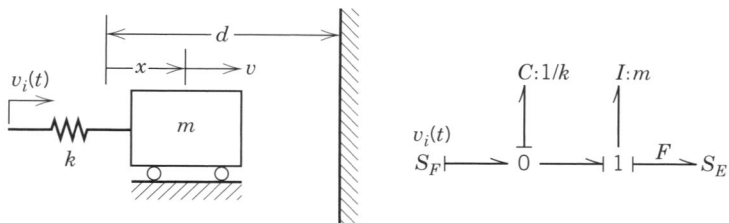

One way of simulating this phenomenon is to specify a boundary impulsive force, F, that acts over a specified but very short amount of time (perhaps one simulation time step) and accelerates the mass such that it comes off the wall with the same speed as the approach, but in the opposite direction.

For an impact duration of Δt seconds, calculate the value of this force, assuming knowledge of the approach velocity; set up the state equations for simulation, including the logic statements needed to handle the boundary impulse.

13-10 For Problem 13-9, another way to handle the impact problem is to assume there is a spring between the mass and wall with the constitutive behavior shown.

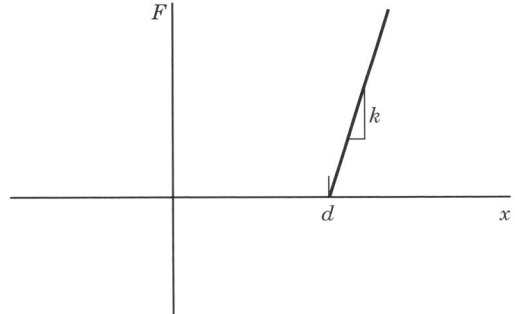

The spring constant, k, must be "very stiff" to simulate the originally intended system, and this may require shortening the simulation time step during contact with the wall. On the bond graph for Problem 13-9, put a C-element in place of the effort source and derive equations ready for simulation. Include the logic necessary to handle the "stiff" compliance.

13-11 Two unequal-length pendulii are attached by a spring as shown. Construct a bond graph model that would account for large angular deflections. Derive a

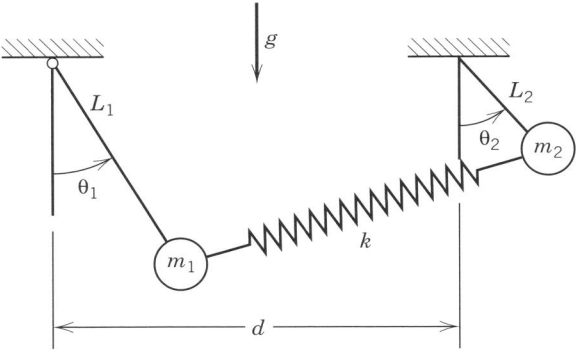

complete state representation and set up for computer simulation. The spring is relaxed when both pendulii are vertical.

13-12   In the slider–crank mechanism, assume the only important inertial element is the connecting rod of mass m, moment of inertia J, and length L. A modified device is shown here where the horizontal sliding constraint has been replaced by two springs, k_H and k_V. If k_V was very stiff, then the horizontal sliding constraint would be approached. Construct a bond graph model of this system and note the use of *MTF*s involving angles θ and α. Derive a complete state representation and organize them for simulation. (*Hint:* Transfer the c_g motion to the end points of the rod; then enforce the velocity constraint at the crank, and derive the spring velocities at the other end.)

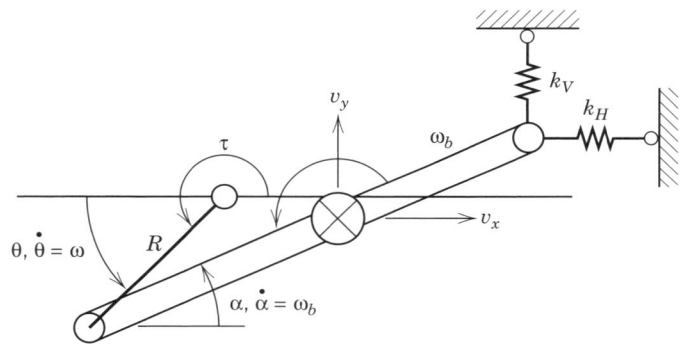

APPENDIX: TYPICAL MATERIAL PROPERTY VALUES USEFUL IN MODELING MECHANICAL, ACOUSTIC, AND HYDRAULIC ELEMENTS

Mass Density, ρ, $[\text{kg/m}^3]$

  Solids

Aluminum:	$2{,}700 \text{ kg/m}^3$
Copper:	$8{,}900 \text{ kg/m}^3$
Rubber, hard:	$1{,}100 \text{ kg/m}^3$
Rubber, soft:	950 kg/m^3
Steel:	$7{,}700 \text{ kg/m}^3$
Titanium:	$4{,}500 \text{ kg/m}^3$

  Liquids

Hydraulic oil, well de-aerated:	900 kg/m^3
Water, fresh, at 20°C:	998 kg/m^3
Water, sea, at 134°C:	$1{,}026 \text{ kg/m}^3$

  Gases

Air at 1 atm. and 20°C:	1.21 kg/m^3
Air at 1 atm. and 0°C:	1.29 kg/m^3
Hydrogen at 1 atm. and 0°C:	0.09 kg/m^3

Modulus of Elasticity, E, $[\text{Pa} = \text{N/m}^2]$

Aluminum:	$71{,}000 \text{ N/mm}^2 = 71 \times 10^9 \text{Pa}$
Copper:	$122{,}000 \text{ N/mm}^2 = 122 \times 10^9 \text{Pa}$
Hard rubber:	$2{,}300 \text{ N/mm}^2 = 2.3 \times 10^9 \text{Pa}$
Soft rubber:	$5 \text{ N/mm}^2 = 0.005 \times 10^9 \text{Pa}$
Steel:	$206{,}000 \text{ N/mm}^2 = 206 \times 10^9 \text{Pa}$
Titanium:	$110{,}000 \text{ N/mm}^2 = 110 \times 10^9 \text{Pa}$

Bulk Modulus, B, $[\text{Pa} = \text{N/m}^2]$

Water at 20°C:	$2.18 \times 10^9 \text{Pa}$
Hydraulic oil (well de-aerated):	$1.52 \times 10^9 \text{Pa}$

Speed of Sound, c, [m/s]

Air at 1 atm. and 20°C:	343 m/s
Air at 1 atm. and 0°C:	332 m/s
Hydrogen at 1 atm. and 0°C:	1, 269.5 m/s
Water, fresh, at 20°C:	1481 m/s
Water, sea, at 0°C:	1500 m/s

Coefficient of Shear Viscoscity, μ, [Pa · s = Ns/m^2]

Air at 1 atm. and 20°C:	1.8×10^{-5}Pa · s
Castor oil:	0.96 Pa · s
Water, fresh, at 20°C:	1.0×10^{-3}Pa · s

Ratio of Specific Heats, $\gamma = c_p/c_v$

Air	1.40
Carbon dioxide	
(low frequency)	1.30
(high frequency)	1.40
Nitrogen	1.40

INDEX